108

STUDYING TEMPERATE
MARINE ENVIRONMENTS

STUDYING TEMPERATE MARINE ENVIRONMENTS

A handbook for ecologists

Edited by

MICHAEL KINGSFORD

and

CHRISTOPHER BATTERSHILL

CANTERBURY UNIVERSITY PRESS
Christchurch, New Zealand

CRC PRESS
Boca Raton London New York Washington, D.C.

First published in 1998 by
CANTERBURY UNIVERSITY PRESS
University of Canterbury
Private Bag 4800
Christchurch
NEW ZEALAND

mail@cup.canterbury.ac.nz
www.cup.canterbury.ac.nz

This edition published in 2000 jointly with
CRC PRESS LLC
2000 Corporate Blvd, N.W.
Boca Raton, FL 33431
UNITED STATES OF AMERICA

Designed and typeset at Canterbury University Press
Printed by SRM Production Services Sdn. Bhd., Malaysia

Publication of this book has been assisted by grants fromthe New Zealand Department of Conservation
and the National Institute of Water and Atmospheric Research Ltd.

Library of Congress Cataloging-in-Publication Data

Studying temperate marine environments / edited by Michael J. Kingsford and
Christopher N. Battershill
    p. cm.
    Includes biographical references (p.)
    ISBN 0-8493-0883-6 (alk. paper)
    1. Marine ecology—Study and teaching—Handbooks, manuals, etc.
    2. Temperate climate. I Kingsford, Michael. II. Battershill, Christopher.

QH541.5.S3 S78 2000
577.7—dc21   00-036073

Front cover photograph: Aggregation of planktivores, Pinnacles,
Poor Knights Islands, northern New Zealand.
M. KINGSFORD

Back cover: A diver recording the number of benthic invertebrates
in a quadrat at Raoul Island, the Kermadecs Islands.
M. KINGSFORD

A biologist processing samples in the laboratory.
NIWA LIBRARY

CONTENTS

10 Identification and treatment of specimens 269

M. J. Kingsford and C. N. Battershill

FOREWORD

The sustainable management of land resources is now part of the mainstream objectives of government and of science, but its equivalent discipline for the marine environment is much more recent. This book is at the forefront of this urgent and important work, setting a new standard against which all future temperate marine studies will be measured.

The authors are all highly respected scientists who have used their theoretical insights, field experience and technical skill to illuminate and explain ecological complexities beneath the sea. The result should be not only better-designed research but better understanding and appreciation of marine ecology, to provide better quality advice to decision-makers. These are essential first steps towards conserving and sustainably managing the rich and diverse marine environments that surround New Zealand and other temperate countries.

The New Zealand Department of Conservation has supported production of this book from its inception and, together with the National Institute of Water and Atmospheric Research, has recognised the need for a comprehensive guide to aid those who are responsible for the health and functioning of coastal marine environments. It will permit them to plan and carry out focused research and monitoring programmes to address the many issues confronting coastal environments today. In addition, the book provides an invaluable resource for use by lay persons interested in their marine environment as well as a key text for use in education at a wide range of levels.

Nick Smith
Minister of Conservation

The experiences of each of us become a shared resource for all only if the stories are told, recorded and accessible.

PREFACE

This book is a synthesis of technical and practical information, both written and otherwise, with the aim of providing a concise first source of approaches, examples and suggestions to aid the study of temperate marine environments. There is an urgent need to understand the dynamics of temperate marine environments, how organisms associated with them interact with each other, and how they respond to changing physical conditions. The demand for reliable information for predicting trends in marine environments is fundamental to good management of marine resources. Human impacts are regular concerns in coastal marine environments, and are frequently accompanied by passionate requests for immediate attention. The issues relevant to management and further research are usually relatively simple to define, but where do you start? The knowledge base required to identify pertinent questions and design effective sampling programs is intimidatingly large. Sources of relevant background information are often scattered in published and unpublished forms (the 'grey' literature), and often encompass topics such as sampling theory, analytical aspects of sampling design, sampling methods, descriptive and experimental ecological studies, life history, anatomy, taxonomy and methods of preservation. There is also a great deal of practical information obtainable only from the experience of colleagues, to which not all of us are privy. In short, we believe that a book of the type presented here is much needed. Although special emphasis is given to coastal environments in New Zealand and temperate waters of the east coast of Australia, the principles of the book are relevant to coastal environments worldwide.

The objective of this book is to provide a quick start that addresses the likely queries of readers and provides useful leads. The book covers a wide variety of disciplines, providing information on approaches for setting up a study, sources of spatial and temporal variation that should be considered in sampling designs, analytical procedures that should be considered for sampling designs, types of sampling equipment and methods of applying them to particular environments (e.g. softsubstrata) and the organisms associated with them. Advice is also given on preservation of organisms, and taxonomy. The book is written with the philosophy that a specific question or hypothesis will determine the most appropriate sampling design and methodology. Accordingly, no single method is advocated as the answer to all sampling problems: rather, examples of descriptive and experimental studies are provided that should help in decision-making. In many chapters specific case histories are given where we articulate the purpose of the work with one or more specific hypotheses. In these examples, the situation the investigator encountered prior to the study is described (i.e. utility of background information), the sampling designs are detailed and the conclusions reached by the investigators are summarised. Some emphasis is given to the study of impacts, including contemporary issues such as effects of marine protection. Descriptive studies and the study of processes using experimental approaches are also covered in detail.

The book is extensively cross-referenced and has a comprehensive list of citations. The references lists included in the book are intended to give readers a rapid and focused access to the broader literature facilitating further exploration of the specific problems of each reader. It could be argued that CD-ROMs, the World Wide Web and other databases can provide all the reference material you need, and indeed we agree that databases are very useful and should be used; but access is only as good as the words used to search and the completeness of the data base. The names of key players in the field are useful for searches and the book should give access to other material that is difficult to obtain on many computer databases. We have pooled all references into one list to reduce repetition from the different chapters and give a complete list of contributors in temperate marine ecology that we have cited.

Our intention is to target principal investigators, decision-makers and managers concerned with marine environments, and people who educate students in environmental and marine science. Specifically, readers who may benefit most from this work include professional ecologists, academics, conservation officers and senior students studying coastal marine environments. All of these persons must set objectives, and design or have input to the design of sampling programs, while taking into account logistical constraints, and consider aspects of sampling that are relevant to the organisms, the habitats in which they are found and the prevailing physical environment. The book should be a useful guide to all aspects of a marine studies, from design to completion. Furthermore, it is intended to provide useful reviews as resources for many non-field-related objectives. Because the book advises on approaches to studying marine environments and is rich in case histories, lecturers should find the book useful for courses in environmental and marine science.

Clearly, the emphases we and the other contributors to the book have put on approaches to sampling, specific methods and case histories are based on a combination of our interpretation of the information available, communications with colleagues, and personal experience. If the reader's thoughts deviate from the conclusions presented, we hope the book has been useful in providing a different perspective. Our hope is that our book will promote a greater understanding of marine environments, provide a succinct document that gives insight into what is known, and offer a vision for the planning of new studies and new marine legislation. Conservation and sustainable marine resources can only happen through the comprehensive transfer of knowledge.

Michael Kingsford

Christopher Battershill

ACKNOWLEDGEMENTS

Special thanks to Bob Creese, Alistair MacDiarmid, Don Morrisey, Rob Murdoch, Stephanie Turner and Kathy Walls for their contributions to the book – the long and winding road reaches an end! Also to Kaye Green of the New Zealand Department of Conservation (DoC) for supporting the book project and for remaining enthusiastic about its concept throughout.

The final stages of this book required a great amount of energy and focus from others associated with the project. Rose-Marie Thompson of NIWA took on the herculean task of checking every reference in the book. Margie Atkinson compiled the index, and members of the Fish Ecology Laboratory at the University of Sydney assisted in final checking. We thank them all for their energy and professionalism. For their help with the figures we thank Megan Oliver, Peter Bennet, Deborah Palmer, Vivian Ward and Graeme MacKay.

Thanks to Tony Ayling, Mike Bradstock, Malcolm Francis, Mathew Galetto, Steve Mercer and Kim Westerskov for providing colour photographs, and to those who supplied black-and-white photographs and line drawings. We also thank Michael Uddstrom, Nils Oien, Jonathon Sharples and Kim Babcock for the images showing sea-surface features.

In a technical book of this type, refereeing is essential. Some brave individuals refereed more than one of the chapters, and for this we give special thanks to Bronwyn Gillanders, Mick Keough, and Duncan Worthington. Our publisher, Mike Bradstock, was the only person, other than us, to read the book in its entirety more than once, and he provided many useful comments based on his own previous experiences as a professional biologist with the New Zealand Ministry of Agriculture and Fisheries.

Specific acknowledgments from the authors of each of the chapters are as follows:

Chapter 1: We thank Sally McNeill for constructive comments and her knowledge of relevant conservation issues in Australia.

Chapter 2: Useful comments were provided by Nicole Gallahar, Bronwyn Gillanders, Mick Keough and Don Morrisey.

Chapter 3: Special thanks to Tim Glasby, who read the text for clarity and thoroughly checked all of the calculations more than once. Other constructive advice was provided by Marti Anderson, Bronwyn Gillanders, Mick Keough and Tony Underwood.

Chapter 4: We thank Duncan Worthington for his comments.

Chapter 5: Maria Byrne, Catherine King and Duncan Worthington. We also thank Maria Byrne and Cathy King for providing photographs for the text.

Chapter 6: Russell Cole and Bronwyn Gillanders provided detailed comments on the manuscript and other comments were made by Howard Choat, Euan Harvey and Vicky Tzioumis.

Chapter 7: Thanks to George Branch and Greg Skilleter for their comments.

Chapter 8: We are grateful to Graeme Inglis and Greg Skilleter for their constructive comments on an earlier draft of this chapter, and to Bob Creese for advice and information.

Chapter 9: We thank Kylie Pitt, Betty Pentury and Iain Suthers for their detailed comments. Many thanks to Sue Tong for advice on methods for studying protists.

Chapter 10: Angela Low and Dennis Gordon gave us many useful comments on preservation and experts in the field.

Our personal communications with people involved in ecological assessments of a variety of types were an invaluable contribution to the synthesis and awareness of the need of a book of this type.

For their open discussion, over the years, and sincere belief in field-orientated research we thank Neil Andrew, Tony Ayling, Bill Ballantine, Greg Cailliet, Howard Choat, Bob Cowen, Mike Foster, Doug Ferrell, Malcolm Francis, Christine Handford, Sally Holbrook, Geoff Jones, Mark McCormick, John McCoy, Russ Schmitt, David Schiel, Greg Skilleter, Laura Stocker, Tony Underwood and Kathy Walls, members of the Fish Ecology Laboratory at the University of Sydney, the research groups at NIWA, Greta Point, the Marine Ecology Group at the University of Alicante, and DoC in Wellington.

This book evolved from an earlier version (Kingsford 1986). The Poor Knights Management Committee had the foresight to designate funding so that a book such as this could be written; as recipients we thank them. The 1986 version was only released as an internal book with a great focus on marine protected areas (MPAs). The Department of Conservation recognised the importance of a more generic document of this type and contributed substantial financial support toward the publication of the book.

The National Institute of Water and Atmospheric Research (NIWA) also substantially supported production of this book, providing facilities, resources and staff time.

Much of the text was written at NIWA, Greta Point, and we thank the Regional Manager, Rob Murdoch for his support. Mike Bradstock nursed us through the process of turning a manuscript into a book. He and the book's designer, Richard King, at Canterbury University Press were always enthusiastic about the project. We thank them both for their perseverance, professionalism and humour.

Finally, our families often suffered as a result of this project; all we can say is thank you for your support and it is over!

CONTRIBUTORS

Dr Christopher N. Battershill
National Institute of Water and Atmospheric Research Ltd, PO Box 14-901, Kilbirnie, Wellington, New Zealand

Chris has 18 years of research experience in marine benthic ecology, predominantly in temperate Australasian waters but with recent experience in both polar and tropical marine ecosystems in the southern hemisphere. Research has focused on sponge biology and ecology, and has expanded into chemical ecological work on marine invertebrates and algae that elicit compounds with pharmaceutical applications. Additional work includes environmental effects assessment associated with industrial developments and marine reserves. He completed a PhD at the University of Auckland and has had postdoctoral fellowships at the University of Canterbury and the University of Wollongong (but based at the Australian Institute of Marine Science). A return to New Zealand enabled him to do marine conservation research at the Department of Conservation, and since 1991 he has been based at the New Zealand Oceanographic Institute, now NIWA. Here his primary research focus is on dynamics and stability of deep-reef ecosystems, sustainable resource development of species that produce biologically active secondary metabolites, and biosystematics of the spones and ascidians. Very recent research initiatives include the assessment of effects of invasive exotic marine species and the mechanisms for their introduction.

Dr Bob Creese
Leigh Marine Laboratory, University of Auckland, PO Box 349, Warkworth, New Zealand

Bob has 22 years' research experience in the ecology of intertidal species and assemblages, with special interest in molluscs. His initial work was on rocky shores, focusing on grazing gastropods and chitons, but the scope of his research has been extended in recent years to include soft-sediment systems, especially filterfeeding bivalves and mangroves. Most of his work has been on temperate coasts in Australia, New Zealand and the western USA. After gaining his PhD from the University of Sydney, Bob held postdoctoral positions at the University of Washington (Friday Harbor) and the University of Auckland (Leigh Marine Laboratory). Bob gained a permanent position at Leigh in 1986 and is now a senior lecturer in the newly formed School of Environmental and Marine Sciences. His research group studies a wide range of ecological and environmental issues around the coasts of northern New Zealand.

Dr Michael J. Kingsford
School of Biological Sciences A08, University of Sydney, NSW 2006, Australia

Mike has 19 years of research experience in the ecology of reefs, with a special interest in reef fishes, larval fish and the biological oceanography of temperate and tropical waters. Although most of his research in temperate latitudes has been done in New Zealand and New South Wales, Australia, he has also worked on reefs in California and southern Spain. He completed a PhD at the University of Auckland on the early life history stages of fish and came to Sydney in 1987 to commence a postdoctoral fellowship on fishers as predators of intertidal organisms. He is now a senior lecturer at the University of Sydney, with long-term interest in approaches to sampling and communicating concepts of sustainable marine resources to the wider community. His research group is active in the areas of biological oceanography, factors determining the abundance of reef fish, novel methods for determining the movements of fish, fisheries and factors determining the detection of reefs by fish larvae.

Dr Alistair B. MacDiarmid
National Institute of Water and Atmospheric Research Ltd, PO Box 14-901, Kilbirnie, Wellington, New Zealand

Alistair has 18 years of research experience in the ecology of reefs, with a particular interest in spiny lobster behavioural ecology and marine reserves in temperate and tropical waters. Most of his research in temperate latitudes has been done in New Zealand, but he has also worked on reefs in Australia and Florida. He completed a PhD at the University of Auckland and moved to Wellington to work as a scientist, first with the Fisheries Research Centre and then with NIWA. He is now programme leader of a project investigating aspects of spiny lobster reproductive and juvenile ecology.

Dr Donald J. Morrisey

National Institute of Water and Atmospheric Research Ltd, PO Box 11-115, Hamilton, New Zealand

Don began his research on the ecology of marine soft sediments with a PhD at the University of Cambridge, UK. This was followed by postdoctoral positions at the University of Bristol, working on the ecology of the Severn Estuary, and at the University of Sydney, working on the effects of heavy metals on the faunas of soft sediments. He currently works as a marine scientist with NIWA in New Zealand, where his particular research interests are the impacts of urban development on estuaries and the environmental impacts of aquaculture.

Dr Robin C. Murdoch

National Institute of Water and Atmospheric Research Ltd, P O Box 14 901, Kilbirnie, Wellington, New Zealand.

Rob has 17 years of research experience in biological and descriptive physical oceanography. His research interests are primarily macrozooplankton ecology and interactions with physical oceanographic features, fish larvae ecology, and pelagic food web processes. He has also undertaken research and consultancy associated with soft-bottom and reef benthos and fishes, molluscan taxonomy, and aquaculture. He moved to Wellington to take up a position as research scientist at the New Zealand Oceanographic Institute (now NIWA) in 1984 after completing his PhD at the Portobello Marine Laboratory, University of Otago. Rob is currently a Regional Manager for NIWA, overseeing oceanographic and aquaculture research.

Dr Stephanie Turner

National Institute of Water and Atmospheric Research Ltd, PO Box 11-115, Hamilton, New Zealand

Stephanie is a marine scientist with NIWA, where her main research interests include the ecology and restoration of coastal soft-sediment and seagrass communities. Previously she worked in Western Australia after completing her PhD at the University of St Andrew's in Scotland. She has seven years' research experience in marine ecology and her interests include the ecology of epifaunal communities (in particular bryozoans), corallivores and coral reef ecology, and the early life history of gastropods.

Kathy Walls

Department of Conservation Te Papa Atawhai, PO Box 112, Hamilton, New Zealand,

Kathy is a marine biologist with New Zealand's Department of Conservation, where she has been co-ordinating the national marine reserves programme for over 10 years. In that time she has overseen the rapid expansion of marine reserves around the country. An important component of her work has been to ensure that marine reserves provide a focus for research and monitoring and she has played a key role in providing support for this book. Her two previous positions saw her working as a marine planner in Northland's harbours and as an adviser to Maori on aquaculture and marine resources, also in Northland. Kathy's academic career began at the University of Auckland's Leigh marine laboratory where she completed a Master of Science degree in 1982 in coastal current patterns and biological oceanography.

ONE

INTRODUCTION TO ECOLOGICAL ASSESSMENTS

M. J. Kingsford, C. N. Battershill and K. Walls

1.1 Background

A large percentage of the population in temperate regions of New Zealand, Australia and, indeed, much of the world is concentrated along the coast. Approximately two-thirds of the population of New Zealand and 90% of that of Australia live on or near the coast. Thus, many commercial, recreational and traditional activities impact on coastal marine environments (Fig. 1.1). What constitutes an appropriate use of these environments is often a subject of heated debate because

many activities can be in conflict. For example, commercial and recreational fishers want access to good fishing grounds, while recreational divers and tourist operators may want the areas to be free of commercial harvesting. These uses can, in turn, conflict with demands for coastal developments such as subdivisions, marinas, causeways, aquaculture, ports and military bases.

In addition to physical impacts from coastal developments, pollutants are released into coastal waters from many sources. These include sewage, pesticides, heavy metals, hydrochlorides and stormwater, which is often contaminated with lead and petrochemicals from artificial surfaces such as roads (e.g. Beder 1989; Glasby

Figure 1.1 Examples of natural and anthropogenic influences on coastal environments.

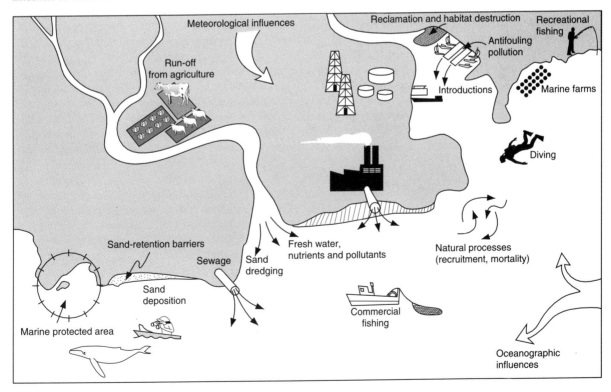

1991; Phillips & Rainbow 1993). The potential and actual impact of pollutants is of concern to users of all marine environments. This is particularly important for managers who are charged with maintaining the quality of habitats and the sustainability of harvestable organisms.

Other areas of conflict that require biological information in decision-making include the burgeoning interest in aquaculture, which has the potential to influence substantial areas of coast. For example, the New Zealand Fishing Industry Board has identified a target of $3 billion in exports by the year 2000, which is a three-fold increase on current levels (NZFIB 1994). Australia has similar objectives, e.g. in South Australia the coast is currently being zoned for actual and potential aquaculture activities. There is also a rapidly growing interest in the creation of marine protected areas (MPAs) and the potentially positive impacts that these areas have on organisms and local fisheries. The existence of threats to assemblages of indigenous organisms by introduced species necessitates a knowledge of the biology, ecology and control of these species. In all coastal marine environments, management of marine resources requires a thorough knowledge of habitats and the biology of organisms involved.

There is great diversity of coastal marine environments in temperate latitudes of New Zealand and Australia. Long stretches of coastline (about 16,000 km in New Zealand and a similar distance at temperate latitudes of Australia) are characterised by rocky reefs, ocean beaches, cliffs, exposed and sheltered bays, drowned valleys and numerous islands of varying sizes (Morton & Miller 1973; Bennett 1987). Each marine environment supports a wide array of marine habitats (Walls & Dingwall 1995). For example, in rocky reef environments habitats may include kelp forests, algal turf, sponge gardens, urchin-dominated barrens (e.g. Ayling 1978; Underwood et al. 1991; Plates 2–19) and unique assemblages such as the black coral of Fiordland in the South Island of New Zealand (Grange pers. comm.). Not all unique assemblages are on natural substrata: for example, the wharf pilings of South Australia have a rich fauna and flora that has been the subject of many biological investigations (Butler 1995).

Those who study and/or manage the biological attributes of temperate coastal waters, therefore, have to deal with a broad range of environments and habitats, and diverse assemblages of organisms associated with them.

With an increase in concerns about conflicts of use, conservation, sustainability and knowledge of marine assemblages, amendments to existing legislation and the introduction of new legislation are inevitable. Coastal-management approaches vary greatly, and laws are regularly revised among countries and states; Australasia is no different in this respect.

In New Zealand, the key statutes involved in sustainable management of marine resources and planning for coastal activities and conservation of the marine environment are the Harbours Act 1950, the Wildlife Act 1953, the Marine Reserves Act 1971, the Marine Mammals Protection Act 1978, the Resource Management Act 1991 (New Zealand Coastal Policy Statement, Regional Plans and Regional Coastal Plans), the Treaty of Waitangi (Fisheries Claims) Settlement Act 1992 and the Fisheries Act 1996 (Walls 1995b). In almost all cases, compliance with the conditions of these acts requires biological information.

In Australia, each state is responsible for coastal waters from MLWM to three nautical miles offshore, and the Commonwealth Government is responsible for waters from three to the 200-mile offshore limit of the Australian EEZ (Kailola et al. 1993; McNeill 1994). Management of the coastal marine environment is divided between three levels of government (national, state and local), with administrative systems varying between the states and each having differences in administration and objectives (Holmes & Saenger 1995). For example, in Western Australia marine-related legislation includes the Fisheries Act 1905, the Wildlife Conservation Act 1950, the Marine Sea Dumping Act 1981, the Conservation Land Management Act 1984, the Marine Harbours Act 1981, the Environmental Protection Act 1986 and the Petroleum Act 1990.

Legislation often requires managers to obtain biological information in order to describe 'what is there', to report on changes in abundance of populations of marine species (due to natural and anthropogenic causes), to further knowledge of the processes that influence patterns of abundance, and to predict what may happen in the future. Biologists, therefore, are asked a broad range of questions so that they may provide appropriate information for management purposes and/or contribute to our understanding of the processes influencing organisms in coastal marine environments. Although the emphasis in this book is Australasian, the biological issues that we outline are typical of marine environments in temperate latitudes all over the world.

Information on marine environments is scattered throughout a diverse literature, and it is difficult therefore to get guidance on how to obtain, process, and interpret information on marine environments (Ward 1993; DOC 1994; Ward & Hedley, 1994; Holmes & Saenger 1995; Walls 1995b).

The intention of this book is to provide a single information source that can help investigators and managers of coastal resources advance our knowledge of temperate coastal marine environments.

1.2 Types of ecological assessments

We have categorised biological assessments into four general types (modified from Green 1979):

1. **Baseline studies**, in which data are collected to define the present state of a biological assemblage. In this book we emphasise the limitations of one-off baseline studies.
2. **Impact studies**, the purpose of which is to determine whether a particular perturbation causes a change in a population or assemblage, together with the spatial and temporal scales of that change; this is given that the nature of the impact is always known (e.g. sewage outfall, effect of the establishment of a marine reserve), but the nature of the response of the organism is not. An impact study allows the type and magnitude of any changes in the organisms (e.g. an increase or decrease in the abundance of organisms) to be described in response to short (pulse) or prolonged (press) impacts, as well as series of impacts.
3. **Monitoring studies** are designed to detect any changes from the present state. Monitoring is often an extension of another category of study (e.g. an impact study) and it is crucial that sampling be designed to detect effects of predetermined importance (e.g. 30% change in abundance). By definition, monitoring studies involve repeated sampling in time. In this book, it is emphasised that if general statements are to be made on temporal change, then the study should also be replicated in space (e.g. two locations or more).
4. **Patterns and processes**. In an ecological context, the study of patterns and processes involves describing distribution and abundance patterns of organisms, with the intention of identifying the processes responsible for them. Descriptive ('mensurative') and experimental approaches should be used (see Andrew & Mapstone 1987).

In this book the word 'study' is used as a general term to encompass all of these categories.

1.3 Types of ecological concerns

Most of the four categories of studies described above are relevant to the following examples of concerns and interests in the biology of marine environments.

1.3.1 Impacts

Coastal environments are subjected to a broad range of impacts. Some of these are potentially deleterious (e.g. sewage outfalls, dumps of toxic waste, dumping of dredgings, introduction of exotic species), while others are potentially beneficial (e.g. MPAs, protection of species). High-profile potentially deleterious impacts in New Zealand include the introduction of the laminarian alga *Undaria pinnatifida* (Hay 1990; Adams 1994) and Asian mussels (Creese pers. comm.) in ships' ballast water or as 'hitchhikers' from Asia, industrial waste in estuaries (e.g. Manukau Harbour) and sewage outfalls (e.g. Wellington). Australian examples include the sewage outfalls off Sydney (Beder 1989; Fagan et al. 1992), the proposed expansion of naval facilities in Jervis Bay, industrial waste (McLean et al. 1991) and unintentional introductions (e.g. northern Pacific starfish to Tasmania, from ships' ballast water). In cases dealing with large-scale impacts such as oil spills, environmental considerations will need to be taken into account in the clean-up process (Paine et al. 1996). Decisions for mitigation are often made more difficult in these situations because there is no information available on the pre-impact state.

Impact studies that have focused on deleterious or potentially deleterious impacts have received much scrutiny in recent years. The inadequacies of studies that profess to quantify or describe environmental impact frequently become apparent in the review process of hearings, court cases and public discussions. Either the data collected proves to be spurious to the key issues (i.e. the wrong question was addressed) or the design of studies was fundamentally flawed. It is not uncommon to see lawyers on either sides of an environmental case using different statistical interpretations of the same data set to advance their cause.

There is a clear need for the relevant questions to be identified in studies of this type so that appropriate sampling designs can be used. In addition to these perhaps self-evident principles, there is also the need for full data sets to be made available for audit and for methods and materials used to be described accurately to enable repetition of the work. It is often appropriate that samples be correctly preserved, catalogued, registered and stored in a suitable repository.

Increasingly, studies are being commissioned to examine impacts or predict disturbances yet to occur. It is a challenge to properly design studies to detect impacts. Sampling should, for instance, be replicated both in time and space, before and after an impact occurs. Furthermore, the estimated size of the impact on coastal environments should be decided before the study, and target species that are likely to be affected should be identified and monitored.

It is clear that the optimal situation for an impact study would be to obtain information before the impact has occurred in both the area(s) to be affected and unaffected areas (controls). This would enable natural

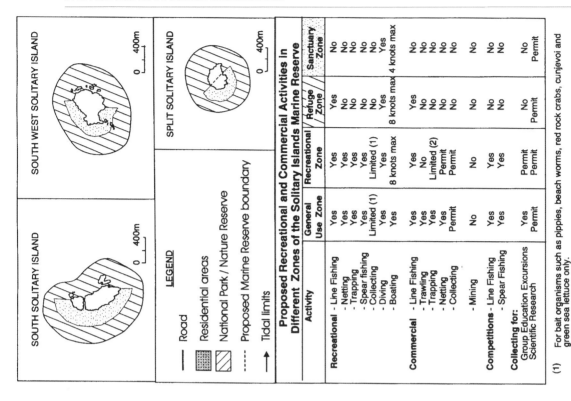

Proposed Recreational and Commercial Activities in Different Zones of the Solitary Islands Marine Reserve

Activity	General Use Zone	Recreational Zone	Refuge / Zone	Sanctuary / Zone
Recreational - Line Fishing	Yes	Yes	Yes	No
- Netting	Yes	Yes	No	No
- Trapping	Yes	Yes	No	No
- Spear fishing	Limited (1)	Limited (1)	No	No
- Collecting	Yes	Yes	No	No
- Diving	Yes	Yes	Yes	Yes
- Boating	8 knots max	8 knots max	8 knots max	4 knots max
Commercial - Line Fishing	Yes	Yes	Yes	No
- Trawling	Yes	No	No	No
- Trapping	Yes	Limited (2)	No	No
- Netting	Permit	Permit	No	No
- Collecting	No	No	No	No
- Mining	No	No	No	No
Competitions - Line Fishing	Yes	Yes	No	No
- Spear Fishing	Yes	Yes	No	No
Collecting for: Group Education Excursions	Permit	Permit	Permit	Permit
Scientific Research	Permit	Permit	Permit	Permit

(1) For bait organisms such as pippies, beach worms, red rock crabs, cunjevoi and green sea lettuce only.

(2) Rock lobsters and crabs only. Subject to MSB consideration.

Note: These activities must comply with existing State and Local fisheries regulations e.g. restrictions on collecting and fishing at North Solitary Island; restrictions on trapping at North and South Solitary Islands; restrictions on Spear fishing, netting and trapping in estuaries; and the current ban on fishing in Coffs Harbour Creek.

LEGEND

— Road

Residential areas

National Park / Nature Reserve

---- Proposed Marine Reserve boundary

→ Tidal limits

SOUTH SOLITARY ISLAND — 0 400m

SOUTH WEST SOLITARY ISLAND — 0 400m

SPLIT SOLITARY ISLAND — 0 400m

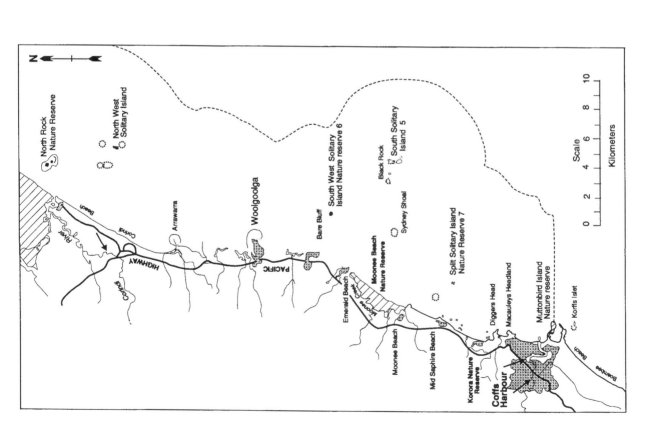

temporal and spatial variation to be distinguished from changes caused by the impact (Green 1979). If the impact has already occurred, and the area influenced is known, the effects can only be inferred from spatial pattern where comparisons are made with adjacent control areas. If there are no appropriate control areas (an approach that should not be encouraged), however, the impact can only be inferred from temporal change. In all situations it is desirable for the sampling design to be sufficiently rigorous to be repeatable. In this way it may be possible for subsequent biological monitoring to detect future changes.

1.3.2 Marine protected areas

In recent years there has been an increasing focus on the study of marine protected areas (MPAs) because of their potential and realised benefits for the management of coastal resources (e.g. Battershill et al.1993; Agardy 1994). We use the generic term 'marine protected areas' (Kelleher & Kenchington 1992; Gubbay 1995) unless areas have a more specific classification. Studies on the influence of MPAs are treated as impact studies and, in general, it is assumed that the impact will be positive (cf. Jones et al. 1993). In both New Zealand (Ballantine 1991) and Australia (Lowe 1996) there is increasing pressure to protect coastal marine environments. In addition, new emphasis has been placed on the identification and protection of coastal marine resources in response to the Convention on Biological Diversity, which has been ratified in many countries including New Zealand and Australia, as well as other international obligations (e.g. UNCED, Agenda 21; UNCLOS). Politicians, scientists and the public of both countries have recognised the importance of gaining information on and protecting coastal marine environments, and this mood is a worldwide phenomenon (Gubbay 1995).

At the time of writing, 14 marine reserves were established under the Marine Reserves Act in New Zealand. In addition to these, two marine parks have been implemented under the Fisheries Act, and a marine protected area under its own Act of Parliament (DOC 1995).

Figure 1.2 An example of zoning that is used for a MPA on the east coast of Australia, Solitary Islands. Details are given of the activities that are allowed in each zone. 'Sanctuary zones' are the only areas in which there is virtually total exclusion. All types of zones are colour-coded for clarity. The whole area is entitled a 'multi-use reserve'. There are many other types of reserves in Australia, including protection for specific habitats (Fish Habitat Reserve), historical sites (e.g. Historic Shipwreck Protected Zone); National Parks (Marine National Park) and conservation of a representative area (Conservation Park). Few of these reserves are zoned as 'sanctuary zones' (Ivanovici 1984; Zann 1996).

In Australia (including tropical latitudes) there are over 200 areas that have various degrees of protection (Ivanovici 1984; Zann 1996).

The degree of protection and size of MPAs varies considerably. For example, in New Zealand most marine reserves are totally protected from commercial and recreational fishing and other forms of harvesting (e.g. Cape Rodney to Okakari Point Marine Reserve – Ballantine 1991), while others permit some recreational fishing (e.g. Poor Knights Islands Marine Reserve). In temperate waters of Australia the situation is even more complicated, where fully protected areas (called sanctuaries) and many other MPAs are open to recreational and limited commercial fishing activities, exploration and mining (Ivanovici 1984; McNeill 1994; Fig. 1.2).

MPAs of the world range greatly in size from those in excess of 50 km long and 100 km^2 in area (e.g. South Africa – about 350 km^2: Buxton & Smale 1989; Thackway 1996) to sanctuaries in Australia that may be as small as 200 m long and about 0.02 km^2. Clearly, the objective of MPAs need to be considered before decisions are made on the degree of protection and size.

The justifications for MPAs are numerous (Ballantine 1991; Kelleher & Kenchington 1992; Inglis 1993; Jones et al. 1993; Walls & McAlpine 1993; McNeill 1994; Gubbay 1995; DOCC 1995) and include arguments that focus on social needs (e.g. education), conservation (e.g. representative areas of coastline) and commercial interests, in particular the sustainability of fisheries (see Table 1.1, p. 22). The arguments vary greatly from criteria such as protecting unique or representative assemblages from exploitation, to more complex justifications that are attractive but for which more information on their validity is required. For example, it is often assumed that MPAs provide a reservoir of commercial and non-commercial species that diffuse into areas that can be fished. Although recent studies described positive effects of protection (Bell 1983; McCormick & Choat 1987; Buxton & Smale 1989; Cole et al. 1990; Rowley 1992; MacDiarmid & Breen 1993), more evidence is required to demonstrate a spillover of commercial species to adjacent areas used for fishing (but see Russ & Alcala 1989; Kelly 1997). Although there has been active debate about the merit of these ideas, the perceived success of established marine reserves and parks in Australasia and other parts of the world has prompted increasing pressure from the public and from scientific groups to establish more MPAs.

Politics aside, decisions concerning the selection and investigation of MPAs will frequently be concerned with a biological resource; namely, the plants and animals that make up the living assemblage within a reserve (Choat pers. comm.). Although it has been argued that we should create reserves and ask questions later

(Ballantine 1991; NOAA 1995), biological assessments of marine protected areas help justify MPAs. For example, the choice of regions for MPAs (Walls 1995a; Thackway & Cresswell 1995), the number of MPAs within regions, the size of protected areas, recording the changes that result from protection, and studying organisms in areas where they are not exploited by humans.

Initiatives to characterise biologically relevant regions are underway in Australia (i.e. IMCRA – Interim Marine and Coastal Regionalisation of Australia: Thackway & Cresswell 1995) and are increasingly part of the argument for the location of reserves in New Zealand (Walls & McAlpine 1993; Walls 1995a). Considerable research effort, however, is still required for the initial characterisation of candidate areas and habitats, together with ongoing monitoring of MPAs and control localities. Further, the diversity of biological questions that relate to MPAs will involve a variety of approaches to sampling (e.g. Battershill et al. 1993). Approaches to the study of MPAs and other types of impacts are emphasised in this book.

1.3.3 Importance of habitats

In marine and terrestrial ecosystems it is well recognised that the biotic and physical attributes of habitats have a major influence on the distribution, diversity and survivorship of marine organisms (Mushinsky & Gibson 1991; Sebens 1991). Many environments (e.g. estuaries and rocky reefs) are made up of mosaics of habitats. Changes in the nature of these habitats can cause rapid changes in the composition of the associated flora and fauna. The relative importance of habitats varies with taxa, and some habitats are identified as being especially important for species diversity and the sustainability of populations of species of commercial interest. For example, seagrass beds within estuarine environments, and in some cases on open coastal environments (Sanchez-Jerez & Ramos-Espla 1996), have been demonstrated to have a great influence on local species diversity and are thought to be important as nursery areas for fishes (Bell & Pollard 1989; Ferrell & Bell 1991; Bell & Worthington 1993; Gillanders & Kingsford 1993, 1996; Underwood & Chapman 1995).

Table 1.1 Commonly used justifications for the implementation of MPA status for an area and the criteria that are used. Some of these criteria can be used to articulate hypotheses that can be tested concerning the 'success' or otherwise of protection. Different approaches to sampling are required for many of these justifications. (Sainsbury 1988; Buxton & Smale 1989; Salm & Clark 1989; NOAA 1990, 1995; Ballantine 1991; Buxton 1993; Carr & Reed 1993; Dugan & Davis 1993; Inglis 1993; Jones et al. 1993; Walls & McAlpine 1993; Jones 1994; Suarez de Vivero & Frieyro 1994; McNeil 1994; Gubbay 1995, Botsford et al. 1997; Agardy 1997).

Social

Education
Research
Beauty and enjoyment for tourists
Preserve natural heritage in the public interest
Culture (historic, religious, traditional, etc.)
Public health (protect from contamination)
Reduce conflicts between user groups on areas of coast
Accessibility and safety for users

Conservation

Maintain or enhance species diversity
Protect rare species
Protect areas of high endemism (= dependency of species on a specific area)
Ecological fragility or vulnerability
Uniqueness (biology and geology)
Increase stability of assemblages
Maintain and increase habitat complexity

Sustainability of habitat for commercial species
Protect areas of high productivity
Representativeness (may include replicate areas of biogeographic provinces)
Buffer adjacent to terrestrial parks
Benchmark for comparison with impacted areas
Critical habitats (e.g. feeding, breeding, roosting, shelter areas)

Commercial, recreational and traditional interests (fisheries)

Refuge for commercial and non-commercial species
Maintain genetic diversity (especially with respect to maximum size)
Refuge for different life history stages (e.g. fish that change sex)
Increase in abundance of commercial species
Increase size and age of commercial species
Maintain an unexploited ecosystem
Enhanced fishery yields in adjacent fishing grounds
Reservoir for commercial and non-commercial species
Prevent obstacles to migration over points and bottlenecks
Restocking

Commercial – reproduction and recruitment

Protect spawning/mating areas
Refuge for fishes of highest fecundity (e.g. a large fish may equal the fecundity of tens to hundreds of small fishes)
Increase reproductive output within MPAs and to non-MPAs
High larval supply within MPAs and to non-MPAs
Increase recruitment within MPAs and non-MPAs

Changes in the nature of habitats can be caused by natural processes (Jones 1991) or humans (Ray 1996). The loss of organisms that characterise habitats (e.g. the seagrass of seagrass habitat or the kelp of kelp forest) will cause rapid changes in the composition of the associated fauna. A range of impacts can influence habitats. Habitats may be destroyed through infilling (e.g. harbour developments, causeways, runways, etc.) or organisms can be affected by an increase in sediment load or other change in water quality (Underwood & Chapman 1995). Impacts on a single member of an assemblage – for example, from disease (Lessios 1988) and extraction by harvesting – can cause a cascade of effects that can fundamentally change the nature of habitats (Hughes 1994; Ray 1996). The impact on extraction of single species will depend on whether organisms are 'habitat formers', 'determiners' or 'responders' (*sensu* Jones & Andrew 1993).

The importance of habitats and the way they influence organisms is a recurring theme throughout this book, particularly how biological questions often demand investigations that go beyond descriptions to the investigation of processes. Recognition that habitats are important has also placed demands on ecologists to mitigate the effects of impacts on organisms, by providing recommendations for alternative or 'replacement' habitats such as artificial structures (e.g. Hair et al. 1994).

1.3.4 Population dynamics, assemblages and harvesting

The effects of extraction of organisms from coastal environments are of concern to managers. They need to know how exploited and unexploited populations vary in size (by number or weight of organisms; e.g. J. B. Jones 1992). It is common for exploitation to be viewed as a single-species problem (Gulland 1988; Hilborn & Walters 1992). For example, how many sea urchins can be removed from rocky reefs without jeopardising the viability of the reproductive stock, rates of catch per unit effort, and the magnitude of recruitment? Important factors that influence variation in the size of a population include rates of recruitment, growth in size to maturity and maximum age, movement (i.e. immigration and emigration) and mortality (harvesting and natural): see Figure 1.3, p. 24.

The relative importance of each factor in influencing the size of a population will depend on the types of organisms (e.g. algae, ascidians, fishes) and type of population that is the focus of the study: the entire biogeographic range (= meta-population); a stock unit within the metapopulation that is largely self-regulating through reproductive output and recruitment (= mesopopulation); or a local population (e.g. a reef or MPA).

An assemblage is a collection of species that have overlapping distributions over space and time within an area (e.g. habitat, depth range of a stretch of coastline), and is often defined as a 'community' (Fauth et al. 1996). Organisms interact with each other directly or indirectly through processes such as predation, interference (competition) and changes in the nature of habitats (Fig. 1.3). Thus an understanding of the dynamics of many populations requires knowledge of the relevant interactions with other species.

It is often appropriate to consider the multispecies aspects of exploitation. The removal of some species (e.g. through harvesting) can cause a 'cascade' of changes in the nature of assemblages (Sloan 1986). These so-called indirect effects can have major consequences for the persistence of other organisms (Wootton 1994). For example, sea urchins have an important role in determining the standard crop of macroalgae (e.g. kelp *Ecklonia radiata* and coralline turf: Andrew 1988). The grazing activity of urchins has a major contribution to the removal of algae in so-called 'barrens' (*sensu* Underwood et al. 1991). The distribution and survivorship of many reef-associated organisms are influenced by habitats (e.g. fishes: Connell & Jones, 1991; Holbrook et al. 1994). Thus, the removal of urchins not only has the direct effect of extraction, but a cascade of effects ranging from a reduced level of grazing and subsequent change in the distribution of habitat types, to changes in the survivorship of other reef-associated organisms. Similar arguments of indirect effects have been given for extraction of rock lobsters from reefs (which may influence numbers of urchins: Elner & Vadas 1990) and large bivalves and crustaceans which can live in soft substrata (can change the nature of sediments and sediment chemistry: Riddle 1988). In a managerial context, therefore, it is important to have a knowledge of processes influencing the abundance and sizes of organisms in coastal marine environments. Moreover, processes can only be elucidated by using manipulative experiments (see section 2.6).

1.3.5 Introductions

Introductions are another way that people can change the nature of coastal marine environments (Carlton 1989). The introduction of exotic marine organisms and disease can affect the structure and 'health' of natural assemblages, as well as have impacts on commercial enterprises (e.g. fisheries and aquaculture). It is well recognised that there are, and have been, multiple vectors that facilitate the introduction of organisms that are either of little concern biologically, or have an

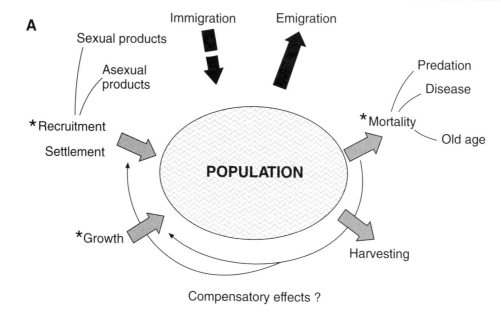

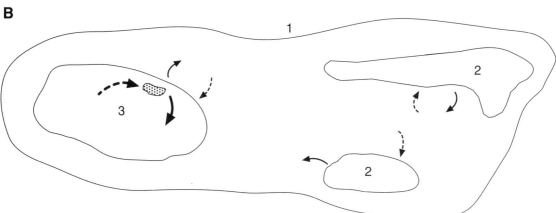

Figure 1.3 A diagrammatic representation of the processes that influence the size of populations and the different types of populations (e.g. a single reef through to the entire biogeographic range) that are often the focus of ecological studies.

A Population

The bubble indicates the size of a population (in weight or number of individuals). The relative importance of each arrow for influencing population size will vary according to the type of organism and the scale of the investigation; i.e., a study on the entire biogeographic range of a species (metapopulation = population of populations *sensu* Hastings & Harrison 1994; where by definition emigration and immigration = 0), a stock within a metapopulation or a local population such as an individual reef (see B). It has been argued for some species in 'open populations' that growth and levels of recruitment are influenced by the size of the population (= 'density dependence' Cushing 1975). So called compensatory effects are thought to exist for some species, but little good evidence

is found for most exploited populations. For example if population size is reduced it is predicted that an increase in recruitment will compensate for this loss. Many other investigators, however, believe that populations are kept below levels at which density dependence is likely to occur, mainly due to the vagaries of recruitment (e.g. Sinclair 1988; Doherty 1991). An assemblage represents multiple populations that are contiguous in space and time; * indicates those factors of each population that may be influenced by interactions with other species (e.g. predation, competition).

B Different types of populations

1. Metapopulation = a group of populations that may equal the biogeographic range of a species.
2. Mesopopulation or stock (usual terminology for fisheries).
3. Local population.

Dashed arrows refer to immigration and the input of reproductive products. Solid arrows indicate emigration and the export of reproductive products. Immigration and emigration should be minimal from a mesopopulation.

ability to cause major changes to local assemblages. A wide range of phyla including protists, algae, cnidaria, bivalves, crustaceans and echinoderms have been accidentally introduced to countries around the world on the hulls of sailing ships (Griffiths et al. 1992; Carlton & Hodder 1995) or modern-day vessels (Carlton 1985; Rainer 1995) as larvae, juveniles and/or adults. Others have been introduced as live bait or have been released intentionally (e.g. Cohen et al. 1995).

Some fouling organisms on ships (e.g. barnacles) may be sexually active and reproduce in ports of call, thereby introducing exotic larvae to surrounding waters. Many exotic taxa have been introduced as larvae into coastal waters through discharge of plankton-rich ballast water pumped aboard ships at foreign ports. Carlton and Geller (1993) found 367 taxa in the ballast water of Japanese ships docking at ports in Oregon, USA, most of them organisms that have a planktonic phase (see Chapter 8). The transport of exotic marine organisms has been a two-way process between hemispheres (cf. Skerman 1960; Gosliner 1995) and although it is recognised that natural assemblages have been altered greatly by introduced species in many parts of the world, evidence suggests that introductions continue unabated (Carlton 1989). For example, Japanese kelp (*Undaria pinnatifida*) has invaded rocky reefs of Australasia, and Asian mussels and tubeworms have invaded habitats of soft substrata (Clapin & Evans 1995). Southern Hemisphere contributions to the Northern Hemisphere include barnacles to rocky substrata (*Elminius modestus*, Bennett 1987) and gastropods to sediment flats (Gosliner 1995). Some invasions have immediate and devastating impacts on local assemblages. For example, the invasion of the Asian mussel (*Pontamocorbula amurensis*) to northern (Carlton et al. 1990) and southern waters (R. Creese pers. comm.), and the northern Pacific sea star (*Asterias amurensis*) to southern waters (Johnson 1994; Byrne 1996), is of great concern because of their rapid proliferation (facilitated by high fecundity, a bipartite life cycle and high survivorship) and resulting impact on local endemic populations.

The biological consequences of invasions can include the following: predation that can have an important (*sensu* Weldon & Slauson 1986) influence on patterns of abundance of local organisms; competition for space or resources (e.g. food); provision of a different food source for predators, thereby altering natural trophodynamics; disturbance and a host of indirect effects (e.g. altering the suitability of substrata for the settlement of other organisms). Introductions can, therefore, alter biogeographic patterns, change local assemblages, cause local extinctions, alter patterns of gene flow between otherwise isolated populations and make it more difficult to understand the origins of assemblage diversity (i.e. a mix of natural and human-mediated dispersal: Carlton & Hodder 1995). Biologists may, for example, be asked to determine the geographic extent of invasions, understand the nature of interactions with other species, or make recommendations on the control of introduced species (e.g. Lafferty & Kuris 1996).

1.3.6 Understanding biological processes

Soundly designed and biologically relevant studies on processes are fundamental to our understanding of the dynamics of coastal marine environments. An understanding of processes requires a comprehensive description of biological patterns (i.e. in space and time) and manipulative experiments. Monitoring required to measure variation in abundance (or other variable) in time (e.g. seasonal change), and a well-designed monitoring study (i.e. which is also replicated in space), should be able to detect changes due to major pulse events (e.g. storms) or slow but unidirectional changes in populations or assemblages (e.g. Holbrook & Schmitt 1998). The investigation of processes, however, requires experimentation and although so-called 'natural experiments' (e.g. natural perturbation such as a storm correlates with a major change in abundance or diversity; or natural spatial variation in assemblages) provide insight and a good platform for hypothesis generating, manipulative experiments are best used to test predictions (reviews: Hurlbert 1984; Krebs 1989; Underwood 1993a).

Experiments that have revealed a great deal about processes in the pelagic and benthic environments are described in this book. They should not, therefore, be considered as belonging to the realm of esoteric science: they have contributed greatly to our understanding of the dynamics of coastal marine environments (e.g. Petraitis 1990; Underwood 1993a; Thrush et al. 1995), and they have often allowed predictions to be made on the nature and magnitude of potential impacts.

The ethics of manipulative experiments and destructive sampling is sometimes questioned, especially where sampling involves the destruction of habitat, removal, injury and death of valued organisms (e.g. reef fish). If this is the case, then the value of the sampling, even the need to carry it out at all, should be considered carefully. Experiments are the only valid method of investigating processes that influence organisms and destructive sampling of some organisms is often the only way that valuable information can be obtained (e.g. on the age of fish). The only alternative to sampling is to guess what is happening with little knowledge of biology, and interactions between multiple important factors. Thus the loss of tens or hundreds of organisms that are considered valuable may be justifiable if the information is crucial to understanding biology for management purposes and the numbers of organisms represent an acceptably small proportion of the population of concern.

1.4 Approach and aims

It is clear from the range of purposes for which biological assessments are undertaken that no single set of sampling methodologies and designs will be appropriate to all situations (Chapter 2). The exact procedure adopted will depend on the specific purpose or hypothesis of the study, background information that is available, and the situation (e.g. logistical constraints) faced by the investigator. For example, is the purpose of the study to:

- Select a locality for a MPA where the objective is to protect representative habitats?
- Obtain baseline information on organisms associated with subtidal reefs?
- Detect impact-related changes in numbers and size-frequency of organisms?
- Estimate total numbers of urchins along a stretch of coastline?
- Determine the factors that are influencing the patterns of recruitment for a particular species?
- Determine the relative importance of abalone (paua) and sea urchins (kina) in maintaining the structure of subtidal assemblages?

These examples require approaches varying from a review of existing information to broad-scale studies and manipulative field experiments. Where biological assessments are made, the investigator must be aware of the characteristics of marine organisms, regardless of whether adequate descriptions and interpretation of temporal and spatial patterns are to be obtained, or manipulative experiments are to be designed.

The reproductive products of most marine organisms disperse into the water column as plankton (Chapter 9) and, as a result, the reproductive output of organisms, or the number of residents, may have little bearing on subsequent recruitment into an area (Holbrook et al. 1994).

The longevity, growth rate, age to maturity, whether or not the animal moults and patterns of movement, and other attributes need to be considered in the sampling design and interpretation of data. Moreover, many marine organisms also have specific habitat requirements. Thus, abundances may vary markedly among habitats and over time. This knowledge should be considered in descriptive and manipulative experimental designs.

Different biological problems as well as temporal and spatial heterogeneity in the distribution of organisms can make the prospect of carrying out an ecological assessment daunting. A series of questions come to mind:

- What sort of precision and biases can be expected from the different sampling techniques?
- Is there a need for preliminary sampling?
- What is a good sampling design?
- How many localities should be sampled?
- How many replicates should be taken?
- What sort of data needs to be collected (e.g. densities, total numbers, occurrence, size)?
- Which species and physical variables need to be considered?
- Will the data be sensitive enough to allow measurement of differences over space and time?
- What degree of change (e.g. 25% change in abundance) are we required to detect?

There is an extensive literature both published and unpublished, concerning sampling design (e.g. Green 1979; Andrew & Mapstone 1987; Krebs 1989; Keough & Mapstone 1995), statistics (e.g. Winer et al.1991); monitoring (e.g. Dahl 1978; UNESCO 1968; Omori & Ikeda 1984; Underwood 1992), sampling methods (e.g. Baker & Wolff 1987; Mapstone & Ayling 1993), marine surveys of specific locations (e.g. Ayling 1978; Underwood 1991; Schiel et al. 1995) and assessing the stock size of organisms for harvesting (e.g. McCormick 1990), as well as literature reviews on marine environments (e.g. Gordon & Ballantine 1976; Morton & Miller 1973; Kelly 1983; Bennett 1987; Underwood & Chapman 1995). However, no explanatory document has been published in Australasia that integrates this information for application in temperate coastal environments.

The aim of this book is to provide a resource for principal investigators and managers to facilitate high-quality ecological assessments that consider the biology of organisms and rigorous quantitative methodology. The principal investigator (or co-ordinator) of a study should be a person with appropriate qualifications for this type of work. The book is not supposed to be a substitute for formal biological training, but is intended to provide a useful document that will help you to make decisions on appropriate approaches, sampling designs, target habitats and organisms.

Although our focus is on temperate marine coastal environments, with an emphasis on organisms of reefs, soft substrata and pelagic environments, the book has broad applicability to descriptive and experimental investigations at other latitudes and in other environments. General procedures for establishing a study and considerations for sampling designs and the treatment of data are described. Approaches to sampling particular environments (e.g. soft substrata) and specific types of plants and animals are also outlined. Specifically, this involves an overview of sources of variation in space and time that should be considered in sampling designs, descriptions of the utility of different sampling devices and methods, special considerations for different types of organisms (e.g. life history), the utility of experiments, the categorisation and treatment of specimens recorded or collected in a study and examples of relevant literature. Case studies are given in each chapter with the purpose (objectives/hypotheses), situation (e.g. logistical constraints, background information on appropriate sample unit sizes etc.), and approach taken (especially the sampling design) clearly indicated and an overview of results and links to related studies are given.

The book is organised in the following way:

Chapter 2 is a general chapter outlining procedures and considerations for establishing a study. Kingsford and Battershill emphasise the importance of identifying the biological question or nature of the problem before describing appropriate sequences for planning. This ranges from obtaining background information and deciding on a sampling design (descriptive or experimental), to choosing equipment. It also lists the priorities and common pitfalls that should be considered in these situations. For example, one-off baseline studies are commonly undertaken, largely owing to constraints of time and money. It could be asked, what is the minimum requirement for such a study? We argue, however, that one-off baseline studies are of limited value because they are often done without a biological question/hypothesis in mind. Biologists are often asked to measure impacts. Support is given to Underwood's (1992) extension of the BACI (Before, After, Control, Impact) philosophy of Green (1979) and the similar Multiple Impact (MBACI) approach of Keough and Mapstone (1995), in that multiple spatial controls should be used and multiple times sampled if impacts (positive and negative) are to be detected. The nature of impacts is discussed. It is emphasised that investigators should decide whether their chosen design is capable of detecting 'pulse' or 'press' type responses to an impact. The nature of and utility of experiments is discussed with special emphasis given to considerations for sampling designs (especially the nature of controls).

Chapter 3, by Kingsford, describes analytical aspects of sampling design in relation to the various problems described in Chapters 4 to 9; great emphasis is given to stratified sampling. This chapter should be consulted for a description of the principles of different analytical procedures and a guide to published examples. Information is also given on methods for determining appropriate sample unit sizes and number, allocation of sampling units in space and time, types of sampling designs, and treatment of the data.

Chapters 4–9 examine various methods used to study assemblages of organisms associated with reefs, soft substrata and planktonic assemblages. Examples of previous studies are presented in each chapter.

Subtidal habitats and benthic organisms (**Chapter 4**, by Kingsford and Battershill) are discussed first because an understanding of the distribution of different habitat types is often a prerequisite for work on other organisms, and the subtidal part of reefs (cf. intertidal) generally encompasses the greatest area of reef. Moreover, changes in the presence or cover of habitats can cause great changes in the distribution and survivorship of organisms (Bell et al. 1991). There are strong links between Chapters 4 and 5 because the large invertebrates discussed in Chapter 5 often have a significant influence on the structure of habitats.

The organisms that are most heavily exploited and are important for the dynamics of reef assemblages include large gastropods (e.g. abalone), urchins, rock lobsters, and reef fish. **Chapter 5**, by Kingsford and MacDiarmid, focuses on large mobile invertebrates, and **Chapter 6**, by Kingsford, on reef fish. It is emphasised in both of these chapters that a priority for any study setting out to assess sampling these animals should be impact on reefs.

Many of these organisms also have important ecological roles where they are found, and investigations of the processes influencing their abundance and their impact (i.e. 'indirect effects') on other organisms is also important. For example, sea urchins are harvested and are often used as bait by divers to attract fish. The urchins, however, have important roles in subtidal assemblages because they graze on macro-algae and can alter the nature of subtidal habitats (e.g. Andrew 1988, 1993). Similarly, should 'top-end' carnivores flourish within a MPA, their impact may be significant on urchin densities, which in turn influences the dynamics of algae (Jones et al. 1993).

A review of methods and approaches to studying demersal fishes of shelf and deep-water environments was beyond the scope of this book, and we advise readers with an interest in fisheries to see 'Fisheries' in Selected Reading at the end of each chapter. We do, however, consider chapters on procedures for setting up a study, sampling designs and specific aspects of study (e.g. age and reproduction of fish, Chapter 6) relevant to studies in these environments.

Chapter 7, by Kingsford and Creese, deals with the organisms of intertidal reef and soft substrata environments and points out that these environments are relatively easy to work in and may be particularly prone to impacts (e.g. Kingsford et al. 1991a). Accordingly, inter-tidal environments should be identified as a priority for sampling in many biological assessments. Many advances in sampling methodology and design have been made in intertidal areas and this environment should always be considered for testing different approaches to sampling. Related chapters include Chapters 4 and 5 for hard substrata, and Chapter 8 for assemblages of soft substrata.

Soft-substrate subtidal environments are common in coastal areas, often the basis of important commercial and recreational fisheries, and are subjected to a variety of anthropogenic impacts. Impacts include changes in riverine influence and regimes of sedimentation (e.g. through developments, dredgings, sewage). More recently, new species introductions have caused major concerns for the health of these environments.

Chapter 8, by Morrisey, Turner and MacDiarmid, describes methods used to sample soft bottoms and experimental methods for investigating processes.

Chapter 9, by Kingsford and Murdoch, deals with planktonic assemblages. Most marine organisms associated with rocky and soft substrata spend time in the plankton as larvae. Many biological assessments require a knowledge of processes influencing survivorship in the plankton (natural or anthropogenic factors) until pre-settlement forms reach potential settlement sites. Moreover, a knowledge of planktonic assemblages can help us to understand processes that influence local production, which in turn can have an impact on pelagic and benthic assemblages (positive or negative). For example, toxic or non-toxic blooms of plankton can result in large sinks of live and dead planktonic organisms that can be either beneficial (as food) or, in some cases, kill benthic organisms through smothering.

Chapter 10, by Kingsford and Battershill, includes sources of expertise on taxonomy for the Australasian regions and gives some advice on the preservation of specimens. All chapters are extensively cross-referenced and a comprehensive index, glossary and literature section is provided.

SELECTED READING

The references listed are examples of research that will lead the reader to a broader literature, also see the text of the chapter. * indicates a multi-authored book or collection of proceedings; references to volumes of a journal refer to a set of relevant papers.

Ecosystems and management: Ward & Jacoby 1992; Norse 1993; *Pacific Conservation Biology* 1995, vol. 2*; Christensen et al. 1996; Caughley & Gunn 1996.

Introductions: Carlton 1989; Carlton & Geller 1993; *Ecology* 1996 vol 77(7)*; Hilborn et al. 1995.

Fisheries: Gulland 1988; Hilborn & Walters 1992; Kailola et al. 1993; Tilzey 1994; Sainsbury 1996.

Habitats (importance and mitigation): Larkum et al. 1989; Battershill et al. 1993*.

Impacts and EIA: Rowley 1992*; Schmitt & Osenberg 1996*; Harvey 1998; see also Chapter 2.

Management: Kenchington 1990; Gubbay 1995*; Zann 1996.

Marine ecology of temperate regions: Keough & Quinn 1990*; Battershill et al. 1993*; Hammond & Synnot 1994; Underwood & Chapman 1995*.

Marine protected areas: Salm & Clark 1989; NOAA1990, 1995; Ballantine 1991; Kelleher & Kenchington 1992; Battershill et al. 1993*; Jones 1994*; McNeil 1994; Ramos Espla & McNeill 1994; Gubbay 1995*; Clarke 1996; Botsford et al. 1997; Schmidt 1997.

Marine and terrestrial links: Glasby 1991; Grimes & Kingsford 1996; Ray 1996.

Natural history: (Australia) Bennett 1987; (New Zealand) Morton & Miller 1973; Adams 1994; Cook in press; see also Chapter 10.

Pollution: Clarke 1989; Miskiewicz 1992*; Phillips & Rainbow 1993; Phillips 1995; Paine et al. 1996; Sindermann 1996; *Marine Pollution Bulletin* (in press, volume)

Sampling design: Green 1979; Andrew & Mapstone 1987; Krebs 1989; Underwood & Chapman 1995*; Keough & Mapstone 1995; see also Chapters 2 and 3.

TWO

PROCEDURES FOR ESTABLISHING A STUDY

M. J. Kingsford and C. N. Battershill

2.1 Introduction

Thorough planning of any study is crucial to its success. The purpose of this chapter is to aid in the identification of a specific biological problem and then to suggest steps that should be taken to ensure the successful completion of a study. Particular emphasis is given to preliminary studies and decisions concerning sampling designs. Impact studies and experimental studies are common, and their utility is discussed, with guidance on common pitfalls in design.

Specific advice is also provided on correct approaches to monitoring and baseline studies. In general, investigators are urged to avoid baseline studies where sampling is only done at one time, or without specific objectives in mind. Many studies require more than one investigator, in which cases we urge principal investigators to train collaborators, to minimise bias among observers and practise procedures to ensure consistency in activities.

The range of gear required for a study can be considerable and because an expensive field trip may founder from overlooking a weight belt or similar item, we have listed various types of equipment that should be considered for a variety of descriptive and experimental studies.

We agree with Caughley and Gunn (1996) in that the nature of the output of the project should be considered at the planning stage. Outputs may include public information, CD-ROMs, unpublished internal reports (which are eventually lost), and papers in refereed books and journals. If the data are an important building block for a programme, or are the basis of informed management for a species or environment, then the source of the information should be recorded carefully. The results of research, and reviews of existing information, should be published in easy-to-access journals or books. It may also be appropriate to provide raw data, reports and publications on the World Wide Web for easy access.

2.2 Identify the problem

The first priority, before beginning any fieldwork, or even planning, must be to identify the question or questions that are to be addressed. The best ways of addressing these questions can then be considered. This initial focus will help you determine the most appropriate sampling designs. If you cannot succinctly outline the objective of your work to a colleague, you are probably about to waste time and money.

Examples of the biological problems that require field assessments are shown in Figure 2.1 (p. 30) and a variety of descriptive and experimental studies are described in more detail in the example section of each chapter. These examples point out the need to ensure that the data collected are strictly relevant. There are many instances where a great deal of information has been collected but proves to be not relevant because of poor design or a poor choice of a subject species. Many organisations request proposals before they will fund a study. The process of refereeing by experts in the field is the best way of ensuring that the study has a sound justification (biological, social, financial, etc.) and the methodology and suggested forms of analyses that are most appropriate for the problem.

There has been a crescendo of pleas for greater rigour in ecological studies. The logic and philosophy necessary for the design and completion of ecological studies are discussed by Underwood (1990, 1997) and Scheiner and Gurevitch (1993). From a set of observations, or understanding from patterns and processes that have been documented in the literature, hypotheses may be generated. The subsequent steps that result in the decisions on sampling design, the collection of data and analyses, may prove challenging. However, the formal constructs of hypothesis testing, or the Popperian approach, described by Underwood and given in Figure 2.2 (p. 31), should result in unambiguous sampling designs. A model (i.e. explanation) is articulated from

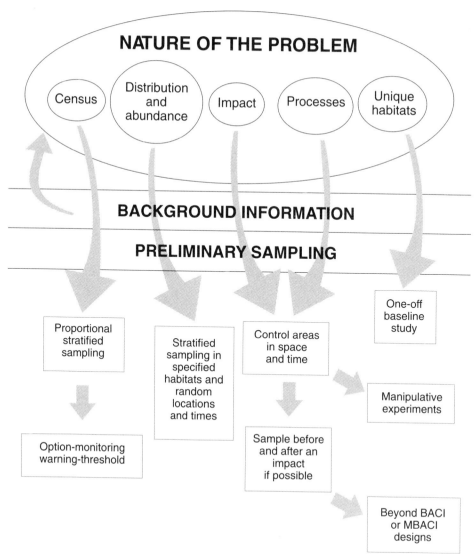

Figure 2.1 Examples of ecological studies that require field assessments. Common problems are given in bubbles, while the figure shows the variety of approaches that have been used to address them. Good planning always requires a knowledge of relevant background information. Background information often leads to the identification of problems that are properly expressed as hypotheses. Preliminary sampling is often required, but not obligatory if background information is adequate. Boxes refer to common options for sampling designs, and analyses that may be appropriate to the original problem and related hypotheses.

Census (= stock assessment)
- What are the total numbers and size-frequency distribution of species A in the study area?
- What is the biomass of species A within the study area?
- What influence will marine reserve status have on the total numbers and size-frequency distribution of species A (usually a target species: see Chapters 5 and 6)?
- If fish species A reaches a predetermined threshold density, recreational fishing will be banned within the study area.

Distribution, abundance and monitoring
- Are there differences in the abundance patterns and size

structure (of target species) among islands at different latitudes?
- What are the seasonal patterns of recruitment of species A along a coastline?

Impact studies and monitoring
- What influence will marine reserve status have on the abundance of organisms (and size-frequency distribution of target species) within the study area?
- Will the dumping of dredged sediment have an impact on populations of commercial shellfish?

Processes
- What influence does the clearance of kelp have on subsequent recruitment of kelp?
- Do subtidal gastropods depend on the grazing of sea urchins to keep algae at a suitable level for grazing?

Identification of unique and representative habitats
- Requires a general description of the marine environments and organisms that are found there.
- Has been requested to choose area for marine reserve status.

Note: we caution the use of this option (see section 2.8).

an observation or ecological question. Based on this model a formal hypothesis is described and this is tested using an appropriate and unambiguous design.

Because it is logically impossible to prove a hypothesis, the alternative approach is to seek disproof or falsification. Thus, a null hypothesis is constructed within which are identified all possibilities except the actual proposition in the hypothesis. If when the data are tested the null hypothesis has to be rejected, the findings are consistent with the predictions from the model.

Example

An investigator observes that abundances of urchins appear low on reefs in the vicinity of a boat ramp with adjoining workshops (potential source of pollution). The model would be that this is so. The hypothesis is that urchin abundance will be lower in areas that have boat ramps (and associated workshops) than at control reefs (without boat ramps), within a restricted latitudinal range. The null hypothesis is that urchin abundance will be no different in the impact areas than at control reefs within a restricted latitudinal range. Samples are collected at multiple impacted and control areas, and the results analysed using ANOVA. The null hypothesis will only be rejected at $P < 0.05$ (i.e. the hypothesis is accepted). Based on the findings and observations from the first sampling design it may be noted that a specific pollutant is released from each boat ramp. Thus, a second model may be that pollution from the ramps is responsible for the observations. The hypothesis will be: if I remove pollution from ramps, numbers of urchins will increase at these ramps compared to ramps that continue to be polluted.

Krebs (1989) cautions the biblical acceptance of only rejecting the null hypothesis using statistical inference. In particular, he states that the major difficulty in ecological statistics is that statistical methods can only cope with random errors. 'In real situations systematic errors, or bias, may be more important, and no statistical test can detect biased data.' In general he argues that ecologists try to minimise bias (i.e. where bias means that the wrong data were collected), but that there should be some scepticism about the results of statistical tests of significance. It should be noted, however, that a crucial link between hypothesis testing and the critical test, is that data should be collected to separate alternative hypotheses unambiguously.

When a statistical test identifies differences among samples, it is always important to consider whether the significant difference is ecologically relevant (Krebs 1989). For example, a 5% difference in the survival rate of male and female fish may be biologically relevant, but may be difficult to detect statistically unless there

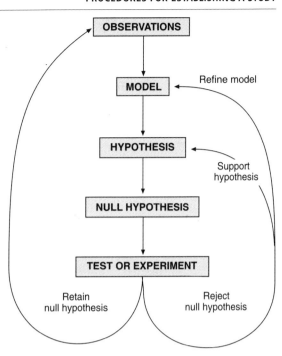

Figure 2.2 Hypothesis testing (adapted from Underwood 1990). The logical components of a falsificationist experimental procedure and their relationships (see section 2.2).

are very large sample sizes. In contrast, by measuring 20,000 abalone at two locations a statistically significant difference may be found between the two populations even though the difference in weight is only 0.05 g and biologically unimportant.

Hypothesis testing with appropriate sampling designs and robust statistical tests has clearly made ecological research more rigorous, but this should be tempered with a biological perspective of what is relevant. The justification of relevance (and sound hypothesis testing) should be determined *a priori* rather than in a subjective fashion *a posteriori* (Green 1994). For example, in the abalone case given above, managers could have decided before a survey that they would only be concerned about a difference in abalone weight among locations if that difference exceeded 10 g. Similarly, fisheries managers may be concerned about detecting changes in densities of fish that exceed 25% of the existing population. This is an example of background information preceding the identification of the question (see Fig. 2.1). With an adequate knowledge of variances in a population, sampling can be designed to detect differences at whatever desired level of precision (see Chapter 3).

2.3 Background information

The results of previous studies will provide information about the organisms that may be found, including some indication of their distributions. If data are available on abundances and variances, they may help in the choice of a sampling unit size and number as well as particular species or species groups (see section 3.3). In some circumstances background information may provide data on the magnitude of variation you are likely to encounter at different spatial and temporal scales, which can be of great use in the planning of descriptive or experimental studies.

Information on the physical and biological attributes of the study area should be obtained; this will help at the design stage of many studies (see Chapter 3). Considerable insight on physical aspects of the area may be gained from aerial photographs, tide tables, charts of bathymetry, topographical maps, swath mapping (see Fig. 4.6), geological maps, and even satellite imagery. If aerial photographs are not available, aerial photography could be included with preliminary sampling for the choice of sampling sites and even the mapping of habitats. This would depend on the budget for the study and the potential relevance of obtaining such information. In most countries, there are key marine research institutes and museums that have Geographical Information Systems (GIS), from which useful information may be readily obtained.

Previous studies can help in the context of manipulative experiments. Likely problems with the design of treatments that may be resolved with background information include: problems in maintaining removals and suitable controls (that were done or should have been done); cage artefacts; and the choice of densities of target organisms (see section 2.9). Moreover, other studies can significantly improve choices of appropriate spatial and temporal scales for experiments (i.e. duration, number, and location of experiments). For example, on rocky reefs the dispersal of some algal spores is predictable. The spores of *Ecklonia radiata* are released over a few months in winter (Kennelly 1987a; Jones & Andrew 1990) and this seasonality may influence the outcome of clearance experiments, depending on when clearances are made. The settlement of many organisms is variable in space and time (e.g. Jernakoff 1983) and experiments will need to be repeated in time and space before the generality of the results can be assessed.

Discussions with colleagues who are working in related fields never fail to be productive. Electronic news-groups can also respond quickly to your queries. This feedback can help with practical suggestions, references you have missed, common pitfalls, and unpublished findings that are relevant to the planning of your study. Most researchers have, at one time or another, laboured over a practical problem only to find it had been resolved some time ago and the solution is well known to a colleague.

2.4 Preliminary sampling

Preliminary sampling is for minimising logistical and methodological problems you could encounter in the field and for making best guesses on sample unit size, number, etc. Because preliminary sampling, or a pilot study, is generally a subset of what you will encounter in the full sampling design, the results should be treated with caution.

Sampling before a study involves a short visit to the proposed study area. For many studies, a plane flight over a study area can give you a much greater insight into the difficulties to expect. This can help greatly in planning the logistics of different sampling designs, as an impression of different scales of habitat structure (or other relative environmental grain) can be obtained. Still photography or a video record of the study area, during such a flight, is useful and may be used as archival material. During preliminary sampling, logistical problems will be identified that are of value in working out methods and designs to be used in the main study, e.g. time taken (and hence cost) to sample at a locality. In this way, you may work out optimal sampling designs and analytical options (see Chapter 3), determine the appropriate size of sample unit, and check that your sampling device or method is sampling the population you think it is sampling (e.g. the mesh size of a sieve, the filtration efficiency of a plankton net). You may refine your original question or change your target organisms after preliminary sampling if the situation you find is different from expected. If it is not possible to carry out preliminary sampling for the best method, sample unit size, number of replicates, rate and ways in which units are searched, etc., then educated guesses may be made from the recommendations in this book, or from other background information.

Within the logistical limits of your study there are two main empirical features that need to be considered in pilot studies to identify the most appropriate sampling units: accuracy or relative bias; and precision (Andrew & Mapstone 1987; Mapstone & Ayling 1993; Fig. 2.3). 'Accuracy' refers to the extent of departure of the sampling mean from the true population mean. 'Bias' refers to the consistency of that inaccuracy (i.e. do repeated estimates tend to differ from the 'truth' in the same direction?). In many field situations it is impossible or difficult to determine the true value of the target being sampled, measured, or counted. In many cases you can only hope that bias remains consistent through time or among locations, so that real biological patterns of relative abundance can be detected. Pilot studies are vital to determine methods with the least bias, but a large pilot study is often necessary to avoid being misled (McArdle & Pawley 1994). 'Precision' is a measure of expected variation in repeated estimates of the same population and is generally expressed as the ratio of the standard error to the mean (see section 3.3). Precision is independent of bias. In other words, it is possible to have a very precise estimate of abundance that is very biased.

Preliminary sampling can help to reduce, or avoid, contamination among samples. Contamination by organisms from non-target sampling strata, or that have not been removed from the sampling gear prior to the next sample, is a problem in many studies, particularly where very small organisms are sampled. For example, it is critical to rinse a plankton net several times to remove all of the organisms that are stuck to the mesh. This is simple to check, by collecting rinses and checking the cod-end for organisms. A pilot study will allow calculation of the degree of contamination and choice of an acceptable level of contamination for the main study. You may, for instance, need to swim demersal plankton traps to the substratum before they are opened, or use a closing mechanism of unknown efficiency on a plank-

ton net. You can easily check that you are not contaminating your sample with organisms from the water column, or non-target-depth strata, by doing a pilot study. A simple improvement of your technique or the gear can completely remove the threat of serious contamination. The behaviour of animals can also pose problems. When counting mobile invertebrates under cobbles and boulders, or counting bivalves in soft sediments, animals are often disturbed when counting commences, and they may escape by burrowing or moving laterally. If the magnitude of escapement is known, then efforts can be made to minimise the loss of individuals.

Preliminary sampling can be essential to the success of manipulative experiments. If an ambitious experiment is planned and you are not sure how well the cage or some other exclusion or settlement device will work, it would be judicious to do a pilot study. You may be convinced that your experimental device will work, but nature doesn't always follow your predictions. For example, a colleague of ours made dozens of concrete clam shells because he was convinced that small damselfishes would find them irresistible as spawning sites. The fish ignored them completely and the concrete clams became another piece of ecological jetsam. A pilot study with a few of the concrete clams would have helped enormously. Most institutions that do marine research have a cemetery of gear that should have worked, so try to keep your contribution to a minimum.

Measuring the size of some organisms may pose problems if preliminary sampling is not done. For example, foliose algae may prove particularly difficult because

Figure 2.3 The distinctions between accuracy and precision. Frequency plots indicate replicate samples from a population. Bias is the distance of the sample mean from the true value.

ADAPTED FROM KREBS 1989

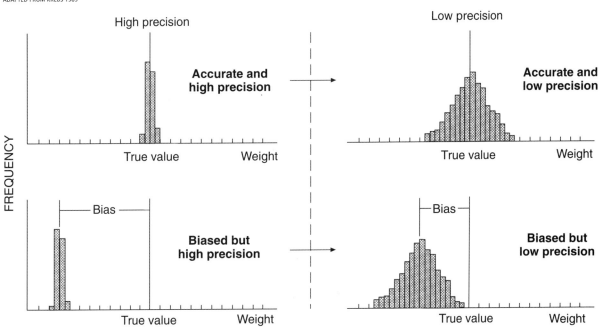

of their complex pattern of growth (McCormick 1990). In such a situation a number of measurements should be taken during preliminary sampling, and the best descriptions (e.g. of plant shape that contribute to an accurate measure of size or weight) should be identified in some exploratory analyses (e.g. principal components analysis, see Pimentel 1979). Some invertebrates (e.g. sponges) pose similar problems due to their irregular shapes and patterns of growth. Estimates of volume and weight may be needed in these situations using procedures such as recording dimensions or using stereo photography with appropriate calibration.

Preliminary sampling may also be used for 'pattern seeking' where you are having difficulty in identifying any biological or physical variation in the environment. Although soft substrata (over a narrow range of depths) may be homogenous, most environments have some biological or physical grain that will require stratified sampling.

Preliminary sampling, therefore, is of value to:

- Assess the logistics of sampling (time, cost, personnel, gear, hazards).
- Determine the feasibility of addressing your original question and the suitability of your target organisms.
- Determine that your sampling device/method will adequately sample your target population.
- Optimise methods to reduce contamination.
- Determine that your sample unit (size and number) is appropriate.
- Check that the rate and way in which a sample unit is searched does not influence results.
- Obtain some idea of the variation in abundance or size of target organisms or other variables at different levels of your sampling design.
- Determine likely problems in the design of treatments for manipulative experiments.
- Determine the best way to measure biological and physical variables.
- Focus the training of personnel (see section 2.10).
- Check for problems with quantities of preservative, the type of preservation and longevity of labels in your chosen preservative (see section 2.11).
- Gain an understanding of the grain (e.g. spatial pattern of habitats) of your study area so that more precise measures of variability in abundance (or other variable) can be obtained for comparisons between different spatial and temporal scales.

2.5 Sampling design

The analytical rationale for the use of particular designs and allocation of sampling is given in Chapter 3. In general, however, it should be noted that:

- An effect (e.g. changes resulting from protection or other type of impact) can only be demonstrated by comparison with controls. It is important to have controls both in time and space. It is preferable to have multiple controls in space and time when the purpose of a survey is to determine the effect of an impact (i.e. beyond BACI, MBACI, see 2.6). A design with a single 'control' and 'impact' is pseudo-replicated (*sensu* Hurlbert 1984). See section 2.6 for more details on the detection of impacts.
- Every level of sampling should be replicated (e.g. sample unit, sites, localities, times). For example, if comparisons are made among 'seasons' sampling should be done at replicate times within seasons. The only exception to this may be where multiple single samples are taken over a gradient for analyses using regression or Analysis of Covariance. However, multiple sets of data (i.e. for each regression) should be collected in space or time for this type of analysis.
- When a study area is being assessed, in general it is better if effort is not concentrated within one locality. Differences in the species composition and abundances of organisms (e.g. Choat & Bellwood 1985) are generally greater among localities than within localities, particularly as spatial scale increases (e.g. Kingsford 1989). Consider fewer replicates and more locations rather than the opposite, but beware of this approach if small-scale variation is high. It is possible to make more general statements about patterns and processes if studies are done at multiple locations.
- Always consider variation on short temporal and small spatial scales. Variation with time of day or on different days within a season can be great (e.g. Holbrook et al. 1994). For instance, regular sampling at monthly intervals (an interval based on tradition, not biology) can miss real changes in abundance, especially when stochastic or relatively regular recruitment or spawning events are measured (Underwood 1989, Fig. 2.4). Stochastic fluctuations in the abundance of organisms can be great and occur over short timescales (Coull 1985; Morrisey et al. 1992a, b).
- The methods used to analyse data should be determined at the design stage. It is no use putting a lot of effort into fieldwork, only to find that the factors you are most interested in are difficult or impossible to analyse. At this stage consider the form of your

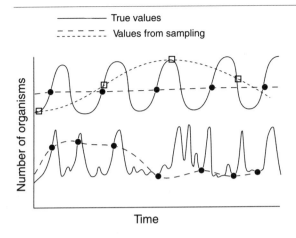

—————— True values

- - - - - - Values from sampling

Number of organisms

Time

Figure 2.4 The problem of monitoring at regular times without an understanding of variation in numbers over short periods of time. 'True values' of variation in abundance in nature and two different regular sampling designs (= values from sampling) that give an erroneous picture of natural patterns.

ADAPTED FROM KELLEY 1976, COULL 1985 AND UNDERWOOD 1989

data (e.g. number per unit area, number per unit time, biomass, species diversity).

- It is often better to have a larger number of small sample units (e.g. quadrats) than a few large sample units. Smaller sampling units are likely to be searched more thoroughly than large quadrats (although it is possible to divide up large sample units). The scale of sampling can, however, influence the measures of patchiness (Andrew & Mapstone 1987; Levin 1992). Preliminary sampling is useful to determine relationships between precision and sample unit size and number (see Chapter 3)

- It is critical that you take replicate samples that are independent of one another, and random samples should be taken if you expect to generalise. If sampling is not done randomly, or haphazardly (where replicates are considered independent) then it will violate the assumptions of many statistical methods (e.g. ANOVA). When this happens the results of analyses may be unreliable (exception: fixed sample units analysed with repeated measures).

- The design should be discussed with investigators who are familiar with aspects of sampling design and marine studies; they may save you time and money. Colleagues may discover that you have chosen a design that is confounded ('pseudo-replicated') and, therefore, is incapable of resolving your original question. Krebs (1989) was even more frank: 'Garbage in, garbage out'.

- Avoid reporting any ecological estimates without some measure of variability of the data (e.g. standard error).

- Recording data only as species presence or absence without some measure of abundance will often give results of poor resolution. Presence/absence data (= binary data) give very little information for the amount of time spent sampling. Furthermore, the records will vary considerably depending on the observer. In some cases it may be appropriate to augment abundance data with casual observations of species presence and abundance (e.g. studies on biogeography where some species are very rare).

- Not everything that can be measured, should be. It is useful to rank information in terms of its usefulness to resolving a biological problem. Ultimately, your hypothesis should determine what you measure.

- Consider the biology of the organisms you are sampling. Relevant considerations may include movements, timing of reproduction, age to maturity, maximum age, and whether your organism moults (e.g. lobsters).

Excellent reviews of approaches to sampling are given by Green (1979), Snedecor and Cochran (1980), Underwood (1981, 1991), Andrew and Mapstone (1987), Krebs (1989) and Schmitt and Osenberg (1996). Many of the considerations above have been adapted from these sources. In addition, see Figure 2.1 for general approaches to sampling based on a particular question.

Ecologists generally use 'statistical inference' in decision making. Biological data are inherently variable and, in essence, statistical inference involves deciding how large the difference among two or more samples must be before it is reasonable to assume that these differences are real and not merely a consequence of differences within samples. By definition this involves a black or white decision whether to accept or reject the 'null hypothesis' (i.e. that samples are not different, or accept the alternative that they are different). Where the null hypothesis is rejected it is generally at probability $P < 0.05$ (i.e. there is a 5% chance that the null hypothesis has been wrongly rejected and that differences among samples were, in fact, due to chance [this is referred to as Type I error]; see Table 2.1, page 36. In special cases this critical level (0.05) may be adjusted (e.g. to $P < 0.01$) where the investigator is being more conservative, for example because the data are heterogeneous (Underwood 1981) or when it is more responsible to consider risks at a lower value of P. For example, Mapstone (1995) argued convincingly that in many studies concerned with impact, critical error rates should be set for each case (e.g. critical $P = 0.1$ rather than 0.05). If trends strongly suggest an impact, it would be prudent (i.e. environmentally responsible) of managers to respond accordingly. For

Table 2.1 The four alternatives of hypothesis testing and the decision-related errors that can be made from statistical testing. Example of an impact study (adapted from Mapstone 1995; Keough & Mapstone 1995). The relationships between the rejection of H_0 and Type 1 errors and non-rejection of H_0 and Type II errors.

Decision	Reality	
	H_0 True (No impact)	H_0 False (Impact)
Reject H_0 (= Impact)	Type I error (α) Impact detected	No error: Power ($1 - \beta$) Impact detected
Not Reject H_0 (= No impact detected)	No error ($1 - \alpha$) No impact found	Type II error (β) Failure to detect real impact

example, the findings suggest there are positive changes in subtidal assemblages in MPAs and a trickle effect of commercial species to adjacent fisheries. These findings, therefore, may justify the gazetting of more reserves. Conversely, where there appear to be negative effects from dumping dredged material close to reefs, it would be judicious to be concerned about impacts that are detectable at $P < 0.1$, because if there is indeed a causal link the consequences of not responding to the impact could be dire.

In situations where an impact has occurred and legislation dictates that whoever is responsible must return the location to its natural state, it may be best to reverse the burden of proof (*sensu* Peterman 1990) by testing for 'bioequivalence'. This is a case of being 'considered guilty until proven innocent' (McDonald & Erickson 1994), and they must demonstrate recovery beyond a predetermined level (e.g. 80%, see section 2.6).

A number of criticisms have been raised about the use of established probability testing (= 'Frequentist approach', Dennis 1996) and the hypothesis-testing framework that has been described in this chapter. The 'Bayesian' philosophy is to use a subjective probability approach based on so-called 'priors', which are *a priori* beliefs about the relative strength of a parameter (by constraining the range of possible values). In our opinion, most of the criticisms of frequentists are based on the investigators who have used the hypothesis-testing approach erroneously. Furthermore, 'there is no alternative framework, in which to evaluate competing conceptual models about how the world works'. (Underwood 1997).

We leave this subject with a quote from Dennis (1996): while staying firmly in the frequentist camp, 'the burden of proof is still on Bayesians to show how ecology can continue its progress with subjective probability approaches. Frequentism, like the peer-review system, is imperfect but has a proven track record.'

2.6 Detection of impacts

Many studies are designed to detect impacts. An impact is essentially a perturbation causing the alteration of the population's density, size, frequency, or behaviour of some members of an assemblage of plants and/or animals. (Broader definitions of perturbations may include physiological variables: see Underwood & Peterson 1988). There are a wide variety of impacts that can affect coastal environments, some of which are designed to promote positive changes in assemblages of organisms (i.e. MPAs) while others are known to cause measurable reductions in the density of organisms (e.g. fishing, sewage outfall effects on fish, habitat destruction, oil spills on intertidal organisms).

The persistence of a measurable impact can vary greatly. Bender et al. (1984) identified two different kinds of perturbations, 'pulse' and 'press'. A 'pulse' perturbation is a relatively instantaneous alteration, after which the system returns to its previous state. Underwood (1989) described such a 'recovery' as 'the mean abundance of replicated, manipulated populations becoming no longer different from that of replicated controls, regardless of the abundance to which they converge.' In other words, the impacted site does not necessarily have to return to its former condition: if the control sites have changed during the time since the perturbation, it can be assumed that the impacted site would have too, even if the perturbation had not occurred. Underwood (1991) illustrated four outcomes of a single environmental disturbance (Fig. 2.5). Some disturbances may cause a pulse response in populations, while others may cause short-term but continued fluctuations in numbers around the previous mean.

Longer-term fluctuations are also possible. A 'press' is a sustained alteration of species densities, that in some cases can lead to the complete elimination of a particular species. Multiple disturbances can cause different patterns (Fig. 2.5F, G). Some populations may be able to recover between stresses, while multiple stresses may exceed the ability of organisms to recover. With an increase in mortality rates the population may slide to extinction. The example given by Underwood focuses on the responses of the organisms, but care should be taken to distinguish between the nature of the impact (press or pulse) and the response of local organisms (press or pulse; Glasby & Underwood 1996). A pulse impact (e.g. oil spill) can potentially result in a press response of the organisms (abundances may be greatly reduced for a considerable time after the impact, even if the perturbation itself only lasted a short time).

So far we have discussed the potential impact of a disturbance. It should also be remembered that there is great variation in space and time at sites that are unaffected by a particular disturbance. If an impact is to be detected, therefore, special attention must be directed toward the sampling design and adequate replication of control areas in space and time. It is unusual to have replicate disturbances (but see Gray et al. 1992). In most cases there is one impact site (e.g. an outfall), but variation at the impact site must be compared with multiple control areas. The inadequacy of one control area is shown in Figure 2.5. This type of sampling design is 'asymmetrical' because the number of impact sites does not equal the number of control sites (see 3.5.7).

Roger Green (1979) coined the term (BACI = Before, After, Control, Impact) in reference to sampling designs where sampling was done both before and after the impact and in both a control and an impacted area. The importance of sampling before and after the impact and in a control area was accepted by environmental scientists, but considerable philosophical and technical improvements have been made to the approach since that time. Stewart-Oaten et al. (1986) argued that sampling should be done at paired (BACIP) and later multiple random times, before and after the impact. They argued that there was no need for more than one control location, because the investigator is not making inferences about populations. Underwood (1991, 1992, 1993b)

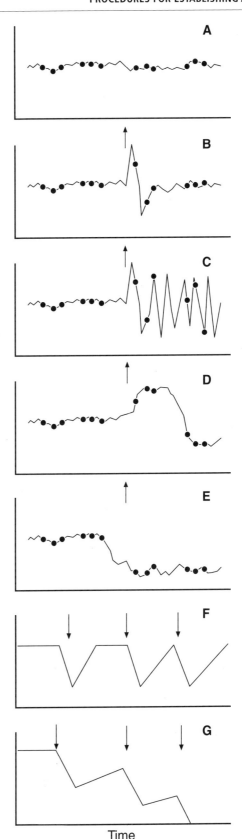

Figure 2.5 The nature of impacts: pulse and press. Lines represent true mean abundance of a population (or other variable) through time at one site that has been subjected to various environmental disturbances; arrows indicate timing of disturbances, dots indicate time of sampling.

A: No disturbance. **B:** Impact causes a pulse in the response of the population. **C:** Impact causes short-term but continued fluctuations in numbers around the previous mean. **D:** An impact causes longer-term fluctuation. **E:** An impact that causes press response where there is a sustained period of low abundance in the population after an impact. Consequences of repeated stresses on **F**, a stable population, and **G**, an unstable population.

ADAPTED FROM UNDERWOOD 1989, 1991 WITH PERMISSION

Time

37

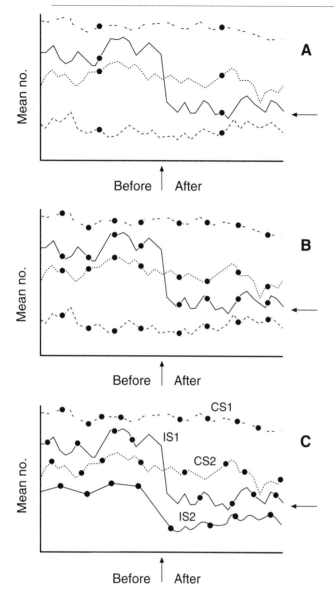

Figure 2.6 The inadequacy of a single control site for an impact study.

A: Principle of multiple controls: here sampling (= dots) is only done before and after the impact and does not allow for temporal variation as for B and C.

B: Four times of sampling before and after an impact (Beyond BACI).

C: Four times of sampling before and after an impact at two impact sites and two control sites (Multiple Before After Control Impact, MBACI).

Solid lines indicate impact sites. CS = control site, IS = impact site Horizontal arrows indicate the population that is subjected to an impact.

ADAPTED FROM UNDERWOOD 1994A WITH PERMISSION AND KEOUGH & MAPSTONE 1995

wrote a series of papers on 'beyond BACI' designs which make use of multiple controls in space as well as multiple samples in time, both before and after.

We follow the 'beyond BACI' philosophy in this book because it is intuitively likely that two places may potentially differ in their patterns of change in the abundance or types of organisms through time, whether or not one place has been subject to perturbation. By having more than one control, we obtain an estimate of the likely size of such differences against which we can then judge the difference between the controls and impacted site (Fig. 2.6). In cases where there is more than one impact site (e.g. several sewerage outfalls), multiple impact and control locations can be sampled at multiple times before and after the impact = MBACI (Multiple Before After Control Impact, Keough & Mapstone 1995).

The general conclusions for impact studies are as follows (adapted from Green 1979; Underwood 1993a; 1994a; McDonald & Erickson 1994):

- An 'optimal' sampling design should commence before the impact has occurred. There should be a knowledge of where the impact will occur and sampling should be done at the impact area(s) and multiple control areas. Multiple times should be included in the design both before and after the impact to provide an estimate of temporal variability and of any change in this variability from before to after.
- If the impact has already occurred, it can only be inferred from spatial patterns. Asymmetrical designs i.e. with more than one control site, should also be used in this type of study (see 3.2.7). Spatial inference of impacts is strongest where an effect is measured at multiple impact sites and compared with multiple controls.
- If sampling is done before and after the impact and there are no controls, an impact can only be inferred from temporal change. You should never intentionally design an impact study without controls because you cannot then eliminate the possibility that the observed changes were due to some extraneous factor that was simply coincident with the impact.
- Impacts can be complicated where the exact size of the impact is unknown. For example, where an impact occurs within a bay it is possible that 'control' sites that are adjacent to the site of impact will also be affected (e.g. due to pollutants, dispersion of organisms, sedimentation etc.). Underwood (1994c) gave an example of this type for bays along the coast of New South Wales, where he recommended that additional control bays and replicate sites be sampled within each of these (Fig. 3.3). In this case there may also be a hierarchy of impacts where there are

differences between the impact area and controls within the impacted bay, but there may also be a bay-wide impact that can only be detected by comparisons with controls from other bays. The need for considerations of this type will vary with local geography and the nature of the perturbation.

- Consider the relationship between alpha and beta in statistical tests for impact studies so that the sampling design does not favour the proponent of a development or the environment (Keough & Mapstone 1997).
- 'Tests for bioequivalence' (see below) are appropriate where a company must return the location to its natural state after the impact. In this case the developer is 'considered guilty until proven innocent' and they must demonstrate recovery over a predetermined level (e.g. 80%; McDonald & Erickson 1994).

Where a company must return all or part of an area to its 'natural state' through mitigation (e.g. after development or mineral extraction), there are major pitfalls to accurately measuring recovery. Sloppy sampling of low statistical power can demonstrate that there is no significant difference between the area before the perturbation and after mitigation, even though there are clearly large biological differences between the two (A in Fig. 2.7). Protection authorities need to regulate mitigation in such a way that an impact study of great precision could detect no difference in selected biological variables from before the impact and after mitigation (B in Fig. 2.7). McDonald and Erickson (1994) proposed a 'testing for bioequivalence' technique. In essence, the regulatory authority must specify limits within which the treated site will be considered bioequiv-alent to the impacted area before the perturbation and with the reference sites(s). For example, the limits may be set at 80% (allowing for some natural variation), in which case if the mean of the impacted

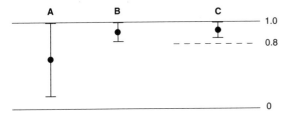

Figure 2.7 Testing for bioequivalence. The line labelled 1.0 indicates identical biological equivalence of the impacted site and reference site. Points indicate the results of studies with 95% confidence limits (CL).

A: Low precision. **B:** high precision. **C:** highly precise study, the 95% CL of which does not overlap with an *a priori* critical level of recovery, 0.8.

ADAPTED FROM MCDONALD & ERICKSON 1994

site can be shown to be significantly greater than 80% (i.e. the lower limit of the 95% confidence interval is above 0.8; C in Fig. 2.7) then we can conclude that the two sites are bio-equivalent. It would be prudent to set these criteria for the abundance of different species and measures of assemblage structure (e.g. MDS, Chapter 3).

Example of an impact study
The procedure for designing an impact study for a marine reserve should be some adaptation of the following example. The purpose of the study is to measure the impact of an MPA by determining whether the abundance and size frequency of a target organism (e.g. abalone) changes as a result of protection. It is hypothesised that the abundance and average size of abalone will be greater inside the MPA than in control areas (i.e. not MPAs) of similar reef type. In this case, the impact can only be inferred from spatial comparisons if no sampling has been done before the impact.

The investigator informs their manager that the proposed MPA should be surveyed a number of replicate times before the impact of protection. Ideally, this would be done for a period encompassing 12 months or more.

Consider a modification of the asymmetrical design discussed by Underwood (1992), the rudiments of which are illustrated in Figure 2.6 and the nature of the analyses in Chapter 3.

In this example, the investigator has been asked to determine whether or not marine reserve status influences the size, frequency, and abundance of abalone. Background information from a previous survey has suggested that 95% of the abalone are in less than 4 m of water and that six 1 x 5 m transects gave adequate precision at the replicate level. It is therefore not considered necessary to do any further preliminary studies on the depth distribution of this species, as this would be irrelevant to the question posed and a waste of effort. The impact study targets this shallow depth range.

It is well known that there is great variation in abalone abundance over small and large spatial scales. Multiple locations are chosen for comparisons with the area of impact (i.e. the MPA) so that the design is not pseudoreplicated (see section 2.5). Because it is known that there is great variation at small spatial scales, the investigator would choose to sample replicate sites within each location and six replicate transects within each site (see Chapter 3). Information from previous surveys suggests that six sites would be adequate within a location to detect a change of 25% in the MPA. Although a minimum of two control locations should be chosen, strong consideration is given to having more as well as multiple times (i.e. greater than two) before and after the impact. In this way the investigator will

39

have adequate degrees of freedom for assessment of impact-related factors at the analysis stage (see Chapter 3 and Underwood 1993b). Sizes of animals will be measured underwater in terms of shell length. Based on the recent conclusions of Worthington et al. (1995), measurements of shell width should also be considered.

Our insistence on reiteration of the importance of replication in space and time, before and after impacts and at several control sites may seem pedantic and unnecessary. Its aim, however, is to provide the strongest possible identification of cause (perturbations) and effect (impact) based on logical argument. Incomplete and confounded studies help no one and may compromise the environment.

Further information and examples of specific impacts are given throughout this book. Schmitt and Osenberg (1996) present current debates on studies of ecological impacts.

2.7 Monitoring

Monitoring generally refers to repeated sampling over time and is a vital component of many sampling designs. A more precise definition would be as follows: sampling in time with adequate replication to detect variation over a temporal range from short and long time periods, done at more than one location. For example, monitoring control locations and impact locations (or location) both before and after an impact is critical in a well-thought-out impact study designed to detect a particular effect size (*sensu* effect size; see 3.5.10); see also section 2.6). Studies on patterns of distribution and abundance generally require sampling through time to accurately describe patterns that could potentially vary on timescales ranging from short-term changes (e.g. state of the tide and time of day) to variation among seasons, years and tens of years (Holbrook et al. 1994).

Because variation in time is often great, the conclusions from studies that do not replicate in time can range from weak to useless. The only exception to this would be where the effect size is huge. For example, large numbers of the bluefinned butterfish *Odax cyanoallix* are found only at the Three Kings Islands, northern New Zealand (Francis 1988; Choat et al. 1988); a one-off study at the Three Kings would reveal that the location is unique with respect to the distribution of this fish. One-off studies of remote islands can, therefore, reveal unique assemblages of species (e.g. Kingsford et al. 1989) or unique habitats (see section 2.8).

In general, however, replication in time is crucial. In many studies it is necessary to monitor changes in biological variables (e.g. number, biomass, diversity) over periods of greater than one year. It is well accepted that seasonal patterns have a great influence on assemblages of organisms (e.g. Jernakoff 1983) and phenomena of longer timescales can cause 'pulse' or 'press' responses. For example, cyclones (Carter et al. 1995) and changes in the southern oscillation (Glynn 1988) can have major impacts. It is typical for investigators to monitor populations at regular monthly, seasonal, or yearly intervals, but there is no replication of these time intervals if you only sample once per month, season, etc. Sampling at these time intervals, therefore, is only legitimate if it has been demonstrated that short-term changes in abundance, biomass, or other variables, are trivial compared to longer-term changes. This scenario, however, should not be assumed because short-term variation can be great for both small organisms (e.g. meiofaunal crustacea, Coull 1985) and large (e.g. fish, Holbrook et al. 1994). If you are sampling every month, additional sampling should be done within some months to demonstrate that variation within a day, between days, and between weeks is less than the variation you are finding among months, seasons and years; and ideally this procedure should be done at more than one location (Morrisey et al. 1992b).

In some studies monitoring is required at very short intervals of time (e.g. descriptions of patterns of settlement and spawning). For example, male damselfish guard eggs for 5–12 days on the substratum (Tzioumis & Kingsford 1995) and intervals of monitoring need to be less than the minimum time of incubation to obtain measurements of spawning periodicity and output. Detailed temporal patterns of this type may be analysed using time-series analysis to test hypotheses concerning the periodicity and duration of events (Table 3.11, p. 71). In all cases, results should always be discussed in the context of the timescales examined.

2.8 Baseline studies

'Baseline studies' simply refer to data that are collected to define the present state of an assemblage. In many cases they refer to a 'before' part of an impact study. If you are doing a baseline study it is worth reconsidering your initial question. For instance, if you are trying to measure changes in abundance, biomass, or diversity from a baseline prior to an impact (e.g. pollution, exclusion of fishermen) you should be asking a question about impacts and considering sampling designs with appropriate controls (see section 2.6). Of relevance in this context is Underwood and Atkinson's (1993) comment: 'It is imperative that sufficient information is collected to be able to estimate the magnitude of natural fluctuations in densities of populations.'

Baseline studies should have replication in time and

space. One-off baseline studies are of limited value, and investigators should argue for more money and/or time to replicate in time.

If a one-off baseline study is your only choice, there are aspects that should be considered. For example, a one-off baseline study of an area may be considered for an MPA in a rocky reef environment. The purpose of the study may be simply to provide baseline information and a typical approach would be as follows:

- The time spent planning the study should be greater than actual sampling time in the field. The time in the field is the most expensive part of the exercise. Read the rest of Chapter 2 and consider how this advice applies to your study.
- Design the sampling in such a way that it will provide good baseline information for formal comparisons at a later date. The design should include multiple locations within the defined area and, depending on the nature of the subtidal environment, it may be appropriate to stratify sampling by habitat type and/or depth. If the purpose of the study is to detect an impact, it is crucial that the sampling design should include appropriate spatial and temporal controls. In other words, perhaps you should reconsider the purpose of the study and change to one that is designed to detect an impact (see section 2.6).
- Obtain a picture of the vertical distribution patterns of subtidal habitats at a number of sites, preferably using one of the quantitative techniques given in Chapter 4.
- Depending on the initial question, priority should be given to sampling organisms that are likely to have been exploited. In this way their abundances and size-frequency distributions can be compared at a later date. If the influence of protection is being assessed, it is more important that good information be gained on these organisms than a very general description of all of the organisms observed. Organisms such as oysters, mussels, lobsters, abalone, urchins, and fish, for example, should be targeted in surveys if there is some concern for human impact on these organisms (see Chapters 4, 5 and 6).
- The intertidal zone may need to be sampled where it has been, or is likely to be, subjected to human activity (see Chapter 7).
- Quantitative data may be supplemented with subjective assessments of large areas by the use of rapid visual approaches. These are outlined in Chapters 4 and 8. Underwater video of habitat profiles or the detail within habitats can be used to obtain quantitative information and a good visual record of the initial study. Replicate video transects should be done over multiple depth profiles at each location.
- Identify unique biological and physical attributes of the area. These can be recorded on video. For example, rare species, rich encrusting fauna associated with caves (see 4.7.4). Video footage is also useful for showing unique aspects or changes in an assemblage to a lay audience. Furthermore, it may be used as a record for later scrutiny of change in comparison with subsequent video images. It may also be important to note species that have been introduced (e.g. the laminarian algae *Undaria* to New Zealand, or northern Pacific starfish to soft-sediment environments of Tasmania).

It should be noted that one-off baseline studies are often only slightly better than no study at all. Managers often demand information on change (e.g. after some type of impact) and often a one-off study has very limited power to provide this. As a result, therefore, cost savings in the field may cause greater expense in court.

2.9 Investigating processes: the usefulness of experiments

The description of patterns of distribution and correlative methods can only be used to infer causality. Gut feelings on causality that are based on descriptive methods are often incorrect (e.g. McGuinness 1988). Models concerning causality should be tested using manipulative experiments. Experiments have contributed greatly to our understanding of processes that influence organisms in temperate coastal environments. This has largely been because the manipulative approaches once restricted to the laboratory have been applied to field situations (Connell 1974). In 1993, Underwood reviewed the literature on the proportion of intertidal studies from 1955 to 1985 that used manipulative experiments. He found that until 1970 less than 30% of the papers were experimental, while 78–80% of papers for the period 1974–85 were in this category.

Underwood (1993a) stated that experiments could be considered as the only epistemologically sound procedure to distinguish among competing models to explain a set of observations. 'An experiment (or test) is any logically derived procedure used to test unambiguously a proposed null hypothesis. The null hypothesis is logically derived from an hypothesis or prediction derived from a model or theory.' (See section 2.2).

There are two forms of experiments (Underwood 1993a): mensurative experiments, which test hypotheses about patterns, where the selection of sites is not by random procedure (e.g. Hurlbert, 1984), and manipulative experiments, which employ an intrusive approach by manipulating biological systems to test the predictions in the hypothesis (e.g. Andrew 1993)

Most investigators seek to make general statements about biological systems. Only when experiments have been repeated several times can consistency of the results be understood (Connell 1983). This should be extended to repeating experiments at multiple locations, because the magnitude or type of change may vary in space (Foster 1990). In some cases, however, the resources available to an investigator do not enable repetition in space or time. In such cases, the generality of the results must be viewed with great caution. Such results are, however, still worthy of publishing because they provide others with the opportunity to repeat the experiments and test the generality of the results.

Underwood (1993a) cites a series of papers that demonstrate the contribution of experiments to our understanding of intertidal systems. For example, limits of distribution of intertidal species are not solely determined by physical processes; competition between and among species of limpets is an important source of mortality; and intense grazing by gastropods has a major impact on intertidal assemblages of algae. Similarly, a series of experiments in New South Wales (Fletcher 1987, Jones & Andrew 1990, Andrew 1993, Andrew & Underwood 1993a) and New Zealand (Andrew & Choat, 1982, Andrew 1988, 1993) have demonstrated that urchins (*Centrostephanus rodgersii* and *Evechinus chloroticus*) have a major impact on algal assemblages on subtidal rocky reefs. The grazing activity of these echinoderms has a great influence on the distribution of kelp forests.

There are many pitfalls with manipulative experiments. Consider the following:

- We urge investigators to have good descriptive information (to an appropriate taxonomic level) on their study organisms before they leap into experimental mode. Many experiments have foundered because the design and implementation of the experiment was ill conceived, based on a lack of information or pilot studies (e.g. for cage design, etc.).
- Use appropriate replication of experiments. Without proper interspersing or use of replicates of the treatments in space and time, no logically valid interpretation of an experiment is possible (Krebs 1989; Underwood 1990). If, for example, all of the replicates of one treatment are in one corner of an experimental site, differences among treatments cannot be separated from differences among the different parts of the site.
- Beware of having only one experimental area and

one control area with replicates within. This is a pseudoreplicated design (*sensu* Hurlbert 1984; see section 3.4), meaning again that differences among treatments cannot be separated from any spatial differences that may exist between areas.

- Be aware that cages, or other exclusion devices, can cause changes (e.g. in water movement, shading, sedimentation, and algal cover) that are unrelated to the effect of an excluded target organism, but could be interpreted as a consequence of exclusion. Have controls for structure of cages (e.g. half cages) and in some cases even the cement you use (Fig. 2.7).
- Use natural substrata where possible. For example, in settlement experiments there can be major differences in patterns between different rock types and artificial substrata (Caffey 1985; McGuinness 1989). Moreover, raised artificial surfaces can also affect results (Kennelly 1983).
- Ensure independence of treatments. Make sure the proximity of one treatment does not affect others. (Underwood 1993a).
- Make sure that the data are collected independently. The replicates must be independent (i.e. replicates not correlated with each other). The investigator may, for example, avoid this problem by sampling small random quadrats within a cage. For considerations on replication, see section 3.4.
- Consider the scale of your experiment. The outcome of experiments can vary greatly with scale (Levin 1992). Some important scales may be too large for experimental studies and conclusions about effects operating at such scales may only be based on observational studies.
- Ensure that nested factors are included in sampling designs when subsampling occurs.
- Use asymmetrical designs if your design is fundamentally orthogonal but there is a non-orthogonal set of controls. For example, Druce and Kingsford (1995) experimented with two different colours and sizes of fish-attraction devices (this was the orthogonal part of the sampling design), but results had to be compared with the densities of fish in open water that might have been captured by chance alone (the asymmetrical parts of the design); see 3.5.7.

For reviews of experimental approaches, with a particular emphasis on field situations, see Hurlbert (1984) and Underwood (1990, 1993a). Experiments that address specific hypotheses on competition generally involve the manipulation of densities – see the review by Underwood (1986). For details of asymmetrical sampling designs see Chapter 3.

In some studies, replicate organisms, plots, or

quadrats are used at multiple times. For this reason analyses using 'repeated-measures' are commonly done (Krebs 1989). Analyses of variance and similar parametric tests that include time as a factor cannot be used because a basic assumption of ANOVA is that each observation is independent. In some cases repeated measures are used when there is a fault in the sampling design that could have been avoided in the planning stage of the project. There are, however, many hypotheses of ecological interest that require comparisons of the same sampling units through time (e.g. removal and recolonisation experiments) and repeated measures ANOVA is a legitimate method of analysing data of this type.

A number of methods are used to manipulate natural assemblages so that processes can be investigated. Three approaches are common (1–3), and some are only used on certain types of organisms (4 and 5):

1. Caging.
2. Removal addition (or organisms/objects).
3. Transplants (by removing and adding organisms or objects).
4. Tethering.
5. Grafting/damaging.

One of the greatest challenges in experimental marine ecology is avoiding experimental artefacts. Greatest problems are generally encountered with caging where investigators attempt to minimise effects that are related to cages. Kennelly (1991) argued that the best form of cage that can measure the nature of impacts is a cage within a cage, where it is assumed that the additive effects of an additional cage would indicate that cage artefacts are a problem. It is possible in some circumstances for the area between the two cages to cause artefacts in their own right, e.g. harbouring small predators. A range of cage designs is shown in Figure 2.8 (p. 44). The design of cages and controls must ultimately be determined by the environment, type of organism, and ecological problem that you are addressing. In many cases study organisms will need to be replaced in cages if they die. This is because changes in density among treatments will confound factors in the experimental design.

The removal and addition of organisms and objects is a common approach in ecological experiments and has the major benefit of eliminating artefacts that relate to the use of cages and similar structures. Experiments of this type have included the removal of intertidal algal and other benthic organisms at different times of the year (Jernakoff 1985), the removal of intertidal kelp (e.g. Andrew & Jones 1990), and the manipulation of topography (e.g. addition and removal of boulders: Connell & Jones 1991). Because the results of clearances may vary with spatial scale, clearances of differ-

ent sizes should be considered in experimental design.

Transplants can be done on many spatial scales. This type of experimental method can minimise the problems of artefacts, but controls still need to be considered carefully (Fig. 2.9, p. 44). Chapman (1986) described a particularly comprehensive set of controls in an experiment on the movement of small intertidal gastropods. Chapman had disturbance controls (Fig. 2.9) where animals were replaced in dispersed and clumped arrangements (the results showed that animals all moved in the same direction only when they were initially clumped) and tagging controls where individuals were tagged in the field and in some cases taken back to the laboratory to be tagged (these two controls had no effect). Transplants have even involved the shifting of large items (e.g. boulders), which have been relocated among habitats and some treatments have been included to assess the effect of disturbance (Andrew 1993; see also 5.4.5).

Tethering experiments are generally used to assess the intensity of predation (Connell 1994). Tethering of course is far removed from a natural situation, but in some cases it may be the only option for exposing study organisms to predators in different habitats (Beck 1995; Aronson & Heck 1995). Some animals such as crustaceans and fishes may become terminally tangled in their tether and exhibit alarm behaviour that may attract predators. In contrast, gastropods which do not move great distances may be amenable to this approach.

Grafting may be used on modular organisms such as sponges and ascidians which have great powers of regeneration and can survive massive incisions. This approach is useful in studies on border effects, ecological succession, and chemical ecology (Hildeman et al. 1979). The experimental disturbance of organisms has sometimes been used. For example, Stocker & Bergquist (1987) compared damaged and undamaged colonial ascidians to determine the influence of urchin damage on size and longevity of colonies.

It is important not to feel constrained by what we have written here: other experimental methods may be appropriate to your ecological problem, particularly in environments where experimental approaches are not often used. Experiments are relatively hard to do in some environments, e.g. mobile sediments and exposed places (see Dayton & Oliver 1980 and Chapters 7 and 8). Similarly, experiments on planktonic and nektonic organisms in the pelagic environment are difficult (particularly over periods of greater than one day) because of the dynamic three-dimensional nature of these waters (see Chapter 9). Nevertheless, the limitations of correlation and description apply in these environments as well and the challenge is to find ways of doing useful experiments.

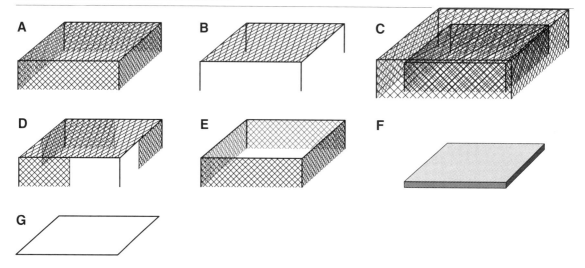

Figure 2.8 Design of cages for exclusion/inclusion with appropriate controls.
A: Total exclusion. **B:** Roof only – this design has been used as a treatment to exclude certain categories of fish (e.g. Choat & Kingett 1982) or control for the structure of cages (Bustamente *et al.*1995). **C:** Cage within a cage, to assess cage artefacts – the assumption is that there will be an additive effect if

artefacts are a problem (Kennelly 1991). **D:** Cage to assess cage artefacts – e.g. half of the walls and a roof to control for potential shade effects (Connell 1997). **E:** Wall to allow access to fish, but exclude grazing gastropods – a hooked edge can increase the efficiency of this control. **F:** A control for artificial substrata (e.g. Caffey 1985) or the topographic anomaly of a plate. **G:** An open control of natural substrata.

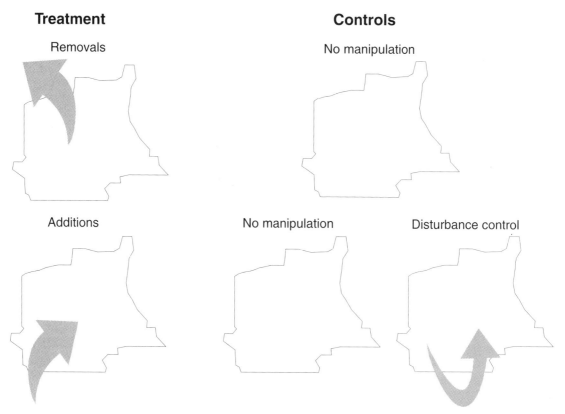

Figure 2.9 Examples of experimental treatments and controls used in transplant experiments. Note that for the disturbance control in cases where multiple locations are used there should

be a disturbance control that involves translocation from one site to another within a location (e.g. tens of metres between sites versus kilometres among locations).

2.10 Training personnel

If more than one individual is involved in a study, it is crucial that everyone involved in the fieldwork is familiar with the sampling design and general procedures. Furthermore, they should have adequate knowledge of the target organisms for the problem being addressed. It should be made clear at the planning stage how plants and animals are to be counted (e.g. as number, percentage cover, or both). Preliminary sampling is invaluable for training personnel in procedures and identifications. At the very least, detailed discussions should be held with all involved before a study begins. Failure to familiarise personnel with procedures can lead to total disintegration of an otherwise good sampling design (see Kenchington 1978). In detailed comparisons of the ability of divers to count fish and 'discrete organisms' (e.g. crown of thorns starfish) on coral reefs, Mapstone and Ayling (1993) concluded that estimates of abundance differed systematically between two observers. 'Thorough training and periodic recalibration of observers will be essential for the utility of data stemming from ongoing monitoring programs.'

Ideally, the same personnel should be used throughout a study so that temporal and spatial patterns of abundance are not confounded by differences among observers (see Inglis & Lincoln-Smith 1995). If large numbers of persons are to be used in a survey that encompasses a broad geographic range, it would be judicious (for reasons of accuracy) to target the most easily and unambiguously identifiable organisms (e.g. abalone and sea urchins, Chapter 5).

2.11 Equipment for studies

The equipment required will vary, but this list may provide a useful reminder of generic categories of gear. The chapters on relevant assemblages of organisms and environments that relate to your study should also be consulted for gear.

Lists, logs, permits and tables
- Lists of gear are essential. For example, if your ecological question requires a large vessel and a cruise lasting a number of days; you may not get a second chance to get equipment such as nets, oceanographic depressor, flowmeters, patching gear, spare cod-ends, conductivity, temperature and depth device (CTD), closing mechanisms, jars and preservative, inclinometer, spare clips, shackles, computer equipment, etc. Also check that all of the gear you need that is resident on the vessel is in working order.
- Log books are an important record. Write down the details of what happened each day (i.e. collections, counts, etc.), locations, times, file names, and names of the disks used. It may even be appropriate to sketch maps of the study area, if this helps with navigation at levels not provided by a GPS. When you write up the log, check that all data have been transcribed from slates before they are accidentally cleaned, and check that you have filled in the necessary details on waterproof data sheets and stored them safely.
- Permits may be required for your collections. You may need some or all of the following: animal ethics approval, fisheries permits, permits from departments of conservation and other authorities. Make sure you think of this six months ahead of time. Many authorities require that you carry a copy of these in the field.
- Licences: In waters of many parts of the world you have to carry your boat licence and registration documents of the vessel.
- Tide tables are useful for most types of studies in coastal environments, and a nautical almanac for a record of the hours of darkness. Tide tables are available on disk in some parts of the world .

Safety
- Safety gear for studies that use boats: check distress flares, V-sheet, radio, water, raincoats (to reduce the chance of exposure), first-aid gear, toolkit, boat spares, food and drink (a stainless-steel Thermos is good for keeping liquids hot or cold), sunblock or other sun protection.
- Advise all members of the party on useful items to bring. Leave details of location and estimated time of return with a contact back in the laboratory.
- Radios and mobile phones are useful for regular 'skeds' or irregular communications.
- If you are travelling long distances in a vehicle, check that is has maps, a toolkit, spares, and a torch.

Sampling equipment
- Diving gear, including a dive watch with a stopwatch function. Dive computers are invaluable for surveys where divers are swimming transects, etc., that cover a broad range of depths. You may consider extra weights if you are working in very shallow water, where surge is a problem, or for drilling underwater. Always have one or more spare regulators for your group. Other spares that are useful include straps (flippers and masks), knives, masks, and a buoyancy compensator.
- Diver catch-bags are useful for work on the shore, and underwater, for holding gear and samples.
- Slates made from white polyethylene with a roughened surface (use sandpaper) and a supply of pencils.

We find H and 2H pencils are soft enough to write, minimise smudging, and are hard enough to maintain a point for the duration of a dive. Consider the organisation of your slate before you get underwater and reassess your first try after a pilot study. Slates can be cleaned with household abrasive cleaner or a rubber. Always transcribe data immediately after your excursion to the field to avoid accidental erasing. Waterproof paper may be better to reduce the handling time of raw data.

- Underwater data sheets, which may be made so that they can be clipped on to a board.

- If data are not recorded directly on to a waterproof sheet, data sheets should be available for transcription from a slate. Data sheets should include pertinent information such as date, locality, site, depth, habitat, method (i.e. sample units, size and number), data recorded as percentage cover, densities, sizes, values for each species. You may want to add the abbreviations for each species that are used in your database: this may make the task of transcribing data easier.

- Tape recorders: Regular recorders, or underwater tape recorders, are an alternative method for recording data. It should be remembered, however, that the tape has to be replayed to transcribe the data, which in some cases can greatly increase the time taken to record data. Underwater tapes are particularly useful if divers are counting or recording percentage cover of organisms in complex assemblages (e.g. Kennelly 1995). The diver can have both hands free for other things and can keep their eyes on the subject organisms. Badly fitting mouthpieces, however, can result in flooded face masks as well as lots of gurgling and rude noises on tapes.

- Measuring tapes: 30–50 m lengths are useful. Those that are shaped like doughnuts are best for placing over stipes of algae and unreeling. Many tapes have clips on the end of them. These invariably get caught in objects as the tape is wound in, whether in the intertidal or the subtidal.

- Plastic bags and sample jars are always useful for storing samples until they are preserved. Take some in case you encounter species you cannot identify and need to bring them back to the laboratory for identification. Write numbers on containers in advance and note what goes in them against the number. This helps to minimise writing in the field.

- Preservatives and labelling: Take a supply of these for many studies, especially descriptive surveys; 10% formalin and 95% ethanol are best for most organisms (see Chapter 10). Take preservative-proof paper (note that waterproof acetate data sheets generally delaminate in alcohol). If frozen samples are taken, check that your labels are freeze-protected. Some alcohol-proof labels disintegrate when frozen, while some parchment-type waterproof paper is not rot-proof. Do a pilot study to check your labels. In special cases you may want to take samples for chemical analyses. Find out if your containers will cause contamination in a pilot study. Even the pen can make a difference. For example, writing on a label with an alcohol-based pen can contaminate the sample. Once again, this type of problem can be checked in a pilot study.

- Vernier callipers or other devices for measuring the size of organisms.

- Quadrats: Also take equipment for measuring percentage cover (see 4.3.3). Large quadrats are best if they can be collapsed for easy transport and storage.

- Intertidal studies: It is advisable to take foam rubber pads (greater than 50 mm thick) to kneel on. Periscopes are useful for studying organisms in rock pools and can be easily made in a workshop.

- Differential GPS is rapidly becoming invaluable for a variety of studies. It can be used to find isolated dive sites where it is difficult to leave surface markers because of the weather or pilfering. On beaches with little in the way of topography a great deal of time can be saved in finding study sites. In studies of planktonic assemblages, GPS can be used to set a course that will take a vessel to multiple locations for plankton tows.

- Temperature data loggers and loggers of other environmental variables can be very useful because they can record environmental variables frequently and over long periods of time (a year or more) with little maintenance. Some probes are relatively cheap (under $300 at the time of writing). Always check the calibrations and delays between ambient and what is recorded before you trust them to the field. Don't forget the software that is used to launch and download the devices.

- Extra rope, small and medium-sized floats are useful for the temporary marking of sites.

- References on taxonomy may be useful (see Chapter 10).

- On soft substrata airlifts, corers, grabs, and sieves may be required (see Chapters 7 and 8).

- Computers are required for many field excursions. Make sure you have all of the necessary hardware (power cord, double adapters etc.), software and blank floppy disks for the backup of your files.

- The transport of samples should be considered where appropriate. Think of how they will be transported. Do you have enough room? Will the samples have to be sent as freight? Where samples have to be frozen, a portable freezer may be needed.

Experiments

A variety of hardware has been used for subtidal experiments on hard and soft substrata, but caging material may include plastic mesh (from garden centres), stainless mesh, netting, or even plastic milk crates. Ayling (1981) used perspex sheeting attached low to the substratum to exclude monacanthids (which feed while orientating vertically) from areas of substratum. Substratum type can have a major influence on the means or ease of attachment. Bolts and/or masonry nails are generally used to attach experimental devices on rock. An underwater drill (connected to a high pressure air source) and masonry bits (lots of them if the rock is very hard) are generally required to drill holes for bolts (Dyna-expandable base type). Bolts are stronger than nails: bolts or a combination of bolts and nails should be used in areas that are subjected to strong swells or if an experiment is expected to last a long time. Water-resistant adhesive (e.g. Expocrete) can also be used to help everything stick, or to block holes. Check that materials are not susceptible to corrosion or toxic to your study organisms. In soft sediment, steel star pickets or plastic tubes can be rammed deep into the substratum, but experiments of this type are difficult if there is great swell activity in the area. Experiments in the pelagic environment can be marked with long-line flags that are anchored or free drifting. Free-drifting flags should have a large drogue to reduce windage. Radar reflectors are useful if you are on a large vessel.

Cameras

- Land: On land an SLR camera with a 24–28 mm lens would be ideal for scenic pictures, and a 55 mm macro lens would be useful for close-ups of the intertidal zone.
- Underwater: Amphibious cameras, such as the various Nikonos models, are extremely versatile for the purposes of a survey. Wide-angle lenses (e.g. 15 mm) provide excellent 'habitat' photographs. The standard lens (35 mm) is useful and with extension tubes can be used for close-ups. Clip-on close-up lenses allow greater flexibility, but beware of bubbles getting caught behind the lens; they are also prone to scratching unless you are particularly careful. Where photographs of quadrats are taken (see Chapter 4), a frame should be constructed. With the camera mounted at 0.9 m from the bottom of the frame, the difficulties of focusing will be minimised. The frame may be labelled with a scale and colour code – the latter is useful for colourful organisms such as sponges. Nikonos cameras can also be used on land with certain lenses (e.g. 35 mm), but not lenses of focal lengths less than 28 mm. A wide range of SLRs can be put in housings and will give excellent results. Although they are more bulky than amphibious cameras, the results are often spectacular. Splash-proof cameras that are waterproof to about 10 m are cheap and may be suitable for your study, but the quality of the slides is likely to be below average
- Digital cameras can be used on land and underwater. Although they lack the flexibility of conventional cameras, especially with respect to lenses, this situation is rapidly changing. The great benefit of this type of photography is that images can be stored and manipulated on computer. A possible limitation is the number of images that can be fitted onto a disk.
- Film: 64 to 100 ASA film would be the most commonly used colour slide film. Despite this, if natural light photographs are taken underwater or in gloomy conditions on land, 200 to 600 ASA film would be better. Black and white film could be considered where large amounts of film are required, but it should be remembered that the cost of film is relatively trivial compared to the costs of diving or even intertidal work. Important images should be digitised and saved on computer disk or CD. Saved images are useful for archives and for other applications such as estimates of percentage cover, number of organisms, and presentations.
- Video: Small video cameras can now give very high-quality results in a wide range of light conditions (Rumohr 1995). For example, Hi8 Sony Handycam cameras can be put in a housing for underwater work. (See section 4.10 for other details on the utility of video). You will often need a lamp for subtidal work. It is always judicious to have a back-up lamp and spare batteries. Don't forget to charge the batteries before the dive. Low-light cameras (sensitivity 1–2 lux), with colour correction filters can give good results underwater and are more compact without the need for a lamp (M. Keough pers. comm.). If you want more than some general footage of the nature of the study area, make sure you do preliminary sampling to check that you are getting the quality of information you require. Digital tapes are ideal, because transferring images to computer is comparatively easy. Images can be stored and manipulated on the computer. Video resolution, especially through computer frame grabbers, is generally of lower resolution than for 35 mm film.
- Storage: storage of photographic gear and film/tapes should be considered. Most films will produce best results if they are stored in a fridge; check packaging for recommendation on storage. Still and video cameras should be stored in dry conditions

otherwise there is a risk of fungi growing on the lenses and this may cause irreversible damage. Colour slides may also be damaged by fungi (see 'Film' above re CD-ROM storage). Airtight containers and silica pellets (to absorb moisture) will avoid most of these problems.

SELECTED READING

The references listed are examples of research that will lead the reader to a broader literature, also see the text of the chapter.
* indicates a multi-authored book or collection of proceedings; references to volumes of a journal refer to a set of relevant papers.

Bayesian inference: Dennis 1996; *Ecological Applications* (1996) 6*: p. 1034–1377.

Experiments in ecology: Chapman 1986; Hairston 1989; Montgomery 1991; Scheiner & Gurevitch 1993; Underwood 1993a, 1997.

Equipment: Baker & Wolff 1987; Coyer & Witman 1990; English et al. 1994; see also Chapters 4–10.

Hypothesis testing: Underwood 1990, 1997; Mapstone 1995; Keough & Mapstone 1997.

Impacts: McDonald & Erickson 1994; Underwood & Chapman 1995*; Schmitt & Osenberg 1996*; Keough & Mapstone 1997; Harvey 1998; Herendeen 1998; Lincoln-Smith 1998.

Photography: George et al. 1985. This field changes very quickly: seek advice from dive shops, large bookstores, libraries and camera shops.

Restoration: Thayer 1992*.

Sampling: Green 1979; Andrew & Mapstone 1987; Krebs 1989; Fletcher & Manly 1994*; Keough & Mapstone 1995.

Statistical methods: see Chapter 3.

THREE

ANALYTICAL ASPECTS OF SAMPLING DESIGN

M. J. Kingsford

3.1 Introduction

This chapter reviews analytical aspects of sampling designs and statistical methods that are appropriate to hypotheses concerning biological assessment of temperate marine environments. It is organised to follow the sequence of decisions that have to be made once the investigator has focused on a problem. This problem should be expressed as an hypothesis or related series of hypotheses (see section 2.2) so that sampling can be designed with an appropriate and unconfounded allocation of sample units.

Before and during the design phase of a study, decisions often have to be made regarding the appropriate size of units to sample, the number of replicates that need to be taken, how to allocate sampling effort within a study area, and how the data should be treated once they have been collected. The latter includes checks of raw data, the presentation of data, and statistical analyses. A priority in this chapter is to present information on types of analyses and the principles on which they are based; appropriate practical applications are cited. Strong emphasis is given to different types of stratified sampling that may involve the random allocation of samples in sampling designs that are orthogonal, nested, partially hierarchical, or have asymmetrical components (stratified sampling with proportional and optimal allocation is also discussed).

An environmental mosaic of different habitats or depths can be recognised in most temperate marine environments, and many hypotheses require stratified sampling designs. In this book, many examples are given of how abundances of organisms (and in some cases other variables such as growth) are influenced by habitat types, depths, locations, and different water masses. Moreover, experiments usually require different levels of sampling that may include factors such as treatment (e.g. + and − grazers); depth (e.g. shallow and deep) and location (1–x). Although some knowl-

edge of statistics is assumed (especially for more complex analyses), where complete accounts are not given most of the sections should provide a sound platform to explore a plethora of more complex technical papers and examples of applications.

Some basic perspectives on sampling design, statistical inference, the nature of impacts and experimental approaches are given to expand on the information in Chapter 2. Comprehensive reviews of approaches to sampling are given by Green (1979), Underwood (1981, 1986, 1992, 1993b, 1994a, 1997), Andrew and Mapstone (1987) and Krebs (1989), who discuss many issues beyond the scope of this work. A guide to other sources is given at the end of the chapter, for readers interested in topics that are not covered such as growth, experiments on competition, bootstrapping, time-series analysis, and multivariate statistics.

3.2 Terminology

Andrew and Mapstone (1987) described terms for sampling and analyses that are relevant to biological studies and their terminology is used in this chapter (Table 3.1, p. 50).

Studies of the organisms associated with coastal environments often require estimates of population size. Although it is possible in some situations to count all members of a population (e.g. seals in a colony), this is unusual in most marine studies.

Three types of populations are worth considering in a biological sense. A 'metapopulation' is a group of populations that may correspond to the geographic range of the species. A 'mesopopulation' is a distinct population nested within the metapopulation (Sinclair 1988; see also Fig.1.3) and this is generally referred to as 'a stock' in fisheries terms. Field ecologists often study local populations (e.g. on a single reef) that are parts of

Table 3.1 Symbols used in Chapter 3. Modified from Andrew & Mapstone (1987).

n	Sample size, i.e. number of replicate sample units
N	Total number of individuals in a population
x	Single measure of a variable e.g. number of fish in a transect
μ	Population mean
$\bar{x}$	Sample mean = an estimator of μ
σ	Population standard deviation
SD	Sample standard deviation = estimator of σ
σ^2	Population variance
s^2	Sample variance = estimator of σ^2
SE	Standard error

a mesopopulation or stock. In a statistical sense, I refer to a 'population' of samples from which sample data are drawn (Andrew & Mapstone 1987). These data are invariably a subset of the total biological population, and inferences are made based on this subset.

In this book I refer to a 'sampling unit' as a device or method that is used to sample organisms or some component of the physical environment (e.g. quadrats, belt transect). The way in which they are deployed in space and time is the 'sampling design'. For example, samples are taken within three depth strata on a reef (upper, mid, and lower) and this procedure is repeated at six locations. At each location and depth stratum on the reef, organisms are counted in six random 1 m² quadrats (see randomisation, section 3.1.5). Sometimes it is necessary to 'subsample' the catch of a sampling unit, particularly where small organisms are being sampled (e.g. epiphytal invertebrates and plankton), and replicate subsamples are required to economise on sampling time.

'Factors' means sources of variation in the sampling design. For example, in the two-factor sampling design above I have the factor 'depth' on the reef with the 'treatments' upper, mid, and lower, and the factor 'location' with the treatments locations 1–6 (Table 3.2). The factors of this design are symmetrical (cf. asymmetrical, section 3.5.9) or 'orthogonal' in that all treatments of 'depth' are sampled at each location. The term 'treatment' is often used in the context of a manipulative experiment. I use it for all categories of design: mensurative experiments, manipulative experiments and descriptive studies (see section 2.9).

'Variables' are one or a combination of the following: the characteristics of organisms within a population (e.g. you may have estimates of density and the size of different species in your study area); species within an assemblage that are being measured; and measurements of the environment (e.g. water temperature, turbidity, salinity). Each set of data is a single variable and you have values for all levels of sampling (i.e. for each factor and each treatment of each factor). Because you will generally have more treatments than variables (especially where sampling is done at multiple times), you will generally find it easiest to organise your database array as 'columns' for variables and 'rows' for levels of sampling; most major statistical packages require the data in this form (Table 3.2). In other words, each sample unit (or subsample) should be a row.

'Parameters' are mathematical or numerical values that are used to describe some characteristic of entire populations of a variable. Field ecologists generally only get estimates of the 'true' measures of an entire population, e.g. the arithmetic mean and variance of densities of limpets at a particular height on the shore. Andrew and Mapstone (1987) used the term 'statistics' in two senses: 'sample statistics', synonymous with the sample estimates of population parameters, and 'test statistics', meaning the derived numerical value (e.g. F-value) used in a statistical test of a hypothesis (see section 2.2).

3.3 Procedures for obtaining the best sampling unit

3.3.1 Principles

The relevant hypothesis of a study will determine the main features of a sampling design, and part of the decision-making involves choosing sampling units. Before any study begins, questions concerning sample size and number should be carefully considered, and ideally these should be assessed during a preliminary sampling exercise (see section 2.4). This procedure may be cost-effective by saving a considerable amount of time and money when it comes to conducting the main study. It is possible that this step may not be necessary if choice of sample unit size and number is based on rigour of previous studies that have studied the same or similar organisms (see Chapter 2), or a precisely specified hypothesis points to the appropriate units.

Major considerations for sample units include the following:

- What size of sample unit gives the greatest precision (smallest standard error) for a given total area searched (i.e. many small or a few large sample units)?

- Sample units that are too small may result in patchiness (Green 1979) and this may give erroneous results for calculations of variance components, cost-benefit analyses, and power analysis (McArdle & Pawley 1994).
- Are the numbers obtained in the sampling unit suitable for analyses and graphics, given a variety of sampling localities and times?
- Have you built in some redundancy to your design to allow for loss of replicates (especially relevant to manipulative experiments);
- Are the counts from each sampling unit accurate (e.g. are organisms being missed)? Accuracy may be low in large sample units that are difficult to

search. You may want to decide how accurate you need to be. Should you count a few units very carefully or do a lot more quickly?

- How long does it take to search a replicate and how does this relate to the total amount of time you can spend at the sampling location? For example, consider the finite area that can be searched in the period of time in which you can sample the lower intertidal or the number of transects that can be sampled in one dive.

In some cases the maximum size of a sample unit may be determined by the dimensions of habitats or other strata. For example, Kingsford (see 6.4.4) found that

Table 3.2 A database showing records of levels of sampling (= factors) and values for biological variables (genera of grazing gastropods, number per 5 m², replicates n = 3) and values for environmental 'variables' that were measured in each 5 x 1 m transect. Date = date of sampling, Obs = initials of observer who collected the sample. Factors in the design were: Dep = depth strata with the treatments 1= shallow, 2 = mid, 3 = deep; Loc = Locations 1–6 where all depth strata were sampled at each location. The variables measured at each level in the design were species of gastropod (= species variable) and the following measurements of the environment: talgae = % cover of tufting algae; paint = % cover of *Lithothamnion* paint; sedi = mm depth of sediment found in each transect. The data matrix is generally described in terms of rows and columns. Factors are often referred to as 'independent variables' and the species and environmental variables as 'dependent variables' (e.g. Tabachnik & Fidell 1983). Dots indicate gaps in the data set that are not shown.

Sampling details		Factors		Species variables			Environmental variables		
Date	Obs	Dep	Loc	Turbo	Haliotis	Cellana	talgae	paint	sedi
12697	PE	1	1	2	25	34	10	80	0
12697	PE	1	1	0	40	25	0	100	0
12697	PE	1	1	1	17	70	20	75	0
12697	PE	1	2	14	10	10	0	100	0
12697	PE	1	2	20	15	30	10	75	0
12697	PE	1	2	16	30	34	10	80	0
.	.	.	.	.	.	.	.	.	.
.	.	.	.	.	.	.	.	.	.
12697	PE	1	6	10	16	15	0	70	1
12697	PE	2	1	4	4	14	0	80	2
12697	PE	2	1	5	3	12	0	100	1
12697	PE	2	1	7	6	16	0	100	1
12697	PE	2	2	7	2	11	0	95	1
12697	PE	2	2	6	0	12	0	75	2
12697	PE	2	2	3	1	10	0	90	0
.	.	.	.	.	.	.	.	.	.
.	.	.	.	.	.	.	.	.	.
12697	PE	2	6	3	12	12	3	60	1
12697	PE	3	1	0	1	6	10	75	4
.	.	.	.	.	.	.	.	.	.
.	.	.	.	.	.	.	.	.	.
12697	MK	3	6	0	0	0	0	80	5
12697	MK	3	6	0	0	1	0	50	4
12697	MK	3	6	0	0	2	2	74	6

a 25 m transect was best for counting fishes in kelp forests along the coast of New South Wales. Transects of greater lengths frequently exceeded the length of the longest axis of the forest. The size of an optimal sampling unit and the number of replicates that are taken will vary according to the organism being studied. Thus some compromises will have to be made when a wide range of organisms is to be sampled with the same sample unit size (e.g. algae, urchins and encrusting animals, Chapter 3).

Descriptions of sample units used in other studies are given throughout the book; these may be useful if it is impossible to carry out a pilot study.

3.3.2 Applications

The procedure should be carried out wholly within habitats or other physical clines (see Andrew & Mapstone 1987). These authors suggested that a minimum of three sampling-unit sizes should be assessed. The examination of more than six units of each size is likely to be too costly in a pilot study. The largest sample unit, at least, should be large enough so that it minimises perceived aggregation by being large enough to bridge aggregations.

If multiple habitats or locations are to be sampled, it is best if the procedure is repeated in more than one area (e.g. Mapstone & Ayling 1993). Information from a pilot study with small sample sizes and from one site is likely to give a misleading estimate of the variation that is likely to be encountered in a more comprehensive design (McArdle & Pawley 1994; see extension of this argument in 3.5.9).

The relationship between number of replicates and sampling precision can be calculated by altering n in the following formula, once data from a large number of sampling units have been obtained:

$$\frac{SD}{\sqrt{n}} \quad \ldots \ldots \quad (3.1)$$

This estimate, called the standard error (SE), is a measure of how precise an estimate of the mean is (the smaller the SE, the more precise the estimate). The results of these calculations can be plotted in a number of ways; the most common methods are as follows (the x-axis is given first).

1. Number of replicates versus $SE/\bar{x}$.
2. Area searched versus $SE/\bar{x}$.
3. Time spent searching versus $SE/\bar{x}$.

These will always result in a negative decay curve (see Fig. 3.1), and these methods are the best plots for data that are normally distributed.

Andrew and Mapstone (1987) present a formula that is useful if a desired level of precision is required.

$$n = [SD/p.\bar{x}]^2 \quad \ldots \ldots \quad (3.2)$$

where p = desired precision (e.g. 0.15 as a proportion of the mean), $\bar{x}$ = mean of the samples, SD = standard deviation of the samples.

It would be conservative to choose more replicates than indicated from formula 3.2 to allow for differences in space and time. There is a risk, however, that this will be wasteful of resources and the cost of that risk can only be determined on a case-by-case basis.

Other advice concerning sample units and the scale of spatial pattern are given by Green (1979) and Andrew and Mapstone (1987).

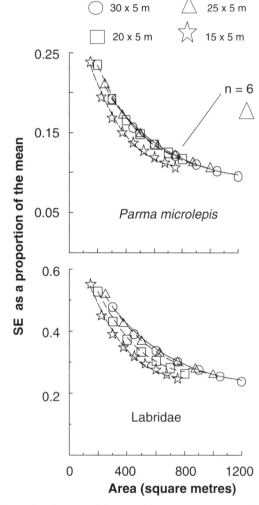

Figure 3.1 Example of data used for optimisation of sample unit size and number. The data are number of territorial pomacentrids (*Parma microlepis*) in barrens habitat, and total labrid fishes in kelp forest, New South Wales. The number of sample units chosen was six, and the sample unit size 25 x 5 m. See text for a detailed explanation on the decision-making.

For the purpose of comparing the precision obtained from sample units of different sizes, source information on the spatial distribution of the organisms may be gathered by:

- Mapping the positions of organisms for a static picture of the study area.
- Sampling the study area repeatedly with the largest sampling unit size subdivided into subunits corresponding to the other units being considered.
- Using information already available on the distribution of an organism.

Ideally any one of these methods should include data from different times and places.

Mapping
Once a map is obtained, sample units of different sizes can be positioned on it by use of random co-ordinates (within the same habitat). In cases where the sampling unit is asymmetrical (e.g. a transect), a random orientation (0–360°) from a random coordinate should also be derived. An example of maps showing the static distribution patterns of an organism is given in Figure 5.6. Problems resulting from the sample unit lying over the edge of the map can be overcome by using the wrap-around technique. This technique involves arranging copies of the map abutting the original. Organisms are counted on the additional maps if the sampling unit extends beyond the original map. Once large numbers of samples (usually 20 or more) are obtained for units of different dimensions, a mean and standard deviation can be derived.

Sampling the study area with units of different sizes
This procedure is identical to that described for mapping, except that sample units of different sizes are tested in the field instead of being superimposed on a map. For example, 100 m transects may be divided into units of 10 m so that information on other sample-unit sizes (e.g. 30, 50 m) can be examined. Avoid taking multiple contiguous small units from the same large transects to avoid problems of independence (i.e. randomly select subsamples of smaller units).

Using information that is already available
Snedecor & Cochran (1980, p. 441) describe an algorithm for calculating the number of samples required to obtain an abundance estimate with an allowable error, in terms of confidence limits.

$$n = 4s^2/L^2 \quad \ldots \ldots \quad (3.3)$$

where L = predetermined allowable error (size of 95% confidence limits) on sample mean, s^2 = sample variance.

Example
Information is available to show that an organism is found at densities of 20 per m^2 with a standard deviation of ± 6; the variance equals 36. You decide your 95% confidence limits must not be greater than ± 5 = L.

$$n = 4\,(36)\,/\,5^2 = 5.76$$

You decide, therefore, that six replicates are required. *Note:* Be careful of spatial and temporal changes in variance.

Green (1979, p. 134) described a technique that is useful for determining whether one species, from a group of species being studied, will be adequately sampled by a sample unit of a given size (e.g. relatively rare sponges in your quadrats), but it does assume a vaguely normal distribution. The approach uses binomial sampling theory which is generally appropriate for organism abundance data (McArdle & Pawley 1994). According to Figure 3.21 in Green, if a species x represents one of the following proportions: 0.3, 0.2, 0.1, 0.05, 0.01 of the total number of organisms collected in sample units, then the probability of detecting that organism (i.e. a 95% chance of getting it) will require sample sizes of 8, 14, 30, 60, 300 respectively.

Example
Background information indicates mean density of all fish is 50 per 100 m^2. The species of interest is present in proportion p = 0.1; this value is then looked up in Green's Figure 3.21. I have decided to use a Y value of 0.05 – the probability that the species of interest will be missed when a given total number of fish is sampled. The graph indicates that there is a 5% chance of missing the species if I obtain a sample containing 30 fish.

30/mean density of 50 = the appropriate sample unit size that is unlikely to miss the species of interest = 60 m^2 (= 0.6 of 100 m^2)

Note: This method should not be used where $P > 0.7$. It is at least useful for determining the minimal sample unit size that should be considered.

A final alternative for the use of background information on abundances is to consult densities described in other habitats, locations, etc. This will enable a crude assessment of whether the sample unit size you or others have used is appropriate in a variety of situations.

Shape of sampling unit

In some cases, the shape of a sampling unit may need to be assessed. The concerns are: the influence of border effects (e.g. square versus round quadrats); and where broad areas are being searched, organisms may be missed – a major concern in transect counts for fish and other cryptic or mobile animals (Caughley 1977; Sale & Sharp 1983; Mapstone & Ayling 1993). Biases are likely to vary according to the organism being studied (e.g. mobility, degree of crypsis, response to investigators, etc; see Chapters 4–9).

Other considerations for sample units

- 'One of the advantages of a well thought out and balanced design is that relatively few replicates per treatment combination can go a long way in producing robust and powerful overall statistical tests.' (Green 1979); see also 3.5.3 and 3.5.4.
- Cost-benefit analyses will calculate how many replicates of the chosen sample unit size are required, given *a priori* decisions of required precision at different levels of a sampling design and the logistical constraints of costs known by the investigator (e.g. time and money; see 3.5.9).
- You will always have to make compromises where you are studying multiple species, because one sample unit size and number may not be ideal for all species. It may be best to optimise your sample unit to functional groups (e.g. grazing gastropods, Chapter 4), target species that are vulnerable to exploitation (Chapters 5 and 6), species that are particularly abundant/rare within the study area, or other criteria of importance (Fig. 3.1).
- Power analysis (see 3.5.10) can be used to calculate how many replicates are needed (and replication at other levels of the design), to detect a significant result for a predetermined effect size.
- Finally, in some situations, logistical constraints may preclude the possibility of assessing sample units of different sizes (e.g. sampling isolated offshore islands). In these cases choices will have to be based on background information, if available, on the densities of the organisms, and experience. Moreover, some sample units may have proved particularly appropriate for the organisms or group of organisms you are concerned with, at a number of localities and times. Examples of these are given in Chapters 4–9.

Examples

- Mapping approach to record position of organisms: rock lobsters (MacDiarmid 1987; see Chapter 5).
- Sampling areas repeatedly with a large sample unit size that was subdivided into subunits: Abundance of drifting algae: (Kingsford and Choat 1986; abundance of agar seaweed on rocky reef (McCormick 1990; large and small tropical reef fishes, large discrete organism (e.g. crown of thorns starfish: Mapstone & Ayling 1993)

- Using background information: Schiel et al. (1995) in a survey of the Chathams Islands (a remote group of islands off the coast of New Zealand) used 1 m^2 quadrats to sample large gastropods and urchins. Replication of n = 5 was based on past experience in temperate waters of New Zealand. The data that resulted were acceptably precise and could be compared with studies from mainland New Zealand.

3.4 Replicates and randomisation

3.4.1 Principles

For many types of sampling design it is critical to take random samples (but see grid sampling; e.g. Zeldis et al. 1995), particularly at the replicate level, because it makes the data more likely to be independent (e.g. Winer 1971; Krebs 1989). If sampling is not done so that data are independent (random, haphazard or representative sampling), then the assumptions of many common statistical methods (e.g. ANOVA) are more likely to be violated. Random samples, for some hypotheses, may need to be taken in the entire study area (e.g. for estimates of total numbers of a target organism) or within strata (e.g. stratified sampling in different habitats). Random allocation of sample units or higher levels of a sampling design (e.g. choice of random location), can be achieved by coding replicates (or defined intersections in space) for replicates of each treatment (i.e. 1, 2 . . . x) and using random numbers (see 3.4.2). Care has to be taken that replicate samples of different treatments do not become 'segregated'. Similarly, clustered samples through random allocation may not be representative subsampling of a larger spatial or temporal scales of the design. In these cases, the randomisation process may be repeated to avoid segregation or promote more representative sampling.

In some studies, replicates too close to each other may influence the results. For example, Kennelly and Craig (1989) found that pairs of traps caught more spanner crabs than replicate single traps separated by 60 m (i.e. had greater collective attraction than single traps).

It is recommended that balanced sampling designs be used (i.e. n is equal for each treatment). Unequal replication in an otherwise balanced design leads to difficulties in analyses and the interpretation of results (see 3.5.3). An exception may be where very large samples of organisms are collected for analyses of size versus age, or similar relationships in the latter case. If, however, analyses if covariance (ANCOVA), or similar forms of analysis are used, you may need to restrict your comparisons to organisms of a similar range of sizes.

Where estimates of size frequencies are required it is also necessary to collect random samples of specimens. This should be done with adequate consideration of potential variation in size among times and at different spatial scales.

3.4.2 Practical application

It is critical that you have a randomisation procedure that reduces subconscious passions to include or exclude organisms from sample units based on *a priori* visions of rejecting or accepting the null hypothesis. Random numbers are required for many of these procedures and they can be obtained from random number tables, many computer programs (e.g. Excel and SYSTAT), digital stopwatches (by using numbers at the level of 1/100ths on the stopwatch, Fulton 1996), a phone book, lotto cards, or dipping for numbers in a hat.

Methods for the allocation of sampled units include:

- Grid up an area (e.g. with tapes) and choose random intersections, that were predetermined using random numbers, for the placement of your replicate samples. However, gridding can be cumbersome, particularly underwater, and other methods (as follows) may be easier to use, but equally rigorous.

- A tape can be run out and predetermined random intersections sampled on the tape; a variation on this could be to sample at random distances to the right or left of the intersection.

- Random distance from a spot, you may have coded, North, South, East, West as numbers 1–4; you can then choose a random direction and distance from the sample spot to place your sample unit. Underwater, flipper kicks may be used to estimate distance without the problem of getting tangled in more tapes; for example, with a given kick rhythm you may do two kicks per metre; a predetermined number of kicks should be generated from random numbers prior to the dive.

- Haphazard sampling may be a necessary substitute for random sampling (perhaps based on time taken) in some environments, but care should be taken that sampling is not biased.

Randomisation may be required at other levels of your design. For example, it may only be logistically possible to sample six out of 20 possible locations. Simply code the locations 1–20 and use random numbers to choose your sample of six. See 'A. randomised' in

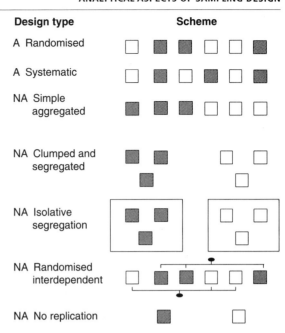

Fig. 3.2 Acceptable and non-acceptable organisation of replicate sampling units. Shading = different treatments, A = acceptable (i.e. interspersed sample units), NA = not acceptable (sample units not interspersed).
MODIFIED FROM HURLBERT (1984) WITH PERMISSION

Figure 3.2, where the six boxes could be treated as six locations, out of which sampling could possibly be done at three.

If data on size frequency are collected, it is crucial that an unbiased sample be collected. Although it may be appropriate to collect large, medium, and small organisms for assessments of reproductive status, the data on sizes in this case are useless for analyses of size-frequency of a population. Options for a random sample include the following:

- Sampling all organisms within random sampling units. However, problems may arise with this method if some size classes or taxa have a very aggregated pattern of distribution. If you are measuring abundance at the same time, in some cases (e.g. where abundance is low) you may require sampling in additional sample units to gain adequate data on size frequency.

- Subsampling organisms from a sample unit (e.g. a plankton tow or trawl).

- Sampling organisms from multiple schools or aggregations, because individual schools are commonly of one size class. Subsamples can be taken from each aggregation.

- Numbering all specimens and measuring those selected by random numbers. Discard those that have already been sampled (R. Cole pers. comm.);

55

although this avoids some of the problems listed above, it is generally impractical.

See also Underwood (1997, pp. 38–44) for special cases of problems in trying to obtain a random sample; for example, random estimates of the sizes of plants with vegetative growth.

Manipulative experiments and impact studies
A number of randomisation procedures have been suggested for the spatial arrangement of replicate treatments within an area. Investigators should be careful that experimental units are far enough apart that they do not influence each other (Underwood 1993b, 1994a). Accepted methods of randomisation have been reviewed by Hurlbert (1984) and are illustrated in Figure 3.2. In a 'randomised design' treatments are allocated randomly to x possible areas. Although this design is legitimate, it can lead to unintentional spatial segregation of treatments.

If spatial segregation of treatments is obvious, the process of randomisation should be repeated. Hurlbert (1984) supported the regular interspersing of treatments because it reduces the chance of a random separation of treatments. Hurlbert considered a 'randomised block design' acceptable where experimental units are grouped together in blocks, be they in kelp forests, rock pools, or tank rooms. The treatments are well interspersed and if one block is destroyed (e.g. a storm destroys a kelp forest) the entire study is not lost. Although this design has been used in many studies, it is vulnerable to the criticism that there is no replication within each block.

A simple linear 'systematic design' also achieves a high degree of interspersion of treatments. It is possible, however, that systematic errors may occur in a periodic environment (e.g. regular differences in rock strata facilitate different settlement patterns for invertebrates, see Fig. 2.4). 'Latin square designs' are an extension of a randomised block design, but are limited in their application because the design must be symmetrical (Krebs 1989). Further, they have restricted applications in space. Investigators should be aware of the assumptions of Latin squares, because the analysis assumes that all interactions will be zero, which is unlikely in ecological experiments. The solution is to not use Latin squares and design the experiment with proper independent replicates (Underwood 1997). In some cases a cline of density may be recognised in the study area. In this situation Hurlbert (1984) recommends maximum interspersion of treatments and maximum similarity (i.e. similar mean values for treatments at the beginning of the experiment).

'Segregated designs' have been used by many investigators, but these designs are often 'pseudoreplicated'

(*sensu* Hurlbert 1984) and should be avoided at all costs (Fig. 3.2). The replicates are not independent and violate the assumptions of statistical methods. For example, an investigator may want to study the influence of urchins on the standing crop of algae. All of the controls are segregated from the experimental removal treatments by 100 m. Clearly, there is no control for the urchin removal because the standing crop and growth of algae may vary greatly on a scale of 100 m. In this case it is not possible to determine whether the urchins or natural spatial variation in algal abundance and growth was responsible for the resultant patterns. A more complex example is 'randomised' versus 'randomised interdependent' (Fig. 3.2). In this case, replicates of the different treatments may be on different water supplies, so water supply and treatment are confounded.

The problems with segregated designs are also shown for impact studies ('NA' in Fig. 3.3) and how this problem can be resolved. Designs marked 'A' are acceptable designs for impact studies.

See Chapter 2 for a more general comment on impact studies.

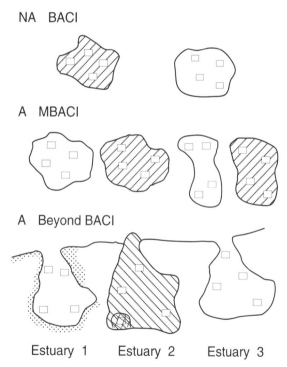

Figure 3.3 Examples of replication in space for BACI, MBACI and Beyond BACI designs.

A = acceptable, NA = not acceptable, Beyond BACI adapted from Underwood 1994c with permission. If the impact is small (cross-hatched), results may differ from multiple-control sites in estuary 2. If the impact is large, results for estuary 2 may differ from control estuaries 1 and 3. Squares = sample units or sites.

Repeated measures

In some studies replicate organisms, plots, or quadrats are used multiple times. For this reason analyses using 'repeated measures' are commonly done (Krebs 1989). Analyses of variance cannot be used because a basic assumption of ANOVA is that each observation is independent. Although there are legitimate ecological reasons why repeated sampling of the same sample units should be done (e.g. Joe Connell's 35-year+ time-series of corals in fixed quadrats at Heron Island), and plotting the data from multiple times is legitimate, analytical problems can often be avoided with careful consideration of the sampling design at the planning stage of the project.

The major problem when the same sample units are resampled is that data are not independent; i.e. what happened at time 1 may influence the outcome at time 2. If repeated measures (a type of ANOVA) is the only option for an hypothesis (e.g. recolonisation experiments: Beck in press), a multivariate approach should be used. Although the multivariate method is less sensitive than the univariate method (Gill & Hafs 1971), it is more robust to different forms of variance-covariance matrices. The analysis is designed to correct for autocorrelation in the data set (because of a lack of independence from time x to time x + 1). Assumptions for the nature of the correction can be violated, and care should be taken with this type of analysis. It should be noted, however, that with only two levels of time in a sampling design the mean square ratios have exact F-distributions (i.e. sphericity is not a problem: Huynh & Feldt 1970). Investigators should also consider how they will deal with the loss of potentially important interactions with the repeated-measures approach (Underwood 1997).

Repeated measures are sometimes used where sites are sampled multiple times and sampling units are allocated randomly within these. The problem with time in the analyses of the same sites is that times are not applied in a random order (are not stationary in a time-series sense) and this could potentially make interpretation difficult.

Although repeated measures are widely accepted, the approach is a topic of active debate (e.g. Winer 1971; Green 1993a; Underwood 1997): and for some hypotheses (and related sampling designs) the literature does not give clear advice on the application of randomised models or repeated measures. (See 3.5.6 for more detail.)

3.5 Allocation of sampling effort

3.5.1 Introduction

Once the investigator has clearly stated the biological problem to be addressed, has chosen a suitable sampling unit (see section 3.1) and is aware of the methods and pitfalls of randomisation, further decisions must be made on the allocation of sampling effort. The questions may be similar to the following:

- How do I allocate my sample units to get an accurate and precise estimate of the total number of sea urchins on rocky reefs along a 100 km stretch of coastline that encompasses a local fishery?
- I have hypothesised that the abundance of species x varies with depth; how many locations should I sample within a 100 km^2 area to make a general statement about depth-related pattern?
- I have hypothesised that the abundance of fish will increase in experimentally disturbed areas when compared to control; how many locations should I sample within a 10 km^2 area to make a general statement about the disturbance-related responses of fish?
- How many locations, times and sample units should I use for my statistical test to be sufficiently sensitive to detect an impact without confounding the spatial and temporal effects?
- How am I going to analyse the data?

Examples are given for descriptions of patterns, mensurative experiments (see section 2.7) and manipulative experiments.

There are two general approaches to sampling temporal and spatial patterns of organisms. The first is to identify patterns with no *a priori* assumptions of structure (potential sampling strata) in the environment. The second is to test mensurative hypotheses based on identifiable strata (e.g. depth, habitat types) in the environment. Where no structure is identified, as often occurs in open ocean or sediment flat environments, regular (e.g. grid) or random (on random co-ordinates) sampling may be appropriate. However, where an environmental mosaic is known from previous work by others, or from your own preliminary sampling, the appropriate sampling design would be a stratified or structured one (Green 1979). This is clearly the case with open coastal reef environments, where subtidal and intertidal habitats can be demarcated according to physical and biological features (see Chapter 4). For this reason, the examples of designs, and the types of hypotheses they address, will be restricted to stratified designs.

3.5.2 Stratified sampling

Principles

Stratified sampling is useful for three general purposes:

1. The determination of density variation over the study area;
2. A reduction of error variance (due to an environmental mosaic) to estimate treatment effects more clearly;
3. The estimation of total abundance of an organism within the study area (Green 1979).

If a study area has a large-scale environmental pattern, it may be broken up into relatively homogeneous sub-areas, which are referred to as strata (Caughley 1977; Green 1979). Homogeneous strata are readily identifiable in open coastal reef environments. On rocky reefs, for example, different habitat types (e.g. kelp forest, urchin-grazed barrens) may be treated as strata. Strata can correspond to physical or biotic variables and the choice of strata will depend on the hypothesis that is to be tested. In some cases, observations or guesses based on the literature may lead to an hypothesis concerning patterns of abundance among depth strata.

In all forms of stratified sampling, sample units are allocated to different strata. Sampling is random within a stratum (i.e. deployment of sample units, see Fig. 3.4).

1 Simple stratified allocation   **2 Optimal allocation**

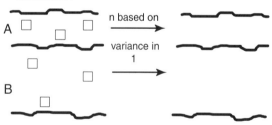

3 Proportional allocation

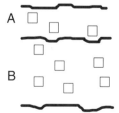

Figure 3.4 A diagrammatic representation of the three methods of stratified sampling used for the allocation of sample units The arrows from 'simple stratified allocation' (Phase 1) to 'optimal 'allocation' (Phase 2) indicate that information on sample sizes in each strata for Phase 2 are based on the findings in Phase 1.

There are three methods of doing stratified sampling, samples being allocated as follows (Cochran 1977):

1. Stratified simple random sampling: Equal numbers of random samples taken from each location and stratum (= randomised design). This type of design may be used to test hypotheses concerning variation in density among strata, or to gain information on local populations within strata for additional procedures such as optimal allocation (see below)
2. Proportional allocation: The allocation of samples is made in proportion to the area of each stratum (i.e. more samples in a larger area). No knowledge of the population is needed before the major sampling program, but a map is required.
3. Optimal allocation: Uses information on the area of each stratum and the variation in abundance of the study organism within those strata; preliminary sampling is required (stratified simple random sampling) for an estimate of variances within each stratum.

Practical application

Simple random sampling requires the identification of strata and where hypotheses are concerned with the generality of patterns of abundance within a defined area (e.g. a 10 km stretch of coastline), sampling should be done at multiple locations and within each strata at each location (e.g. fish are counted in kelp forest and urchin-grazed barrens at all locations).

All three methods can be used to estimate the total abundance or biomass (often used in the context of harvesting) of an organism within the study area. Proportional and optimal sampling, however, are likely to result in more precise estimates of total abundance. Analytical methods used to calculate estimates of total numbers are identical for all methods of stratified sampling. The only exception is that simple random and proportional allocation does not require the calculation of optimal allocation of sampling units to each stratum. If estimates of total abundance are to be obtained, then a detailed knowledge of the area of each stratum in the study area is required. This will have to be done using existing maps (e.g. subtidal habitats of the Leigh marine reserve, Fig. 4.8) or your own map generated from one or a combination of the following: profiles (e.g. Fig. 4.8); triangulation from known survey points; aerial photographs (or photographs from a vantage point), and casual observations. Sampling units are distributed randomly within each stratum. In the case of diving exercises this may be achieved by sampling all strata at multiple random locations along a stretch of coastline. Although optimal allocation will generally give a more precise estimate of total abundance, preliminary sampling is required for estimates of the density and variation in density of the study organisms within each stratum. Hence this procedure will require more time. A comparison of the total population estimate from optimal allocation can be compared with that from simple random sampling. Formulae for the calculation of population estimates can be found in Cochran (1977). The main sources of error with optimal stratified sampling are likely to be the calculation of sampling error on variance estimates from the pilot study, and the estimate of area for each stratum. McCormick and Choat (1987) found that six profile transects along a 2 km stretch of coastline gave estimates of area, for each habitat type, that were not significantly different from those measured from Ayling's (1978) map (Fig. 4.8). Further simulations from maps, like that of the marine reserve, may indicate that fewer than three profiles per kilometre give adequate estimates of area. The optimal allocation of sampling units to each stratum does not indicate how the sampling units should be distributed within the study area. Clearly, samples should be taken randomly over the whole area under study (i.e. representatively), not crammed into one small part of the study area. A good method for achieving this is to divide the entire area of each stratum into large contiguous grid squares and allocate random sampling units to each.

Examples
- Simple stratified allocation: Kingsford et al. (1989) tested a null hypothesis that there were no differences in the abundance and composition of reef fishes at deep and shallow sites at multiple locations around the subantarctic Auckland Islands (for a detailed account, see 6.4.1). This was a biogeographic study and deep and shallow sites were sampled in case some species were only found in deep or shallow water. For other examples, see Jones (1984a), a study of recruitment of labrid fish.
- Proportional allocation: Haddon et al. (1996) used stratified sampling with proportional allocation to estimate biomass of shellfish associated with beaches in New Zealand. There was also a haphazard optimal component to this study, where the authors added sample units to strata based on perceived variances as they were sampling.
- Optimal allocation: McCormick and Choat (1987) tested a null hypothesis that there were no differences in the total number and size frequency of large cheilodactylid fish along the 5 km marine reserve coastline at Leigh, New Zealand, compared with an adjacent 5 km stretch of coastline without protection. Optimal allocation was used (for more details on transect size, replication and ecological interpretation see section 6.4.3). An improvement

to this study would have been the inclusion of more than one stretch of coastline that was not protected. For other examples, see Siniff and Skoog (1964), McCormick (1990).

3.5.3 Factors, degrees of freedom and planning

The focus of this section is on sampling designs that are relevant to analysis of variance. The principles of the design phase, however, are relevant to many other forms of analysis. If using other forms of analysis (e.g. multivariate methods), consider degrees of freedom for the specific test. Degrees of freedom for multivariate methods, for example, may change with number of variables.

Fixed and random factors
Before examples of sampling designs are given, some aspects of design protocol and analytical procedures require clarification. Factors are described as fixed or random (Underwood 1981, 1997).

Fixed factors indicate that you have randomly allocated sample units in all of the possible treatments that are available and are relevant to your hypothesis. For example, you identified three depth strata on a reef (shallow, mid and deep) and sampled all of these; there were five identifiable habitats in your study area and all of these were sampled. You should always replicate within a fixed factor, e.g. sample shallow, mid and deep at more than one location within your study area. Samples taken from only one location may be unrepresentative of depth-related patterns in the study area.

Random factors: you choose your treatments as representative of a much larger set of possibilities. This enables investigators to make more general statements about the outcomes of tests for random factors. Examples are as follows: of a large number of islands you could only sample three; of a large number of accessible rocky reefs along a stretch of coastline you could only sample five; of a possible 90 days within a season you only sample on five. A reliable procedure for choosing these treatments is to number all possible options and then designate your treatments by random numbers. In a situation where locations are sampled at regular intervals along a coastline, location should still be treated as a random factor (if the study was done again it would be unlikely that the same locations would be chosen). Furthermore, if locations or sites are chosen arbitrarily (no *a priori* reason for choice), treat them as random factors. Unlike fixed factors, the parameters of interest are not means, but variances (Winer 1971).

Winer (1971) provides a simple method for determining whether your factors are fixed or random, where the ratio of the number of treatments of a factor to the potential number of treatments is called the 'sampling fraction' of a factor.

> N = total possible treatments for a factors
> n = number of treatments sampled
>
> If $n/N = 1$, then the factor is fixed
> If $n/N < 1$, then the factor is random

For example, in your sampling design you sample five of a possible 20 locations.

Thus, $5/20 = 0.25$, therefore the factor 6 'location' is random.

Exceptions

Some authors treat factors as fixed, when you may think that designation as a random factor was more appropriate. For example, Keough and Mapstone (1995) treat time as fixed in their MBACI design for impact studies. Their argument is threefold:

1. Sampling for impact studies of this type is rarely done over more than a couple of years; time therefore is not representative of variation over a wide range of times (also see example 3.5.6);
2. They are trying to make management decisions about the particular time when the impact occurred, rather than one broad section of time;
3. If you sample five consecutive years, that cannot be considered a random sample of all possible years.

By extension, an argument for fixed factors in space can be made. For example, if you make your sampling universe small (e.g. one degree of latitude) and you sample all estuaries within that area, then the factor 'estuary 'can be treated as fixed. The conclusions, however, must be restricted to that narrow stretch of coastline. In rearing experiments, if organisms are sampled at fixed times (e.g. 0, 5, 10, 15, 20 days), then time is a fixed factor (Underwood 1981).

The distinction between fixed and random factors has a great influence on the calculation of F-statistics in ANOVA because this determines the appropriate denominator mean squares (MS) that are used to calculate F-statistics (F = numerator MS/denominator MS). An example, is given in Table 15 of Underwood (1981) for a two-factor orthogonal model: if factors A and B are fixed factors, then the denominator MS for A and B and the A x B interaction will be the residual MS. If, however, B is random then the denominator MS of A will be tested over the MS of the A x B interaction and B and A x B will be tested over the residual MS (for a similar design see 3.5.4).

Methods for calculating the expected values of mean squares for straight forward and specialised designs are shown in Appendix 3.1.

In some cases multiple random factors can result in a 'no-test' for a factor that is directly relevant to the test of the hypothesis. This may be resolved in the following ways:

1. It may not matter if there is expected to be a significant interaction that involves this factor because the significant interaction would have to be interpreted and it would be illogical to interpret the main factor in isolation.
2. If factors are not significant at $P > 0.25$ (a value of P for which the factor is thought to explain a negligible amount of the total variation in the analysis) then they may be pooled and the model simplified so that there is no longer a factor without a denominator MS because non-significant terms have been removed (Winer 1971). In many cases, however, this will not be possible (i.e. P is not > 0.25).
3. Revisit the reasoning of your design and construct a different hypothesis from the underlying model Underwood (1997). It is preferable that this be done *a priori*.
4. Quasi F-tests (Winer et al. 1991) may be used as a last resort if you do not take options 1–3. In this case, mean square components that are known from other parts of the model are extracted to determine the confounded components of the mean square estimates of the factor with a 'no-test'. This method is frowned upon by some, because the estimates of associated probabilities are considered inaccurate (Underwood 1997). It does, however, appear similar to the determination of resolvable MS estimates in complicated asymmetrical sampling designs.

Degrees of freedom and planning

Unless stratified sampling using optimal or proportional allocation techniques is adopted (see section 3.3), it is best to have an equal number of replicates in each treatment (Underwood 1997). Where replication is unbalanced, it can result in:

- Invalid F-tests for some sources of variation, although this could potentially be solved with a randomisation test or similar procedure (Manly 1991);
- Approximate degrees of freedom that make it difficult to use tests based on probabilities associated with an F-ratio;
- Violations of assumptions of normality that may increase the chances of a Type I error.

Lost data often occurs even with the best of planning. If you lose a replicate from balanced design data, through demonic intrusion, then you can replace the missing cell in the analysis with the mean of the others and lose a degree of freedom from the residual. Note that this procedure should only be done when you have a reasonable number of replicates (i.e. n ≥ 3). The use of 'dummy values' becomes more complicated with loss of multiple replicates and your options are limited if there are interactions in the data set (see Underwood 1997).

Even great differences in abundance of organisms (or other measures) are unlikely to be detected if you have poor degrees of freedom for the factors that are of greatest interest to your hypothesis. The distribution of F values (see an F table, e.g. Winer et al. 1991) shows very high values at low degrees of freedom [e.g. df = 1,1; $F_{0.05}$ = 161], followed by a rapid decline with increasing degrees of freedom. Once degrees of freedom are greater than 1,4 (where critical F = 7.71, at P = 0.05) for subsequent increases in degrees of freedom there are very small concurrent drops in F values. The aim of any design where the hypothesis to be tested is not based on an *a priori* size effect, is to obtain degrees of freedom (for factors important to the hypothesis) that are in this plateau of F-values. Low degrees of freedom greatly increase the chances of Type II error.

Beware of 'pseudofactorialism' (*sensu* Hurlbert & White 1993). In this case an investigator has made multiple measurements on a single experimental sampling unit and called them replicates. This results in the erroneous inclusion of a factor (at the level of experimental units in the correct design) and an incorrect increase in the number of degrees of freedom. Hurlbert and White (1993) used an example of a tank experiment where two measures of capture time of prey by predators (= sequence of capture) – time to first capture, and mean time between subsequent captures – were made in each replicate tank and then included as the factor 'sequence' in the experimental design. This resulted in double the degrees of freedom of the correct experimental design (i.e. where the two measures of sequence of capture are analysed separately).

An *a priori* knowledge of the 'effect size' (see Power Analysis, 3.5.10) of biological or social interest will help with the development of an hypothesis and the allocation of sampling effort at different levels of the design. Ultimately, however, you will have to work out the best sampling design to cater for logistical constraints of time, weather, and money.

When you articulate your hypothesis and make decisions on the allocation of sample units, always consider variation on short temporal and small spatial scales. Variation with time of days, or on different days within a season can be great (e.g. Chapters 6 and 9). Regular sampling at say monthly intervals can miss real changes in abundance, especially when processes such as recruitment are measured (Underwood 1989; Fig.

2.1). In other situations preliminary sampling and background information may indicate that short-term temporal changes (e.g. variation with time of day) are unlikely to confound comparisons among longer time periods and locations, i.e. you can count organisms at any time of the day in a study (e.g. Kingsford 1989). In cases where long-term temporal changes are of primary interest, always replicate in space to justify claims of generality of temporal changes.

The sampling designs that follow are examples based on carefully stated hypotheses. Correct F-tests for a factor require a knowledge of the denominator mean square. The denominator mean squares were calculated according to the procedure shown in Appendix 3.1. In all the examples that follow, where values of F statistics for different levels of P are presented they were taken from Winer (1971, Table C.3). Note that the hypothesis determines the detail of each design, so the designs should not be 'pasted' into another biological problem. Variations on multifactorial designs are numerous. For example, Underwood (1997) stated that there are more than 100 unique alternatives with four-factor designs. The section that follows assumes you have chosen an appropriate size and number of sampling units (replicates) from background knowledge or the procedures given in section 3.3.

3.5.4 Orthogonal sampling designs

Orthogonal (= factorial experiment) designs enable an investigator to evaluate the effects of two or more factors. The combined effects of the factors can be evaluated from the interaction effects (see Fig. 3.5, p. 62).

Example 1
An orthogonal design that could be used for a mensurative hypothesis concerning the distribution of sea urchins. A diagrammatic representation of the allocation of sampling is given in Figure 3.5.

Hypothesis
That abundance of urchins will vary with depth and that patterns of abundance will be consistent among locations.

The degrees of freedom in the design (and related statistical power) change greatly with changes in the number of treatments for each factor. If location = 2, for example, then (b – 1) = 1, and in this case factor depth (a – 1) = 2, then the critical value of F (at P = 0.05) will be based on degrees of freedom 2,1 ($F_{1,2}$ = 200; i.e. the difference would have to be huge to detect a significant effect). Since the hypothesis is primarily concerned with depth and the generality of these patterns, then it would

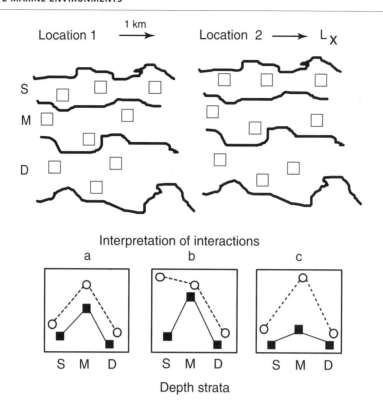

Interpretation of interactions

Depth strata

Figure 3.5 A diagrammatic representation of the allocation of sample units for an orthogonal sampling design. Sample units are shown as quadrats within strata that were sampled at each location; S = shallow, M = mid, D = deep. Diagrammatic representations are given of three outcomes of the results (a–c). A significant interaction would not be expected for plot a, but would be for b and c.
Solid line = location 1; broken line = location 2.
See text for the mensurative hypothesis and a discussion on interactions; also ANOVA Table 3.3.

Table 3.3 An orthogonal sampling design for a mensurative experiment. F = fixed factor, R = random factor (see section 3.5.3); df = degrees of freedom; the denominator mean square for each factor is used for the calculation of an F-statistic in ANOVA. See also diagrammatic representation in Figure 3.5. 'Depth' is fixed because all depth strata were sampled. Location is 'random' because the treatments are representative of many possible locations along a stretch of coastline. In column df, a = treatments of factor depth, b = treatments of factor location and n = number of replicates.

Row	F/R	Source of variation	df	Row of denominator MS
1	F	Depth	$(a - 1)$	3
2	R	Location	$(b - 1)$	4
3		D x L	$(a - 1)(b - 1)$	4
4	R	Res	$ab(n - 1)$	

be logical to increase the number of locations so that there is greater statistical power (see section 3.5.10) to test the factor depth. For example, if there are seven locations, then the critical value of $F_{2,6} = 5.14$. It is not possible to change levels of depth (and related df) because it is a fixed factor with only three possible treatments. Possible outcomes of the design are indicated as plots of urchin abundance in Figure 3.5a–c. Figure 3.5a would not result in a significant D x L interaction. Differences in the abundance of urchins were found among depths. Abundance was greatest at mid depths and the patterns of abundance were consistent among depths at all locations. In Figure 3.5b there were differences in urchin abundance among depths at both locations, but rank abundance among depths varied for each location and this would result in a significant D x L interaction. In Figure 3.5c patterns of urchin abundance with depth varied with location. Although a trend for highest abundance at mid depths was found in all locations, the magnitude of this difference varied with location and resulted in a significant D x L interaction.

A posteriori tests (e.g. SNK and Ryan's tests) could be used to test hypotheses concerned with differences among specific depths. If there was no significant interaction, a single test could be done among depths for the whole design (i.e. pool all replicates for a given depth from all locations). If, however, the interaction was significant then separate tests among depths would have to be done for each location, as the interaction implies. It may be of no relevance to your hypothesis to test for specific differences among locations at the same depth strata.

Example 2
An orthogonal design that was used for a manipulative experiment.

Hypothesis
That abundance of fish will be highest in disturbed treatments and this will be consistent among locations.

This design was used by Glasby and Kingsford (1994) to study disturbance by reef fish. They sampled at six locations, thus the critical value of $F_{1,5} = 5.79$ at $P = 0.05$ for the factor 'disturbance' and $F_{5,48} = 2.45$ for the interaction, which allowed the authors to comment on the generality of the plus and minus disturbance treatments. A detailed description on the biological interpretation of the work is given in section 5.4.6.

For other applications of orthogonal sampling designs that use manipulations, see Jernakoff and Fairweather (1985); Kennelly (1987a); Hair et al. (1994). Mean square estimates for many mixed-model orthogonal designs are given in Underwood (1997), but these can be derived easily using the methods given in Appendix 3.1 (p. 83).

3.5.5 Nested sampling designs

Nested, or so-called 'hierarchical' designs, are a potent approach to dealing with spatial and temporal variation on a variety of scales. In all of the chapters on specific groups of organisms that make up assemblages in coastal environments (Chapters 4–9) it is clear that there is great variation in abundance in space and time. For example, abundance may vary among days within a month, months within a season and seasons within a year. The variation among seasons, however, may be greater than any of the shorter-term sources of variation.

Where hierarchical designs are not used, multiple levels of variation may contribute in unknown ways to the residual variance, rather than be partitioned and testable in a comprehensible way.

Table 3.4 An orthogonal sampling design for a manipulative experiment. F = fixed factor, R = random factor (see 3.5.3); df = degrees of freedom; the denominator mean square for each factor is used for the calculation of an F-statistic in ANOVA. 'Disturbance' was treated as fixed because the treatments were disturbance and no-disturbance. Location is 'random' because the treatments are representative of multiple possible locations along a stretch of coastline.

Row	F/R	Source of variation	df	Row of denominator MS
1	F	Disturbance	$(a-1)$	3
2	R	Location	$(b-1)$	4
3		DxL	$(a-1)(b-1)$	4
4	R	Res	$ab(n-1)$	

Example 1

A nested design that could be used to describe variation in the abundance of organisms at different spatial scales in one habitat or over a whole reef.

Hypothesis

That abundances of sea urchins in barrens will vary between islands and that these differences will be greater than differences on smaller spatial scales (i.e. locations and sites).

There are no logical interactions in this design. It is easy to see from Figure 3.6 that it would be illogical to have an interaction between island and location (i.e. I x L interaction does not exist in the design). Locations 1–3 are placed randomly within each island and potential comparisons in rank abundance between locations 1–3 at island 1 and locations 1–3 at island 2 are meaningless (cf. interactions in orthogonal designs, p. 63); i.e. location 1 island 1 has no relationship with location 1 island 2, etc.

Note that all factors are treated as random in this design. The denominator mean square (for the F-ratio) is always that next factor down the hierarchy (i.e. Island is tested over Location and Location over Site). In this case the hypothesis is primarily concerned about variation among islands.

A more powerful test for Islands can be gained by increasing the number of locations within each location. For example, if Islands = 3 and Locations = 2, then the critical value of $F_{2,3} = 9.55$ at $P = 0.05$; if the number of locations is increased to three, then the critical value of $F_{2,6} = 5.14$. This increase in the power of a test for Islands makes intuitive sense because we are gaining more information about each island by sampling at more locations per island.

A greater number of locations also makes biological sense because great variation in abundance of

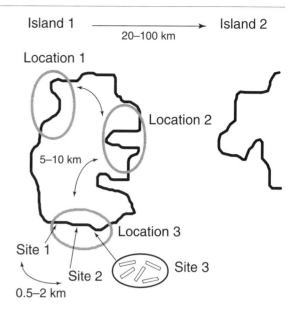

Figure 3.6 A diagrammatic representation of the allocation of sample units for a nested sampling design. Five non-overlapping and haphazardly placed transects were sampled at each site. Sites were located randomly within each depth strata and location. See text for the hypothesis; also Table 3.5.

urchins on reefs is commonly found at scales hundreds of metres to kilometres (Chapter 5).

If all factors in rows 1–3 were significant, the results would be interpreted as follows. Although there was significant differences among sites within locations and locations within islands, significant differences in the densities of urchins were detected among islands.

Although some authors argue that it is obligatory to treat nested factors as random (Underwood 1997), this does assume that the treatments of nested factors are random samples at defined spatial or temporal scales. I support the treatment of factors as random in this case.

Table 3.5 A nested sampling design for abundance patterns of urchins. F = fixed factor, R = random factor (see 3.5.3); df = degrees of freedom; the denominator mean square for each factor is used for the calculation of an F-statistic in ANOVA. See also diagrammatic representation in Figure 3.6. Island was treated as random because a subset of many possible islands was sampled. For column df, a = treatments of factor island; b = treatments of factor location; c = treatments of factor site and n = number of replicates.

Row	F/R	Source	df	Row for denominator MS
1	R	Island	$(a-1)$	2
2	R	Location (Island)	$a(b-1)$	3
3	R	Site [L(I)]	$ab(c-1)$	4
4	R	Res	$abc(n-1)$	

In some cases, however, this may be difficult to argue if you have great restrictions with your options for random samples in time and space. For example, Keough and Mapstone (1995) have argued for fixed nested factors (times) in some MBACI sampling designs (see p. 66, example 2).

If it is relevant to the biological question, investigators may wish to determine the proportion of variation that can be attributed to each factor in a nested design; this can be calculated using variance components (see 3.5.10). For other examples of nested designs, see Morrisey et al. (1992a), temporal variation; Kingsford (1989), Jones et al. (1990), Morrisey et al. (1992b), Thrush et al. (1995) , spatial variation.

3.5.6 Partially hierarchical sampling designs

It is common for there to be orthogonal and nested components to a sampling design. Designs of this type are called partially hierarchical. For example, an investigator knows that, based on the literature, patterns of fish abundance often vary with depth. There is, however, a concern that the depth-related pattern may vary at spatial scales of 10–15 km as well as on scales of 0.5–1 km within a location.

Example 1
The sampling of fish at different depths and spatial scales

Hypothesis
That abundances of fish will vary among depths and that these differences will be consistent among locations (5–10 km apart) and at smaller spatial scales within locations (i.e. sites).

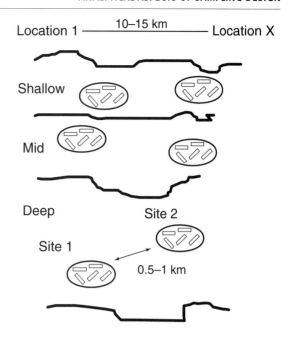

Figure 3.7 A diagrammatic representation of the allocation of sample units for a partially hierarchical sampling design. The sampling units are five non-overlapping and haphazardly placed transects at each site. See text for the mensurative hypothesis and Table 3.6.

As for orthogonal example 1 (p. 61) variation in numbers of locations has a great impact on critical values of $F_{0.05}$ for the factor depth. For some hypotheses the primary concern may be variation in abundance among locations, but fish can only be counted at a finite number of locations within a restricted window of opportunity for sampling. An earlier pilot study may have indicated that variation among sites was low (e.g. $P > 0.25$: Underwood 1981), and therefore it may be possible for

Table 3.6 A partially hierarchical model (i.e. orthogonal and nested components to the sampling design) for abundance patterns of fish. F = fixed factor; R = random factor (see 3.5.3); df = degrees of freedom; the denominator mean square for each factor is used for the calculation of an F-statistic in ANOVA (for calculations see Appendix 3.1). For diagrammatic representation, see Figure 3.7.

Row	F/R	Source of variation	df	Row for denominator MS
1	F	Depth	$(a-1)$	3
2	R	Loc	$(b-1)$	4
3		D x L	$(a-1)(b-1)$	5
4	R	Site [L]	$b(c-1)$	6
5		D x S[L]	$(a-1)b(c-1)$	6
6	R	Residual	$abc(n-1)$	

sites to be deleted from the design and sampling effort redirected to the level of locations (see cost-benefit analysis and the pitfalls, section 3.5.8).

Although it may not be necessary to have sites in the design it may still be important to avoid having all replicates close together within a location, but spread apart at a scale approximating the separation among sites. For a discussion on the interpretation of interactions and considerations for *a posteriori* tests, see 3.5.3.

Example 2
MBACI (Multiple Before After Control Impact) design (partially hierarchical design for impact studies).

For a general discussion of sampling designs that are used to measure impacts, see p. 36.

Hypothesis
There will be a reduction in abundance of snails after impacts in two areas, that are greater than 10% of the original densities measured at multiple times (temporal controls) and compared to multiple times after the impacts, and that there will be no drop in abundance, measured at multiple times before and after the impacts, at two control sites that are assumed to be outside of the area of influence of the impacts.

MBACI designs have the advantage that they are not as complicated as asymmetrical designs (Fig. 3.8, Table 3.7). In many cases, however, it is not possible to have more than one impact location; and an asymmetrical design is the only option.

The factors of greatest interest in the MBACI design are the C x B and the C x Times(B) interactions (Green 1979; Keough & Mapstone 1995). The interactions are of primary interest because impact studies are designed to detect differences (or a lack of them) in temporal trends in abundance (or other variable) at control and impact sites before and particularly after the impact. For this type of analysis there is generally, and unavoidably, a poor test for the C x B interaction (i.e. 1, 2 df). Notes on the interpretation of interactions and considerations for *a posteriori* tests are given in 3.5.4. Differences between locations from control and impact areas (i.e. interactions involving Loc (C)) are also of interest to interpret the generality of patterns for each of the impact treatments.

There are three ways that this design may be analysed:

1. The design may be justified as a partially hierarchical design with the random allocation of sample units within each location at each time of sampling. If the degrees of freedom are calculated using the rules given in Appendix 3.1, then there are no tests for two main factors (C and B) and the important C x B

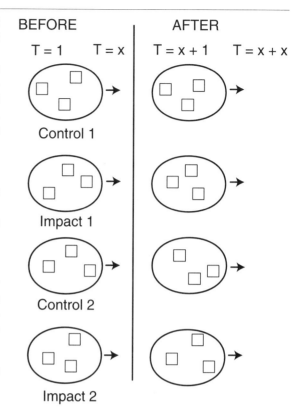

Figure 3.8 A diagrammatic representation of the allocation of sample units for an MBACI (partially hierarchical model) sampling design for the detection of an impact. The sampling units are three non-overlapping and haphazardly placed quadrats at each location. See text for the manipulative hypothesis and ANOVA table. Locations within impact and control treatments are random, and not orthogonal, pairs of impact and control locations. Quadrats are placed randomly at each time of sampling.

interaction. It is possible that after analysis the interactions could be pooled to create a test for previously untestable terms.

2. A partially hierarchical design where the factor time is treated as a fixed factor along with factors C and B. Keough and Mapstone (1995) argue that in a study of this type Times is rarely a selection of random times. It is more likely that the study can only be done in a fixed number of regular periods (e.g. years) and, therefore, time should be treated as a fixed. If time is treated as fixed, then denominator mean squares can be obtained for all factors (Table 3.7a).

3. Keough and Mapstone (1995) argue that the design should be treated as a repeated measures design because it is easier (Table 3.7b). This method, however, results in a no test for location (C) nor the L(C) x T (B) interaction and no residual, all of which they consider is of little interest.

Keough (pers. comm.) comments that it doesn't matter whether you use options 2 or 3 (above): you will get the same answer.

The varied application of randomised and repeated measures designs is based on different interpretations of the assumptions of ANOVA (cf. Green 1993a, b; Underwood 1997). The repeated component of the design is repeated sampling of the same location rather than the same sampling units (see Green 1993a). Green argues that the use of the residual (a measure of spatial variation) is incommensurable with factors that include temporal variation. For this reason, it is assumed that the error (= residual) cannot be measured. It is unclear, however, where this leaves factors that only measure spatial variation (e.g. location that includes both control and impact treatments). I assume the inclusion of

Table 3.7a A partially hierarchical model (i.e. orthogonal and nested components to the sampling design) for a randomised MBACI design. F = fixed factor; R = random factor (see section 3.5.3, and justification for the treatment of Times (B) as a fixed factor, item 2 in text above); df = degrees of freedom; denominator mean square for MS is used for the calculation of an F-statistic in ANOVA. In column df (= degrees of freedom for numerator in F-test), a = levels of impacts versus control, b = levels of location, c = levels of before versus after, d = levels of time, n = number of replicate sample units; the results of these calculations are given as the numerator df in the far right column. The degrees of freedom for F-tests (far right column) are based on times = 4, locations = 2, n = 3. Derived from the example of Keough and Mapstone (1995); see Figure 3.8. Note, a − 1 must be 1.

Row	F/R	Source of variation	df	Row for denominator MS	df for F-test Fig. 3.8
1	F	Impacts vs. Controls = C	$a - 1$	2	1,2
2	R	Loc (C)	$a(b - 1)$	9	2,64
3	F	Before vs. After = B	$(c - 1)$	7	1,2
4	F	Times (B)	$c(d - 1)$	8	6,12
5		C x B	$(a - 1)(c - 1)$	7	1,2
6		C x Times(B)	$(a - 1)c(d - 1)$	8	6,12
7		Loc (C) x B	$a(b - 1)(c - 1)$	9	2,64
8		Loc (C) x Times (B)	$a(b - 1)c(d - 1)$	9	12,64
9		Residual	$abcd(n - 1)$		64

Table 3.7b A partially hierarchical model (i.e. orthogonal and nested components to the sampling design) for a repeated-measures MBACI design. Rows 1 and 2 refer to spatial variation and rows 3–8 refer to temporal variation. $l' = l_c + l_i$ where l_c = number of control locations, l_i = number of impact locations; $t' = t_B + t_C$ Denominator MS for Figure 3.8 is based on times = 4 and locations = 2 as for Figure 3.8. Example adapted from Keough and Mapstone (1995). Note, there is no residual as in row 9 of Table 3.7a.

Row	F/R	Source of variation	df	Row for denominator MS	df for F-test Fig. 3.8
1	F	Impacts vs. Controls = C	1	2	1,6
2	R	Loc (C)	$l' - 2$	No test	–
3	F	Before vs. After = B	1	7	1,2
4	F	Times (B)	$t' - 2$	8	6,12
5		C x B	1	7	1,2
6		C x Times(B)	$t' - 2$	8	6,12
7		Loc (C) x B	$l' - 2$	8	2,12
8		Loc (C) x Times (B)	$(l' - 2)(t' - 1)$	No test	–

a variance component that includes time in the mean square estimators makes location commensurable with the rest. There appears to be no satisfactory resolution of when and when not to use repeated measures. However, if the sampling units are truly random samples that are independent of each other among times, then I can see no reason why the analysis should be done using options 1 or 2 above.

3.5.7 Asymmetrical sampling designs

Asymmetrical designs are commonly used for experiments (Underwood 1986) and impact studies (Underwood 1991, 1994a; Chapman et al. 1995; see also section 2.6) and can be very complex. As the name implies, some parts of the analysis are unbalanced (i.e. cf. orthogonal and nested designs in 3.5.4 and 3.5.5). For example, there may only be one site with an impact, but you have compared the densities of organisms at multiple control sites with the potential site of impact.

This type of analysis should not be confused with 'unbalanced' designs (*sensu* Winer 1971) where different numbers of replicate sample units are used per treatment (i.e. n at the residual level, Glasby 1997).

Principles of the designs
- An unbalanced number of levels within groups of treatments that prohibit orthogonal comparisons among all factors.
- Equal numbers of replicate sample units are sampled per treatment.
- In the design phase of this approach, consider the design as though it were balanced (symmetrical) to determine the mean square estimates of factors (for methods see Appendix 3.1) and then delete the parts of the design that are logically impossible (Glasby 1997).
- In the analytical phase two methods can be used (Underwood 1993b): (1) calculations are based on the results of more than one analysis of variance, or (2) using the formulae for complex models provided in Winer (1971), method 1 is used in the examples that follow.
- *Post hoc* pooling is possible when factors are not significant at $P \geq 0.25$. This can result in more powerful tests for factors and interactions that are important to hypotheses concerning impact.
- At the time of writing, multivariate analyses could not analyse all of the possible interactions in an asymmetrical design (*cf.* ANOVA), but these methods are useful for viewing the 'distances' between samples from different treatments in multivariate space (e.g. using MDS, Clarke 1993).

In many Environmental Impact Studies (EISs), it is common to address problems concerning a single point source of pollution. For example, an investigator may be asked to determine the impact of a sewage outfall, point source of industrial waste, or a single thermal plume from a power plant. The assessment of a Marine Protected Area (MPA) is another scenario where patterns of abundance, diversity or size frequency in one affected area are compared with more than one control area. Asymmetrical designs are necessary to address problems of this type, or the sampling design will be pseudo-replicated (see Chapter 2 and section 3.4). Ideally, a sampling design of this type should include multiple sampling times in the impacted area and at multiple control areas both before and after an impact (Underwood 1993b; Chapter 2).

In a high percentage of impact studies the disturbance has occurred before sampling commences. Glasby (1997) presented the results of a study into the impacts that marinas have on benthic assemblages where it was only possible to make spatial inferences of impacts because it was not possible to sample before and after the potential perturbation caused by the construction and normal functions of a marina. The design was as follows (see also Fig. 3.9): marinas were sampled in three different waterways in the greater Sydney area (factor = places). Within each place sampling was also done at two control locations (private jetties). A total of eight replicate quadrats was sampled by taking photographs of the epibiota on multiple pilings at control and impact (marina) locations.

Hypothesis
That abundance and diversity of the epibiota on pilings will vary according to the presence of marinas and that these patterns will be consistent among places.

In this design (Table 3.8) 'place' was a random factor because any places with marinas and controls could have been chosen. 'Treatment' was treated as a fixed factor, because there were only two possible levels, marina and control. Control locations represented random factors because they were two control locations of a number of possibilities at each place.

The benefits of an asymmetrical design for an impact study of this type are as follows: formal comparisons are possible for the presence of marinas versus controls (Marina versus Control, = Treatment, Table 3.8); the generality of any potential impact can be tested (i.e. is the cover of epibiota always different on the pilings of marinas when compared to controls, Place x Treatment interaction). For notes on the interpretation of interactions see 3.5.4. Formal comparisons of differences in abundance were possible among places

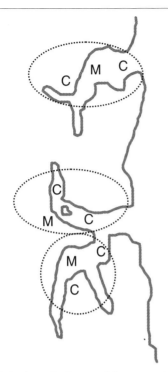

Figure 3.9 A schematic diagram of the asymmetrical sampling design used by Glasby (1997) to study the epibiota of the wharf pilings of marinas and controls. Dashed circles indicate separate waterways (factor = place). Only one marina (M) was sampled at each place and comparisons of the diversity and cover of epibiota were made with two control wharfs (C = controls/non-marinas) at each place (for ANOVA, see Table 3.8). Having a single marina and two controls at each place gives the design asymmetry.

(regardless of the presence or absence of a marina; factor = place) and between locations (marinas and controls) within a place. These factors are, however, generally of little relevance when testing hypotheses concerning impacts. It is important to note that without multiple controls it would have been impossible to separate natural differences from impact related differences and that more than one control location was required within each place.

The calculations for Table 3.8 are based on four analyses. A full explanation of the procedure is provided by Glasby (1997). In summary the analyses were as follows: Design analysis a is a hypothetical symmetrical analysis with Place x Treatment as orthogonal factors, so there is a P x T interaction nested in location, Loc (P x T). The remaining analyses (b–d) were used to calculate sums of squares. Analysis b used all of the data, and there was no differentiation between marinas and controls. This gives an orthogonal analysis (Location, Place and L x P). Analysis c, using data from controls only, controls orthogonal with Place; analysis d using data from controls only, a nested analysis of control locations nested within places; n = 8 in Glasby's design (Place, C(P)).

In experiments it is often appropriate for controls to be an asymmetrical part of the design. For example, in a study by Druce and Kingsford (1995) an experiment was done to assess the influence of colour and size of fish attraction devices (FADs) for attracting small fishes. The design was essentially orthogonal with the factors of size and colour, but these treatments had to be com-

Table 3.8 An asymmetrical model (i.e. unbalanced number of levels of each treatment; in this case there is only one marina location per place that is compared with two control locations) used by Glasby (1997) to study variation in the cover of epibiota at impact and control locations close to Sydney, Australia. The design also has orthogonal and nested components.
F = fixed factor, R = random factor (based on calculations of MS estimates from a hypothetical model). Df = degrees of freedom; the denominator mean square (MS) is required for the calculation of an F-statistic in ANOVA and was calculated from analyses a, b, d. Diagrammatic representation Figure 3.9; n = 8. SS = Sums of Squares. Treatment = Marina versus Controls. Only spatial inference of an impact is possible with this design.

Row	F/R	Source of variation	df	SS	SS from analysis	Row of denominator MS
1	R	Place	2	SS_P	b	4
2	F	Treatment	1	$SS_L - SS_C$	b&c	3
3		P x T	2	$SSP_L - SSP_C$	b&c	4
4	R	Control location (Place)	3	$SS_{C(P)}$	d	5
5		Residual	63	$SS_{(R)}$	b	
6		Total	71			

Table 3.9 Asymmetrical sampling design of Druce and Kingsford (1995): numerals indicate the number of replicates per treatment, the orthogonal section of the design is highlighted with a bold line.

Factor	Size		Asymmetrical cells
Colour	Large	Small	Control
White	3	3	3
Black	3	3	

pared with samples from open water (controls). The controls, therefore, were the asymmetrical part of the design.

Two analyses were used to calculate Sums of Squares in this design. Firstly, all of the levels were analysed using a one-way ANOVA. Secondly, the orthogonal section of the design was analysed as a two-way ANOVA. Appropriate sums of squares (SS) for the source of variation 'Control versus Rest' were derived from the total SS in the one-way ANOVA-total SS in the two-way ANOVA.

For competition experiments that used asymmetrical designs, see Underwood (1986), Fletcher and Creese (1985) and Quinn and Ryan (1989).

3.5.8 Variance components

Investigators often want to test hypotheses concerning the relative 'importance' of levels within a design by measuring the proportion of the variation explained by each factor. Variation can be measured with so-called 'variance components' (Winer 1971).

Principles and pitfalls
- Variance components can only be calculated legitimately for designs that have fixed or only random factors. It is most common for variance components to be calculated on designs with only random factors. In this case all levels of the analysis, including the residual, are commensurable (i.e. comparisons of variances, not means as for fixed factors; see 3.5.3).
- Although variance components have been presented in many publications for mixed-model designs (i.e. a combination of fixed and random factors), Underwood and Petraitis (1993) argue that they are not commensurable and, therefore, are illogical where interactions occur. This is still a matter of debate, given that Burdik and Graybill (1992) show how calculations of this type can be made.

- Variance components should only be calculated for raw data; if the data require transformation for calculations of F then the ANOVA table can be presented for the transformed data and the variance components for untransformed data.
- The reliability of the method has been questioned in cases where confidence limits for the estimates are not provided (McArdle & Pawley 1994); the bootstrapping method is best for confidence limits (see 3.5.9).
- Only talk about significant effects, because the confidence intervals for variance components from non-significant effects generally overlap (Keough pers. comm.).
- Variance components are useful for measures of importance among factors in experiments; for legitimate comparisons of the 'importance' of factors among experimental designs, the designs must be identical in terms of the number of factors, treatments of each factor and the residual variation (Underwood & Petraitis 1993). The assumption for residuals may be hard to meet for many organisms.
- The method of calculating variance components for a nested design is given in Table 3.10.

The proportion of variation explained by each factor has often been used to calculate the relative 'importance' of processes in manipulative experiments (models where all of the factors are fixed). This approach, however, has many pitfalls (Underwood & Petraitis 1993). In comparisons among similar experiments, percentage values for main factors can vary according to the number of replicates used in the design. These differences may be important, depending on the degree to which the amount of replication changes (Underwood & Petraitis 1993). Relative importance is often calculated using the percentage sums of squares (Weldon & Slauson 1986). This approach, however, has major disadvantages compared to the method of Winer shown above, in that it does not consider the relevant denominator mean squares of the ANOVA model.

It should be noted that comparisons of relative importance among similar experiments are impossible if there are major differences in the sampling designs. For example, Underwood and Petraitis (1993) give an experimental case where the factors of the design are predators (plus versus minus) and densities (1–x). If more density treatments are included (and the residual variance remains unchanged), then the percentage variation attributed to predation can change as a result of detecting increased variability for the factor density.

In experimental studies that investigate the importance of processes, Underwood and Petraitis (1993) recommended that comparisons among studies can

ANALYTICAL ASPECTS OF SAMPLING DESIGN

Table 3.10a Calculation of variance components. 'Calculation' column: The letters that are used as the denominator in these calculations are the alpha prefix of the MS estimates (derived using the methods shown in Appendix 3.1). This example is based on the design shown in Table 3.5; all of the factors in this design are 'random'. A = treatments of factor A = islands; B = treatments of factor B = locations; C = treatments of factor C = sites. Lower-case letters equal the number of treatments for each factor; n = number of replicates. σ^2 = estimate of variance and level of the design is indicated, based on the result from column 2. Negative results should be treated as zero. Variance components as a percentage of total variation are given in column 4 and are derived from the Σ given in column 1, row 5.

Variance and level	Calculation	Variance components	% of total variation	Row
$\sigma^2 A$	$\dfrac{MSA - MSB(A)}{bcn}$	i	i / Σ x 100	1
$\sigma^2 B$	$\dfrac{MSB(A) - MSC(B(A))}{cn}$	j	j / Σ x 100	2
$\sigma^2 C$	$\dfrac{MSC(B(A)) - MSRes}{n}$	k	k / Σ x 100	3
σ^2	MSRes	l	l / Σ x 100	4
		$\Sigma_i \ldots l$		5

Table 3.10b ANOVA results used for calculating variance components in Table 3.10c, using the design shown in Table 3.5, where number of: islands, a = 3; Locations, b = 3; Sites, c = 3; replicates n = 5. * = $P < 0.05$; ** = $P < 0.01$. The mean squares were determined from raw data on abundance of urchins among locations. Note that differences were found among islands, despite significant differences among locations within each island.

Row	F/R	Source of variation	df	MS	F	Row for denominator MS
1	R	Island	2	11727.27	18.83**	2
2	R	Location (Island)	6	622.76	4.66*	3
3	R	Site [L(I)]	18	133.73	1.67	4
4	R	Res	108	80.1		

Table 3.10c Calculation of variance components based on methods in Table 3.10a and the results of Table 3.10b. Note that data for the calculation column are from Table 3.10b.

Variance and level	Calculation	Variance components	% of total variation
$\sigma^2 A$	(11727.27 − 622.76) / 45	246.77	66.67
$\sigma^2 B$	(622.76 − 133.73) / 15	32.6	8.81
$\sigma^2 C$	(133.73 − 80.1) / 5	10.72	2.90
σ^2	80.1	80.1	21.64
Total		370.19	

only be made where the following are comparable: the organisms, habitats, processes (e.g. the types of predators involved) and size of experiments (with particular emphasis on spatial scale, e.g. metres versus kilometres is not comparable).

3.5.9 Cost-benefit analysis

Principles

Information from a pilot study may help in allocating effort as economically as possible. These data can be used in a cost-benefit analysis where the aim is to maximise precision and minimise the costs (Underwood 1981). This exercise may be particularly useful where the problem involves monitoring over a number of years. The cost may be expressed in monetary terms or processing time. In many cases there is a limit to the amount of time (e.g. boat time, time that assistants are available), or money, that is available. In such a programme, decisions have to be made on the best way to allocate these costs.

- Information from a pilot study is used to help in allocating effort as economically as possible, while maximising precision.
- Use untransformed data: transformed data will give incorrect results.
- Inadequate pilot studies can give erroneous estimates of variances that are likely to be encountered (McArdle & Pawley 1994). I would argue, however, that in many situations (e.g. 'applications' below and Chapter 2) pilot studies are essential and considerable time and money can be wasted without this step. It is judicious, however, to treat the outcome of results conservatively (see below).
- I conclude that cost-benefit is most useful for decisions on precision that are relevant to your hypothesis. Calculations of costs have a great influence on the outcome of results and may be difficult to calculate. Costs may, however, be easiest to calculate when there are major costs for analyses in laboratories.

Requirements are as follows (modified from Underwood 1981):

Estimated variances at different levels of the design (as variance components, see derivation in Table 3.10. The results of analysis in Table 3.10c are used to demonstrate the use of the formulae below)

Procedure to determine an allocation of samples that maximises precision and cost effectiveness (i.e. where V [= variance] and C [= cost] are minimised):

Determine the cost of each unit of sampling at each level of the design (this may be in dollars or minutes).

Formulae for the calculations of variances (e.g. for $\sigma^2 C$) and an example of calculations is given in Table 3.10a. Costs for the different levels were as follows: $12 per replicate = C_n; $80 per site = C_c; $690 per location = C_b; $2125 per island = C_a

recommended numbers of n
$$= \sqrt{[C_c \times \sigma^2 / C_n \times \sigma^2 C]} \quad \ldots\ldots \quad (3.4)$$

From: Table 3.10c
$$n = \sqrt{[(80)(80.1)/(12)(10.72)]} = 7.05 \approx 7$$

recommended numbers of c
$$= \sqrt{[C_b \times \sigma^2 C / C_c \times \sigma^2 B]} \quad \ldots\ldots \quad (3.5)$$

From: Table 3.10c
$$c = \sqrt{[(690)(10.72)/(80)(32.6)]} = 1.68 \approx 2$$

recommended numbers of b
$$= \sqrt{[C_a \times \sigma^2 B / C_b \times \sigma^2 A]} \quad \ldots\ldots \quad (3.6)$$

From: Table 3.10c
$$b = \sqrt{[(2125)(32.6)/(690)(246.77)]} = 0.68 \approx 1$$

A result of 0.69 was obtained when the cost of islands was increased to allow for the loss of a day through storms at each of the three islands. Note there are still costs for time and petrol and other resources when field activities are compromised. It would be advisable not to reduce the number of locations below two, given that significant differences were found among locations in the analysis.

Note that different optimal sample numbers were derived depending on whether time or money was used as the cost. It is up to the investigator to choose what the most important cost is. In some situations only money will be important, as time taken to do the study is no problem. Alternatively, money may be unlimited compared with the short amount of time a research vessel is available to do a study.

The calculations were repeated where cost was expressed as time (hours) for the different factors in Table 3.10 as follows: 0.69 hours per replicate = C_n; 3.27 hours per site = C_c; 54.81 hours per location = C_b; 176.4 hours per island = C_a. The results were similar to calculations for cost in dollars as follows: n = 5.9 ≈ 6; c = 2.3 ≈ 2; b = 0.65 ≈ 1.

Recommended levels of b remained unchanged when hours were increased to 200 to account for loss of time with bad weather.

Cost-benefit analysis may be used to justify requisitioning more boat time and money in order to reduce the chances of type II error.

Procedure to determine a desired level of accuracy:
For this procedure, you decide on a predetermined measure of error for comparisons among levels of A. Thus the variance of the estimated mean of treatments of factor A is calculated using the following formula (SE = standard error of the mean and MS within levels of A = mean square estimator calculated from procedure shown in Appendix 3.1):

$$SE^2 = V = \underline{\frac{MS \text{ within treatments of A}}{bcn}} \quad \ldots \ldots \quad (3.7)$$

therefore:

$$V = \sigma^2 e + n\sigma^2 C(B(A)) + cn\sigma^2 B(A)/ncb \quad \ldots \ldots \quad (3.8)$$

Using variance symbols from Table 10c:

$$V = \sigma^2 e + n\sigma^2 C + cn\sigma^2 B/ncb \quad \ldots \ldots \quad (3.9)$$

Once you have an estimate of the mean of factor A (μ) and the SE (square this to get required V) required for your *a priori* level of precision, the numbers of treatments of each can be altered to achieve this precision, because values of V will change with any changes to bc and/or n.

Therefore, by transposition of the values in formula 3.9, required levels of b

$$= (\sigma^2 e + n\sigma^2 C + cn\sigma^2 B)/ncV \quad \ldots \ldots \quad (3.10)$$

Maximum mean urchin abundance at an island = 42.

If it is decided *a priori* that standard error will be no greater than 10% then SE should be no larger than 4.2, and thus the variance = 17.64.

$$\therefore b = (80.1 + 5 \times 10.7 + 15 \times 32.6)/15 \times 17.64$$

$$\therefore b = 622.6/264 = 2.38 \approx 2\text{--}3$$

Note that the outcome of this calculation will be influenced by number of replicates (n) and sites (c). Levels of n and c can be altered in the numerator and denominator of formula 3.10.

McArdle and Pawley (1994) noted that estimates of error are particularly sensitive to the presence of aggregation in the population. Although log (x + 1) transformed data are slightly more reliable, they are more affected by mean density, pilot study size and aggregation. The authors concluded that cost-benefit analysis and related techniques (e.g. variance components, 3.5.8; and power analysis, 3.5.10), which are based on pilot studies, will seldom give reliable recommendations, but that the routine use of confidence intervals on variance ratios will protect workers from being misled. They concluded that bootstrap method (re-sampling at all levels of the design) gave the most reliable confidence limits, but even these were sensitive to suboptimal designs and aggregated organisms. Your confidence in the procedures described in this section should be weak if there is little replication for multiple factors in the initial sampling design.

Refer also to Green's (1979, p. 43) method for calculating the number of replicates required to detect a change in abundance of a predetermined percentage (e.g. 50%). This may be especially useful in an impact-control situation. Note, however, that this approach is based on a log scale.

Applications of cost-benefit analysis
Costs of histology for sampling replicate animals, replicate slides (animals) and sections on each slide (Underwood 1981, p.558). An example for a mensurative design is given by Kennelly and Underwood (1985) and a manipulative experiment in Underwood (1997, p. 283).

3.5.10 Power analysis

There are many useful applications of power analysis in ecology and particularly applied ecology for management (Green 1989, 1994; Peterman 1990; Fairweather 1991b; Keough & Mapstone 1995; Chapter 2).

Principles of analysis
Power of a test is the probability that it will yield a statistically significant result (i.e. rejection of the null hypothesis; Cohen 1988). There are two kinds of error in any statistical test: type I error (α), the probability of rejecting the true null hypothesis (H_o); type II error (β), the probability of failing to reject a false null hypothesis (see Table 2.1). If the statistical test estimates that type I error is small then reject H_o. If the significance criterion (e.g. $P = 0.05$, but see discussion of Mapstone 1995) is smaller than the statistical estimate of P in the analysis, then H_o is accepted. It is important to note, however, that investigators rarely use similar criterions for the risk of a type II error (Rotenberry & Wiens 1985).

Statistical power is the probability that a particular test will result in the rejection of the null hypothesis at a particular α level when the null hypothesis is false, therefore power = $1 - \beta$.

If the characteristics of a test are known, power can be calculated. Requirements and considerations for power analysis are as follows (see also Fig. 3.10, p. 74):

- Determination of a significance criterion a (e.g. $P = 0.05$), and whether the test is to be a one- or two-tailed test (two-tailed test = the direction of a difference between the means is not specified).

- Estimate or measure an effect size (ES).
- Sample size (n).
- Estimate of variation in the data set (s); this is used in most cases.
- The use of tables of non-central F-distribution for fixed factors: this procedure requires the calculation of a non-centrality parameter ϕ.

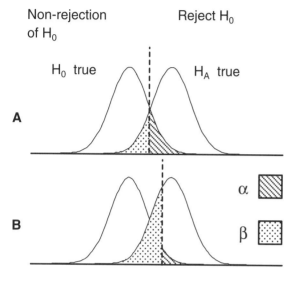

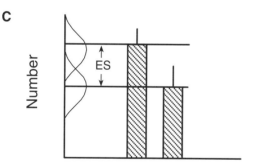

Figure 3.10 Considerations for power analysis. A and B show the relationship between type I and type II error (modified from Winer 1971). Curves represent the distribution of the relevant statistic (compare this normal distribution with that of an F-ratio statistics shown on p. 547 in Underwood 1981). The dashed vertical line represents the critical value at which the null hypothesis is rejected (determined by α; the shaded area for α is the critical region for $P < 0.05$) – note that this critical level has increased in β as the chosen α is made smaller. The value α is equal to the area under the part of the curve that falls in the area of non-rejection. In B the value of α is smaller than that of A, so B has a smaller chance of type I error. A decrease in type I error will increase the potential for type II error. As β increases, and higher critical levels of α are set, power $(1 - \beta)$ decreases. C shows the concept of effect sizes between two samples, normally expressed as the difference between the means (ES) as a proportion of one of the means.

What is effect size? The concept of ES is illustrated in Figure 3.10. Keough and Mapstone (1995) described it as the expected, or known, departure from a nominated null hypothesis. This can be more complicated when there are multiple means (e.g. in many ANOVAs) because the ES may be the product of differences among one or several of the means (Andrew & Mapstone 1987).

Applications of analysis
There are two general applications for power analysis, as follows.

Planning, or *a priori* calculations of power. This involves calculations of power for a proposed design which is expected to be able to detect a significant result for a given effect size (e.g. differences between the means). For example, a manager is concerned that a 25% change in numbers of fish should be detected, so that a suitable managerial response can be developed quickly. The effect size used can be based on what is considered to be acceptable (social or biological criteria), or based on previous studies where biologically important effects (cf. statistical significance, Chapter 2) have been demonstrated at a certain level of change (Underwood 1981; Toft & Shea 1987). As Rotenberry and Wiens (1985) state, 'There is no conventional methodology for estimating *a priori* the magnitude of an ES for ecological problems.' Cohen (1988) categorised effect sizes into operational units of small, medium, and large. Clearly the category chosen will depend on the details of your hypothesis or hypotheses, and the categories may be of limited use, even completely useless, for many biological problems. If using a pilot study for information on effect sizes that are used in the development of a larger sampling design, consider McArdle and Pawley's (1994) criticism regarding inadequate pilot studies. Since power varies with sample size (n), and the ability to detect *a priori* effect sizes varies with n, decisions on sample sizes should also be made at the planning stage (see section 3.3). Many investigators have opted for a power of 80%, i.e. $\beta = 0.2$, with the convention of $\alpha = 0.05$, giving a ratio of 4:1 for error rates used either to calculate n or an expected $(1 - \beta)$, or a minimum detectable effect size (MDES) as a guide for designing experiments when the effect size is uncertain (Keough & Mapstone 1995). The philosophy of a fixed ratio between type I and type II error rates (based on α and β) is discussed in Keough and Mapstone (1997), with emphasis on the problems of dealing with scaling a monitoring programme back for long-term monitoring in impact studies; they argue that a desirable ratio of $\alpha:\beta$ is 0.5. In this way both α and β are changed in a predetermined way before monitoring commences, which favours neither side in an environmental argument.

A posteriori calculation of power. This is particularly relevant for non-significant results. For example, an investigator may claim that no effect of grazing urchins was detected on the standing crop of algae on a temperate reef. A subsequent analysis of power in the investigator's analysis may reveal that the test for 'impact' was weak and a 98% change in algal abundance would have been required before a significant result was possible. If this application is used, it should be applied with caution. Clearly, the original hypothesis (and effect size) of the investigator will influence estimates of power. It should be noted that *a posteriori* calculations of power are a useful double-check of the *a priori* calculations.

Warning: see McArdle and Pawley's (1994) comment re confidence limits in 3.4.9.

See Green (1989) for multivariate considerations on calculation of power. The power of the test in this case greatly depends on the direction of response of each variable (e.g. if only one variable responds versus an equal response of all variables, or different vectors of response for each variable).

Examples

Many of the examples of power analysis found in the literature are of simple designs (e.g. one-way ANOVA = ANOVA with one factor) and there is still debate about the methodology and assumptions of calculating power for complex designs (e.g. mixed model and partially hierarchical designs, see 3.5.6). Cohen (1988) gives an example of two- to three-way orthogonal designs with fixed effects. Some of the problems of partially hierarchical designs may be alleviated if random and nested factors can be pooled (only where not significant at $\alpha = 0.25$). A common approach is to run a power analysis as a one-way ANOVA model on main effects from a more complex model, but be wary of interactions that may be of considerable biological interest (Andrew & Mapstone 1987). It is difficult to calculate power for interactions because the effect size is hard to specify. It is important to note that in complex ANOVA models the error term must be the mean square that is the denominator in the F-ratio of the term (not the residual) for which power is being calculated (Keough & Mapstone 1995, see also Appendix 3.1). Calculations of power for even moderately complex designs are difficult, so it is worth seeking advice.

Examples of power analysis for *a priori* cases: one-way ANOVA (Underwood 1981); t-tests, correlations, chi-square, ANOVA, ANCOVA, multivariate methods (Cohen 1988); multifactorial ANOVA (mixed models and partially hierarchical designs), (Winer 1971, Keough & Mapstone 1995; an effect size measured in a mensurative sampling design that was used to plan a

manipulative experiment (Connell 1996); impacts and detection of predetermined effect sizes (Green 1979).

Example of power analysis for *a posteriori* example: Sweatman (1985); multifactorial ANOVA (mixed models and partially hierarchical), Keough & Mapstone 1995.

A-posteriori calculation of power for Table 3.10, using the formulae provided in Underwood (1997).

Effect size for island:
$$MSA - MSB(A)/bcn \quad \ldots \ldots \quad (3.11)$$
$$= bcn\sigma^2 I/bcn$$
$$\sigma^2 I = (11727.27 - 622.76)/45 = 246.8$$

Note that the result is identical to that for the variance components (Table 3.10b).

Power $(1 - \beta)$ for island
$$bcn\theta = bcn\sigma^2 I/\sigma^2 L(I) \quad \ldots \ldots \quad (3.12)$$
$$bcn\sigma^2 I = bcn \times 246.8 \text{ from above}$$
$$45 \times 246.8 = 1106$$
$$bcn\theta = 11106/622.76 = 17.8$$

F-statistic df 2,6 at 0.05 = 5.14 (where df = that for island in Table 3.10b)
$$F' = 5.14/1 + bcn q = 5.14/1 + 17.8$$
$$F' = 0.27$$

Note: the α to which F' corresponds with 2,6 df is the power. You will need a good set of tables with values of a above 0.5, or you will need to interpolate.

Power $= 0.77 \approx 0.8$

Effect size for location
$$cn\sigma^2 L(I) = MSB(A) - MSC(B(A)) \quad \ldots \ldots \quad (3.13)$$
$$\sigma^2 L(I) = (622.76 - 133.73)/15 = 32.6$$

Power $(1 - \beta)$ for location with df 6,18
$$cn\theta = cn\sigma^2 L(I) /\sigma^2 S(L(I)) \quad \ldots \ldots \quad (3.14)$$
$$cn\sigma^2 L(I) = 15 \times 32.6 = 489$$
$$cn\theta = 489/133.73 = 3.7$$
$$\text{F-statistic df 6,18 @0.05} = 2.66$$
$$F' = 2.66/1 + 3.7 = 0.57$$
$$\text{Power} = 0.75 \approx 0.8$$

Effect size for site
$$n\sigma^2 = MSC(B(A)) - MSRes \quad \ldots \ldots \quad (3.15)$$
$$S(L(I)) = n\sigma^2/5$$

$$= (133.73 - 80.1)/5 = 10.7$$

Power $(1 - \beta)$ for site

$$n\theta = n\sigma^2/\sigma^2 \quad \ldots \ldots (3.16)$$

$$n\sigma^2 = 5 \times 10.7 = 53.5$$

$$n\theta = 53.5/80.1 = 0.7$$

F-statistic df 18, 108 @ 0.05 ≈ 1.7

$$F' = 1.7/1 + 0.7 = 1$$

Power $= 0.47 \approx 0.5$

Note that calculations vary from the above for fixed factors (examples: Underwood 1997). Power, for fixed factors, is determined by the non-central F-distribution and requires the calculation of a non-centrality parameter ϕ.

Power required to detect rare species (see Green 1979, figure on p.134).

Power tables, curves and non-central F-distributions Winer 1971 (ANOVA fixed factors; Table C11; non-central F-distributions Table C14); Cohen 1988 (t-tests, correlations, chi-square, ANOVA, ANCOVA, multivariate methods) and are calculated in some statistical computer packages (review Thomas & Krebs 1997).

3.6 Treatment of the data

3.6.1 General considerations

- Check the data for aberrant data points that were the result of errors in transcription from the original data sheets. It is generally best to enter and check data using two persons. Box plots help as a final check for erroneous data points and give you a preliminary view of variation within and among treatments (though not presented as a normal distribution). Box plots show the median, the ranges in which 50% (box) and 90% (Whiskers) of the data are found and outlying data are plotted as separate points (e.g. Wilkinson 1996).
- Once the data are checked, explore it graphically to get a good feel for it. Study papers (for good and bad examples) and software manuals for different ways of presenting graphics, but always consider accurate presentation of variation in the data set (e.g. SE or 95% CL). Pie diagrams, for example, rarely give measures of variation.
- If the design stage was carried out correctly, it should enable your original question to be tested with formal statistical procedures that were chosen *a priori*.

- Critical levels should have been chosen *a priori*. Although default levels of α are generally used (i.e. $P = 0.05$) it is often appropriate to vary the critical to keep the risk of type II error low (Mapstone 1995). Consider relationships between α and β for impact studies (Keough & Mapstone 1997).
- Check that your data meet the assumptions of your analyses. Many analyses have, for example, assumptions regarding the homogeneity of the data. Where the normal data violate these assumptions a transformation may be necessary. Failure to transform the data may result in an incorrect test of a hypothesis (Hurlbert & White 1993; Underwood 1997). In cases where data are intractably heterogeneous, you may check your result with a randomisation procedure (Manly 1991), adopt a more conservative critical value of a (e.g. 0.01 for ANOVA) or go ahead and do the analysis anyway because the analysis is robust to heterogeneity (e.g. ANOVA, when the design is balanced: Underwood 1981, p. 534).
- Present data in clear graphs, diagrams, and tables. For example, graphs may display abundances and size-frequency distributions of organisms. With modern graphic packages you may be able to superimpose graphics on another image such as a diagram of your reef profile or target organism. The final decisions on the presentation of data and measures of variation for publication will depend on your original hypotheses and the results of analyses.
- If multivariate plots are used where values of the original variables cannot be identified (e.g. PC, CDA, MDS ordinations), I suggest that you augment them with univariate plots. If this is not done, readers cannot extract basic information about relative abundance and variances at different levels of the sampling design. It may not be possible to present data for all variables. In this case, the multivariate analyses should be used to determine which variables are most responsible for variation among factors that are relevant to your hypothesis, and the raw data plotted.
- There are many types of analyses that may be appropriate for your biological problems and data. Table 3.11 provides a guide to a broader literature.
- Always keep the original data sheets; they may contain other information that is needed at a later date.

3.6.2 Multivariate considerations

Many hypotheses require the analysis of multiple variables in the one analysis. Multivariate analyses provide systems for analysis under conditions where there may be several independent variables and dependent variables (Table 3.2) that are correlated with one another to varying degrees (Tabachnick & Fidell 1983). Most multivariate problems can be viewed in terms of 'distances' between samples, where a distance matrix is calculated and the values are plotted in two- or three-dimensional graphs. Many tomes have been written on multivariate statistics and detailed treatment is beyond the scope of this book. I recommend Manly (1994b) as a good primer on multivariate methods. (See Selected Reading for more on these methods.) A problem-orientated guide to multivariate statistics and other approaches is given in Table 3.11 (page 79). For those who are unfamiliar with multivariate methods, it is worth making some generic statements.

Applications include the following:

- Analyses of the composition of assemblages; e.g. variation in the representation of 20 or more species among factors of a sampling design may be adequately represented by two or three 'super-variables' such as MDS axes (Tabachnick & Fidell 1983).
- Analyses of relationships between sets of variables, where data are in the form of two or more distance matrices; e.g. comparisons of biological and physical variables such as the composition of assemblages versus sediment characteristics (two-distance matrices, e.g. Canonical Correlation analysis, Mantel's test), or comparisons of species assemblages, with measurements of pollutant residues and the results of toxicity tests (three-distance matrices, e.g. sediment-quality triad: Green et al.1993).
- Analyses of the morphometrics of organisms from different sites and times, e.g. to contribute to the identification of fish stocks (Haddon & Willis 1995).
- The identification of variables that explain a high proportion of the total variation in a data set. This may be an end point of the analyses or an initial phase of variable choice for other analyses.
- Multivariate scores may be used in univariate analyses, e.g. the body shape or size of an organism may be best described by a combination of body measurements that are well described by multivariate scores (e.g, PC1 of principal components analysis may describe 85% of the variation in size of individuals based on multiple different measurements; see also 'Size', Table 3.2). These PC measurements could then be compared at appropriate spatial and temporal scales.
- Analyses of impact versus control areas at multiple spatial scales. Procedures such as analysis of similarities between groups of samples (ANOSIM, Clarke 1993) can enable difference in the structure of assemblages to be analysed at sites separated by different spatial scales.

Common considerations for multivariate analyses

- Consider degrees of freedom and levels of the design at the planning stage. A number of procedures have been mentioned in this chapter that will assist in the allocation of sample unit with a view to statistical tests that have an appropriate level of power to test hypotheses. In many cases this advice will hold for multivariate approaches, but the use of multiple variables also has an influence on the degrees of freedom for tests, and therefore, power (Green 1989).
- If multiple variables are highly correlated, it is not necessary to use all of them in the analysis. In some cases this can compromise the power of the analysis to distinguish groups.
- It is generally critical to consider the number of variables in relation to the number of rows (= replicate samples), particularly where the number of samples (or cases/rows in computer jargon) is small. For example, Grossman et al. (1991) demonstrated that PCA and associated eigenvalue tests were unreliable when levels of replication were less than twice the number of species. B. McArdle (pers. comm.) argues that investigators should aim at having fewer than 10 variables for canonical discriminant analyses (CDA), even when there are a large number of cases in the data set. The number of variables used for multivariate analyses may be reduced by deleting one or more highly correlated variables; deleting variables based on critical levels of abundance or presence/absence; and finding variation in the data set that describes greatest variation in a first-step analysis before the final multivariate analysis.
- Transformation is often necessary because many multivariate methods are sensitive to non-normal of the data (e.g. MANOVA).
- Large numbers of zeros and missing values can compromise analyses (i.e. invalidate assumption of multivariate normality) and in some cases make calculations impossible. Where there are a small numbers of missing values it may be legitimate to include estimates of these values in analyses (Manly 1994b, p. 144).
- Variables of different units (e.g. abundance of plankton, salinity and temperature) generally have to be standardised in some way (e.g. proportions, residuals) before they are commensurable in analyses.
- Graph ordinations to supplement hypothesis tests.
- In most cases a combination of the univariate and multivariate grouping procedures will extract the most information from the available data (e.g. Anderson & Underwood 1994).
- Present data somewhere in a paper as original variables, rather than just multivariate measures of distance between groups of replicates.

3.6.3 Cusums: a special case for monitoring

In some situations where the abundance of an organism is monitored through time (or other variable of interest) it may be useful to have a warning system which will give a threshold of an impending deleterious change in density. For example, if a particular species of fish reaches a predetermined threshold density, it may be decided that recreational fishing will be banned from the area. A method for presenting this graphically is a cusum chart (Manly 1994a).

Cusums are a way of detecting small deviations away from normality. The method plots cumulative deviations from a preset 'standard' or arbitrarily desired level of μ. Your interest may be a systematic change in the distribution of a variable, expressed as changes in the mean or variance. As samples are taken and sample means ($\bar{x}$) are obtained, cumulative deviations from the standard are constructed; these are called cusums (cumulative sums). The chart consists of plotting these sums against the observation number (Fig. 3.11).

Manly (1994a) stated that the practical advantages of the method are as follows:

- The principles are relatively simple to understand.
- The most important results are displayed graphically.
- The method is able to detect unusual patterns of change that are not necessarily reflected in mean and standard deviation.
- It is easy to accommodate missing values.

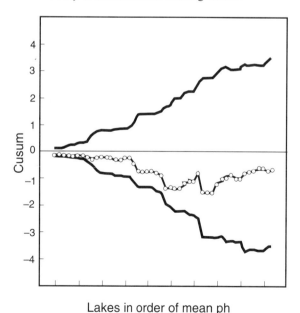

Lakes in order of mean ph

Figure 3.11 Cusum plot. The upper and lower continuous lines are the limits obtained from randomisations of data, and the cusum is the line with open circles.

ADAPTED FROM MANLY 1994A WITH PERMISSION

- Extension of the method to multivariate data is straightforward.
- No assumptions are made about the distribution of the variables being studied.

When variations in density (or some other measure, e.g. heavy metal concentration in the flesh of a mollusc) are 'normal', the plot oscillates about a horizontal line. Any change of inclination is subjective unless critical levels of change are determined. A CUSUM envelope can be created using a randomisation method (Manly 1994a). Maximum and minimum values from the randomisation procedure are plotted on the CUSUM plot for an envelop in which the data should lie. If the real data go outside of the envelope then there is evidence that the null hypothesis is not true.

Example: Acid rain study in Norway (Manly 1994a).

3.7 Statistical software

There are many excellent packages that will enable you to analyse data using parametric (univariate and multivariate statistics) and non-parametric statistics. Many of these packages also give some explanation of the analytical approach and theoretical justification for doing so. Packages include the following: SAS, SPSS-X, BMDP, Minitab, SYSTAT, GLM. Many spreadsheet programs (e.g. Excel, Lotus) provide modules for basic statistics (e.g. ANOVA, regression) and great flexibility of analysing data using a broad range of 'formulae'. Macro-notepads can even allow you to run complicated routines that may involve bootstrapping with multiple iterations of your original data set.

Regardless of whether you are using a package produced by a well-respected company, some public domain software with few details, or some 'in-house' software, *always test the program with a data set with known results of your required statistical method.* Data sets and the results of analyses are often provided in books on statistical methods (see Selected Reading). This is not only useful for checking the accuracy of the program, but it also enables you to interpret the printouts, which are often long and confusing. Never assume anything about a package. Windows-type programs and fast computers may give you results, but they will equally readily give you the wrong result if you do not know what the program is doing. For example, many statistical programs will treat all factors in an ANOVA as fixed factors (see 3.5.3) unless you instruct the computer otherwise. It is clear from the models given in section 3.5, that using the residual as the denominator MS for all factors is incorrect where not all factors are fixed. Advice on the output of analyses in some software, and related theory, is provided in some statistical texts (e.g. Tabachnick & Fidell 1996).

Many of these statistics packages, spreadsheet programs and draw programs (e.g. Freelance, Powerpoint) have graphics modules that provide broad options for data exploration and the plotting of results from tests of specific hypotheses. Modern software also enables the easy transfer of graphics from one package to another, using clipboards. This is often useful if you have done your graphics in a statistical package, but want to add a scanned image to the graph. In this case the graphic can be pasted (or imported using a metafile or similar format) to a draw program (see above). Ask other people in the field if your requirements are particularly specialised. For example, most standard packages do not give you great flexibility with the control of two- and three-dimensional contouring plots. You may need specialist software (e.g. Surfer) for this type of work.

General considerations
- Carefully consider how the data should be entered for the software of your choice, and note that methods for the coding of data (e.g. by time, location, site; Table 3.2) vary between packages.
- Always check that data have been entered correctly and carry out checks for errors with graphical methods such as box-plots.
- Always backup important files.
- Keep a careful record of disks/computers on which you have stored data, file names and the logic of your codes.
- Write relevant details of computer, disk, file name on hard copies of printouts.
- Write the date of the last time of editing on the hard copies (this is provided automatically in some software).
- If you are confused what a program does, analyse a set of data from a text book or some other source where you know the correct result. This can also be useful if you are not quite sure how to interpret the print out from an unfamiliar program.
- Code temporary files consistently (e.g. tmp.sys, tmp1.sys . . . tmpx.sys).
- Code transformations consistently, for example add L to new variables that are the log $(x + 1)$ of the original variable; S to new variables that are the square-root of original variables, etc.

Table 3.11 A guide to a broader literature on statistical methods that are relevant to a range of biological problems. If a source is not given in the 'Theory' column, then it is covered in the references listed for 'Practical Application'. Also see Selected Reading for texts that deal with univariate and multivariate statistical methods of parametric and non-parametric types. Some good advice on statistical methods is provided in statistical packages such as SAS, SYSTAT, and SPSS-X. 'Problem' = biological or statistical problem.

Problem	Method	Practical application	Theory
A *posteriori* comparisons among means	Multiple comparisons SNK, Ryans	Day & Quinn 1989 Underwood 1997	Day & Quinn 1989 Underwood 1997
Assemblages/communities	Multidimensional scaling (MDS)	Clarke 1993 Anderson & Underwood 1994	Clarke 1993
Assemblages and environmental variables	Canonical analyses, Procrustes analysis; Mantel's test	Green 1993b	Gittins 1985 te Braak 1987
Assemblages and environmental variables (MDS)	Multidimensional scaling	Somerfield et al. 1994	Clarke & Ainsworth 1993
Assemblages and environmental variables- quality triad	Multiple multivariate methods	Green et al. 1993	
Bayesian approach	Bayes' theorem	McAllistair et al. 1994 Dennis 1996	
Classification and ordination-assemblage/ communities	Multiple		Rohlf 1974 Pielou 1984 Minchin 1987

Problem	Method	Practical application	Theory
Classification and ordination-assemblage/communities	Bray-Curtis	Abel et al. 1985	
Classification and ordination-assemblage/communities	Correspondence analysis	Syms 1995	Greenacre 1984
Cohorts of organisms	Cohort analysis	Cassie 1954 Worthington et al. 1992b	Grant et al. 1987 Hilborn & Walters 1992
Community (see Assemblages)			
Competition experiments	Asymmetrical ANOVA	Underwood 1992	Underwood 1992
Effect sizes	Power analysis	Cohen 1988, Peterman 1990 Fairweather 1991b Mapstone 1995	Cohen 1988 Fairweather 1991b
Electivity (see Selectivity)	Electivity indices	Hyslop 1980, Lechowicz 1982	Hyslop 1980 Lechowicz 1982
Directionality	Circular statistics	Batschelet 1981	Batschelet 1981
Dispersion		Krebs 1989	Green 1979, Krebs 1989
Distance	Measurements of distance between observations: nearest, furthest, average, centroid distances, etc	Faith et al. 1987	Pielou 1984
Diversity	Indices		Lande 1996 Smith & Wilson 1996
Growth	Multiple	Kaufman 1981, Schnute 1981 Schnute & Richards 1990 Chen et al. 1992	
Growth	Von Bertalanffy model	Maller & De Boer 1988 Ricker 1987	Fabens 1965
Impacts	BACI, BACIP, Beyond BACI, MBACI – usually analysis of variance	Underwood 1991 Keough & Mapstone 1995 Chapters 2 & 9	Green 1979 Keough & Mapstone 1995 Underwood 1994a
Impacts	Principal component biplots	Williams 1994	Williams 1994
Impacts	Multidimensional scaling	Clarke 1993	Clarke 1993
Impacts–compliance monitoring	Response surface	Keough & Mapstone 1995	Montgomery 1991
Mapping		Jones 1983 (see also Fig. 6.10), MacDiarmid 1991 (see also Fig. 5.6)	
Mark-recapture	Multiple methods	Manly 1985a, Burnham et al. 1987, Tsukamoto et al. 1989	Krebs 1989
Mark-recapture	Multiple sightings	Fletcher 1994	Fletcher 1994 Cormack 1994
Mark-recapture	Imperfect observations	Gardner & Mangel 1996	
Mitigation	Bioequivalence		McDonald & Erickson 1994

Problem	Method	Practical application	Theory
Modelling	Multiple	Nisbet & Gurney 1982 Murdoch 1994	
Morphometrics	Multiple		Pimental (1979)
Morphometrics	Principal components analysis (PCA)	Koslow et al. 1985	
Morphometrics and measurement error	Discriminant analysis	Francis & Mattlin 1986	
Mortality		Sale & Ferrell 1988 Connell 1996; Hixon & Carr 1997	
Mortality (Fishing and natural)		Hilborn & Walters 1992	Hilborn & Walters 1992
Mortality (Fishing and natural)	Tagging	Burnham et al. 1987	
Mortality experiments		Doherty & Sale 1985; Hixon & Beets 1993; Steele 1997	
Movement		Manly 1985b	Manly 1985b
Movement–data independence	Power analysis	Swihart & Slade 1986	Cohen 1988
Movement paths	Fractal analyses		Turchin 1996
Multiple studies–generality	Meta-analysis	Adams et al. 1997	Adams et al. 1997
Patchiness		Lloyd 1967, Underwood & Chapman 1996	
Population structure	Life tables	Caughley & Gunn 1996	Caughley & Gunn 1996
Power–statistical		See section 3.5.10	
Randomisation	Randomisation tests	Manly 1991, 1997	Manly 1991, 1997
Rare species	Multiple		Green 1979
Selectivity indices (see also electivity)			Vanderploeg 1978
Simulations	Randomisation, Monte Carlo		Manly 1991
Size	Complex shape described by multivariate scores	Manly 1986	McCormick 1990
Spatial variation	Grid sampling–spatial Autocorrelation and contouring	McArdle & Blackwell 1989	Ripley 1981
Spatial variation	Multiple comparisons	Underwood & Chapman 1996	Legendre & Fortin 1989
Spatial variation	Hierarchical ANOVA	See 3.5.5	Underwood 1997
Stock identification	Canonical discriminant analysis and analysis of covariance	Haddon & Willis 1995	
Stock size	Egg-production method	Zeldis 1993	
Stock size	Stratified sampling–optimal and proportional allocation	McCormick & Choat 1987 McCormick 1990	See 3.5.1
Stock size	Multiple methods		Hilborn & Walters 1992

Problem	Method	Practical application	Theory
Temporal variation	Grid sampling–temporal autocorrelation	McArdle & Blackwell 1989	Ripley 1981
Temporal variation	Hierarchical ANOVA	See section 3.5.5	Underwood 1997
Threshold–warning of change	CUSUMS	Manly 1994	Manly 1994
Time series	Multiple	Milicich 1994	Chatfield 1980 Powell & Steel 1995
Tracking data	Multiple	White & Garrott 1990	White & Garrott 1990
Uncertainty	Bootstrap method		Lepage & Billard 1992 Efron & Tibshirani 1993

SELECTED READING

This list represents selected reading, not a review of each subject area. See also Table 3.11; * indicates a multi-authored book or collection of proceedings.

Good tabular guides to applicable statistics are proved on the inside covers of Sokal and Rohlf (1995, parametric statistics) and Siegel and Castellan (1988, non-parametric statistics). The texts I have listed deal with most standard statistical methods including analyses of the following:

One-sample cases (or one group of replicate samples): frequency of categorical data (e.g. chi-square test, binomial test, cohort analysis); functional relationship between two variables (e.g. regression analysis); statistics of 'association' between two variables (e.g. correlation analysis, Spearman's rank test); statistics of 'association' between three or more variables (e.g. partial correlation), multiple variables measured for the group (multivariate methods: e.g. principal components analysis, canonical correlation analysis, multidimensional scaling).

Two-sample cases (or two groups of replicates): related matched pairs (e.g. paired T-test, Wilcoxon rank test); independent samples (e.g. T-test, Mann-Whitney U-test); comparisons of frequencies (chi-square test for 'class' x 2, Komogorov-Smirnov test); functional relationships between 2 or variables (e.g. analysis of covariance, stepwise regression, multiple regression); multiple variables measured for the group (multivariate methods, e.g. principal components analysis, canonical correlation analysis, multidimensional scaling).

Multiple sample cases (or multiple groups of replicate samples): comparisons of frequencies (chi-square test for 'class' x 2); analyses of multiple groups of replicates that are classified according to one or more factors (e.g. analysis of variance, Kruskal-Wallis one-way ANOVA, Friedman's two-way ANOVA); functional relationships between two or variables (e.g. analysis of covariance, stepwise regression, multiple regression); statistics of 'association' between two or more variables (e.g. correlation matrices); multiple variables measured for the group (multivariate methods, e.g. principal components analysis, canonical correlation analysis, multidimensional scaling).

General parametric statistics: Winer 1971; Huitema 1980; Snedecor & Cochran 1980; Underwood 1981, 1997; Cohen 1988; Winer et al. 1991; Trexler & Travis 1993; Sokal & Rohlf 1995.

Multivariate statistics: Pimental 1979; Dillon & Goldstein 1984; Pielou 1984; Digby & Kempton 1987; James & McCulloch 1990, Rohlf 1990; Grossman et al. 1991; Kent & Coker 1992; Clarke 1993; Palmer 1993; Manly 1994b; Tabachnick & Fidell 1996.

Non-parametric statistics: Conover 1980; Siegel & Castellan 1988; Johnson 1995.

Mathematics for biological sciences: Arya & Lardner 1979.

Sampling design and statistical methods: Green 1979; Underwood 1986, 1997; Andrew & Mapstone 1987; Krebs 1989; Hurlbert & White 1993; Fletcher & Manly 1994*; Keough & Mapstone 1995; Kirk 1995.

Statistical tables: Fisher & Yates 1963; Rohlf & Sokal 1981 (Most statistics books include standard tables).

Appendix 3.1

Methods for calculating the expected values of mean squares for straightforward and specialised designs. Based on rules from Winer (1971) and Underwood (1981, 1997). F = fixed factor; R = random factor.

Mean square estimators for factors in a partially hierarchical design (see Table 3.6). Lowercase letters indicate subscripts for each factor, e = residual error; () = nested term.

Linear model statement:

$$X_{abcn} = \mu + Aa + Bb + ABab + C(B)c(b) + AC(B)ac(b) + en(abc)$$

where X_{abcn} = any replicate (n) from any treatment in the design.

Rules for individual factors when calculating values in table below:

If the column subscript is in the row and if:

x is fixed and not nested, then = 0
x is fixed and nested, then = 1
x is random and not nested, then = 1
x is random and nested, then = 1
If the subscript of a factor is not present, then = subscript in columns.

When determining the MS estimators for each factor below, cover the column for that factor. For example, for **A**a (row 1 below), which is a fixed factor (F), cover column a and only consider factors in rows 1–6 that have subscript a in them. For interaction **AB**ab, cover columns a and b and only consider factors that have a and b in them in rows 1–6.

Note that Underwood (1981, 1997) presents mean square estimates for different combinations of fixed and random factors for some orthogonal models. It is unlikely, however, that MS estimators for factors of all of the hypotheses you test can be taken directly from these examples.

Row	Factor	F	R	R	R	Mean square estimators	Row of denominator
	Column Subscript	a	b	c	n		
1	Aa	0	b	c	n	$\sigma^2e + n\sigma^2AC(B) + cn\sigma^2AB + bcn\sigma^2A$	3
2	Bb	a	1	c	n	$\sigma^2e + n\sigma^2AC(B) + n\sigma^2C(B) + acn\sigma^2B$	4
3	ABab	0	1	c	n	$\sigma^2e + n\sigma^2AC(B) + cn\sigma^2AB$	5
4	C(B)c(b)	1	1	1	n	$\sigma^2e + n\sigma^2AC(B) + n\sigma^2C(B)$	5
5	AC(B)ac(b)	1	1	1	n	$\sigma^2e + n\sigma^2AC(B)$	6
6	en(abc)	1	1	1	1	σ^2e	

σ^2 = variance for each level of the design

FOUR

SUBTIDAL HABITATS AND BENTHIC ORGANISMS OF ROCKY REEFS

M. J. Kingsford and C. N. Battershill

4.1 Introduction

The rocky subtidal is highly heterogeneous in terms of both topography and the organisms in it (Plates 2–19). Different habitats are usually identifiable and can be characterised according to a combination of physical and biological attributes (e.g. shallow broken rock, urchin-grazed rock flats, *Ecklonia* (kelp) forest: Ayling 1978; Cole et al. 1990). Habitats such as these are found at temperate latitudes all over the world (e.g. Australia: Underwood et al. 1991; Maine: Elner & Vadas 1990; Chile: Dayton 1985; South Africa: Buxton & Smale 1989; California: Ebeling et al. 1980; Schiel & Foster 1986). We have adopted the classification of types of organisms described by Jones and Andrew (1993) to discuss the members of subtidal assemblages. 'Habitat formers' are large and/or sedentary organisms that characterise a habitat (e.g. laminarian algae), and 'habitat determiners' are mobile organisms that are capable of influencing the distribution of habitat formers, usually through grazing or predation. 'Habitat responders' are those organisms whose distribution and abundance is strongly influenced by habitats. Some organisms may change status as they increase in size; for example, small echinoids, *Evechinus chloroticus*, may be responders when small and determiners when large.

A description of the distribution patterns of subtidal habitats, and/or proportional cover of the reef, is often an important prerequisite for sampling designs that address questions concerning the abundance of other organisms (e.g. fish, Chapter 6). Sampling, therefore, may be structured to accommodate this heterogeneity in a way that enables more precise estimates of abundance of target species. Furthermore, temporal changes in the abundance of a habitat responder may be linked with fundamental changes in habitat type (Choat & Ayling 1987). For example, if *Ecklonia* forests were removed by the grazing activity of urchins, there would be a large decline in the number of algae-associated fish (e.g. spotties: Jones 1984a) and subcanopy assemblages.

The nature of many habitats is determined by interactions among members of the assemblage, which in turn may be influenced by exogenous factors. A knowledge of habitats (and associated species) and an understanding of processes influencing them is sought by many investigators, and progress requires both descriptive and experimental approaches (see Chapter 2). Processes that influence the distribution and persistence of habitats include:

- Grazing by urchins and gastropods, which can greatly alter algal cover and diversity (see Chapter 5).
- Physical disturbance from storms on short and long time scales, which can alter sedimentation rates either directly (e.g. Cortis & Risk 1985; Battershill & Bergquist 1990) or through run-off (Griffiths & Glasby 1985; G. P. Glasby 1991).
- Mortality of important organisms through freshwater input (Andrew 1991), disease, predation (e.g. urchins: Elner & Vadas 1990; Hughes 1994; sponges: Ayling 1981).
- Direct and indirect human impact on habitat formers and determiners (e.g. impacts on canopy algae and grazers: Dayton et al. 1984; Connell & Keough 1985; Kennelly 1987a, b; Castilla & Bustamente 1989; Fanelli et al. 1994; Culotta 1994; Siegfried 1994).
- Recruitment, which may vary in space and time (e.g. Keough 1983).
- Abundance of other organisms (e.g. Breitberg 1984; Stocker & Bergquist 1986; Kennelly 1987c) and the presence of conspecifics (Tegner & Dayton 1977; Schmidt 1982; Stocker & Bergquist 1987).
- Competition, which can alter the nature of assemblages that characterise habitats (Butler 1995).

- The direction of competitive interactions, which may change with physical disturbances.
- The life history of the organisms themselves, which may influence habitat structure, as great variation in abundance can result from changes due to asexual reproduction, vegetative expansion, rapid growth rates, and longevity of individuals (see 4.2.1).

Processes influencing habitats are often complex and a cascade of effects can result from a particular perturbation. For example, a fishery for sea urchins may have a direct effect on numbers of urchins, and indirect effects may include a rapid increase in algal cover, a loss of small grazers, and changes in patterns of recruitment for invertebrates and vertebrates (Chapter 5). In some cases altered habitats may not revert to the original state for a considerable period of time, or not at all. Fanelli et al. (1994) found that the removal of mussel beds facilitated the development of urchin-grazed barrens. Increased urchin grazing was thought to prevent recolonisation by mussels.

In this chapter, great emphasis is given to considerations for sampling designs and methodology that can be used to describe habitats, the organisms that characterise them, and the organisms that are associated with them. The characterisation of assemblages in habitats that are unique by virtue of their geomorphology (e.g. large caves and fiords) is also discussed. We emphasise the importance of considering the life history of organisms when biological questions concerning the dynamics of populations and assemblages are addressed. Special emphasis is given to the life history of important organisms that are often habitat formers (e.g. macroalgae and modular organisms such as ascidians and sponges). The utility of experimental approaches is also reviewed as the only way to obtain unconfounded information on the processes that influence the dynamics of habitats. Readers are urged to identify their biological questions before proceeding (see Chapter 2).

Greater attention is given to mobile and conspicuous invertebrates such as urchins, lobsters, abalone and other large gastropods (e.g. *Turbo* spp.) in Chapter 5. Some hard substrata in shallow waters at temperate latitudes are not rocky reefs; for example, the pilings of substantial piers. We draw attention to the review by Butler (1995) on organisms associated with piers and the processes influencing the dynamics of populations and assemblages in this environment.

4.2 Considerations for sampling designs

When quantifying the distribution of habitats and considering experimental designs for elucidating ecological processes, it is important to consider aspects of the life history of organisms, because their attributes can have a great influence on the dynamics of populations and assemblages. Sources of variation in space and time should also be considered in sampling designs. In most cases, only a subset of these will be relevant to your study.

4.2.1 Life history

The dynamics of many habitats are strongly influenced by the life history of habitat-forming organisms. Thus, knowledge of the life history of target organisms, or those that make up an assemblage that is representative of a habitat type, should be considered in mensurative and manipulative sampling designs. Most reef-associated organisms (plants and animals) spend part of their lives in the plankton as spores or larvae before they settle back to reefs (not necessarily their natal reef) as juveniles, adults or to form gametophytes (algae: Foster & Schiel 1985; invertebrates: Chapter 5; reef fishes: Chapter 6).

The life history of algae is complex and the sexual phases of plants (gametophytes) are often microscopic (Figs 4.1, 4.2, pp 85–86). Even minute gametophytes are an important source of food to grazers (e.g. urchins, limpets and turbinid gastropods). Sporophytes are the main macroscopic habitat-forming stages of brown macro-algae. They release spores that give rise to gametophytes, and this is usually seasonal (e.g. *Ecklonia*, Fig. 4.1). The distance that spores disperse was once thought to be on a scale of metres, but it has been demonstrated for *Macrocystis* that the spores have energy stores (Reed et al. 1992), are motile, and can travel hundreds of metres, particularly during storms (Reed et al. 1988). In some cases sporophytes are annuals and are shed from the substratum (e.g. *Sargassum*: Schiel 1985a, b) into coastal waters as drift algae (Kingsford 1993). An understanding of the life history of algae and timing of the release of spores is crucial for the planning of mensurative and manipulative experiments as well as the interpretation of the data. This is also relevant to impact studies; for example, the introduction of *Undaria* to reefs in the southern hemisphere may be underestimated if studies are done when sporophytes are rare or absent (Battershill pers. obs.). For the life history of other types of algae, see references on taxonomy in Chapter 10.

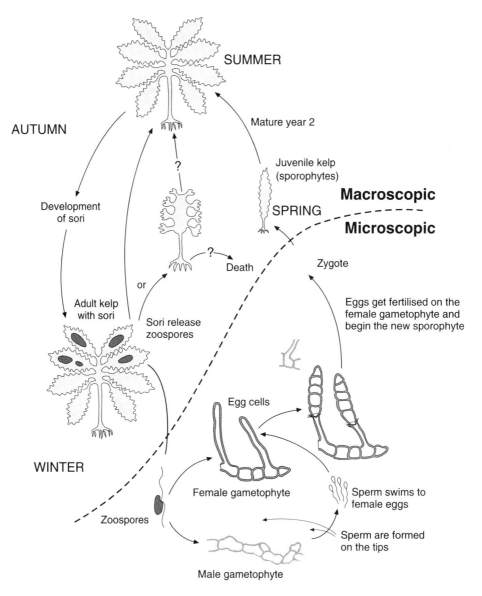

SUMMER

AUTUMN

Mature year 2

Juvenile kelp
(sporophytes)

Macroscopic

Development
of sori

SPRING

- - - - - - -

Microscopic

?

Death

Zygote

Adult kelp
with sori

or

Sori release
zoospores

Eggs get fertilised on the
female gametophyte and
begin the new sporophyte

Egg cells

WINTER

Female gametophyte

Sperm swims to
female eggs

Sperm are formed
on the tips

Zoospores

Male gametophyte

ABOVE: **Figure 4.1** The life history of *Ecklonia radiata*, a laminarian brown algae that is categorised as a 'habitat former' on temperate reefs in Australia and New Zealand. Sporophytes (macroscopic phase) are rarely longer than 1 m. The morphology of these plants varies greatly with exposure and water depth: plants are generally longest in deep water (Plates 4, 5). At some times sporophytes lose many laminae, but it is not known what proportion of these die.

ADAPTED FROM KENNELLY 1995 WITH PERMISSION

OPPOSITE TOP: **Figure 4.2** The life history of *Macrocystis pyrifera*, a laminarian brown algae that is categorised as a 'habitat former' on cold temperate reefs in both northern and southern hemispheres (including Australia and New Zealand: Dayton 1985). Sporophytes (macroscopic phase) are often longer than 30 m. The morphology of these plants varies greatly with exposure and water depth; plants are generally longest in deep water. Sporophytes have floats and are very common as drift algae in nearshore waters (Kingsford 1995). Microscopic zoospores may disperse considerable distances, especially during storms.

ADAPTED FROM FOSTER & SCHIEL 1985

OPPOSITE BOTTOM: **Figure 4.3** Diversity of sexual and asexual products. Sexual products are on the upper side and asexual products on the lower side of the diagram. Sexual reproduction results in the recombination of genes contributed by males and females. Adults may have separate sexes (gonochoristic), be sequential hermaphrodites or simultaneous hermaphrodites (i.e. reproduce as male and female at the same time). The fertilisation of an egg results in a zygote, the beginning of a new individual. Asexual products are genetic clones of the source individual. Fragments knocked off and buds that are pinched off can give rise to a new individual. Sponges sometimes give rise to environmentally resistant gemmules, and the polypoid phase of jellyfish produces similarly resistant podocysts. Asexual reproduction is common in some taxa, including sponges, cnidaria and ascidians, and in some cases clones may represent the majority of new recruits.

MODIFIED FROM GIESE & PEARSE 1974

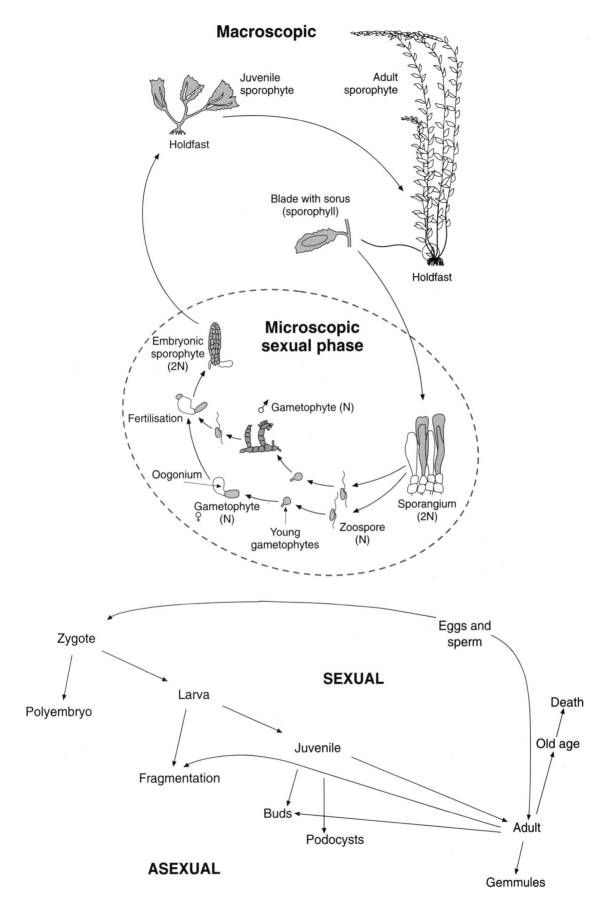

Macroscopic

Juvenile sporophyte

Adult sporophyte

Holdfast

Blade with sorus (sporophyll)

Holdfast

Microscopic sexual phase

Embryonic sporophyte (2N)

Fertilisation

♂ Gametophyte (N)

Oogonium

Gametophyte ♀ (N)

Young gametophytes

Zoospore (N)

Sporangium (2N)

Zygote

Eggs and sperm

Polyembryo

SEXUAL

Larva

Death

Old age

Juvenile

Fragmentation

Buds

Adult

Podocysts

Gemmules

ASEXUAL

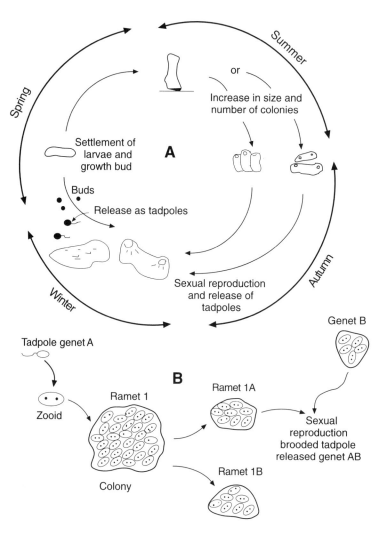

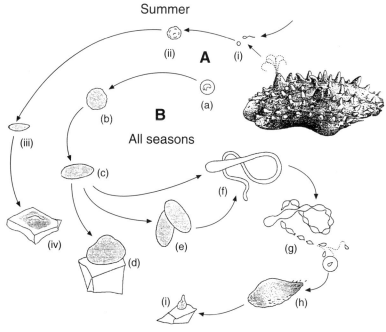

Organisms such as ascidians, urchins, gastropods and crustaceans (Figs 5.1, 5.2, 7.1) generally have external fertilisation (from gonochoristic, sequential or simultaneous hermaphroditic parents) where reproductive products are broadcast-spawned or transferred from males to females as a spermatophore (e.g. lobsters: MacDiarmid 1988). In some cases fertilisation is internal and the young are brooded until they are released as larvae or juveniles (e.g. ophiuroids and asterinids: Byrne 1994, 1995). Where larvae are released into the plankton they may spend anything from a few hours (ascidian larvae: Davis 1987; Svane & Young 1989) to hundreds of days (e.g. teleplanic echinoderm larvae: Scheltema 1986). The mode of reproduction and nature of the larvae can have important consequences in terms of distance of dispersal and the variation in numbers of organisms in local populations (Underwood 1979; Chapter 5).

The life history of many marine invertebrates, however, is complicated because of different asexual products that can have a great influence on numbers of individuals or colonies within a population (especially sponges and ascidians; Figs 4.3–4.5). Many invertebrates are single units (e.g. urchins = unitary organisms), others, however, are joined together in groups (or modules) to form colonies (= modular organisms). Individual modules can live as unitary organisms, which

gives them great resilience to damage as it seldom affects all modules of the colony. Moreover, a high percentage of colonies in an area can be asexual clones (ramets) of the same genotype (genet). Colonies generally increase in size as they grow and may fuse with other colonies or split (by fission); see Figure 4.4.

The Porifera represent one of the most plastic animal groups in terms of reproductive choice (Bergquist 1978; Battershill & Bergquist 1990). Recruitment of sponges can result from sexual products or asexual reproduction in the forms of viable fragments, buds and, in some cases, gemmules (usually at high latitudes). The most variable species found to date is *Polymastia croceus*, which can reproduce sexually (with separate sexes and as a hermaphrodite) as well as having three separate modes of asexual reproduction (Fig. 4.5), which involve different levels of budding.

The ascidians are perhaps the best exponents of budding, which can lead to a massive increase in biomass and area covered (Svane & Young 1989). Ascidians are the temperate-reef equivalent of coral-reef soft-coral groups because of their powers of asexual reproduction. Through budding, ascidians can rapidly alter their representation in a habitat from a few settled individuals to almost a genetic monoculture (e.g. *Didemnum*: Stocker 1986; Fig. 4.4). Both colonial (Cook in press) and solitary ascidians can bud (exception: *Pyura spinifera*: Davis 1996), but the process and habitat-modifying effect are most pronounced in the colonial forms (compound and social ascidian species: *sensu* Stocker & Battershill *in* Cook in press). It is also common for cnidarians to reproduce asexually, which can cause an increase in representation of zooids on a variety of substrata (e.g. *Cyanea*: Grondahl & Hernroth 1987; *Phyllorhiza punctata*, see Figure 9.17).

4.2.2 Population dynamics

Many studies are concerned with variation in the numbers and perhaps size of individual species (e.g. for harvesting). Important factors include recruitment to the population, growth in size to maturity, mortality, and harvesting (Fig. 1.3). Recruitment to a population is more complicated for modular organisms (above) in that a substantial proportion of the population may result from asexual products such as buds, fragments and gemmules (Wulff 1986; Kelly-Borges & Bergquist 1988; Svane & Young 1989; Battershill & Bergquist 1990). Growth characteristics are of great concern to many investigators, particularly growth to maturity and changes in the biomass of a population based on growth, which are relevant to regimes of harvesting. Estimates of growth require information on age, which is difficult to obtain for many benthic organisms (e.g. algae:

OPPOSITE TOP: **Figure 4.4** The life history and seasonality in the size of a colonial ascidian, *Didemnum candidum*.
A: Seasonality in the timing of growth of colonies and the release of ascidian tadpoles. Variation in the size of colonies can have a great influence on percentage cover of the substratum and their relative importance as occupiers of space in assemblages of sessile invertebrates. **B:** Colonies grow rapidly from one tadpole (from sexual reproduction). Individual colonies (ramets) from the same tadpole (genet) split and sometimes fuse with each other. Letters labelling colony indicate relationships of ramets through time. The timing of splitting is an area of active research and may depend on factors such as the size of ramets and proximity of other sessile organisms such as sponges.

OPPOSITE BOTTOM: **Figure 4.5** The life history of a sponge, *Polymastia croceus*. The complicated behaviour of these asexual products in these sedentary organisms has only recently been appreciated Sexual and asexual reproduction in the diagram shows: A (i) fertilisation of an egg from the release of eggs and sperm from different individuals; A (ii) cleavage of the zygote; A (iii) a creeping blastula is formed; A (iv) settlement; B (a) a sponge fragment detached from the sponge surface papilla; B (b–c) free propagules; B (d) direct settlement of the propagule with beginning of oscular tube formation; B (e) fusion of propagules; B (f) extension of propagule; B (g) beading along the length of the propagule and the length and migration of detached elements; B (h) migrating propagule; B (i) settled propagule with oscular tube.

Hatcher et al. 1987; Cheshire & Hallam 1989). Mortality represents loss from a population and can result from a variety of causes, including death through old age; disease (e.g. Elner & Vadas 1990); predation (e.g. Andrew & Choat 1982) and competition (e.g. Keough 1984a, b; Davis 1987). There can, of course, be substantial loss from benthic populations as a result of harvesting. For example, urchins have been all but wiped out on temperate reefs of France through harvesting (M. Byrne pers. comm.). Many organisms that are crucial are habitat formers and determiners are the targets of fisheries and these include algae (e.g. *Macrocystis pyrifera*: Foster & Schiel 1985, and *Ascophyllum*: Vadas et al. 1990) and grazing invertebrates (Chapter 5). Moreover, the extraction of these organisms can cause a cascade of ecological changes that include changes in the presence and distribution of habitats (e.g. Elner & Vadas 1990).

4.2.3 Benthic assemblages

It is difficult to study populations in isolation within habitats because there are many important interactions with other populations. For this reason many studies focus on more than one population (assemblage = a collection of populations). For example, inter- and intra-specific interactions among benthic invertebrates (Kay & Keough 1981; Connell & Keough 1983; Butler 1995) and among algae (Kennelly 1987a, b) can have a major influence on patterns of abundance. In some cases larvae actively avoid other benthic organisms, often for good reason. Davis (1987) followed individual ascidian larvae and found that most of them avoided sponges of the genus *Mycale*. There was a 100% mortality of those larvae that settled on sponges. The recruitment of kelp can be inhibited by the presence of turfing algae (Kennelly 1987a, b). Conversely, settlement of some organisms can be facilitated by the presence of conspecifics or other species (Davis 1996); for example, the abundance of kelp can influence patterns of recruitment of fishes (Chapter 5). Interactions among grazers are strong, and diverse assemblages of grazers (including gastropods, urchins and fishes) are found in many habitats, especially the barrens of rocky reefs. In some cases grazing by larger organisms is essential for providing algae of a suitable size that can be consumed by small grazers (Chapter 6). Moreover, grazers can greatly influence the successful settlement and longevity of other benthic invertebrates (e.g. sponges and ascidians: Stocker & Bergquist 1987). The relative importance of these processes can only be determined with multifactorial experiments (Chapter 3).

4.2.4 Spatial variation

Depth

On most temperate reefs of the world there is great variation in representation of different habitats with depth (e.g. California: Foster & Schiel 1985; Europe: Peres & Picard 1964, Carlos & Calvo 1995; see also Plates 2–19). For example, in northern New Zealand it is common to have a band of macroalgae in very shallow water (e.g. *Xiphophora chondrophylla* 0–1 m; *Carpophyllum machalocarpum* 1–2 m); barrens grazed by urchins (*Evechinus chloroticus*); algal turf (e.g. *Corallina, Delissia* spp.); kelp forests (*Ecklonia radiata*) and in some cases deeper stands of fucoid algae (e.g. *Carpophyllum flexuosum*) over a depth range of 2–20 m (Choat & Schiel 1982); in deep water (> 15 m) it is common to have assemblages of sponges on rock or attached to rocks and cobbles under a shallow bed of sediment (Ayling 1980; Battershill & Bergquist 1990). In addition, on vertical walls distinct assemblages of sponges, ascidians, bryozoans and cnidarians are found (Battershill 1986). On the southeastern coast of New South Wales, in shallow water (0–3 m) patches of the large fucoid alga *Phyllospora comosa* and solitary ascidians (*Pyura* spp.) are found, and a 'fringe' of tufting algae is also common in this depth stratum. Barrens grazed by urchins (*Centrostephanus rodgersii*), kelp forest (*Ecklonia radiata*) and turf habitats (algae: *Corallina, Lobophora, Padina*) are found over a broad depth range (3–25 m), and sponges (e.g. *Mycale*) and ascidians (e.g. *Botrilloides*) are common in deeper reef habitats (Underwood et al. 1991).

At high latitudes in both hemispheres (Dayton 1985), kelp habitat is in the form of giant kelp forests (*Macrocystis pyrifera*). If sampling is done within these forests, it may be necessary to consider the subcanopy separately as it may be composed of several understorey layers of algae such as tufting algae and sponges (e.g. *Ecklonia radiata* forest). In California, the subcanopy algae (under *Macrocystis*) can include *Pterygophora* and *Laminaria* spp. (which are as large as *E. radiata* in southern hemisphere kelp forests). Furthermore, the subcanopy habitat in *Macrocystis* forest may be devoid of algae (apart from corallines) and be regarded as urchin-grazed barrens. Here, we describe relatively broad categories of habitats, but in some studies it may be appropriate to define microhabitats (e.g. sand gutters or cobble patches within kelp forests, crevices, and flat areas in barrens: Syms 1995).

Depth-related patterns of habitats have often been described in terms of 'zonation' (e.g. Morton & Miller 1973), where diagrams often imply that local habitats will always be found along a given depth profile of the reef. In many cases, however, variation in the distribution of habitats is not so predictable. Processes that influence the distribution of habitats may include reef exposure, local geomorphology, sedimentation, patterns of grazing, disease and incidence of disturbance. For many habitats, it can only be said that there is a high or low probability of a certain type of habitat type being found within a certain depth range. Moreover, some habitats are found as a mosaic within a depth range. For example, at the Poor Knights Islands in northern New Zealand, a mosaic of patches of kelp forest and *Carpophyllum flexuosum* forest can be found over the same depth range. Similarly, on the coast of New South Wales urchin-grazed barrens and kelp are generally found over the same depth range.

Horizontal variation

Variation in the distribution of habitats occurs at a wide range of spatial scales from many degrees of latitude and longitude to variation from metres to tens of metres within a depth range. For example, in New Zealand *Macrocystis* is rarely found at latitudes north of Wellington ($41°18'$ S). In New South Wales, urchin-grazed barrens can represent up to 70% of the substratum area of reefs to the south of Port Stephens (approx. $32°$ E), but are absent from reefs to the north, where most of the substratum is covered with solitary ascidians (*Pyura*). It is not surprising that the geographic limits of many habitats coincide with major biogeographic boundaries or transitions of a variety of organisms. Geographic transitions of this type are relevant to many biological questions (e.g. zoning coasts and the positioning of marine reserves: see 6.4.4).

On spatial scales of less than 100 km the distribution of habitat types can vary greatly with respect to exposure (Schiel et al. 1995), local geology (e.g. presence of boulders: Andrew 1993), local oceanography, such as upwelling (Kiirikki & Bloomster 1996) and proximity to freshwater plumes (Andrew 1991; Kennelly & Underwood 1992), sedimentation rates on reefs (Cortis & Risk 1985, Grange 1989), the vagaries of recruitment (Keough 1983; Butler 1986) and other biological factors (see section 4.1). Even on scales of tens of metres the percentage representation of habitats along a reef profile can vary greatly. For example, Underwood et al. (1991) found great variation in the percentage representation of kelp forest, barrens and other habitats in profiles (see 4.7.3) within sites that were only separated by 5–15 m. In special circumstances there can also be quite distinctive horizontal variation in habitat character; for example, between inner and outer locations in fjords (Grange 1990) and with distance into caves (Battershill 1986; see also 4.7.4)

4.2.5 Temporal variation

Proportional representation of habitats can vary greatly on timescales of greater than one year. The size and frequency of storms can cause massive changes on reefs. For example, changes in the El Niño-Southern Oscillation (ENSO) and attendant storms can alter the supply of nutrients to reefs and have been correlated with great changes in algal cover (e.g. *Macrocystis, Gelidium*: Ebeling et al. 1985; Dayton et al. 1992; Foster & Schiel 1993; Holbrook & Schmitt 1996), encrusting fauna (Ayling 1983) and, hence, the nature of habitats. Variation among years in the recruitment of habitat-forming organisms can also change the nature of habitats. The timing of a disturbance can also have a great influence on the nature of habitats. For example, Kennelly (1987a, b) demonstrated that the removal of kelp by storms during winter would result in bare space that was colonised by kelp spores, which are released from sporophytes at this time (see 4.7.6). If the kelp is removed outside of the period when spores are released, the space is colonised instead by turfing algae, among which kelp does not become re-established. Jernakoff (1985) also found that the timing of disturbances had a great influence on the nature of subsequent colonisation. These types of observations have also been made in many studies that have described the colonisation of artificial structures (Butler 1995).

It is well known that there is seasonal variation in resident biomass and settlement patterns of algae (Ayling 1980; Bonin 1981; Kennelly & Larkum 1983; Schiel 1985a; Kaehler & Williams 1996) as well as the recruitment of benthic organisms from sexual (e.g. Elner & Vadas 1990; Andrew 1991) or asexual products (e.g. ascidians: Svane & Lundalv 1982). For example, there are enormous changes in biomass of *Sargassum* spp. found on rocky reefs, because members of this genus shed most of the thallus during the later stages of reproduction, in spring (Schiel 1985a). *Sargassum* spp. are regularly observed drifting in coastal waters at this time of year (Kingsford 1993). Annual red algae are similar in that only a very small amount of tufting algae covers the substratum over winter, whereas a carpet of algae may be found in summer (Bonin 1981).

If seasons cannot be accounted for in the sampling design, it should be acknowledged that seasonal changes will, to some extent, alter the nature of subtidal environments that have been described at only one or two points in time. Care should be taken, therefore, not to confound seasonal changes with those induced by impacts. If monitoring at different times of the year is not possible, the problem may be minimised by repeating studies at the same time of year. Short-term temporal variation as a source of error should also be considered in sampling designs. For example, in New South Wales filamentous red algae can be abundant in barrens habitat for short periods of time (days or weeks) before it is consumed by the abundant invertebrate and vertebrate herbivores in this habitat.

Because many of these sources of variation form a hierarchy, sampling designs often need to incorporate nesting (see Chapter 3). For example, if sampling is to be done on broad spatial scales (hundreds to thousands of kilometres) and locations are separated by large distances (e.g. > 100 km), multiple sites should be sampled within locations (e.g. separated by kilometres) to adequately describe the potentially great variation at this smaller scale. Sampling may have to be done in different habitats (the orthogonal part of the design) at each site and organisms counted in random sample units within each habitat. Hierarchies and the potential for nested designs should also be considered seriously for temporal variation. It is also important to define what is considered a reef, so that continuous reef is not compared with, say, areas of small cobbles in sediment.

Common questions that relate to the distribution of habitat types and the organisms within them are as follows:

- What is the percentage cover of different habitats down the vertical extent of a reef (from the low tide mark, to the seaward edge of the reef profile)?
- How consistent are depth-related patterns of habitats on spatial scales ranging from tens of metres to hundreds of kilometres?
- What is the percentage cover of habitats of different types within an x kilometre stretch of coastline?
- How are different benthic organisms (e.g. abalone) distributed or dispersed with respect to habitat type and depth?
- Does protection (i.e. a MPA) influence the representation of habitats when compared with multiple reefs without protection?
- How does habitat type influence the recruitment of organisms?
- Are unique or rare habitats catered for in sampling designs (e.g. 4.7.4)?
- What processes influence the distribution and persistence of habitats?
- What processes influence the distribution and persistence of organisms within habitats?

Most of these problems will require different sampling methods and designs.

It is crucial to carefully consider your ecological problem before proceeding with the choice of method and the allocation of sampling units in your sampling design (see Chapter 2).

The term 'benthic organisms' is used in this section to refer to algae, sessile invertebrates and mobile invertebrates (see section 4.3). The term 'assemblage' refers to a collection of species that are concurrent in space and time. Readers who are designing sampling programs for coral reef environments should note the 'manual' by English et al. (1994).

4.3 Methods for describing subtidal habitats

The method that is best suited to your biological question will vary according to the scale of your investigation (i.e. metres to thousands of kilometres), whether discrete (e.g. quadrats) or continuous sampling (e.g. video) is considered necessary, whether destructive samples are required, and the cost of sampling. In this section we first describe methods that are used and then protocols for their application. In most published studies you will find one or a few of the following methods are used (e.g. see section 4.7).

4.3.1 Methods

(a) *Satellite images and aerial photography*
Satellite imagery is improving in quality all the time and provides the potential for low-cost, quick assessments of the distribution of some habitats. Imagery is available from a variety of satellites that have photographic images of areas of coastline and digital data on physical (e.g. temperature) and biological variables (e.g. chlorophyll); Plates 53–59. In general, however, the imagery is only likely to be of value for the choice of areas of coastline that have different environments (e.g. rocky reefs, sandy beaches, *Macrocystis* forest). Aerial photography can give high-resolution images suitable for an overview of the study area and the mapping of shallow habitats. On days when the underwater visibility is exceptional and the tide is low, excellent information on subtidal topography and the distribution of different habitats can be gained from aerial photography (Plate 51). Infrared film may also be used if it is necessary to obtain images of intertidal populations (Plate 44). Photographs may be used to provide background information at the sampling design stage (see Chapter 2). Moreover, they provide a useful supplement to information gained while diving for the construction of subtidal maps (e.g. Ayling 1978). It is useful to have good shoreline maps (topographic and marine charts) as supplements to aerial photographs; GIS may also be available to provide bathymetry. In some cases experi-

mental treatments (e.g. the clearance of algae: N. L. Andrew pers. comm.) can be observed from low altitude, and photographs are also useful to aid diver navigation. If water visibility is consistently good and ground-truthing indicates that reliable pictures of subtidal habitats can be obtained to a given depth, contiguous series of photographs may be obtained on spatial scales ranging from tens of metres to hundreds of kilometres. In New South Wales, information of this type is assisting in decision making on zoning (e.g. for MPAs; S. McNeill pers. comm.).

Advantages
✔ Provides broad-scale information on the percentage cover of different environments along stretches of coastline.
✔ Aerial photography can give good quality images of subtidal habitats.
✔ Useful for an overview of study sites and later navigation within the study site (use laminated photos).
✔ Augment information from diver-intensive work (e.g. profiles).
✔ Relatively inexpensive if few flying hours are required.

Disadvantages
✘ Poor quality from high altitude.
✘ Quality dependent on weather, water clarity, angle of the sun and aircraft, and the sea-surface conditions.
✘ Expensive if many flying hours are required.

(b) *Swath mapping*
Modern developments in side-scan sonar can allow investigators to get surprisingly high-quality images of different habitat types in the subtidal (e.g. Arron & Lewis 1993; Fig. 4.6). It is unlikely that vessels will get closer than 50 m to the low-water mark and the beam width is very restricted in shallow water (= 2 x depth). The technique can be particularly useful for the location and mapping of deep reefs and soft substrata (see Chapter 8). Mapping can be very accurate using differential GPS and hydronavigation techniques (e.g. Page & Battershill in press).

Advantages
✔ Search large areas.
✔ Can be used in deep water where it is impossible or difficult to dive.
✔ Continuous data

Disadvantages
✘ Difficult to find vessels that will sample in the near-shore environment.
✘ May require ground-truthing.

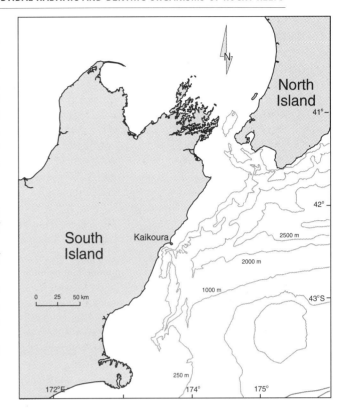

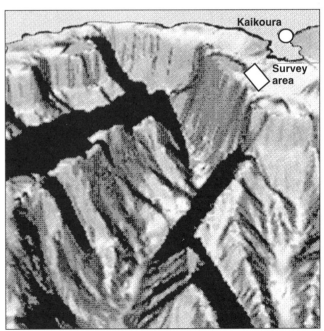

Figure 4.6 Example of a swath map, showing the detail of bottom topography that is useful for the identification of different habitat types. Two scales are shown: the general area of Kaikoura (50 x 50 km) and inset (700 x 500 m). The shadow shown is computer-generated.

(c) *Rapid visual methods*

Several techniques are available, and many have been used on the Great Barrier Reef for making general assessments of reef condition and for locating special areas of interest (e.g. infestations of crown of thorns starfish: Kenchington 1978, 1984; English et al. 1994). McCormick (1990) has also used them to target certain algae for more detailed investigation, while they can also be used for locating potentially unique habitat types such as caves with a rich fauna of encrusting organisms (e.g. anemones, ascidians, and sponges). Because it is not usually possible to identify reef components to species while doing rapid surveys, taxa are generally categorised according to Taxonomic Units (see 4.5.1).

Manta tow: A diver is towed at speeds of up to 2.5 knots for a short distance/time period. This method allows divers to observe large areas of reef on snorkel or scuba while they are towed by a small boat. The diver surfaces and is debriefed by the boat person, who asks a series of questions (e.g. what was the reef slope, substratum type, abundance of *Ecklonia*, approximate size of plants?) For each question a range of broad categories is given to the diver to choose from (e.g. abundance of algae: rare, common, abundant). Items on the question sheet will vary according to the goal of the study.

Dive computers that allow semi-continuous depth recording are invaluable tools for this type of work. Some models allow direct down-loading of depth profiles onto a PC. Voice-communication equipment can enhance this operation by allowing real-time communication from the diver to boat crew.

Advantages

✔ Search large distances over reef (hundreds to thousands of metres).

✔ Find patches of habitat that have large distances (tens of metres) between.

✔ Qualitative assessment of the nature of habitats.

✔ Relatively simple work.

✔ In shallow water there is no need for scuba.

Disadvantages

✘ Little directional control available to the diver.

✘ The diver can become saturated with information, especially in a rich area.

✘ It is not possible to decrease speed or pause to study an interesting area.

✘ Cryptic organisms can be overlooked.

✘ Towing may become difficult in areas of foul ground and reef profiles of varying angles.

✘ The diver may be towed away from reef substrata.

✘ Of little use in poor visibility. There is little time for identification and taxonomic units often have to be used (section 4.5.1).

For more detail on this method (percentage categories, arm signals, etc.), the design of manta boards and their utility, see English et al. (1994) and Chapter 8.

DPV: The method of collecting information while using a diver-propulsion vehicle (DPV) is very similar to that of a manta study: a boat follows the diver for debriefing. McCormick (1990) used a DPV to make a rapid visual assessment of *Pterocladia* biomass along the 5 km coast of the Leigh marine reserve. McCormick found that the DPV allowed considerable freedom of movement and the option of stopping. Moreover, it allowed the contingency of surfacing for debriefing if there was an abrupt change in reef morphology before the completion of the standard time/distance run. McCormick took 50 minutes to an hour to traverse 1 km of coastline.

Although the cost of buying, hiring or making a DPV is relatively high (compared with a manta board), the vehicle overcomes many of the problems encountered with manta boards. The ability to start and stop is a great advantage over a manta board. A current meter could be used to determine the approximate distance travelled, and a compass used to determine direction.

Free-swimming observer: This technique has been widely used (Kenchington 1978). The major advantages are that the observer is free to modify speed and direction, hands are free to record observations on a slate, and the technique requires no specialised equipment. Disadvantages are that the observer frequently has little navigational information to enable sightings to be related to location, and may unwittingly cover the same area more than once (though the use of a compass may minimise this problem) and a smaller area will be covered than with other methods.

The free-swim method could be modified by attaching a line and small float to the diver. While the diver swims along the reef a boat person may plot the position of the float at regular intervals on a map (e.g. every one or two minutes). If the diver also keeps track of the time, observations can be related to position on the map. The diver may surface at regular intervals for debriefing and in this way avoid swimming in circles for over an hour.

Advantages

✔ Cover tens to hundreds of metres.

✔ Cheap because no specialised equipment is required.

✔ Qualitative information.

✔ Finding habitats that are separated by metres to tens of metres.

Figure 4.7 Divers collecting data for a stratified-sampling design where sampling is stratified by habitat. Habitats include shallow *Carpophyllum* (fucoid algae), urchin-grazed barrens, *Ecklonia* forest (kelp), and deep reef with sponges and sand. A tape has been run out perpendicular to the shore to provide a profile of the reef. The borders of habitats are marked by depth and distance from the shore. Other divers are collecting information on the densities of organisms (e.g. gastropods and urchins) in each habitat, using quadrats. Estimates of the abundance of benthic organisms within quadrats are recorded on a slate, using a pencil. Sample for other organisms such as fish and lobsters may also be stratified by habitat.

BRIAN BISHOP AND VIVIAN WARD

Disadvantages

✗ Lack of navigational information that can be related to position on a map.

✗ Only qualitative information.

(d) Reef profiles

A reef profile is continuous line from the immediate subtidal to the seaward edge of the reef, in a direction perpendicular to the shore. A long tape or weighted rope is usually used for profiles (Fig. 4.7), but a DPV fitted with a flowmeter can also be used. Habitat boundaries (kelp forest, barrens, etc.) are recorded in relation to the tape as distance from the start point and depth of the water column (truthed to MLWS or similar datum). In this way, the depth distribution of habitats and the horizontal extent of the reef can be reconstructed (e.g. Ayling 1978). Video can be used to record data from the profile so that it can be examined later in the laboratory. Replicate profiles within a site (three or more)

will give a measure of proportional representation of habitats within an area, with some estimate of error. For example, Underwood et al. (1991) used four 100 m profiles to estimate the representation of habitat types on reefs in New South Wales (see 4.4.5). High-precision estimates of habitat representation may be used for stratified sampling using proportional or optimal allocation (3.5). In many cases one may not reach the end of the reef at, say, 100 m, in which case the profile would have to be defined as a profile to 100 m from the shore. The densities of organisms within each of the identified habitat types can be obtained with other methods (e.g. quadrats and transects).

Advantages

✔ Measurements of the proportional representation of habitats on a reef.

✔ Data may be used to make subtidal maps (e.g. Fig. 4.8, p. 96, from Ayling 1978).

✔ A basis for stratified sampling of other reef-associated organisms.

✔ Distances between habitat types can be measured.

✔ A permanent record of the profile can be kept if video is used.

Disadvantages

✗ Cannot be used to describe generality of patterns without replication.

✗ Lack of quantitative measures of the densities of organisms unless augmented with direct measures of density stratified by habitat.

✗ Time taken to complete each profile may vary greatly with horizontal extent and depth of the reef.

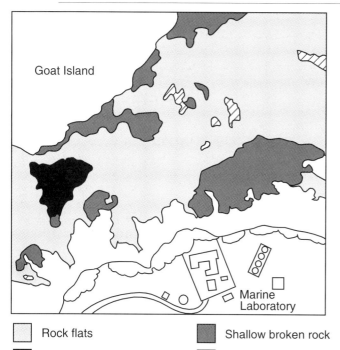

Rock flats

Sediment covered-rock flats

Shallow broken rock

Sand and gravel

Figure 4.8 Extract of a subtidal map of Goat Island showing subtidal habitats. This map was created from diver profiles where the position of different habitats was marked with respect to distance from the shore. This information was augmented with aerial photographs. A map of this type can be very useful for underwater navigation and questions that require stratified sampling (especially of the 'proportional' type: see Chapter 9). If the map is reasonably accurate, the borders or existence of habitats can be studied at later dates. Creating a map is very work-intensive and the cost benefits should be considered seriously before embarking on a project of this type.

(e) *Underwater video systems*

Video has advanced greatly in recent years, and relatively small-sized Hi8 systems are available in housings. Videos with a wide-angle lens and good lighting (it is always useful to have more than one lamp as backup) may be used for some of the methods above. You should choose a video housing that has external control of zoom functions (e.g. Sony Handycam housings). If good records of depth and distance from the low-water mark are kept, video can be used for profile and the tapes may be archived as a record (see 4.3.1d). Video can be used for quadrat sampling and the images 'frame grabbed' on a PC for analysis. In this case it is important that the video be kept at a fixed distance from the substratum with the assistance of a measuring rod, tripod and mount, or laser-positioning lights. Investigators should ground truth their data to determine the

accuracy of the method (e.g. Foster et al. 1991). The accuracy of video for determining habitat features may not be good enough if you want more than a record of large-scale differences of macro-algae between habitats (Leonard & Clark 1993). New techniques and software, however, now enable stereo video imaging and analysis and, with two cameras, can allow estimates of size (Harvey & Shortis 1996). See also comments on digital tapes in 2.11.

Advantages
- ✔ A permanent record.
- ✔ There can be consensus in decision-making among investigators.
- ✔ Data other than that relevant to the current biological question can be extracted at a later date.
- ✔ Among-observer bias can be eliminated.
- ✔ Video can allow large distances (and modest transect widths) to be photographed, which can be useful for permanent records of profiles and organisms within habitats.
- ✔ Images can be scanned or grabbed (for video images), and processed using image analysis programs.

Disadvantages
- ✗ Technical problems in obtaining data (e.g. flat batteries, leaking housings, faulty connections, etc.).
- ✗ Difficult to obtain a good image of the tape and the substratum when doing reef profiles.
- ✗ Variation in skill may pose problems.
- ✗ Lengthy viewing times and analyses escalate the cost.

(f) *Spot-check method*

Spot checks are made at regular intervals around the coastline (e.g. 200 m, 500 m intervals: Kenchington 1978). The check may be made by looking over the side of a boat with a face mask (Kenchington 1978) or by a short dive, which may involve swimming over a given range of depths. The advantage of this technique is that the diver can cover a very large area; the disadvantage is that it is a subsampling technique, by which only a small proportion of the study area is covered. Preliminary information on the spatial distribution of the organisms (or habitats) of interest would be useful for the choice of a sampling interval. Choice of sampling interval may be based on a knowledge of habitat heterogeneity from aerial photography, the literature or casual observations.

Advantages
- ✔ Medium- to broad-scale observations (hundreds to thousands of metres).
- ✔ Qualitative assessment of the reef.

Disadvantages
- ✗ Discrete sampling (cf. continuous of 4.3.1c).

✘ No quantitative data.

✘ Divers may have problems with multiple pressure changes (i.e. ears) in one day.

(g) Quadrats and belt transects

A wide variety of quadrats and transects can be used for discrete sampling. For quantitative methods, decisions have to be made concerning sample-unit size and shape, and number of replicates. Procedures for working out optimal sample-unit size and number that are appropriate to a single species or closely related group of organisms are given in Chapter 3. Accordingly, compromises will have to be made when a variety of organisms (e.g. algae, sponges, gastropods) are to be sampled by use of the same sampling unit. It is also likely that more than one sample unit size will be required to adequately sample target organisms in any study that attempts to quantify the biotic components of habitat types (e.g. Underwood et al. 1991). In some cases, photographic quadrats of encrusting organisms may be desired, and the sample unit size may be limited by the lens of the camera or turbidity.

Many studies that have carried out quantitative assessments of benthic organisms in the subtidal zone have generally used 1 m² quadrats (Ayling 1978; Andrew & Choat 1985; Schiel 1985b; Battershill & Bergquist 1990; Schiel et al. 1986). This sample unit size generally offers an acceptable level of precision (SE as a proportion of the mean) for a wide range of organisms at different latitudes and in a variety of habitat types. In addition, numbers from 1 m² quadrats are generally large enough for analyses.

Quadrats measuring 1 m² will not be the most appropriate sample-unit size in all situations. Sample-unit sizes of 0.25–10 m² have been used for large gastropods and urchins (see Chapter 5). Battershill (1986) used 0.25 m² quadrats to sample the benthic encrusting fauna at the Poor Knight Islands, northern New Zealand, as did McCormick (1990) for *Pterocladia* seaweed near Leigh, New Zealand. In contrast, Battershill (unpub. data) used 250 m² quadrats to sample the incidence of fungal infection on choristid sponges during an El Niño event. A useful approach when there are large differences in the sizes, densities or secretive nature of organisms, is to use more than one size of quadrat. Schiel (1984) used 0.25 m² quadrats for juvenile urchins at the Poor Knights Islands and 1 m² quadrats for adults. Underwood et al. (1991) used 5 m² transects (1 x 5 m) for estimating densities of urchins, and 1 m² quadrats for limpets and other small grazers. Large sample unit sizes may be required for some large organisms. For example, in a comprehensive set of comparisons of different sample units on the Great Barrier Reef, Mapstone and Ayling (1993) concluded that 50 x 5 m transects were most suitable for discrete benthic organisms such as crown of thorns starfish and some corals.

The shape of a quadrat may also be important (e.g. circular, rectangular, square: see 3.3). From a pragmatic point of view, however, McCormick (1990) commented that square quadrats are easy to construct, easily gridded up for percentage cover determination, and enable random sample allocation to be simplified (see 3.4). They are also amenable to estimation of percentage cover of organisms. The clumsiness of a 1 m² quadrat can be reduced by using a 0.5 x 1 m quadrat and flipping it over for the complete 1 m²; a more sophisticated version could include two 0.5 x 1 m quadrats hinged together. A two-sided quadrat may be practical in some habitats, the third and fourth side being treated as imaginary lines. From our experience, however, you can get lost and waste time if you try this method in dense stands of macroalgae. Quadrats taken on vertical cliff faces can be clipped to fauna on the rock wall using nylon cord attached to at least two corners.

The number of replicate quadrats to be taken should be worked out as part of the sample unit optimisation procedure. Alternatively, in general studies of benthic organisms, 5–10 quadrats sampled at each stratum usually give adequate precision (e.g. Schiel 1984; Schiel et al. 1986). Ultimately, the number of replicates will depend on logistic constraints and the precision required (see 3.3).

In general, sample units should be placed randomly, or haphazardly, within a study area or strata (e.g. different habitat types: see randomisation, Chapter 3). There are some biological problems, however, where it is necessary to have fixed or permanent sample units. For example, one of the best-known ecological data sets is that of Joe Connell at Heron Island, where he has photographed fixed quadrats over a period of 35 years (Tanner et al. 1994). It should be noted that use of fixed sample units violates the assumptions of many statistical procedures (e.g. ANOVA: see Chapter 3), but they may be needed where the purpose of the study is to examine something other than estimates of abundance (e.g. patterns of overgrowth in encrusting organisms).

Advantages

✔ Non-destructive.

✔ Accurate and precise estimates of abundance can be obtained.

✔ Some organisms require minimal training to be counted correctly.

✔ Size measurements can also be obtained.

✔ No specialised equipment.

✔ Some sample-unit sizes may be combined with a photographic approach.

Disadvantages

✘ Discrete sampling.

✘ One sample unit size will not be suitable for all types of organisms in a general study.

✘ Energy-intensive.

✘ Some organisms require good taxonomic expertise (e.g. sponges).

✘ Bias/non-independence if the placement of sample units is not random.

(h) *Mapping*

Maps are of great value for many biological investigations on a variety of scales. Maps of habitats have been used to describe the distribution of habitats and organisms within them on scales of greater than 1 km (e.g. subtidal maps of the Leigh marine reserve) and tens of metres where the borders of individual habitats are monitored and the position of individual boulders is noted (see Fig 6.10). Maps may be used to study the distribution and spatial arrangement of habitats, and how these change through time. Moreover, they are often a prerequisite for stratified sampling of other organisms (e.g. mobile invertebrates and fishes: Chapters 5 and 6).

Mapping techniques may be combined with sampling by divers (e.g. with quadrats) to measure densities and sizes of organisms within habitats. In this way, more detailed information is gained on the nature of habitats. Maps of large areas will require detailed information from reef profiles (see 4.3.1d), and this can be augmented with outlines of habitats (e.g. kelp forest and barrens) from high-resolution photography. Small-scale maps (up to hundreds of metres) may be done using a combination of profiles and triangulation (or polygons) to measure the distances between three or more topographic features and/or habitat boundaries. It is useful to adopt drafting methods common to cartography such as different outlines (i.e. continuous and dashed lines) for high points (e.g. boulders) and depressions.

Innovations in GIS software and satellite imagery can greatly enhance the potential spatial scale and detail of maps (Page & Battershill in press).

Advantages

✔ Density data can be extracted from maps (see 5.2.2b).

✔ Show spatial arrangement of habitats.

✔ Useful for stratified sampling.

✔ Useful for the choice of sites for detailed studies of particular organisms.

✔ Study temporal change in the distribution of organisms within mapped areas.

Disadvantages

✘ Areas may have to be re-mapped as the distribution of habitats changes greatly with time, thus investigators should be beware of considering the spatial arrangement of habitats as permanent.

(i) *Underwater photography*

Regardless of whether a subjective or quantitative method is used to describe the biological features of subtidal habitats, good underwater photographs of each habitat are an invaluable supplement, particularly where the results of an investigation are to be presented to laypersons who are unfamiliar with interpreting graphs showing densities of organisms.

In many situations, especially where a case is being made for protection, it is important to recognise and record unique geological and biological aspects of the area. For example, the caves, arches and vertical drop-offs of the Poor Knights Islands (see Battershill 1986, 4.7.4 and Plate 12). Time may not enable detailed investigations of these features, but their biological and aesthetic importance can be recorded on film either haphazardly or in a structured way by taking photographic quadrats at different depths, distances into caves, etc. It is also possible to use stereophotography for the purpose of measuring the sizes of organisms and reef topography. Svane and Lundalv (1982) used stereophotography to study the densities and sizes of *Pyura* (solitary ascidians) in fixed quadrats (see 4.3.1g) for 10 years in a fiord in Sweden. Stereophotography has been used for estimates of size of other organisms underwater (e.g. Klimley & Brown 1983; Harvey & Shortis 1996). For a general reference on photography as a research tool, see Price et al. (1980). Video may also be used in the same context (see 4.2.1e). Photographic techniques in percentage cover or estimation of densities should always be substantiated by ground-truthing (e.g. Foster et al. 1991) and the collection of samples for taxonomy and a reference collection (see Chapter 10 for treatment of specimens). See section 2.11 for information on equipment and the storage of photographs as digitised maps.

Advantages

✔ Enables habitats, species assemblages and dispersion patterns of species to be visualised.

✔ With appropriate ground-truthing, a large amount of information can be obtained in a short period of time.

✔ Estimates of density and percentage cover can be obtained from photos.

✔ Interactions amongst sedentary organisms can be determined by using photography for monitoring.

✔ The fate of individuals or colonies can be followed by monitoring fixed quadrats (beware, analytical debate of the problems of repeated measures).

✔ Images can be scanned or grabbed (for video images) and processed using image-analysis programs.

✔ Scanned images are excellent for long-term storage and reduce problems of damage to originals.

Disadvantages

✘ Sample units generally have to be small.

✘ Mechanical or electronic breakdowns – a backup camera is a good idea for expensive field trips.

✘ Depth of field can be a problem in close-ups of complex surfaces.

✘ Quality of the photographs is influenced by visibility and especially the number of large particles in the water column.

✘ May be time-consuming extracting data from the images.

(j) *Airlifts*

Airlifts are useful for sampling small organisms (as adults or juveniles) in habitats such as algal turf or reef-associated sediments. These devices have been used to sample organisms such as epiphytal and sediment-dwelling crustaceans (Choat & Kingett 1982) and the juvenile stages of molluscs such as abalone (McShane & Smith 1988). These venturi-suction samplers are made of a pipe (2 m long, of plastic or galvanised iron) with an inhalant end that is orientated to the substratum and an exhalant end that has a mesh cod-end on it. The mesh size will depend on the target organisms (Choat & Kingett 1982). Within 10 cm of the exhalant end is attached a lateral high-pressure pipe with the second stage of a diver's regulator attached. With the tank turned on, suction is facilitated and the material from the substratum is deposited into the cod-end. The sample-unit size is usually defined by a quadrat (or core for soft sediment). It will often be appropriate to do pilot studies on appropriate sample-unit sizes (see sections 2.3 and 3.3).

Advantages

✔ It is possible to obtain quantitative samples of small and cryptic organisms.

✔ Sampling can be done on very small (centimetres) and large (kilometres) spatial scales.

✔ Small patches of habitat can be sampled.

Disadvantages

✘ Discrete samples of small areas.

✘ Much time and energy is needed for broad-scale studies.

✘ Uses a lot of scuba air.

✘ Destructive sampling; damage to delicate organisms.

✘ Considerable sorting may be necessary if a great deal of non-target material such as algae and sediment is collected.

✘ May cause a reduction in visibility.

(k) *Plastic and mesh bags*

Bags may be used to collect small organisms associated with macroalgae. The mesh size will depend on the target organisms. In general this method is used for the collection epiphytal invertebrates and small mesh sizes have been used. Gallahar and Kingsford (1993) used a plastic bag that measured 70 x 50 mm, with a 100 mm mesh bag attached to one corner for drainage. The bag was carefully lowered over kelp plants (*Ecklonia radiata*) to within a few centimetres of the substratum. The holdfast was cut and the bag closed with a rubber band. The cod-end allowed water to be drained from the bag without the loss of organisms when the collection was brought on board a boat. The same technique was used by Taylor and Cole (1994), but they used a mesh size of 1 mm. It should be noted that many epiphytal organisms move among plants and into the water column at night (see Chapter 9). Plastic 'zip-lock' bags are also very useful. Slimy or soft fragile species (especially sponges and cnidarians with of potent chemical composition) should be collected into separate bags, otherwise they will contaminate other species (see Chapter 10).

Advantages

✔ Estimates of density and species composition on plants.

✔ Alternative to airlifts and other devices that are difficult to use on macroalgae.

✔ Mesh bags allow drainage.

✔ Haphazard collections of organisms.

Disadvantages

✘ Destructive sampling.

✘ Discrete sampling on small spatial scales.

✘ Variation in abundance is great over small spatial scales and abundances change quickly in time, thus intensive sampling is required to adequately describe variation in abundance over large temporal and spatial scales;

✘ Processing of samples is time-consuming.

✘ Identification, especially of juvenile crustaceans, is difficult or impossible.

(l) *Markings and tagging*

Many ecological questions require information at the level of individuals, and this can be gained by tagging. An understanding of population dynamics demands a knowledge of growth, age and mortality, and natural markings and tagging can assist. Furthermore, interactions among individuals can be studied if identification can be assured. If organisms have unique markings, it may be possible to identify individuals by natural markings, as for lobsters (MacDiarmid 1987), fish (Jones 1983) and cetaceans (Dawson & Slooten 1997), thereby avoiding any problems of tag-artefacts. Alternatively, individuals of some sedentary organisms (e.g. discrete sponges and algae) may be identified at specific locations and fixed quadrats, and their existence and growth monitored, in some cases using photographic techniques (see 4.3.1i). Natural markings in particular are unlikely to be useful if large numbers of individuals have to be identified. It is possible to tag organisms internally or externally (for a more extensive account of conventional and new innovation in tagging see Chapters 5 and 6 for large mobile invertebrates and fish). Colour coding of even very delicate organisms with stains (Levin 1990) or subcutaneous marks (see tagging of fish, Chapter 6) is possible, or marks may be made on hard parts (e.g. file marks on shells).

It is important with any tagging technique to determine if tag/marking artefacts are a problem, tags and marking may alter the behaviour of organisms, make them more prone to disease or cause death. In addition, the life of some tagging and marking techniques are short and monitoring at regular interval is likely to determine problems of short-lived marks. Some organisms have a form that lends itself well to tagging. Algae may be tagged around the stipe with cable ties and dymo markers, but care has to be taken not to damage the plant (Hatcher et al. 1987; Andrew 1994). Damage through tagging may be avoided for some organisms by attaching tags to a bolt or nail adjacent to sedentary organisms.

Advantages
✔ Sessile individuals can be tracked from recruitment so that estimates of age and growth of organisms can be gained, especially those that cannot be aged by other methods.
✔ Estimates of age can be obtained.
✔ Estimates of mortality of discrete organisms or ramets and genets can be obtained (see Fig. 4.4).
✔ Enables determination of the spatial scale of movement of sedentary organisms.

Disadvantages
✘ Energy-intensive.
✘ Damage or death of organisms may result.
✘ Changes in growth, mortality and behaviour that are tag-related.
✘ Tag loss (e.g. through tearing, grazing by urchins).
✘ Overgrowth of tags with algae.

4.3.2 Approaches to sampling

Sampling designs that use any of the approaches reviewed below should consider replication in space along shore in a way that is relevant to the ecological question (see Horizontal variation, 4.2). In general, it is appropriate to have multiple locations and sites within locations to allow claims of the generality of patterns. In some cases it may be possible to restrict sampling for broad-scale studies to a small number of habitats or depths. For example, it may be demonstrated that a high percentage of target organisms – say, more than 90% – are found in shallow habitats – say, less than 3 m deep.

The identification of patterns of distribution of organisms is crucial for the planning of manipulative experiments (see 4.6), for without it the experiment may be of limited ecological relevance. The habitats or depths at which experiments are located will depend on the hypotheses that are to be tested, but a sound knowledge of patterns of distribution may enable experiments to be restricted to one habitat. Moreover, repeating experiments at different locations and times will make general statements about the findings more justifiable.

(a) *Representation and depth of habitats to the seaward edge of reefs*
Information of this type is gathered if the objective is to describe the proportional representation of habitats on reefs, and is usually done by profiling (see 4.3.1d). Data of this type are relevant to a wide variety of biological questions that require a knowledge of important habitat types at a location, and calculations of the percentage cover of habitats along broad stretches of coastline. This type of information is essential if stratified sampling with random allocation is needed, or if total counts of target organisms are required using stratified sampling with proportional or optimal allocation (see 3.5.2). If a map is required, it is often necessary to profile the study area. Where information is required on densities of specific organisms, depth-related patterns of habitats should often be augmented with stratified sampling (4.3.2b).

(b) *Sampling stratified by habitat*
Here the objective is to quantify the densities (possibly size and other morphometrics) of benthic organisms in sample units within a selection of habitats or in every habitat. This will generally follow a subjective or quantitative identification of the habitats present within an area (see 4.3.1 for methods). Sample units may be placed haphazardly, but non-overlapping, within a habitat; for example, by swimming up in the water column, dropping the quadrat and sampling where it falls. More correctly, sample units should be placed using a randomisation procedure (see section 3.4) such as random co-ordinates (e.g. Schiel 1984). Two tapes, placed at right angles to each other, serve as axes for co-ordinate sampling. Pairs of co-ordinates are selected from a random-number table before the dive. A similar method of randomisation involves placing replicate

quadrats at predetermined random intersections along a 25 m transect and repeating the procedure along multiple transects. This procedure may be modified further by rolling a sample unit (e.g. 1 m²) end over end to measure the distance to a predetermined position perpendicular to the tape (e.g. 0, 1, 2 m each side of the tape). This tape method may be more appropriate if investigators are relating habitat type to the distribution patterns of organisms such as fish that have to be counted over a relatively large area. Where a tape is not used, random sampling within a habitat can be achieved by swimming in a random and predetermined compass direction and distance (measured in quadrat flips as above) from the first quadrat (x) and the procedure is repeated from each subsequent replicate quadrat (x + 1 ... x$_j$) until data from the required number of replicate quadrats is collected (short-cut method).

The tape and short-cut methods are generally quicker than co-ordinate sampling. A disadvantage of stratified sampling by habitat is that if habitats of a particular type are found over a broad range of depths, then some of the variation in the abundance among locations may simply relate to maximum depth of a particular habitat. This may be addressed by stratifying sampling by depth and habitat (see 4.3.2d).

(c) Sampling stratified by depth

Quadrats are sampled at regular depth intervals to the edge of the reef (e.g. every 5 m). Unwind a tape measure from the immediate subtidal to the seaward edge of the reef, in a direction perpendicular to the shore. The placement of replicate quadrats should be done using a randomisation procedure (see section 3.4). Quadrats may be placed randomly along a depth isobar by choosing random distances from the tape. For example, within a 10 m wide lane (5 m each side of the tape), allocate 11 possible intersections (one per metre); if replication = 5, then use random numbers to select five positions from a possible 11. Roll the quadrat end over end to measure the predetermined distance. It may be relevant to note the distance from shore that each depth stratum is sampled.

In addition to quadrat sampling at regular depth intervals, the depth and distance along the tape that habitat boundaries are found should be recorded. You may decide to use a sampling protocol that is structured according to depth until habitats are identified and you can justify the choice of habitats for method (see 4.3.2b).

Sampling that is stratified by depth is likely to yield data with high variances because sample units may end up in different habitat types at fixed depths of sampling. This type of sampling design has commonly been used at remote islands to describe patterns of abundance of

organisms on rocky reefs where there is no information on the distribution and number of habitats on a reef that could enable sampling stratified by habitat (e.g. Schiel et al. 1986, 1995). Figure 4.9 shows data collected using depth sampling. There is, however, a high probability that some habitats will be over-sampled (e.g. Ecklonia forest or urchin-grazed barrens) and others will be under-sampled, particularly in habitats with a narrow, deep distribution (e.g. bands of algae in the immediate subtidal).

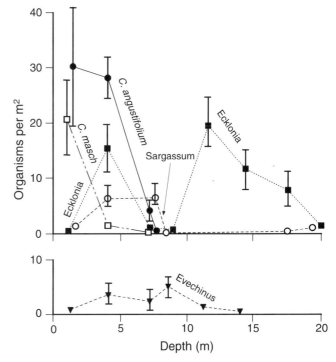

Figure 4.9 Data collected by Schiel (1990) using a depth-stratified design. The objective of this study was to determine whether the abundance of macroalgae and urchins varies with depth.

(d) Sampling stratified by depth and habitat

Some habitats (e.g. Ecklonia forest) may be found over a broad depth range, in which case a combination of depth-stratified and habitat-stratified sampling may be used. For example, some organisms may only recruit to kelp forest in shallow water. Furthermore, some habitats may only be found over a narrow depth range, but cover a high percentage of the profile of a reef (where the gradient is gradual). Methods of the random placement of sampling units would be as for 4.3.2b.

(e) Structured according to the linear distance down the shore

This procedure is identical to 4.3.2b except that quadrats are sampled at regular distances from the shore (e.g.

every 5 m), rather than with depth. On reef with a long and gentle depth profile, this method may be more cumbersome than methods 4.4.2b–d. You may decide to use a sampling protocol that is structured according to linear distance until habitats are identified and you can justify the choice of habitats for further sampling (see 4.3.2b).

(f) Belt-transect method

Contiguous quadrats are sampled from distance zero to the seaward edge of the reef (e.g. Dodge et al. 1982; UNESCO 1984). This method could also be used to justify the choice of habitats. This type of design is energy-intensive and is unlikely to be practical on reefs that extend a great distance from shore.

4.3.3. Information recorded from each sample unit

There are four major considerations:

1. The density of organisms per unit area.
2. The percentage cover of organisms.
3. The size-frequency distribution of organisms.
4. The morphometrics of organisms.

The information recorded will vary according to the purpose of the investigation. In studies of assemblages, measurements of size frequency distribution may be limited to a few target organisms (e.g. large gastropods, urchins; see Chapter 5). Some organisms may be counted as size or age categories (e.g. juvenile, adult). In some cases it may be relevant to study the morphometrics of organisms. For example, the morphology of *Ecklonia radiata* is quite variable (Morton & Miller 1973; Plates 4, 5), and comparisons of stipe length and size of the laminae may enhance an understanding of variation in numbers of epiphytal invertebrates or associated fishes. In some cases the area and number of units in colonial organisms may have to be measured (e.g. modular organisms; see 4.2.1).

An example of the procedure used in a descriptive study of habitats would be as follows (adapted from Schiel 1984):

- Record the number of juvenile and adult laminarian algae (e.g. *Ecklonia*) and fucalean algae (e.g. *Carpophyllum*); plants less than 250 mm in total length are categorised as juveniles. In addition, juveniles are distinctive in that they have one lamina (Fig. 4.1).
- Count numbers of urchins, gastropods, starfish and any other conspicuous invertebrates, and measure the size of target organisms (e.g. urchins).
- Measure percentage cover of the various layers (e.g. subcanopy and canopy) of encrusting organisms and vegetation.

The percentage cover of organisms within a quadrat is usually estimated with a point-count method. Methods include the following:

1. Gridding up the quadrat with cord so that intersections are used as points.
2. Putting regularly spaced dots on a clear plastic board that is laid over the substratum (Battershill 1986).
3. The point quadrat bar method (see Schiel 1984; Leonard & Clark 1993).
4. Using a 25 x 25 cm grid of garden mesh with random intersections marked on it (Underwood et al. 1991).

The sampling device used for method 3 is a metal bar with a string attached to the ends, like a loose string of a bow (Fig. 4.10A, page 104). Knots are tied at random along the string. The bar is tossed haphazardly into the area of a quadrat, and a sample point is determined by grasping a knot, pulling the string taut and pressing the knot to the substratum. The size of the device will vary depending on quadrat size. In a 1 m^2 quadrat, Schiel (1984) used a 1.4 m long bar and 1.8 m string. Five knots were tied in the string and 10 knots were sampled in each quadrat. A variation is the point quadrat bar method, which entails tossing a notched bar into the quadrat, the notches being treated as points. Leonard and Clark (1993) found that more taxa were detected using this method than in video transects. Some bias may be experienced with the knotted-string method in dense kelp forest, as the string tends to pull around stipes, thus increasing the likelihood that the sample points will end up on holdfasts. Tangles can also slow down the process of measuring percentage cover.

Methods 1 and 2 above are generally used where there is a minimal vertical component with the organisms involved (e.g. encrusting organisms: Battershill 1986). It is extremely difficult, however, to press a gridded quadrat or perspex board close to the substratum in a dense forest of algae. Alternatively, a point-contact bar is easy to use in this situation to assess percentage cover. Underwood et al. (1991) used method 4; they found it easy to use in all habitats, including *Ecklonia* forest.

The number of points sampled will depend on the precision required. For example, having only 10 points will be very insensitive to organisms that have a percentage cover of 10% or less. A pilot study should be conducted to determine the number of points required, especially if precise estimates of percentage cover are to be obtained for rare species. Binomial sampling theory may aid in the choice of the numbers of dots to be tested in the pilot study. For example, where an organism makes up 10% of the population, 30 samples (dots) would be needed for a 95% chance of detecting the target organism (see 3.3.2). Data on percentage cover may be augmented with a list of rare species that were also observed in quadrats.

In many subtidal habitats, investigators are faced with vertical structure within a quadrat, as well as horizontal variation in distributions. For example, in a forest of algae there is an algal canopy, beneath which is a subcanopy flora and fauna. Schiel (1984) used a method that overcame this problem in kelp forest (*Ecklonia*): he recorded the presence of all attached species that intersected an imaginary line from the substratum below the knot to the highest algal canopy. In this way, percentage cover of canopy and subcanopy organisms can be determined. All personnel taking part in the study must be made familiar with the procedure (see section 2.10). Alternatively, sampling can just be stratified according to different levels of the canopy (e.g. primary canopy of *Macrocytis* and subcanopy of encrusting or foliose algae). In some cases, it may be appropriate to have different sample-unit sizes for different levels and assemblages of organisms.

An alternative to point-count methods is to trace the outline of organisms on to acetate sheets with a Chinagraph pencil. Sheets can be taken back to the laboratory for detailed measurements of percentage cover; graphics tablets are useful for this procedure. The method would be uneconomical for broad-scale studies, and its use is probably limited to detailed investigations of small study areas with minimal vertical relief. Traces would be particularly useful where permanent quadrats are being monitored (Polderman 1980; Battershill & Bergquist 1990). Photographs of quadrats may provide a quick alternative, but some ground-truthing of the accuracy of species identification would be required (Plate 15).

4.4 Physical measurements

4.4.1 Topography and substratum type

Many studies require information on topographic complexity, substratum type, and the organisms on the substratum. For example, you may want to relate variation in the bottom topography of an area with the abundances of fish (Leum & Choat 1980; Jones 1988), or gain precise estimates of total abundance of organisms where sampling is stratified according to the proportional representation of different types of substrata. Information on topography and substratum type may be required at spatial scales ranging from millimetres to kilometres Methods for assessing the substratum have been explored in detail on coral-reef environments

(UNESCO 1978, 1984; Dodge et al. 1982; English et al. 1994; McCormick 1994b), and are briefly mentioned here.

(a) *Variation in bottom topography*
There are two components to measuring reef topography: the frequency and amplitude of corrugation, and the degree of angulation (McCormick 1994b). The 'rope-and-chain' method has often been used to measure topography, as described by Luckhurst and Luckhurst (1978). A rope and chain, or two tapes, are used: the chain (or a tape) is moulded over the bottom contours and the rope (or other tape) is pulled straight to determine a linear distance. The ratio of surface distance to linear distance is expressed as a measure of rugosity. Replicated estimates should be taken within each study site (Leum & Choat 1980). The length of tapes used will depend on the scale of topography the investigator is concerned with. For example, Leum and Choat (1980) used 10 m tapes, and Kingsford (1980) and MacDiarmid (1981) used 25 m tapes. Topographic complexity, as for many other ecological patterns (Wiens 1989; Cullinan & Thomas 1992), may vary with spatial scale and you may consider using tapes of different lengths if you want to describe a hierarchy of variation in topographic complexity. A problem with this technique is that it does not describe the actual topography (e.g. small holes versus gutters). The diver can, however, note this as a supplement to the technique.

McCormick (1994b) concluded that the profile-gauge method was a rapid method of quantifying surface topography on a coral reef. He used it in a study that compared the abundances of fishes with measures of cover of benthic organisms and topography. The profile gauge has multiple calibrated rods that can be pushed onto the substratum, and the depth to which rods can be pushed is measured in relation to a horizontal frame with a level (Fig. 4.10B, page 104). Replicate 3 m transects took approximately five minutes to complete. He compared six methods of quantifying surface topography to determine which was best for differentiating among types of reef architecture. The Consecutive Substratum Height Difference (CSHD) and two other measures were best for differentiating between schematic profiles on the reef. The CSHD was the summation of the squared differences between heights of consecutive bars used. The data were then square-root transformed to linearise the index. McCormick acknowledged that topography measures are dependent on the scale of measurement and urged investigators to measure the substratum variable at the scale that is important to the organism of interest (i.e. the profile gauge can be used in transects of different sizes). It is

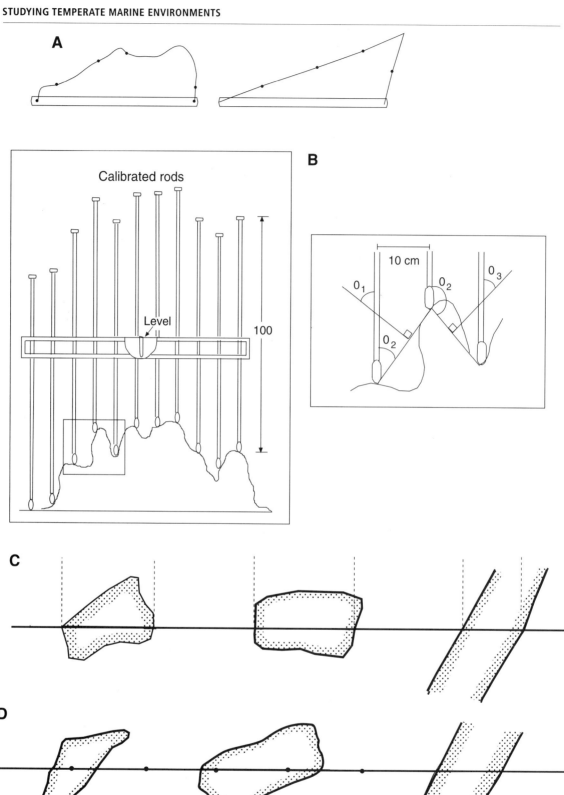

likely, however, that the profile gauge would be inadequate for measuring large-scale changes in topography (e.g. small valleys, peaks and drop-offs that have differences in height in the order of metres and are separated by tens of metres). McCormick did note, however, as did Wilkins and Myers (1992), that combinations of habitat descriptors may be required for some ecological questions.

Stereophotography should also be considered as a method for the study of surface topography, although this would probably be most useful for measurements of topography on small spatial scales (centimetres to tens of metres).

(b) *Quantitative estimates of substratum type*

Intersected-length method (ILM): A transect is laid over the reef area being studied and the intersected length is then recorded to the nearest millimetre (Dodge *et al.* 1982; Fig. 4.10C). The method assumes that the total length of substratum intersected on the transect length equals the proportional cover (= Line Intercept Transect of English et al. 1994). The length of a particular substrata type is often expressed as a percentage of total length. Multiple transects should be done for replication and to provide a measure of variance of percentage cover (see section 3.3 for determining sample-unit number). The appropriate length of transect will vary according to the ecological problem. English *et al.* (1994) recommended replicate 20 m tapes (see this source for other references) and listed some of the following advantages and disadvantages. This is a smaller-scale version of the profile method (4.3.1d).

Advantages
- ✔ If life-form categories are used (e.g. coral massive; see also 4.5.1), then persons with limited experience may collect data.
- ✔ The method is reliable for quantitative information on percentage cover.
- ✔ Little equipment is required.
- ✔ Can be augmented with photo-quadrat or video-quadrat methods.

Figure 4.10 Examples of methods and devices used to gain estimates of percentage cover of substratum type and the organisms on it and measures of topographic complexity.

A: Point quadrat bar method: organisms or substratum type are recorded under each knot as the string is pulled taut while the bar is held.
B: Profile-gauge method, a device that is used to quantify surface topography (adapted from McCormick 1994b, see also section 4.4).
C: The intersected-length method of measuring the cover of different substratum types.
D: The point method of measuring percentage cover.

Disadvantages
- ✘ Difficult to standardise some life forms (though some investigators may be able to use the technique to finer taxonomic distinction).
- ✘ Questions must be limited to percentage cover.
- ✘ The data are of little or no use for measures of growth or mortality.

It is also possible for this type of information to be extracted from video tapes, where sample units can be frame-grabbed from the tape for analysis using image-analysis software.

Point method: Random or regularly spaced points are arranged along a transect line. The type of substratum falling beneath each point is characterised (Fig. 4.10D). The method assumes that the number of points/total points equals the proportional cover. A pilot study may be needed to determine the appropriate number of dots to get the accuracy required (see section 3.3).

Advantages
- ✔ Faster than ILM.
- ✔ Others as for ILM.

Disadvantages
- ✘ Likely to be less accurate than ILM.
- ✘ Other problems as for ILM.

Other methods are available that are pertinent to coral reefs rather than temperate rocky reefs; these include the quarter-point method (Dodge et al. 1982; UNESCO 1978, 1984) and Choat and Bellwood's (1985) method for assessing substratum identity, percentage cover, height and degree of shrubbiness.

(c) *Sediments*

Sediments can have a great influence on reef-associated organisms (Battershill & Bergquist 1990; Floderus & Pihl 1990; Culotta 1994). Even modest changes in rates of sedimentation can affect suspension feeders, and a cover of sediment over a reef may inhibit the recruitment and growth of algae and other benthic organisms, as well as affect the ability of herbivores to feed. Sediment traps can be used to compare sedimentation rates among sampling sites (Bloesch & Burns 1980; Asper 1987). English et al. (1994) describe and illustrate a trap made of an 11.5 cm long PVC tube with an internal diameter of 5 cm. Baffles (grid) placed inside the tubes reduce the chances of sediment being resuspended and the likelihood of organisms resuspending sediments. Arrays of these traps can be attached to stacks and deployed in the study area. If formalin reservoir is added to the base of the trap, differences between biogenic and sediment fallout can be calculated (Bloesch & Burns 1980). It is sometimes possible to gain historical records of sediment transgressions onto reefs from aerial photographs, and aerial photography can be continued during the project. Neil Andrew (pers. comm.) has used this method to document sediment intrusions onto reefs along the coast of southern New South Wales. These events can result in the destruction of habitats and cause substantial changes to the percentage representation of organisms on reefs.

Sediment depth can be measured, and this may be done with random and replicated (e.g. n = 10) depth recordings per quadrat, using a ruler or calibrated probe. Where deep soft sediments overlie more compacted material, a metre-long probe with a blunt end can be rammed into the sediment. Where it is important to quantify sediment structure, grain size may be quantified using a series of sieves (Buller & McManus 1979); advice from a sedimentologist would also help (see Chapter 8).

4.4.2 Sea water

In many studies it is relevant to collect information on physical and biological attributes of the water column. A broad range of cheap data loggers are now available that can be used for measuring seawater temperature with great accuracy and at multiple depths and locations (e.g. HOBO probes: Doherty et al. 1996). Data loggers can also be used to measure salinity and turbidity, but they are generally more expensive than temperature probes (see Chapter 9). Turbidity can also be measured with a Secchi disk if data loggers are not available, but the data will be discrete rather than continuous and of less quality where there are great changes in light conditions. Changes in the supply of nutrients to reefs can influence the growth and survivorship of algae and in some cases (e.g. South Africa and California) a lack of upwelled and nutrient-rich water can result in the widespread destruction of stands of algae. The nature of planktonic assemblages can also influence benthic organisms, by changes in the quality and quantity of food and sedimentation rates of live and dead cells to the substratum (see Chapter 9 for methods to measure nutrients and plankton).

More comprehensive descriptions of oceanography will require CTDs (conductivity, temperature and depth devices) and other equipment that can be used to obtain temperature, salinity and density profiles of the water column and hence provide models of local oceanography that can be related to the biology.

4.5 Categories of organisms

The purpose of this section is to outline categories of organisms that are useful to consider while recording data on slates (Table 4.1). Furthermore, these categories may be used to summarise results and provide broad categories for organisms before they can be properly identified from taxonomic references or by consulting experts (Chapter 10). Underwood et al. (1991) used slightly different categories of algae as follows: foliose, filamentous, encrusting and *Ecklonia*. Detailed species lists are available for some areas if required (e.g. Australia: Middleton Reef: Hutchins 1988; New South Wales, Millar & Kraft 1993, 1994; New Zealand: Poor Knights: Kelly 1983; Schiel 1984; Battershill 1986; Leigh: Stocker 1985; Pritchard et al. 1985; Kermadec Islands: Nelson & Adams 1984). If unknown benthic organisms are found, they should be placed in a plastic bag for further examination and recorded descriptively (e.g. red encrusting sponge). Recommendations for preservation and identification are given in Chapter 10.

4.5.1 Taxonomic units

'Functional Groups' or 'Operational Taxonomic Units' (= OTUs: *sensu* Hooper 1994; Steneck & Dethier 1994) describe groupings of key taxa for which identification to species within them is either difficult or irrelevant to the biological questions posed. For example, Steneck and Dethier (1994) argued that for the benthic algae assemblages they studied, relatively few species attributes are of overriding importance to the overall structure of algal assemblages. Species attributes are often shared among taxonomically different species (e.g. relative productivity). Fletcher (1987) used functional groups to refer to different types of grazers that were ecologically similar in terms of body shape, size and the way in which they consumed algae (e.g. limpets and turbinid gastropods). The results of manipulative experiments were presented for the different functional groups (see 4.7.6).

Functional groups can also allow for the rapid description of benthic assemblages by non-expert personnel in cases where the objective of a study is to gain general descriptions of habitat type only. The concept has been extended formally to descriptions of assemblages. We say 'formally' because the technique of pooling taxa has been common since the inception of subtidal marine research. The method is generally used in tropical ecosystems (English et al. 1994) because of the high species diversity of organisms (particularly corals) and the difficulty of identifying all species in the field. Functional groups have often been used in temperate environments, particularly for quantitative descriptions of habitats (Table 4.1). The approach can also be combined with 'ground-truthing' where the species compliment of each unit is described in detail (Ayling 1978; Schiel 1984; Battershill 1986; Battershill et al. 1992). It is also possible that a subset of species can be easily identified and used to test the generality of the group.

OTUs enable ready interpretation through procedures such as community score analysis (see Singleton 1992) which use canonical discriminant analyses and other similar analytical procedures. In addition to enabling rapid identification of major habitat distribution, particularly in heterogeneous areas, they also allow for preliminary interpretation of trends in species assemblage/morphological groupings with respect to other physical or biological factors. For example, Battershill (1986) found definite patterns in the presence of encrusting and turfing morphologies (e.g. encrusting algae, sponges and ascidians) with respect to depth and distance into caves and archways. Descriptions of this type can lead to the design of testable hypotheses that address questions about which biological or physical features are important in organising the observed structure of an assemblage.

4.6 Experimental methods

Experimental methods have given great insight into processes that influence the nature of habitats on subtidal reefs and the organisms in habitats. A general account of the philosophy, utility and pitfalls of experiments is given in Chapter 2. Moreover, a review of approaches to experimentation is given, including caging, removal/additions (by removing and adding organisms, natural substrata or objects), transplants and grafting. There are excellent examples of each of these types of experiments on subtidal reefs, where in some cases the ecological questions have focused on processes influencing the structure and persistence of habitats and how the nature of habitats influences associated organisms, while others have studied interactions between species of mobile and sessile organisms and intraspecific processes. In most cases, experiments have

Table 4.1 Categories of benthic organisms; adapted from multiple studies in north eastern New Zealand.
* = a functional group of organisms (also called OTUs, operational taxonomic units).
The groupings that are chosen depend on the biological question that is asked, the groupings given here are examples only.

Large brown algae*

Fucoids e.g. *Carpophyllum, Sargassum, Cystophora*

Laminarians e.g. *Ecklonia, Macrocystis*

Understorey algae*

Tufting algae*

Red algae e.g. *Asparagopsis, Gigartina*

Brown algae e.g. *Colpomenia, Leathesia, Zonaria, Padina*

Green algae e.g. *Caulerpa, Codium, Ulva*

Thin encrusting algae* e.g. *Lithothamnion, Hildenbrandia*

Mobile invertebrates

Gastropods* e.g. *Cellana, Haliotis, Trochus*

Chitons e.g. *Cryptoconchus*

Echinoids* e.g. *Evechinus, Centrostephanus, Heliocidaris*

Nudibranchs e.g. *Glossodoris, Dendrodoris*

Sessile fauna

Sponges*

Thin encrusting e.g. *Cliona, Microciona, Pronax*

Finger e.g. *Callyspongia, Raspailia, Axinella*

Massive e.g. *Ancorina, Polymastia, Aaptos, Geodia, Cinachyra*

Bryozoa

Thin encrusting e.g. *Steginoporella, Cellaria, Baenia*

Turfing e.g. *Bugula, Margaretta, Amathia, Costaticella*

Ascidians*

Thin encrusting e.g. *Aplidium, Didemnum, Lissoclinium, Leptoclinides*

Bulbous e.g. *Pseudodistoma, Podoclavella, Aplidium*

Solitary e.g. *Styela, Asterocarpa, Cnemidocarpa*

Hydroids

Turfing e.g. *Obelia, Theocarpus*

Bushy e.g. *Solandaria*

Anthozoa

Gorgonians e.g. *Primnoides*

Soft corals e.g. *Alcyonium*

Anemones e.g. *Actinia, Corynactis*

Hard coral e.g. *Flabellum*

Black coral e.g. *Antipathes*

Zoanthids e.g. *Zoanthus*

Worms

Sabellid

Tube builders e.g. *Pomatoceros, Galeolaria*

Brachiopoda e.g. *Terebratella*

Barnacles e.g. *Balanus, Chamaesipho*

been initiated based on hypotheses that have been developed from the literature and/or from comprehensive mensurative studies on spatial and temporal variation in abundance and other relevant variables.

Algae and herbivory

Algae are one of the main types of cover on subtidal reefs and important descriptors of habitat types. In many cases, therefore, it is appropriate to consider manipulating algae to measure responses (e.g. changes in recruitment rates, number, weight/biomass, growth) to disturbance, intra- and inter-specific interactions (Kennelly 1983) and the influence of grazers. For example, Kennelly (1987a) found that the timing of kelp removal had a great influence on subsequent colonisation of the area. In winter, areas cleared of *Ecklonia* were colonised by kelp, while in other seasons, turf grew and inhibited subsequent settlement by *Ecklonia* (see 4.7.6). Grazers (especially urchins and gastropods) have a great impact on algal assemblages, and this has been demonstrated by experimentally removing grazers or excluding them (with cages) from 'barrens' devoid of algae. This generally results in a rapid recolonisation of filamentous, turfing and/or canopy algae such as kelps to the area. (Examples of these experiments are given in sections 4.7, 5.3, 5.4.) In some cases, grazing pressure on algae may vary with topography and the nature of grazers. For example, Fletcher (1987) and Andrew and Jones (1990) showed that the urchin *Centrostephanus rodgersii* is crevice-bound during the day and grazes at night. Kelp forest in central New South Wales, is generally found on areas of reef with few crevices. By translocating boulders into kelp forest and barrens, Andrew (1993) demonstrated that urchins recruit into crevices and that grazing pressure was greatest in the vicinity of crevices. Details of the experimental design and controls for disturbance are given in 5.4.5.

The impact of grazers on algae (as determined by experiments in New Zealand), and the consequences of this grazing on the distribution of subtidal habitats, is reviewed in Andrew (1988) and Schiel (1990).

Settlement and recruitment

Settlement plates are sometimes used for experimental investigations into the settlement and recruitment of algae and invertebrates. Keough and Downes (1982) used 150 x 150 mm clay tiles (20 small pits were drilled into each tile, often used as refuge by invertebrates; also note 'suitable settlement sites' of Raimondi 1990) to study the early mortality of settling larvae. Cages were used to exclude fish from half of the tiles, and panels were placed at three experimental sites to allow for variation in patterns of settlement and to enable more general statements to be made about the patterns on

experimental plates. The entire experiment was done twice, giving further weight to the generality (or lack) of patterns. A caging effect was detected for *Tubulipora* (bryozoan), where fewer recruits were found in the presence of fish, but only on exposed surfaces of tiles. The authors concluded that the mortality of invertebrates could be high immediately after settlement, and that variation in recruitment might be due to a combination of planktonic events, active choice by larvae, and subsequent mortality (see also Keough 1984a). The size of experimental units can influence the outcome of experiments, and the use of experimental units of different sizes should be considered in experiments (e.g. Keough 1984b).

Kennelly (1983) demonstrated that some care should be taken to consider experimental artefacts. He found that thick settlement plates lead to an overestimation of initial algal growth compared with growth on natural patches in the presence (uncaged) and absence of grazers (caged). Moreover, thick plates can exclude some grazers. Kennelly also found that the presence of cages could influence assemblages of algae by attracting small snails. Other cage artefacts included edge effects, attracting fishes, increased sedimentation, and changes in water flow and light (Schmidt & Warner 1984; Stocker 1986; see also Fig. 2.8).

The presence of algae can influence the settlement of other organisms. For example, Battershill and Bergquist (1990) and Stocker and Bergquist (1987) demonstrated that the presence of coralline turf enhanced recruitment of sponges and ascidians (*Pseudistoma novaezelandiae*); they recorded lower recruitment rates in the absence of turf. Recruitment was five to seven times higher in areas with turf. Breitberg (1984) demonstrated that crustose coralline algae could inhibit the settlement of other sessile organisms. Boulders with an upper surface area of 15–74 cm^2 were used as experimental units. Some boulders used were completely covered with crustose coralline algae, and some were without (a scalpel was used to remove all organisms visible under 12 x magnification, and the rock was then scrubbed with a brush). Two to four boulders were attached to concrete stepping-stones (30.5 x 30.5 x 5 cm), some of which were caged to exclude grazers, while others were left uncaged to provide 'grazed' treatments. Breitberg found that total invertebrate recruitment was reduced even in the absence of grazers (i.e. in caged treatments). She argued that grazing resistant organisms such as crustose coralline algae may contribute to the persistence of coralline algae/sea urchin assemblages (= barrens).

Grazers can influence the settlement of encrusting invertebrates, but Stocker and Bergquist (1987) demonstrated experimentally that this is not always the case. Removal of the large gastropod *Cookia sulcata* and the echinoid *Evechinus chloroticus* was found to have no effect on recruitment of ascidians when compared with controls where the densities of grazers were not manipulated. Grazers can cause mechanical damage to benthic organisms, and therefore influence the composition assemblages that characterise a habitat. Some workers have damaged organisms experimentally to check their powers of survival and regeneration (e.g. sponges: Ayling 1983).

Larval supply
Differences in larval supply (e.g. Shanks 1983; Gaines & Roughgarden 1985) can influence spatial patterns of recruitment. A number of other factors that relate to larval choice of settlement sites and larval mortality can also influence the abundance of recruits. Most of these factors have been the subject of experimental investigations, including surface topography (see above), surface films (Todd & Keough 1994; Keough & Raimondi 1995, 1996), residents (e.g. Davis 1987), post-settlement mortality (see above) and the presence of conspecifics. Many benthic organisms are gregarious at settlement (e.g. ascidians: Stocker & Bergquist 1987; Davis 1989; polychaete worms: Pawlik 1986; urchins: Dayton & Tegner 1984) and some of the best evidence has come from barnacles where it has been demonstrated experimentally that a combination of surface texture and the presence of conspecifics influences patterns of recruitment (Raimondi 1990).

Competitive interactions
Many benthic organisms exhibit swift cellular and chemical responses to protect their integrity by rejecting encroaching allogenic ('non-self') cells and for repairing damage. This is particularly true for modular animals such as sponges. Reciprocal grafting techniques are often used to investigate inter- and intra specific interactions, in that sponges may accept or reject transplanted tissues. For example, Smith and Hildemann (1986) compared responses by *Callyspongia diffusa* to 'allografts' (non-self) and 'autografts' (self, generally used as a control for the graft procedure and ability to respond to damage). Small fingers of sponge (3–5 cm) were cut for grafts and allowed to heal for 28–48 hours before they were used in experiments.

There were three experimental treatments: allografts, autografts and injuries induced on animals to check responses. Responses to grafts were checked in multiple time periods from 0.5 to 48 hours and were studied using histology. In the case of allografts, the cells around the graft rapidly became necrotic and the allogenic tissue was released. Autogenic grafts initially showed signs of rejection, but eventually the tissues of the graft and sponge merged. Damaged tissue healed quickly and no scars remained. Experiments of this type have been done to induce chemical responses where the chemicals are of great interest to scientists for their potential medical applications and antifouling properties (Battershill 1990; Blunt et al. 1990; see also 10.4.1).

Competitive interactions can influence patterns of distribution, abundance, survivorship and growth of organisms. Organisms compete for resources, and to demonstrate competition it must be shown that a resource is in limited supply. Food and space are two potentially limited resources (Butler 1995). Experiments are essential to understand the consequences of changing resources on interactions between species. Studies have included competition for space by sessile invertebrates. For example, work by Butler, Kay, Keough and Butler on pier pilings has demonstrated many competitive interactions with the use of experimental clearances. Butler (1995), however, concluded that single species are unlikely to dominate in this environment because many interactions are stand-offs and they lose as many as they win. See also Andrew (1989) for a discussion of the consequences of food limitations for gastropod assemblages, and Underwood (1986) for recommendations on the design of competition experiments.

Impacts
Humans use reefs for recreation and harvesting (e.g. Kingsford et al. 1991). The exclusion of humans or the restriction of their activities (e.g. harvesting) has been a successful way of determining impacts on intertidal (e.g. Castilla & Bustamante 1989) and subtidal assemblages (Ballantine 1991). MPAs, with good policing, are essentially experimental exclusions of fishers (see Chapters 1, 5 and 6).

Trophic interactions
For interactions between fishes and the habitats and the impact of fishes on benthic assemblages, see Chapter 6.

4.7 Examples of sampling designs

General perspectives and analytical aspects of sampling designs are given in Chapters 2 and 3. This section provides examples of sampling designs that have been used to describe subtidal habitats and the benthic organisms within them as well as factors causing changes in habitats (by experimentation). The purpose of each of these studies was different and, consequently, different sampling designs were used. We have used the term 'purpose' in a general sense, as some examples test more than one hypothesis.

4.7.1 One-off baseline study of the Chatham Islands

Example: The structure of subtidal algal and invertebrate assemblages at the Chatham Islands, New Zealand (Schiel et al. 1995)

Purpose: To obtain baseline information on rocky subtidal assemblages in terms of the benthic organisms and aspect of the reef, and to aid in the management of the abalone fishery and inshore areas.

Situation: No background information was available on the distribution patterns of benthic organisms, and detail on marine charts was poor. The amount of fieldwork that could be completed was determined by logistical constraints of time and weather; four divers involved.

Approach: Get to as many localities as possible (based on weather), encompassing a wide range of exposures (e.g. bays and points on different sides of the island). Benthic profiles (by use of the depth-stratified method, see 4.3.2c) were completed at nine sites encompassing sheltered harbours and exposed points. Although reefs extended to depths of over 20 m at some study sites, deepest sampling levels were 20 m because of the physiological limits of divers. With good visibility, however, casual observations of the subtidal area past 20 m supplemented the quantitative information. The following patterns of abundance of conspicuous benthic organisms were found. The immediate subtidal was occupied by the bull kelp *Durvillea* spp. (see Plates 34, 35). A suite of 11 species of fucoid algae were found to a depth of 10 m, while another fucoid, *Carpophyllum flexuosum*, was mostly found in deep water. *Macrocystis pyrifera* was abundant at 2.5–15 m at sheltered sites, while the only other laminarian, the endemic *Lessonia tholiformes,* was abundant at 2.5–15 m. The commercially valuable abalone *Haliotis iris* was very abundant in shallow water. Although the sea urchin *Evechinus chloroticus* (which is responsible for urchin-grazed barrens in northern New Zealand, Plates 6 and 7) was common, no urchin-dominated barrens were observed (this is typical for latitudes south of 40° S).

For information on the generality of depth-related patterns of distribution of benthic organisms in New Zealand, see Schiel (1990).

4.7.2 Detailed survey map of a stretch of coastline

Example: Survey of the Leigh marine reserve (Ayling 1978).

Purpose: To obtain detailed baseline information on the distribution of subtidal habitats. This would be useful for investigations to focus on the organisms in particular habitats and for precise estimates of total numbers of organisms in the area (e.g. fish, Chapter 6).

Situation: No background information was available on the distribution of benthic organisms. Numerous volunteer divers were available, access to the study areawas easy, and considerable logistical support and time was available.

Approach: Completed a large number of depth profiles in the study area for drawing habitat isobars. The reef-profile method (see 4.3.1d) was used and supplemented with spot dives and aerial photographs. Quantitative information on the densities of benthic organisms was also obtained within subtidal habitats (e.g. large brown algae and mobile invertebrates). Part of the completed map is shown in Figure 4.8. Investigators should show some caution with the use of maps, particularly in places where there is great temporal change in the distribution of habitats.

Other large-scale subtidal maps from New Zealand include Mokohinau Islands (W. Ballantine and A. Ayling with RNZAF) and Mimiwhangata Peninsula (R. Grace), Kapiti Island (Battershill unpubl. data). In Spain, an example is the Isla de Tabarca MPA (Ramos 1985).

4.7.3 Survey of a 750 km stretch of coastline

Example: Survey of the coast of New South Wales (Underwood et al. 1991).

Purpose: Provide a description of subtidal habitat types along a 750 km stretch of coastline.

Situation: Some preliminary information was available on appropriate sample unit size and numbers from work done by Andrew and Underwood (1989). Unpublished data by Underwood, Kennelly and Scanes indicated that four or more profiles gave accurate estimates of habitat cover.

Approach: Six locations (tens to hundreds of kilometres apart) were sampled over 750 km of coastline, and within each location two sites were sampled that were separated by 0.2–2 km. The entire sampling design was done in summer and winter in case algal abundance, and therefore the nature of habitats, changed between times. All of the work was done from a small boat and required three divers in the water. A 100 m tape was used for profiles of the reef. At each site a diver swam the tape out four times, where each profile was separated by 20–30 m. The proportion of the tape that lay over each habitat type was measured and the depth at which habitat transition occurred was recorded (reef profile, see 4.3.1d). In some cases the reef extended more than 100 m, and beyond feasible diving depths. Hence the description of habitats was limited to the first 100 m of reef rather than the entire reef profile.

Detailed information on each habitat was obtained using stratified sampling (see 4.3.2d). Five replicate 1 m^2 quadrats were done in each habitat at each site. Number of macroalgae (e.g. *Ecklonia* and *Phyllospora*) and large mobile or sedentary invertebrates (e.g. *Pyura* and *Cellana*) were measured. Percentage cover was measured using method 4 in 4.2.4. The grid (25 x 25 cm), with 10 random dots, was placed haphazardly within the 1 m^2 quadrat so that percentage cover was estimated from a total of 30 points. This number of points gave a 95% chance of detecting organisms that are present at the 10% level or greater (see 3.3.2). Algae were characterised as macroalgae (e.g. *Ecklonia*), foliose (e.g. *Pterocladia* and *Lobophora*), encrusting (e.g. 'coralline paint') and filamentous (identified by colour). The density of very small invertebrates (e.g. *Patelloida mufria*) was estimated within each 25 x 25 cm grid quadrat within each 1 m^2 quadrat. The densities of these small limpets were too high to estimate in the large quadrat. Large invertebrates (e.g. *Centrostephanus rodgersii*, *Cabestana* spp.) were counted in five replicate 1 x 5 m transects within each habitat.

Major differences in the representation of habitat types were found in the study area. At the three southern locations, major habitats were an algal 'fringe', urchin-grazed 'barrens', *Ecklonia* forest, and deep reef dominated by sponges. Although the fringe and deep reef were found in shallow and deep water, *Ecklonia* forest and barrens were found as a mosaic over a range of depths (Plates 15–17). At the two northern locations, the substratum was dominated by high densities of ascidians (*Pyura* habitat). The study was successful in identifying major geographical differences in the nature of subtidal reefs. This type of information may be useful for justifying regional areas for the choice of representative reserves (see section 2.8) and to generate hypotheses for testing the influence of habitat types on the distribution of other organisms (e.g. fish).

Potential improvements to the design for more detail on broad transitions in the representation of habitat types could include a greater number of loca-tions, and sites within locations, but this would greatly increase the time taken to complete the design. In this study, diving was weather-dependent, and it took 1–1.5 months to sample all locations. Aerial photography with appropriate ground-truthing could also help greatly to determine the generality of patterns of habitat distribution along the coast. Obviously, the utility of aerial photography will depend on visibility and the depth of habitats. It would be possible to augment information on deeper habitats by using rapid visual techniques (see 4.3.1c). This would be useful if there were greater focus on organisms such as sponges that are found in deeper waters.

4.7.4 Detailed investigation of marine benthos of caves, archways and vertical reef walls

Example: Study of the Poor Knights Islands (Battershill 1986).

Purpose: To quantify the distribution and abundance patterns of the benthic flora and fauna, especially in unique habitats, caves and archways.

Situation: Preliminary information collected by Ayling (1978) and Schiel (1984) helped in the development of a suitable sampling design. Because the survey was located offshore, considerable logistical support was required for the study; one or two divers carried out the work on each trip.

Approach: Preliminary information from casual observations and photographic quadrats indicated that sponge communities varied according to levels of light and exposure. Accordingly, these factors were catered for in the detailed study. Figure 4.11 (p. 112) shows the design used by Battershill (1986) to sample caves. Positions on the wall (10 x 5 m) were subsampled at two depths (5–10 m and 15–20 m) at three different distances (= penetration) into the cave. The whole procedure was carried out at each side of the cave (east and west). Within the wall positions, six replicate 0.25 m^2 quadrats were sampled. The difficulty of holding a quadrat on to a vertical face was overcome by clipping the quadrats on to any protruding organism on the wall. A similar procedure to the above was adopted for the sampling of archways and open walls.

Major differences were found in the species composition and size-frequency distributions of sessile benthic organisms. Battershill related these to the intensity of light and relative exposure of the sites. This sort of variation in distribution is of particular interest to recreational divers as well as scientists. Sponges and ascidians were previously regarded only in terms of their aesthetic

111

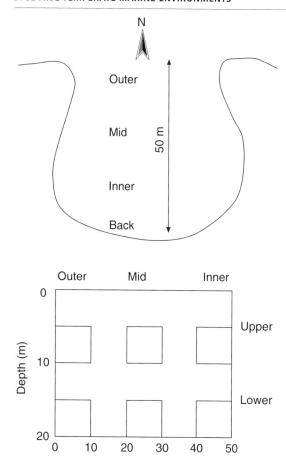

Figure 4.11 Sampling design by Battershill (1986). The investigator's question was: does the abundance, percentage cover and patch size of sessile benthic invertebrates vary with distance into a cave and with depth? The sampling design, therefore, had orthogonal factors (cave penetration and depth). Random sampling was done within each of the squares indicated. Multiple other caves of similar geomophology were also studied for generality of patterns.

attributes, and some were useful as bath sponges. Recent research, however, has demonstrated that compounds from these sessile organisms show great promise for antitumoral and antiviral properties (Munro et al. 1993). Hence these animals are worthy of careful management with respect to harvesting and further investigation into the ecological context of the production of compounds (e.g. the presence of encrusting organisms; see also bioprospecting in Chapter 10).

Although considerable amounts of time, money and taxonomic expertise were required for this study, information can be obtained on these unusual or isolated habitats by other methods. Photographic quadrats can be completed quickly and analysed in the laboratory. The significance of the habitats may then be brought to the attention of scientists and managers by use of photographs.

Also note McCormick's (1990) investigation of algae (*Pterocladia*) with a very patchy distribution. He used a DPV (see 4.3.1c) to search large areas of coastline and identify those with significant stands of algac. The second phase of the work was a detailed examination of algae in areas where they were abundant.

4.7.5 Experimental study on kelp and algal turf

Example: Physical disturbances in an Australian kelp community (Kennelly 1987a).

Purpose: Determine the timing, frequency and severity of natural disturbances that lead to the removal of canopies in a particular kelp forest. Use manipulative experiments to determine the consequences of disturbance on the structure of the assemblage.

Situation: Pilot studies on sample-unit sizes had been done in previous studies by Kennelly. He also had a good working knowledge of the study area (i.e. location of kelp forest) from his previous work and had made casual observations of the physical effects of storm on kelp cover.

Approach: Information on wave height was obtained from waverider buoys. Counts of kelp were made every two months to determine frequency of disturbance. Sampling involved placing three 20 m transects randomly in kelp using a tape. Each metre along the tape, adult (> 150 mm) and juvenile kelp (< 150 mm) plants were counted within a 1 m^2 quadrat. Also recorded were the numbers of kelp plants that lacked laminae and those lacking both laminae and stipes. The frequency of gaps was determined by recording the sizes of openings in the canopy (> 0.5 m^2) that occurred along each transect during sampling. The amount of light reaching the understorey was estimated using a photometer. Experimental clearances were made at four sites (2 x 2 m) and four were left uncleared in each season. Only the holdfasts of plants were left where kelp was cleared.

Experimental treatment were sampled using two techniques. Macroscopic benthic organisms and cover of sediment were sampled using a 30 x 30 cm, 100-point quadrat to record percentage cover. Microscopic organisms were also counted using an underwater microscope. All data were recorded underwater, on a tape recorder. Kennelly found that kelp plants were only removed in significant numbers during very rough weather, and natural clearances that ranged in size from 0.5–60 m^2 were found. New kelp recruits appeared in August of each year. Encrusting algae decreased in abundance regardless of the season in which kelp was removed. There was a rapid increase in the number of kelp holdfasts in clearances that were made in winter, but in other seasons there was a rapid increase in the cover of tufting algae. Species richness of algae was found to be less in cleared than in uncleared sites. As Kennelly acknowledged, ideally the experiment would have been done in multiple years to determine the generality of the patterns found. It is common for algal turf to inhibit the growth of kelp (e.g. Dayton et al. 1984; Reed & Foster 1984), and it is only during the recruitment period of kelp (winter) that this is unlikely to happen. Kennelly (1987b) concluded that even partial damage to kelp canopies will lead to similar effects on benthic species as for total clearances. Where kelp was experimentally thinned (cf. clearance), the removal of at least 50% of the canopy was required before the effect on benthic assemblages was noticeable.

4.7.6 The influence of gastropods and urchins on the structure of subtidal habitats

Example: Interactions among subtidal Australian sea urchins, gastropods and algae: effects of experimental removals (Fletcher 1987)

Purpose: The aims of the study were to examine spatial relationships and the influence of the sea urchin *Centrostephanus rodgersii*, three species of limpets and two turbinid gastropods on an assemblage of subtidal algae on the eastern coast of Australia. Experimental removals were used to test hypotheses that sea urchins were responsible for the maintenance of areas of crustose coralline algae (barrens), thereby facilitating the survival of limpets. The generality of any results was examined by repeating the manipulations at two sites and two times.

Situation: Pilot studies on sample-unit sizes had been done in previous studies by Fletcher. He also had a good working knowledge of the study area (i.e. location of urchin-grazed barrens and boulders) and had observed that grazers appeared to aggregate in the vicinity of crevices.

Approach: Grazers were sampled in 259 quadrats (0.25 m²) throughout the study area to compare relationships between the density of each type of grazer with other grazers and algal cover. The influence of crevices on patterns of abundance was also examined. Ten areas were chosen where crevices were at least 10 m from other crevices. From a crevice in each area ten 50 x 50 cm quadrats were sampled contiguously. The density of grazers and percentage cover of algae was

recorded in each. The results of this descriptive work showed that there were positive relationships between the density of urchins and that of gastropods. There were also significant positive relationships between the cover of coralline algae and the abundance of grazers of all categories. With increasing distance from crevices, the percentage cover of encrusting algae decreased as the cover of foliose algae increased. The density of gastropods decreased with increase in distance from crevices.

A series of experiments was designed to determine interactions among grazers and the consequences of grazing for cover of algae; here we discuss one of them in detail. Experimental units were stable boulders (2–4 m²) that were originally covered with crustose coralline algae, with a large number of gastropods present on the surface and urchins sheltering around the base. Manipulations involved removals and units were monitored weekly to remove intruding animals. Manipulations were maintained at 10% or less of grazers in nonremoval boulders. Urchins, limpets and turbinid gastropods were considered as functional groups, and thus eight combinations of presence or absence of each group was required (e.g. + urchins – limpets; – urchins + limpets). There were two replicate boulders for each of the eight treatments (= 16 boulders), and this arrangement was repeated in both shallow (1–3 m) and deep (7–10 m) water at two sites about 500 m apart, so giving the investigator a greater ability to make general statements about the results. Moreover, some of the treatments were repeated a second time, giving the author greater scope to discuss generality of the outcome. Percentage cover of algae was sampled in five 0.25 m² quadrats on each boulder every 3–5 weeks for up to 24 months. Turbinids were found to have little impact on algae, and after six months these treatments were discontinued. Cover of algae increased quickly in the absence of urchins and limpets (80–100% after 12 months). In areas where only urchins were removed, this recovery was slower and took 18–24 months. Where only limpets were removed, cover of algae was only slightly more than in controls (i.e. + all grazers). Urchins, therefore, can essentially maintain the characteristics of barren habitat in the absence of limpets, but the converse is not true. Other experiments found that the recruitment of some limpets was initially higher in the absence of urchins, and the recruitment of small limpets was higher in the absence of large limpets. Numbers of all limpets, however, tended to zero as the cover of tufting algae increased. The persistence of limpets in barrens, therefore, depended on the presence of urchins. Importantly, the results of experiments were similar at all sites and times.

See also Chapter 5, especially example 5.4.5.

SELECTED READING

The references listed are examples of research that will lead the reader to a broader literature, also see the text of the chapter. * indicates a multi-authored book or collection of proceedings; references to volumes of a journal refer to a set of relevant papers.

Algae: Shepherd 1981; Schiel & Foster 1986; Clayton & King 1990; Santelices 1990; Lobban & Harrision 1996; see also Ecology and Taxonomy.

Aquaculture: Barthel & Theede 1986; Adams et al. 1995, Duckworth et al. 1997.

Black coral: Grange & Singleton 1988; Parker et al. 1997.

Gastropods: Underwood 1979; Chapters 5 and 7; see also Taxonomy.

Ecology
General: Keough et al. 1990*; Underwood & Chapman 1995*; see also Chapter 5.
Chemical: Blunt et al. 1990; Hay 1991, 1995; Steinberg 1994; Cronan & Hay 1996; Hay 1996; Miller & Hay 1996.
Disturbance: McClintock 1987; Connell & Keough 1985; Keough et al. 1990*.
Competition: Underwood 1986; Butler 1995;
Algae and grazers: Foster & Schiel 1985; Jernakoff 1985; Schiel 1985a, 1988, 1990; Trenery 1986; Schiel & Foster 1986; Johnson & Mann 1988; Andrew 1988; Reed et al. 1988, 1992; Hay 1990, 1991; Underwood & Kennelly 1990b; Keats 1991; Leonard & Clarke 1993 (Chapter 4); Bruhn & Gerard 1996; Hay 1997.
Encrusting organisms: Ayling 1980, 1983; Svane & Lundalv 1982; Battershill & Bergquist 1990; Butler 1995; Davis 1996
Predator-prey relationships: Elner & Vadas 1990; Estes & Duggins 1995; see also Chapters 5 and 6.

Epiphytal invertebrates: Edgar 1983a, b, 1991; Coull 1985; Edgar & Moore 1986; Jacoby & Greenwood 1988; Gallahar & Kingsford 1993; Taylor & Cole 1994; Chapter 9, see also demersal zooplankton.

Fisheries and harvesting: Storr 1964; Hay & South 1979; Schiel & Nelson 1990; see also Chapter 5.

Habitat dynamics: Dayton 1985; Schiel & Foster 1986; Keough et al. 1990*; Underwood & Chapman 1994*; see also Chapter 5.

Larvae: Anderson et al. 1976; Lindquist & Hay 1996; see also Chapters 5 and 9.

Lobsters: See Chapter 5.

Marine protection and impacts: Cole et al. 1990; Siegfried 1994*; Reed et al. 1994; Reed & Lewis 1994; Underwood & Chapman1994*; Gubbay 1995*; Schmitt & Osenberg 1996; see also Chapters 1 and 2.

Methods: Dahl 1978; UNESCO 1978; 1984; Coyer & Whitman 1990; English et al. 1994; sampling designs, see Chapters 2 and 3.

Octopus: Cortez et al. 1995a, b.

Pollution: Clarke 1989; Phillips & Rainbow 1993; Reed & Lewis 1994; Chapman et al. 1995; Anderson & Kautsky 1996.

Settlement and recruitment of invertebrates: Tegner & Dayton 1977; Battershill & Bergquist 1990; Keough & Downes 1982; Olsen 1985; Keough et al. 1990*; Davis 1987, 1996; Svane & Young 1989; Young 1990; Johnson 1992*; Graham & Sebens 1996.

Surveys: Ayling 1978; Choat & Schiel 1982; Bell 1984; Battershill et al. 1985, 1992; Choat & Bellwood 1985; McCormick 1990; Underwood et al. 1991; Beckley & Branch 1992; Cole et al. 1992; Schiel et al. 1995.

Taxonomy: See Chapter 10.

Urchins: See Chapter 4.

FIVE

LARGE GASTROPODS, URCHINS AND CRUSTACEANS OF SUBTIDAL REEFS

M. J. Kingsford and A. B. MacDiarmid

5.1 Introduction

Large, mobile invertebrates (e.g. sea urchins, abalone and lobsters) are an important component of the fauna on subtidal rocky reefs. At all temperate latitudes in all parts of the world, large herbivorous and carnivorous invertebrates are thought to have important direct and indirect effects on the distribution and abundance of 'habitat formers' such as kelp (review: Elner & Vadas 1990). For example, in New Zealand (Andrew 1988) and Australia (Fletcher 1987; Andrew 1993), Chile (Dayton 1985) California (Schmitt in press) and Nova Scotia (Elner & Vadas 1990), the grazing activity of gastropods and urchins creates characteristic habitats, often called 'urchin-grazed barrens' (Underwood et al. 1991), that are largely devoid of macroalgae. Experimental removal of these animals often results in a rapid recolonisation of the area by algae. In the northwestern Atlantic, it was argued that the removal of predatory lobsters (*Homarus americanus*) resulted in a population explosion of herbivorous urchins that caused fundamental changes in the nature of subtidal habitat types, although the evidence for this is rather weak (Elner & Vadas 1990). Changes in the distribution and abundance of these organisms, therefore, can potentially cause a cascade of affects on reef assemblages. For these reasons many of these organisms have been coined 'habitat determiners' (Jones & Andrew 1993).

A number of large invertebrate species are heavily exploited by recreational and commercial fisheries, and are excellent candidates for assessing biological changes that occur as a result of protection. A prime objective of any study that sets out to assess the influence of protection, or other anthropogenic impact, should be to target organisms such as these for sampling (see also fish as target animals, Chapter 6). In addition to measuring the densities of target species, a description of the size-frequency distribution (by sex, if possible or appropriate) of the resident population is important. Because exploited invertebrates are easy to identify and are slow-moving, the use of a good sampling design and competent observers will enable very precise information on abundance and size to be obtained. Clearly, animals that are over the legal size limit are likely to be exploited in the greatest numbers and show the greatest changes in abundance in protected or unprotected areas. Measurements of size in one dimension may not be adequate for some estimates, such as biomass. For example, Worthington et al. (1995) found that slow-growing abalone were wider and heavier at a given shell length. Hence, in this chapter we give some attention to the utility of measurements of size, growth and methods for tagging invertebrates.

An understanding of the dynamics of populations of invertebrates is the focus of many studies, and this requires information on all aspects of life history (see Chapter 4). Most large and mobile invertebrates have a bipartite life cycle (e.g. lobsters, Fig. 5.1, p. 116) where larvae spend days to months in the plankton. (This is not exclusive; for example, the young of *Heliocidaris erythrogramma* are direct developers; that is, they have no planktonic pluteus stage: Williams & Anderson 1975; cf. Figs 5.2, 5.3.) Some gastropods also have direct development (Underwood 1979). Planktonic processes and methods for study are discussed in Chapter 9. An understanding of post-settlement processes that influence the dynamics of invertebrates requires information on population size and composition (size and age); inputs to populations, including recruitment and growth (need information on age and size); immigration; and outputs. Outputs include mortality (from old age, disease, predation, disease and harvesting) and emigration (see Fig. 1.3).

We provide information and literature on estimates

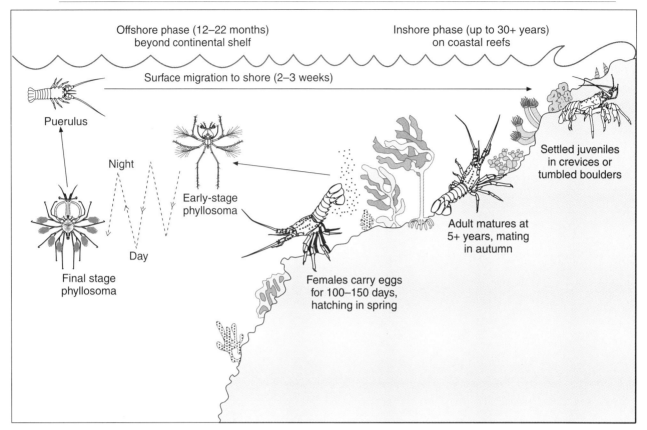

Offshore phase (12–22 months)
beyond continental shelf

Inshore phase (up to 30+ years)
on coastal reefs

Surface migration to shore (2–3 weeks)

Puerulus

Night

Day

Final stage
phyllosoma

Early-stage
phyllosoma

Females carry eggs
for 100–150 days,
hatching in spring

Adult matures at
5+ years, mating
in autumn

Settled juveniles
in crevices or
tumbled boulders

Figure 5.1 Life history of rock lobsters (*Jasus edwardsii*). A depth profile of a temperate reef is shown with adults and juveniles. Females walk to deep water (below 15 m) at the time of hatching of the eggs. Eggs hatch near dawn, and larvae commence a pelagic existence that lasts a year or more before the non-feeding puerulus migrate inshore to settle.

JOANNE MOORE AND PETER BENNETT

of abundance, age and size structure, and measurements that are needed to elucidate inputs and outputs of populations.

Some biological questions demand an understanding of the reproductive biology of large invertebrates, particularly issues or problems relevant to harvesting practices, population dynamics, phenology, the parts of a reef that merit protection, resilience to starvation, and interactions with other grazers (e.g. Andrew 1989; Shepherd et al. 1992b). Conclusions and methods from studies of reproduction in gastropods, urchins and lobsters are also given.

The removal of large invertebrates from an areas, or conversely their protection, can potentially cause a cascade of effects to the resident assemblage of organisms (e.g. Elner & Vadas 1990; Jones et al. 1993). If investigators are to determine the processes influencing interactions between organisms on a reef, then experiments are essential (see Chapter 2). We have therefore given

some emphasis to the utility of manipulative experiments.

Two general categories of invertebrates are discussed in this chapter: large gastropods and echinoids, especially abalone and urchins, all of which are important herbivores; and large crustaceans, with special reference to rock lobsters, that consume invertebrates, including herbivores, as prey.

It is often necessary to stratify sampling by habitat, or complete mensurative or manipulative experiments within the same habitat. For this reason Chapters 3 and 4 are essential companions to this chapter. Moreover, the separation of discussions to subtidal (Chapters 4, 5

OPPOSITE: **Figure 5.3** The early life-history stages of *Heliocidaris erythrogramma*, a species with lecithotrophic and direct development, which therefore does not have a planktonic pluteus stage. (Approximate time from fertilisation in cultures reared at 20° C is given.)

A: Wrinkled blastula (15 hours) 450–460 µm.
B: Gastrula (21 hours) approx. 480 µm.
C: Late larva, at the onset of metamorphosis, showing five primary podia at the vestibular opening and spines within the echinoid rudiment (4 d). The ciliated band around the larvae is not visible in this view.
D: Juvenile, oral view showing definitive spines. Those on the aboral side are bifurcated and lost later in development (6 d).

CATHERINE KING

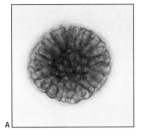

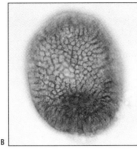

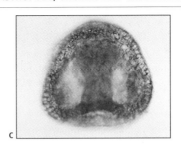

ABOVE and RIGHT: **Figure 5.2** The early life-history stages of *Centrostephanus rodgersii*, a temperate sea urchin with a planktotrophic development stage. (Approximate time from fertilisation in cultures reared at 20° C is given.)

A: Unhatched blastula with fertilisation membrane (12 hours).
B: Blastula (22 hours).
C: Gastrula with invagination and blastopore at the vegetal pole (40 hours) approx. 150 μm.
D: Pluteus larva with long oral arms and functional digestive system (70 hours) 200–400 μm.

CATHERINE KING

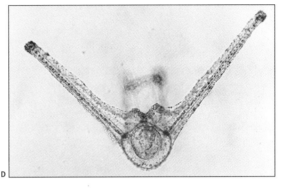

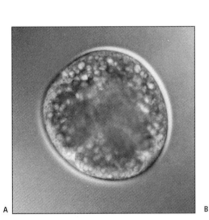

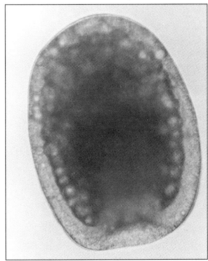

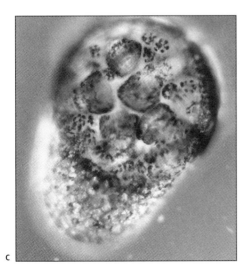

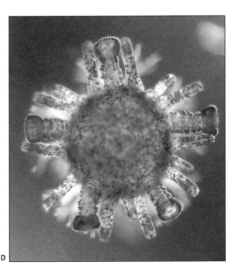

117

and 8) and intertidal (Chapter 7) parts of reefs has some biological relevance (i.e. many organisms spend their lives in one or the other), but note that some organisms may move between the two (e.g. urchins: Kerrigan 1987).

Many of the sources of variation in the abundance, behaviour and composition of species assemblages are similar in intertidal and subtidal environments, so approaches to sampling and concerns with respect to sampling designs are applicable to both environments (see Table 6.1).

5.2 Methods and considerations for sampling large mobile invertebrates

5.2.1 Planning

There is great variation in abundance of gastropods, urchins and large crustaceans in space (e.g. among habitat types, depths, sites separated by metres, kilometres and greater than 10 km) and time (e.g. days, months, years, more than 10 years, stochastic events such as disease (Hughes 1994) and storms). Thus it is often important to stratify sampling and/or nest sampling units according to different spatial and temporal scales. If this is not possible, or appropriate to the hypothesis of interest, investigators should clearly define the sampling area according to habitat type, depth, spatial extent and time of sampling. For a more extensive discussion of sources of variation in benthic assemblages of rocky reefs, see Chapter 4. An account of the principles and applications of stratified, nested and other designs is given in Chapter 3.

Carefully consider the nature of your sampling problem and relevant sources of variation before proceeding further (see Chapter 2).

Examples of problems that investigators may address:

- How are turbinid gastropods distributed with respect to depth and habitat type?
- What is the size of the abalone stock in latitudinal bands of 1 degree along a temperate coastline?
- What is the impact of a marine protected area on abundance and size-frequency of sea urchins?
- Is the size/age to maturity of sea urchins different between fished and protected areas?
- What is age to maturity and maximum age of abalone over a biogeographic range of 10 degrees of latitude?
- Are urchins habitat determiners; that is, do they change the distribution patterns of 'habitat formers' such as kelp plants?

- Do lobsters have an important impact on numbers of sea urchins?

Although multiple methods can be used to study large invertebrates, most studies restrict themselves to one or a small number (see examples in section 5.5).

5.2.2 Sedentary herbivorous invertebrates

Approaches to addressing questions on the distribution and abundance of mobile invertebrates have been described in Chapter 4 (p. 92). Information may be obtained on patterns of vertical distribution on a reef by the use of stratified sampling, with quadrats (or transects) as replicates sampled within depth or habitat strata. Any individuals found in these sample units can be measured to identify differences in size-frequency distribution. Small animals, for example, may only be found in shallow water.

Restricted distribution patterns may enable subsequent investigations of abundance to be carried out over a limited depth range. This knowledge may be particularly important when detailed comparisons of target animals are being made along broad stretches of coastline.

Distribution of sea urchins and abalone usually shows distinct variations in densities with depth (e.g. Ayling 1978; Andrew & Choat 1985; Jeffs 1986; McShane et al. 1993). Urchins are called 'habitat formers' (Jones & Andrew 1993) because their feeding activity can remove large swathes of algae from an area (Plates 6, 7, 15–18). Moreover, their presence may partly characterise a habitat type, usually from 4 m to 12 m, especially in northern New Zealand and central New South Wales ('rock flats': *sensu* Ayling 1978; or 'barrens': *sensu* Underwood et al. 1991). In southern New Zealand and southeastern Australia, urchin barrens often form part of a mosaic of habitats. In some areas individual barrens may only be a few square metres in extent (e.g. Andrew 1993; Andrew & Underwood 1993a; McShane et al. 1993; Schiel et al. 1995), while others may be tens to hundreds of square metres (Plate 18). Abalone are typically most abundant in shallow water (less than 10 m in New Zealand: Jeffs 1986; Schiel & Breen 1991; McShane et al. 1993), but some Australian species may be found at depths of up to 50 m.

There is no information to suggest that the depth distribution of abalone or urchins changes between seasons, and this stability of these patterns makes them ideal candidates for biological questions that require estimates of stock size (e.g. McShane 1994) or very precise measures of abundance through stratified sampling at different depths or habitats (see Chapters 3 and 4). If information on the size of animals obtained from stratified sampling with quadrats or transects is inadequate, it can be supplemented by a stratified rapid visual technique for additional measurements (e.g. McShane 1994, 1995).

Many grazing invertebrates are conspicuous, which makes it easier to address questions on dynamic and static distribution patterns; the positions of individuals may even be mapped (see 5.2.2, also Underwood 1977; Underwood & Chapman 1985; Andrew & Stocker 1986). Furthermore, some animals can be tagged so that their movements and interactions with other organisms can be recorded.

There are some practical difficulties in estimating sizes and abundances of urchins, particularly for cryptic urchins that are difficult to extract from crevices (e.g. *Centrostephanus rodgersii*). An urchin hook (a metre-long brass tube with an urchin-sized hooked end that is flattened to easily slip under animals) greatly speeds up visual estimates of abundance; use gloves for protection.

When planning a field study on urchins, or large gastropods, consider the following points:

(a) *Size of quadrat*
It is often necessary to determine the best sample unit size during preliminary sampling (see Chapters 2 and 3). In general, the lower the abundance and the greater the patchiness of the target species, the larger the sample unit size needs to be. If time is not available for adequate preliminary sampling of the target species, make a 'best guess' of sample unit size and replication based on local knowledge and published information. Quadrats measuring 1 m^2 have generally been used to study densities of sea urchins (*Evechinus*), abalone (*Haliotis*) and other large gastropods in New Zealand (e.g. *Cookia, Turbo, Patella kermadecensis*: McShane 1994; Cole et al. 1992; Schiel et al 1995). Jeffs (1986) used 4 m^2 quadrats to sample abalone but provided no evidence that they gave greater precision than 1 m^2 quadrats. Andrew and Underwood (1989) estimated abundances of Australian urchins and abalone in 10 x 1 m transects in areas of 'fringe' habitat. Each transect was started at a randomly selected point along a 30 m tape measure. Tapes were laid out in haphazard directions from their starting points. Fifteen 10 x 1 m transects were sampled in each locality. Underwood et al. (1991) used five replicate 1 x 5 m transects to estimate abundance patterns of *Centrostephanus rodgersii* in subtidal habitats on temperate reefs of New South Wales, Australia. This sample unit size gave useable means (mistakenly presented as number per m^2 in the paper) and standard errors that were generally 20–30% of the mean. Kingsford (unpub. data) sampled *Heliocidaris erythrogramma* in waters less than 2 m deep in New South Wales using 1 m^2 quadrats and measured average densities of 20 m^2; clearly a small sample unit is more appropriate for this species (cf. *Centrostephanus*). Small quadrat sizes (e.g. 0.25 m^2) have been demonstrated to provide more precise estimates of juvenile urchin abundance (see Chapter 4). A Venturi suction sampler may be the most efficient means of sampling very small urchins and abalone within a quadrat (McShane & Smith 1988).

In summary, experience has shown that, for example, for adult abalone and *Evechinus* in New Zealand and *Heliocidaris* in Australia, 1 m^2 quadrats are often an effective sampling unit. For adult *Centrostephanus* and abalone in Australia, however, transect sizes of 1 x 5 or 1 x 10m are adequate. Even longer transects may be appropriate for some discrete organisms. For example, Mapstone and Ayling (1993) concluded that 50 x 5 m transects were best for discrete organisms such as crown of thorns starfish (*Acanthaster plancii*) and poritid corals on the Great Barrier Reef.

(b) *Rapid visual techniques*
Shepherd (1985) and McShane (1994) argued that quadrats and transects are inadequate to estimate densities where the species sought is cryptic, in low average densities and aggregated in shallow, wave-swept habitats, as is often the case for abalone. They suggest that use of a free-swimming observer (see 3.2.2d) is the most efficient way of locating the maximum number of individuals in a given time. The difficulty of this method, however, is that estimates of absolute density cannot be readily derived from such data, unless the power and efficiency of the diver are known. Shepherd (1985) provides a formula that includes a large number of subjective components and, therefore, unknown variances. In view of this, the method is less useful for estimating absolute densities. However, free swimming in urchin or abalone 'habitats' is an effective way of gathering information on relative abundance, size-frequency distribution and patch frequency, especially if animals are clumped and present in low mean densities (McShane 1994, 1995). Moreover, the general advantages of rapid visual techniques make these ideal for locating suitable habitat or high densities of animals for detailed studies.

(c) *Measurements of size*
Check the literature to determine the most appropriate measure of size, especially if you want to compare your data to those collected previously. Is a measure of length or weight needed? Test diameter is often measured in urchins. For abalone, shell length is generally used as a measure of size in 'slow-growing' populations; however, some abalone are comparatively broad for their length so both shell width and length may need to be

measured for underwater estimates of biomass (e.g. Worthington et al. 1995).

Measurements are generally taken as follows:

- With vernier callipers (e.g. Jeffs 1986; Schiel et al. 1995).
- With a 'paua board', adapted from a fish measuring-board shape (S. H. Hooker pers. comm.).
- By the size-class method, using a 'paua board' with a long slate attached under a ruler divided into columns corresponding to 1 cm size-classes; the size of individuals is recorded in the appropriate class. The advantage of this method is that you have immediate size-class plots.
- Using a metal ruler.
- By visual estimate (the crudest method). A useful scale for estimates is a spread hand; for adults this measures about 180–200 mm in width (check with a ruler).
- By the Shepherd method (Shepherd 1985). This is adapted from a caliper design, with a gauge shaped to fit around the abalone. Along the length of the gauge (scale on callipers) is a plastic strip into which a needle is pushed to signify the lengths of numerous individuals. The strip can be rapidly changed by a diver under water. Measurements on the plastic can be read under low power on a binocular microscope back in the laboratory. An obvious problem with this method would be repeated needle holes on the same stretch of plastic, when a year-class of similar size was being measured.

(d) *Measures of age and growth*

Growth may be determined and age estimated by repeated measurement of discrete size cohorts or the same individuals (see cohort analysis, Table 3.11). The counting of increments in the shell or test can be used to age some species, and information of this type can be used to generate age/size relationships (e.g. Erasmus et al. 1994; Shepherd et al. 1995).

Usually the only way to uniquely identify individuals is to tag them, either externally or internally. Any effects of tagging on growth must be first determined before tagging is widely applied. External tag codes can be simply painted using fast-drying nail polish, or etched using hand-held engraving tools. Plastic or ceramic tags can also be attached using quick-drying adhesives, e.g. super glues. Tags intended for marking bees have also been used but, as with other tags, they are prone to removal through grazing by the urchins that sometimes feed on algae attached to the shells of some gastropods (Ettinger-Epstein 1996). Gastropods can also be tagged by drilling through the shell close to the opercular opening and attaching a spaghetti tag.

Internal tags such as microwire tags or microchip tags may be better suited to echinoderms and soft-bodied gastropods. These tags can be detected using a hand-held underwater 'wand' (Hagen 1996). Microwire tags are relatively cheap (less than NZ$2 each), but to read the unique code they must be excised, usually leading to the death of the host. In circumstances where animals must remain alive, cohorts may be tagged using microwire tag, and repeatedly identified over a period where change in mean size is determined. In contrast, microchip tags are expensive (NZ$10 each in 1995) but can be read *in situ* using the 'wand' detector. Chemicals such as tetracycline and calcine (Ebert 1982; Ebert & Russell 1992; Pirker & Schiel 1993; Day et al. 1995; N. Andrew pers. comm.) or shell notching (Pirker 1992) may be used to insert a datum mark into the shell or test at the time of initial tagging. Incremental growth can be measured at recapture, or verify the rate of deposition of ring structures (see also 6.2.10 for the importance of validation).

(e) *Reproduction*

Research on the reproduction of target species is important for understanding demographic processes that are relevant to a wide variety of biological problems that may relate to one or a combination of the following: harvesting practices (e.g. Sainsbury 1982; Shepherd et al. 1992b), population dynamics, phenology, sex change (e.g. *Patella kermadecensis*: Creese et al. 1990), the parts of a reef that merit protection, and resilience to starvation and interactions with other grazers (e.g. Andrew 1989). The focus of such studies may be on the timing of reproductive processes, age and size of grazers at maturity, reproductive output per recruit, the influence of food quality and quantity on the size of gonads, and the possible use of gonads as a storage organ (urchins). Exploitation rates of gastropods and urchins by humans are often high, and the focus of exploitation is often gonad tissue of the animals, particularly for those animals with gonads of a desired colour (Dix 1977).

A knowledge of the timing of reproduction, therefore, is important for decisions concerning the harvesting of species and associated management policies that facilitate a sustainable fishery. The timing of reproduction will often vary greatly with latitude. It cannot be assumed, therefore, that the timing of reproduction for a particular species in one area is indicative of patterns along the entire coast. For example, in studies of the timing of reproduction of abalone (*Haliotis iris*) at the northern (Hooker & Creese 1995) and southern (Wilson & Schiel 1995) regions of New Zealand, it was found that animals to the north had a spawning period of much longer duration (eight months versus four). Patterns of this type have also been found for urchins. It has been revealed (N. Andrew & M. Byrne pers. comm.) that *Centrostephanus rodgersii* along the coast of New South Wales have a prolonged period of spawning at high latitudes (37° 06' S) and a short period of intense spawning at low latitudes (30° 08' S). The number of spawning events may also vary with latitude, as for abalone (Hooker & Creese 1995). Within a latitudinal range, there is often great variation in the timing of reproduction among species. The proximate and ultimate factors (*sensu* Pearse et al. 1986) that determine timing, therefore, vary greatly among species (e.g. Underwood 1974).

Size of gonads is not always a good indication of quality of reproductive tissue for invertebrates (see also p. 156 for a discussion on gonad indices). It is common for urchins to have substantial quantities of nutritive tissue in gonads outside of the reproductive periods. Thus changes in gonad indices (GSI) or absolute weight of gonads should be interpreted with caution in the absence of matching data on histology. Where comparisons are made among times and locations it is important that animals of a similar size are compared. Urchin gonads often contain gametes at various stages of development during the reproductive period. While urchin gonads are in the resting phase they are characterised by early-stage gametes around the periphery (Fig. 5.4). This pattern is found in *Evechinus chloroticus* (Dix 1970; Walker 1982), *Centrostephanus rodgersii* (Fig. 5.4) and *Heliocidaris erythrogramma* (Dix 1977; Laedsgaard et al. 1991). *H. tuberculata* is unusual in that individuals can be found with gametes at various stages of development at all times of the year (Laedsgaard et al. 1991).

The gonads of urchins are sampled by holding urchins the oral side down (specimens with long spines will need a 'haircut' first) and cutting around the equator. The five sections of the gonads can then be scraped out for histology (e.g. Laedsgaard et al. 1991) and for measurements of their total size. Gonad size can be measured in a volumetric cylinder or by weighing on a balance after drying. Urchin gonad weight is measured

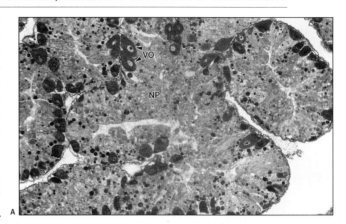

A

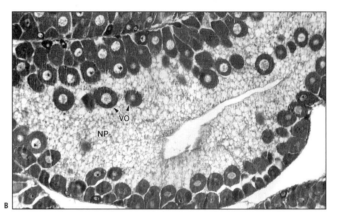

B

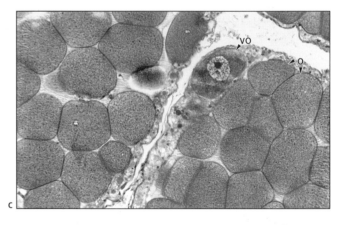

C

Figure 5.4 Histology of the gonads of the temperate sea urchin *Centrostephanus rodgersii*. This urchin is common along the southeastern coast of Australia and is rare in northeastern New Zealand.

A: Female gonad, recovering from spawning with some growing oocytes. **B:** Female gonad maturing with large numbers of growing oocytes. **C:** Mature ovary filled with fully developed ova. **D:** Male maturing with columns of developing spermatocytes. NP = nutritive phagocytes; O = ova; SC = spermatocyte columns; VO = vitellogenic oocytes.

MARIA BYRNE

D

either as wet weight (Rogers-Bennett et al. 1995) or dry weight; the latter is less likely to be influenced by variable water content (e.g. Laedsgaard et al. 1991).

Gonad tissue of gastropods is obtained by cracking the shell and removing the visceral mass. The arrangement of the gonads of most gastropods is similar (Joll 1980). Gastropods may be gonochoristic or hermaphroditic, and ovaries and testes are generally a different colour. For example, in *Turbo intercostalis* (a gonochorist from Western Australia) the ovary is green, the testis is creamy white and there is sexual dimorphism of the genital papilla. Connective tissue separates the gonads from the digestive gland in both sexes. Photographs of the anatomy of genitalia are shown in Joll (1980), and examples of histological sections of gonads are given in Underwood (1974), Joll (1980) and Wilson and Schiel (1995). Histology is generally essential because the external appearance of male and female gonads is unreliable for estimating phases of the reproductive cycle (Underwood 1974). Indices of the size of gonads have generally been obtained from analyses of cross-sections of the 'conical appendage' (gut and gonad: Poore 1973). The conical organ is fixed in Bouin's fluid and then sectioned at four or more equally spaced points between its tip and the end of the gonad. The area of each section that is gonad and digestive gland is then measured and compared to total cross-sectional area as an index. If you only want to obtain data on gonad indices, preservation in 10% gives acceptable results and may avoid some of the problems of dealing with saturated picric acid in the field (M. Byrne pers. comm.).

A knowledge of size, age at maturity, and egg production is often important, and this can be achieved by sampling invertebrates of a wide size range. Although size to maturity has been documented for a number of urchin species (e.g. Dix 1977), you should be cautious about simply using information from the literature, because size to maturity may vary with location and latitude, as with abalone (Hooker & Creese 1995). Appropriate morphometric dimensions to measure size should also be considered. For example, Worthington and Andrew (1998) found that quick-growing abalone (in terms of length) had relatively low egg production per recruit compared with slower-growing counterparts. To optimise the fishery, they argued that a minimum size measured as width rather than length would increase the minimum length at which fast-growing abalone could be removed. This would increase the exploitation of slow-growing populations by reducing the intensity of fishing at sites where abalone grow quickly.

The fecundity of invertebrates (e.g. the urchin *Strongylocentrotus franciscanus*) can vary greatly with position on a reef. For example, differences in fecundity have been found between intertidal and subtidal urchins from the same location (e.g. *Paracentrotus lividus* from Ireland: Byrne 1990). A small proportion of the stock, therefore, may contribute disproportionately to the reproductive output of local populations. Rogers-Bennett et al. (1995) found that gonad indices were greatest for *S. franciscanus* in shallow water. Although no histological work was done in this study (i.e. to compare the proportion of nutritive tissue to actual gametes), they argued that 'shallow beds of urchins would be ideal candidates for harvest refugia', promoting the production of larvae to replenish deeper harvested habitats. Because the morphology of urchins (e.g. variation in spine length) varies with depth, the protection of shallow refugia could be ensured by checking catches.

5.2.3 Large crustaceans

Large crustaceans are usually found in low average densities, and where abundant they are often clumped. Accordingly, the behaviour of the animals and their distribution patterns will require a different sampling approach to that described for sedentary invertebrates. This section will focus on rock lobsters as target organisms (see review by Breen & McKoy 1988), but the general principles will also be relevant to other crustaceans (e.g. crabs). Traditionally, information on the distribution of large crustaceans has been derived from catches in pots (e.g. Morgan 1974; McKoy & Esterman 1981). Although this method will provide some data on numbers removed from different areas, the usefulness of catch statistics from pots for providing information on distribution is limited for several reasons. Rock lobsters may be attracted to bait from long distances (e.g. Jernakoff & Phillips 1988) and will not enter pots during periods of moulting (mature male *Jasus edwardsii* late spring; mature females late summer–autumn) or reproduction (June in northern New Zealand: MacDiarmid 1991). Furthermore, they may move in and out of traps, and very small animals are secretive and rarely caught in traps.

Excellent information on rock lobster distribution and abundance can be gained while diving. Crevices and holes within a defined area are searched and all lobsters counted. The carapace length of individuals can be estimated (see 5.2.2d) and sex determined by use of a torch to illuminate the back walking leg, which in females is chelate (has a pincer), but in males is similar to the other walking legs (Fig. 5.5). Additional information can also be obtained from individual lobsters where appropriate; for example, tag number and the presence of eggs on females (MacDiarmid 1989a, b, 1991, 1994; Pitcher et al. 1992). Female lobsters are

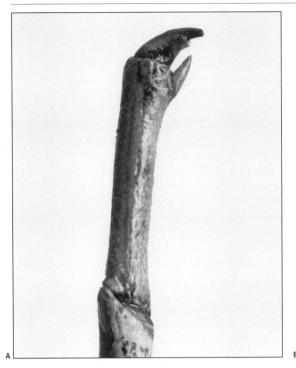

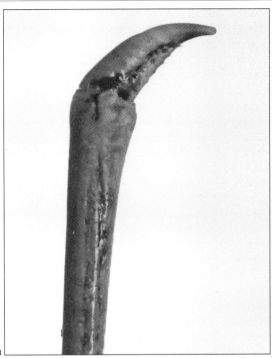

Figure 5.5 Sexual dimorphism in the rock lobster, *Jasus edwardsii*.
A: Claw on rearmost walking leg of a female. **B:** Claw on rearmost walking leg of a male. **C:** Ventral surface of female abdomen showing large paired pleopods. **D:** Ventral surface of male abdomen showing the small unpaired pleopods.

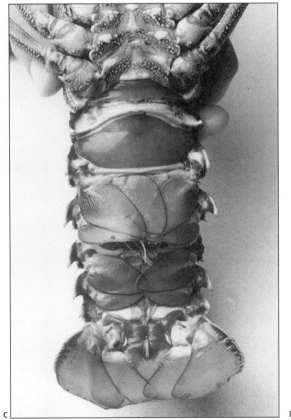

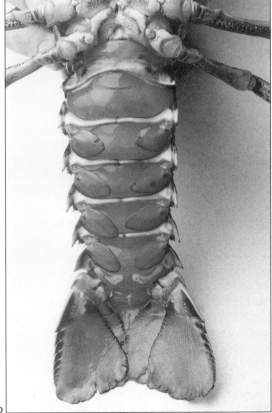

particularly obvious when brooding eggs ('in berry').

Methods for describing patterns of distribution and abundance are as follows:

(a) *Belt-transect method*

Belt transects (for other types, see Chapter 6) have been the most common method for measuring abundances of rock lobsters (e.g. Ayling 1978; Battershill et al. 1985; MacDiarmid 1991; Pitcher et al. 1992). Sizes of transect range from 10 x 10 m (MacDiarmid 1991) to 4 x 500 m (Pitcher et al. 1992). MacDiarmid (1991) tested transect units of different sizes, by use of the mapping technique (see Chapter 3) on a high-density population of lobsters, and concluded that highest precision was obtained when 10 x 10 m transects were used, with 20 replicates. This transect size proved most appropriate at different times of the year and varying depths on the reef. An examples of the maps that MacDiarmid (1994) used to choose the best sampling unit is given in Figure 5.6. The transect method may be incorporated with counts for fish, especially when two divers are involved (see 5.4.1). Pitcher et al. (1992) used pilot studies to determine that using two 4 x 500 m transects per loca-

tion provided the most cost-effective method to search for tropical lobsters that were found in low densities.

(b) *Mapping*

In situations where a detailed study of rock lobsters is required (e.g. MacDiarmid 1991, 1994), the positions of individuals may be marked on a map of the subtidal zone (e.g. Fig. 5.6). This method allows detailed information to be obtained on the distribution patterns of males and females, without measures of dispersion being influenced by the size of sample units (see Andrew & Mapstone 1987). Furthermore, mapping is invaluable for behavioural work. Although this is an excellent method for detailed investigation, the time required to map an area means it is unlikely to be a useful method for studies where sampling is done over spatial scales of kilometres or more.

(c) *Rapid visual techniques*

A free-swimming method (see also 4.3.1c) would be useful where average densities are low and the habitats in which lobsters are found (e.g. broken rock) are scattered over a large area. If very few animals were found in detailed transect counts, supplementary information on the size-frequency distributions in the study area could be obtained by free swimming. (For general problems that would have to be accounted for while using this method, see 4.3.1c.)

Figure 5.6 Map showing positions of lobsters in an area of reef. Data on densities and the spatial arrangement of animals can be extracted from maps of this type.

FROM MACDIARMID 1987

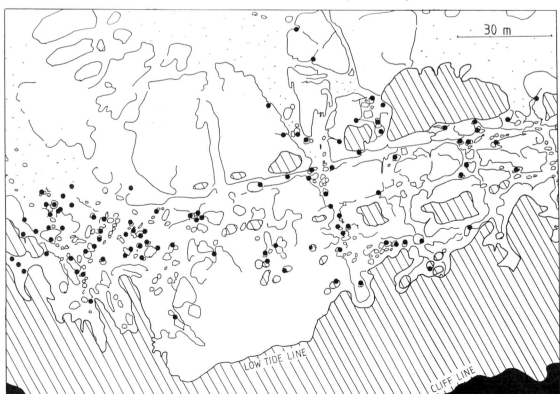

Because of the secretive behaviour of rock lobsters, other rapid visual techniques would be useful only in identifying suitable habitat for further investigation.

(d) *Size*

Size is usually measured using vernier callipers. When a diver has observed a wide size range of rock lobsters and measured them under water, the carapace length of individuals can be estimated by eye (MacDiarmid 1991). Divers require practice in estimating the sizes of objects under water before they can be relied upon to give accurate estimates of carapace lengths or tail widths (see also 6.2.9).

(e) *Age, growth and tagging*

Tagging is useful where information is required on growth or the movements of rock lobsters. Crustaceans regularly moult, and therefore external tags will be lost and are unsuitable for measurement of growth. Nonetheless, these tags may serve some short-term studies of movements (e.g. MacDiarmid et al. 1991). Tags that are retained through the moult are required to determine individual growth rates. Lobsters have been successfully tagged using 'Western rock lobster tags', which comprise an internal toggle anchor (Fig. 5.7),

Figure 5.7 A range of tags that have been trialed on rock lobsters. From top: microchip tag (length 15 mm), T-bar tag, sphyrion tag, small streamer tag, western rock lobster tag, dart tag, large streamer tag. The streamer tags are inserted using the attached steel needles, but all others require the use of a special tagging gun or modified surgical syringe.

inserted into the abdominal muscle block using a special tag inserter, connected by a nylon filament to an external spaghetti tube that carries the unique code. T-bar tags similar to those used to attach price tags to clothing have been used with mixed success. Tags may remain in animals for considerable periods of time; for example, Booth (1984) described tagged *Jasus verreauxi* that were caught 950 days after liberty. Microwire and microchip tags (e.g. Buckley et al. 1994) are ideal for tagging rock lobsters because of their small size and high rates of retention.

Rock lobsters do not have hard parts that can be used for ageing (cf. fish, p. 151). Estimates of age may only come from a knowledge of planktonic duration (Fig. 5.1), cohort analysis, tagging and, perhaps, recent advances in the use of physiological age marks such as lipofuscin (Sheehy et al. 1994). A knowledge of age is crucial for any demographic study and other studies where distribution, behaviour, foraging, movement, and even susceptibility to pollutants, may vary with age.

(f) *Moulting and mating*

There are generally strong links between moulting and mating in large, reef-associated crustaceans such as lobsters and crabs. Females often moult before brooding eggs on the exoskeleton. It should also be noted that catches of crustaceans will often drop at the time of moulting, because the animals generally hide and do not eat while the exoskeleton hardens. Lobsters and other decapods are gonochoristic, and fertilisation is generally external; see experiment of MacDiarmid (1988) for *Jasus edwardsii*.

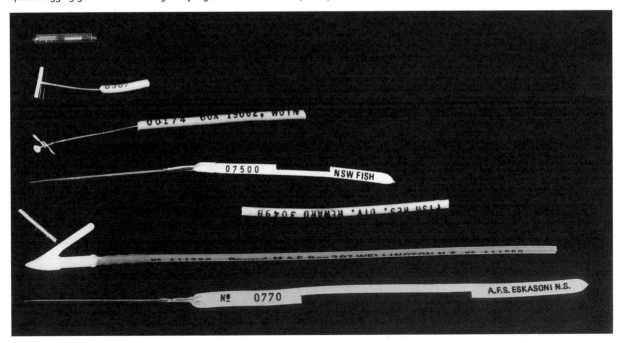

Although gross abundances and size-frequency distributions of rock lobsters (*Jasus edwardsii*) in a rocky reef environment may show little variation over a year, there are large movements of males and females up and down a reef profile at different times of the year. MacDiarmid (1989a, b 1991) confirmed earlier suggestions (e.g. McKoy & Leachman 1982) that movements of mature males and female *J. edwardsii* up and down the reef relate to moulting and reproductive cycles. Females move into habitats of shallow broken rock during autumn, over which period they moult. After moulting they spread over all depths of the reef to court and mate. Females carry eggs on their newly hardened exoskeletons for 103 days in northeastern New Zealand and about 150 days in southern New Zealand. Towards the end of this period they move to the seaward edge of reefs, where the larvae hatch. From October to November, males move into shallow water and moult. In contrast, over summer large numbers of males are found at the seaward edge of the reef. Juveniles do not make these movements. It is clear, therefore, that a sampling design for estimating the densities and size-frequency distributions of rock lobsters should take these movements into account. Moreover, the movements should influence the interpretation of data on densities, sizes and sex ratios collected from each depth stratum.

The reproductive output and quality of the eggs (based on egg diameter) of decapods can be measured relatively easily because they are brooded externally by females.

(g) Survivorship

Survival of highly mobile benthic invertebrates is often difficult to determine because losses from an area may be considerable as a result of either emigration or mortality. One way to separate these effects is to use Jackson's method (Manly 1985a), where the study area is divided into four equal subsquares and data on recaptures of tagged individuals is used in two different ways to estimate survival. First, recaptures (or resightings) made within the subsquare of release are considered. Second, all recaptures within the study area are considered, irrespective of the point of release. The emigration rate from the study area as a whole should be half that for a single subsquare because, on the balance of possibilities, half the immigrants from a subsquare move into another subsquare and are thus inside the study area. Some preliminary information will be required to help scale the size of the study area to the distance moved by the study animal. Simulations by Manly (1985b) show that this simple method works well in practice.

If relative rather than absolute estimates of survival

are adequate, tethering experiments can be used to determine short-term rates of mortality in the field. With this method study animals are tethered, using a nylon line of appropriate strength, to a fixed point, and numbers surviving after a period are counted. Pilot studies are necessary to determine how long the experiment should be left to run so that survival is neither 100% nor 0% in all treatments. Relative survival can be compared between microhabitats, between areas with contrasting population densities or different predator abundance, and so on. Rates of observed mortality are likely to be very high compared with natural rates, and effort must be taken to investigate potential experimental artefacts that might vary among treatments (Peterson & Black 1994; Aronson & Heck 1995). Nonetheless, the technique can sometimes be the only way to investigate the relative importance of various factors (e.g. exposure, risk of predation, etc.) in determining survival. This technique has been used successfully with clawed and spiny lobsters to investigate the relationship between shelter size and relative survival (e.g. Eggleston et al. 1990, Wahle & Steneck 1992) and with crabs (Beck 1995). For other measures of mortality (e.g. size-based models), see Hilborn and Walters (1992).

5.3 Experimental methods

In many ways marine invertebrates such as gastropods, echinoderms and lobsters are ideally suited to experimental manipulation to investigate aspects of their ecology. They are often abundant, large and ecologically significant either as grazers or predators (e.g. Schmitt 1987). Their abundance and size can be assessed much more easily than most other groups of marine organisms, enabling meaningful comparisons over various spatial and temporal scales. Their densities can be experimentally reduced or enhanced, and they are amenable to exclusion (and sometimes inclusion) cage techniques, enabling density-dependent effects to be readily investigated. They can be tagged so growth, survival and patterns of movement can be determined. One negative aspect is that they are usually most often active at night. They are, however, less likely to show behavioural responses to the presence of divers during daytime sampling (cf. reef fish).

Ecological problems that have been addressed using experimental methods for large mobile invertebrates include interactions between grazers and algal assemblages (e.g. Fletcher 1987; also Chapter 4; Andrew 1988), the influence of topography on the distribution and grazing activities of invertebrates (Andrew 1993;

see also 5.4.5), competitive interactions between grazers (Schmitt 1985, 1996), interactions between predatory invertebrates and grazing invertebrates (Branch et al. 1987; Barkai & McQuaid 1988; Elner & Vadas 1990; Schmitt 1982, 1987) and interactions between predatory fishes and grazing invertebrates (Cowen 1983). See also experiments involving other invertebrates (Chapters 4 and 7) and fish and invertebrates (Chapter 6).

The ethics of experiments on these organisms may be questioned: our perspective is as Kingsford has stated in Chapter 6 for fishes.

5.4 Examples of sampling designs

See also advice on sampling designs (Chapter 2) and, especially, allocation of sampling effort in Chapter 3.

5.4.1 One-off survey of an offshore island

Example: Survey of the Auckland Islands, subantarctic New Zealand (Schiel, Battershill & Kingsford, unpublished data).

Purpose: To obtain baseline information on organisms in subtidal rocky reef environments, including large reef-dwelling invertebrates.

Situation: As for 6.4.1. Owing to the limited amount of time available at each locality, counts of invertebrates were included with counts of reef fish. Background information suggested two types of large crustacean were likely to be found: rock lobsters (*Jasus edwardsii*) and spider crabs (*Jacquinota edwardsii*).

Approach: Crustaceans were sampled at shallow and deep strata down a reef profile to allow for the sort of variable pattern that MacDiarmid (1987) found for rock lobsters (see 5.2.2 f). At each of nine localities, crustaceans were counted in three replicate 10 x 10 m transects at each depth. The procedure required two divers. While one counted fish (see 6.4.1), the second diver followed and sexed all crabs. The carapace length was measured by callipers and the sex determined by examining the abdomen: females had a wider abdomen than males. After completing the fish counts, the first diver counted crustaceans in a 10 x 10 m transect. Thus information was obtained on densities, sizes and sex ratios in deep and shallow water at each locality.

5.4.2 Influence of marine reserves on densities and size-frequency distributions of rock lobsters

Example: Study of rock lobsters along the coast of Northland, New Zealand (MacDiarmid & Breen 1993).

Purpose: To compare the densities and size-frequency distributions of rock lobsters among protected and fished populations.

Situation: Pilot studies by MacDiarmid indicated that accurate and precise estimates of abundance were obtained with 100 m^2 transects. A large amount of background information (e.g. MacDiarmid 1991) indicated that mature male and female rock lobsters undergo directed movements up and down a reef at different times of the year (see 5.2.2f). The study was done in autumn when female lobsters move to shallow water to moult and males are generally in found in shallow water. The study was carried out by two divers and required support from a launch. The exception was around the coast of Leigh, where access was by dinghies.

Approach: To maximise the number of sites that could be examined in a short time, only shallow broken rock habitats were sampled. Stretches of coastline, 5 km in length, were sampled at six localities, which included two marine reserves (Fig. 5.8, p. 128). Five sites were sampled haphazardly (for definition, see 3.2.1a) within each locality, and rock lobsters were counted in fifteen 100 m^2 transects at each site.

A number of interesting points came out of the study: rock lobsters were most abundant in the Leigh marine reserve; although a large number of rock lobsters were found at fished localities (e.g. Hen Island), they were mainly small individuals; despite the fact that the Poor Knights Islands are a marine reserve (i.e. fishing effort was assumed to be zero or very low), only a small number of large lobsters were found. This last point emphasises that natural processes (e.g. settlement patterns and movements) also influence patterns of lobster distribution; for example, lobsters may rarely recruit or immigrate to the Poor Knights Islands. Patterns of this type need to be adequately described before changes induced by humans are claimed (see section 2.6). It is also noteworthy that important information was derived both from estimates of abundance and from size-frequency distributions, not from just one of these. Sampling done in 1978, 1984 and 1992 indicated that the abundance of rock lobsters had increased in the Leigh marine reserve. Greatest increases in abundance were found for female lobsters. It was thought that male density was more 'volatile' because of offshore migrations around the time of sampling.

5.4.3 Broad-scale stock assessment and size structure of rock lobsters

Example: Survey of spiny lobsters (Pitcher et al. 1992) using visual transect methods.

Purpose: To estimate the stock size of the tropical lobster *Panulirus ornatus* in Torres Strait, Australia.

Situation: The usual method of estimating yield by analysis of catch and effort data is not suitable for the Torres Strait diving fishery for *P. ornatus* because only part of the effort is monitored and there is no long-term data. Consequently, an alternative approach using divers to estimate stock abundance over an area of 25,000 km² was adopted.

Approach: Pilot studies were undertaken to assess the feasibility of a full-scale survey and to optimise the sampling design. These showed that 4 x 500 m belt transects were the most cost-effective of the different sizes compared, the optimal allocation of replication was two transects per location, and 300 locations were necessary to achieve a 95% confidence interval of ± 10% of the mean density found in the pilot study. Suitable lobster habitats in Torres Strait were classified using satellite imagery into three broad strata, and the 300 locations were allocated to each stratum in proportion to its area (proportional stratified sampling, see Chapter 3). Locations were randomly positioned inside each stratum. The surveyed population was sampled concurrently using spears, to determine size and sex structure. The resulting estimate of abundance

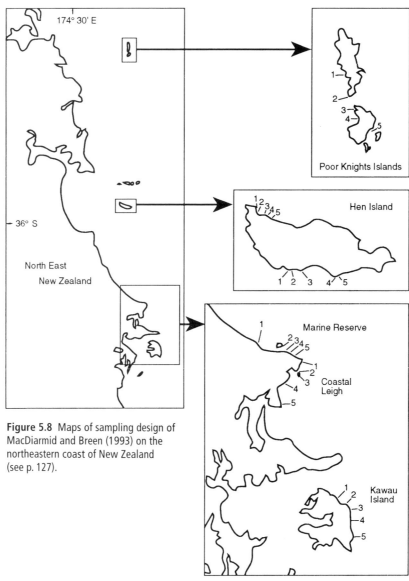

Figure 5.8 Maps of sampling design of MacDiarmid and Breen (1993) on the northeastern coast of New Zealand (see p. 127).

was approximately 14 million lobsters with a 95% confidence interval of $\pm$ 21%.

5.4.4 Broad-scale stock assessment of sea urchins

Example: Pre-fishing surveys of sea urchins (*Evechinus chloroticus*) in Dusky Sound, southwest New Zealand (McShane & Naylor 1991; McShane et al. 1993).

Purpose: To determine the abundance, size structure and population trajectory of the sea urchin *E. chloroticus* in an area prior to the start of commercial fishing in one subarea.

Situation: There are few fisheries that progress with direct information on the population structure of the unfished stock. In a new fishery on sedentary invertebrates such as sea urchins, there is an opportunity to measure the impact of fishing directly and isolate its effects from other factors causing variation in population size and structure. Parties interested in developing a new fishery for *E. chloroticus* in New Zealand paid for surveys of sea urchin stocks and associated reef communities prior to commercial fishing starting in one subarea.

Approach: A series of stratified random surveys of sea urchin populations was conducted in Dusky Sound. The number of sites allocated to each stratum was determined by its relative area. Sites within strata were chosen at random. At each site two replicate 25 x 1 m random belt transects were sampled for sea urchins and other dominant epifauna. Algal floristic composition in each transect was described in terms of percentage cover by taxon and density of large brown macroalgae. All sea urchins in the transect were sequentially collected until 60 had been taken. Test diameter, total weight, gonad weight, gonad volume and gonad colour were measured. The jaw apparatus was removed from a subsample of sea urchins and stored for processing and measurement of jaw length. Sea urchins were found at all sites surveyed and were observed in dense aggregations down to 26 m, although less than 10% of sea urchins occurred below 9 m. The standing stock of sea urchins to 9 m depth in the initial survey was estimated at 3401 $\pm$ 808 tonnes. The results of four pre-fishing surveys showed that there was considerable natural spatial and temporal variation in sea urchin density, recruitment and gonad condition; any fishing-induced change must be judged against this. The studies generated multiple hypotheses regarding the response of sea urchin abundance, growth, jaw length and gonad yield, and algal densities to large-scale fishing, with a realistic expectation that these could be tested by using a beyond-BACI design with multiple controls (see Chapters 2 and 3).

5.4.5 Sea urchin grazing and habitat structure

Example: Spatial heterogeneity, sea urchin grazing and habitat structure on reefs in temperate Australia (Andrew 1993).

Purpose: To determine the role of shelter for sea urchins in causing shifts in reef-habitat structure from macroalgal forest to urchin-grazed 'barrens'.

Situation: By day the sea urchin *Centrostephanus rodgersii* is restricted to crevices, but it emerges at night to graze over the surrounding substratum, creating haloes of barrens habitat devoid of macroalgae around shelters. The removal of urchins causes a shift back to a habitat dominated by macroalgae. The processes that lead to the reverse shift, from foliose algae to barrens habitat, are unclear.

Approach: The hypothesis that shelter was a sufficient precondition for both the creation and maintenance of barrens habitat by urchins was tested by a field experiment. Large boulders were transferred using lift-bags from barrens to algal forest habitats (and vice-versa) and arranged in pairs to either minimise or maximise shelter for adult urchins (see Fig. 5.9, p. 130). Controls for the effects of translocation consisted of moving boulders within the same habitat. There were four replicate pairs of boulders for each combination of experimental factors. The abundance of algae, urchins, and other invertebrates was monitored for 31 months. The fauna and flora on translocated boulders became progressively similar to surrounding habitats. Foliose algae on boulders transferred into barrens habitat were gradually grazed to low densities. Macroalgae became abundant only on boulders within the algal forest which had minimal shelter for urchins. Juvenile sea urchins settled to all boulders in the experiment but survived in greatest numbers as adults where shelter was abundant. The availability of shelter was clearly a sufficient condition for the creation of areas of barrens habitat.

5.4.6 Enhancement of wild stocks

Example: Experimental evaluation of commercial-scale enhancement of abalone (Schiel 1993).

Purpose: To determine the biological and economic viability of enhancing populations of abalone *Haliotis iris* in New Zealand.

Situation: Commercially valuable abalone fisheries have suffered serial depletion in many countries including both New Zealand and Australia. It may be possible to augment depleted abalone stocks to help ensure their long-term sustainability by enhancement with hatchery-raised juveniles. To do so, it is necessary to understand the links between early life stages and adults in

terms of habitat requirements, survival and growth. Enhancement must also be economically viable.

Approach: The biological and economic viability of enhancing abalone populations by placing hatchery-reared juveniles into natural habitats was evaluated at New Zealand's Chatham Islands. Using local broodstock, 80.000 hatchery-reared juveniles of shell length 3–30mm were seeded by hand into eight sites of shallow-water boulder habitat. Densities of abalone prior to seeding were determined by counting and measuring all individuals in 40 randomly placed 0.25 m²

Figure 5.9 Experimental design of Andrew's (1993) study (see p. 129). The balloon is a heavy-duty airlift device for underwater use. F = kelp forest, B = Barrens. Boulders with and without kelp (*Ecklonia radiata*) are drawn.

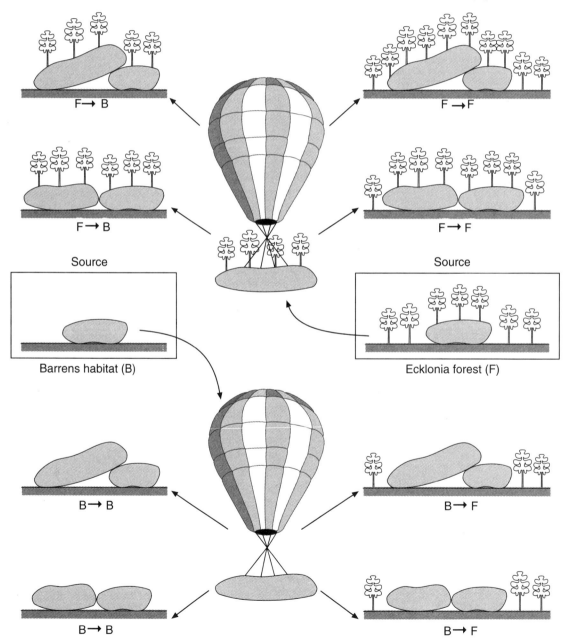

quadrats at each site. Survival and growth of seeded abalone were monitored for up to two years by counting and measuring all juvenile abalone in 40 randomly placed 0.25 m^2 quadrats at each site on every sampling occasion. Hatchery-reared juveniles could be distinguished from wild individuals by the presence of a bright green mark on the apex of the shell. Growth was measured in terms of increase in mean shell length. Percent survival was determined by comparing sample abundance means to the initial seeding density. The results were incorporated into a simple economic model to determine the internal rate of return for each seeding site. Mortality and growth varied among sites. Mortality was greatest at sites affected by sand movement that partially buried juvenile habitat. The overall rate of return of the eight sites was predicted to exceed the original investment.

Another approach to reseeding areas that may be more cost effective and less energy intensive is the translocation of adults to areas of suitable habitat that have no or few abalone: local recruitment can be greatly enhanced as a result (Neil Andrew pers. comm.).

SELECTED READING

The references listed are examples of research that will lead the reader to a broader literature, also see the text of the chapter.
* indicates a multi-authored book or collection of proceedings; references to volumes of a journal refer to a set of relevant papers.

Ecology
Abalone and other gastropods: Underwood 1977, 1979; Schmitt 1982, 1985, 1987, 1996; Shepherd et al. 1982; Shepherd 1985, 1988; Jeffs 1986; Keestra 1987; Creese 1988; Kerrigan 1987; Fielding 1995; McShane 1996; Seki & Taniguchi 1996; Shepherd & Daume 1996.
Rock lobsters: MacDiarmid 1985, 1987, 1989a, b, 1991, 1994; Schmitt 1987; Barkai & McQuaid 1988; Elner & Vadas 1990; Andrew & MacDiarmid 1991; MacDiarmid et al. 1991.
Urchins: Ebert 1982; Andrew & Choat, 1985; Andrew & Stocker 1986; Kerrigan, 1987; Andrew 1988, 1993; Elner & Vadas 1990; Ebert & Russell 1992; Estes & Duggins 1995, Mooi & Telford 1998.

Fisheries
Abalone and other gastropods: Sainsbury 1982; Tegner et al. 1989; Shepherd 1990; Schiel & Breen 1991; Shepherd et al. 1992b*; (see also Reproduction below); Hilborn & Walters 1992.
Rock lobsters: Morgan 1974; Jernakoff & Phillips 1988; Hilborn & Walters 1992; Proceedings of the 5th International Conference on lobster biology, fisheries and management *Mar. Freshwater Res.* 29(1)*.
Urchins: Hilborn & Walters 1992; see Reproduction and 5.4.4.

Larvae
Abalone and other gastropods: Prince et al. 1987; Strathmann 1987*; Morse 1990; Shepherd et al. 1992a, b*; Watanabe et al. 1996*; Moran 1997.
Rock lobsters: Phillips 1981; MacDiarmid 1988; Pearse & Phillips 1988; Lemmens 1994; Booth 1994; Booth & Phillips 1994; Caputi et al. 1996.

Urchins: Williams & Anderson 1975; Pearse & Cameron 1991; Emlet & Hoegh-Guldberg 1997.

Marine protected areas
Abalone and other gastropods: Rogers-Bennett et al. 1995; Gubbay 1995*.
Lobsters: MacDiarmid & Breen 1993; Chapter 1.
Urchins: Jones et al. 1993; Rogers-Bennett et al. 1995.

Movements
Abalone and other gastropods: Underwood & Chapman 1985.
Lobsters: Booth 1984.
Urchins: Andrew & Stocker 1986.

Reproduction
Abalone and other gastropods: Poore 1973; Shepherd & Laws 1974; Joll 1980; Tutschulte & Connell 1981; Sainsbury 1982; Shepherd 1986; Strathmann 1987*; Fletcher 1987; Creese et al. 1990; Shepherd et al. 1992b*; Hooker & Creese 1995; Wilson & Schiel 1995; Moran 1997; Worthington & Andrew 1998.
Lobsters: MacDiarmid 1988, 1991; Kittaka & MacDiarmid 1994; Baur & Martin 1991; Phillips et al. 1994*; Groenveld & Rossouw 1995.
Urchins: Dix 1970; Williams & Anderson 1975; Walker 1982; Pearse et al. 1986; Laedsgaard et al. 1991; Pearse & Cameron 1991; King et al. 1994; Drummond 1995.

Reseeding, enhancement and aquaculture
Abalone and other gastropods: Emmett & Jamieson 1989; Shepherd et al. 1992b*; Schiel 1993; McCormick et al. 1994; Howell et al. in press*; Schiel 1997.
Lobsters: Nonaka & Fushimi 1994; Herrnkind et al. 1997.

Surveys
Lobsters: Pitcher et al. 1992; MacDiarmid & Breen 1993.

Taxonomy: See Chapter 10.

SIX

REEF FISHES

M. J. Kingsford

6.1 Introduction

Fishes are important members of the fauna of rocky reefs. Reef fishes can have an important influence on the dynamics of other organisms through herbivory (Foster 1972; Norman & Jones 1984) and predation (e.g. Cowen 1983; Keough & Downes 1982; Gaines & Roughgarden 1987). Moreover, their feeding and excretion (Robertson 1982; Bray et al. 1981), as well as their role as prey, can make an important contribution to the trophodynamics of reef fishes (e.g. Hixon 1991). As reef fishes are often harvested for food (e.g. Sparidae, Latridae, Labridae, Cheilodactylidae) or sport (e.g. grey nurse sharks, Odontaspididae: Pollard et al. 1996), the population dynamics of fishes are of special interest to many investigators.

An understanding of population dynamics requires information on the following:

- Life history (bipartite life cycle for most; Fig. 6.1, but cf. live bearers, e.g. Embiotocidae of the North Pacific).
- The input of new recruits to reefs (Doherty & Williams 1988).
- Growth (which requires a knowledge of age), number and biomass of fishes in populations.
- The size frequency of populations and how this is related to reproduction (i.e. age/sex to maturity, reproductive output, sex change; see also Fig. 6.1).
- Mortality (natural and fishery related) and movement (Fig. 1.3).

The magnitude of movement of reef fishes varies greatly and will have a significant influence on the dynamics of local populations. Some fishes are territorial and may only move tens of metres during post-settlement life (e.g. Pomacentridae; *Parma microlepis*: Moran & Sale 1977). In contrast, some species show size-related movements between environment, such as sparids and labrids, which may recruit in estuaries and subsequently move to reefs (Gillanders & Kingsford 1993, 1996; Bell & Worthington 1993), while other fishes move among reefs and spend some time on the continental shelf away from reefs (e.g. sparids and monacanthids).

Reef fishes are of considerable interest to divers, and some species are keenly sought after as food. The variety of fish species and their abundance and size-frequency distributions are commonly equated with the health of reefs. For example, one of the main reasons divers travel to the Cape Rodney to Okakari Point Marine Reserve at Leigh (hereafter referred to as the Leigh marine reserve), northern New Zealand, is to observe large fishes under water, especially snapper (Bell 1984; Cole 1994). What is more, the selective exploitation of some species makes them particularly good candidates for assessing changes resulting from positive impacts (protection) or negative impacts (harvesting). In many cases, information on fishes and other large discrete organisms (e.g. urchins and lobsters, see Chapter 4) are used as justification for the choice of areas for marine reserves, their size, shapes and maintenance (Dugan & Davis 1993). Data on abundance, size-frequency, movements, life-history characteristics (e.g. Buxton 1993) and the timing and sites of reproduction are relevant to biological questions of these types (e.g. Ayling 1978; Russ & Alcala 1989; Buxton & Smale 1989; Cole et al. 1990; Bayle-Sempere & Ramos-Espla 1993).

Studies in the temperate waters of California (Ebeling et al. 1990; Ebeling & Hixon 1991), the east coast of North America (Levin 1991; Levin & Hay 1996), New Zealand (Jones 1984a, b; Choat & Ayling 1987; McCormick & Choat 1987; Kingsford et al. 1989; Syms 1995) and Australia (Kingsford unpub. data) have found that abundance patterns of fishes are influenced by the proportional representation of different habitat types (but see Holbrook et al. 1990). Changes in the composition of assemblages of fish, therefore, can often reflect changes in the representation of habitats (e.g.

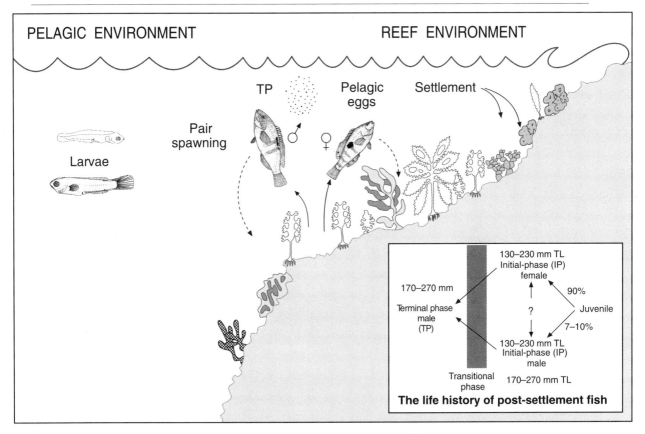

Figure 6.1 Life history of *Notolabrus celidotus* (Labridae), common name 'spotty' (formerly *Pseudolabrus celidotus*). The reef profile shows reef-associated phase of these protogynous hermaphrodites. The initial-phase fish (IP) have the same coloration as the female shown. *N. celidotus* spawn pelagic eggs, and spawning is usually done in male territories in deep water. The early life history of these fish is spent in the plankton; some larval forms are shown. Based on the duration of the larval period of other labrids, they may spend 15–121 days in the plankton (where 30–40 is most typical of the family: Victor 1986a). The inset shows the typical paths of development from initial-phase fish to terminal-phase males (TP). A small proportion of fish live to be TP fish.

ADAPTED FROM JONES (1980, 1983), CROSSLAND (1981, 1982A) AND THOMPSON (1981)

Bodkin 1988; Russ & Alcala 1989), and links between the fish fauna and benthic organisms should be considered in sampling designs.

The variety of sizes, distributions and behavioural characteristics of reef fishes has resulted in the development of a wide range of techniques for recording abundance, presence/absence and the behaviour of species. Methods are reviewed in this chapter. Attention is also given to the importance of studies of age/growth, feeding, biology and reproduction for many ecological problems. The utility of experiments is emphasised for investigating processes influencing fishes and interactions between fishes and other organisms in reef environments.

Sampling fishes is more difficult than for most other large members of reef assemblages (e.g. urchins and abalone). Preliminary sampling is particularly important in work on fishes (see Chapter 2). If different observers are used among times and places, bias among observers can be a major problem. Mapstone and Ayling (1993) carried out detailed comparisons of different types of methods used to measure abundance of fishes and concluded that estimates of abundance differed systematically between two observers. Furthermore, they concluded that thorough training and periodic re-calibration of observers is essential for the utility of the data stemming from on-going monitoring programs. If bias (see 2.4) is consistent for one observer, biological patterns (e.g. differences among depths, habitats, locations, times) can still be detected. In general, therefore, problems relating to differences among observers can be avoided if the same observer is used in a study.

The utility of this review will increase greatly to readers if a biological problem is identified *a priori*. The focus of this chapter is on mensurative and experimental studies on reef fishes, but many of the principles and some of the methods can be used to study fishes in other environments (e.g. seagrass beds and other

Table 6.1 Spatial and temporal sources of variation in numbers of fish on temperate reefs. Habitats are defined according to characteristic organisms (eg. algae, sedentary and grazing invertebrate) and local topography (e.g. presence of boulders/shelter). References listed are examples only. Mechanisms include: habitat destruction (e.g. storms, grazing, changes in water temperature and chemistry), environmental change (e.g. El Niño), feeding, reproduction (e.g. spawning migrations), recruitment, predation (which may alter in space and in time according to the movements and recruitment of predators), interspecific interactions (e.g. interference or resource competition), evolutionary processes, ontogenetic change in behaviour (e.g. increase in home range with age and size). Note: spatial variation in assemblages of reef fish is often greater across the shelf than for similar spatial scales alongshore.

ADAPTED FROM HOLBROOK ET AL. 1994

SPATIAL

	Mechanism		
Scale	**Environmental**	**Fish biology**	**Examples**
Microhabitats ($< 1\ m^2$)	Algae	Recruitment	Moran & Sale 1977
	Prey	Feeding	Norman & Jones 1984
	Shelter	Ontogeny	Holbrook & Schmitt 1984
	Grazers	Predation	Syms 1995
Among habitats	Algae	Recruitment	Choat & Ayling 1987
	Prey	Feeding	Jones 1984a Bodkin 1988
	Shelter	Ontogenetic	Carr 1989, 1991
	Invertebrates	Predation	Holbrook et al. 1990
			Connell & Jones 1991
			Syms 1995
Among depths	Habitats	Recruitment	Larson & Demartini 1984
	Prey	Feeding	Holbrook & Schmitt 1989
	Light	Reproduction	Kingsford et al. 1989
		Ontogeny	McCormick 1989
		Competition	
Among reefs 1–10 km	Habitats	Recruitment	Jones 1984a
	Oceanography	Reproduction	
	Pollution	Feeding	Gillanders 1997a
		Ontogenetic	
		Physiology	
11–100 km	Habitats	Recruitment	Choat & Ayling 1987
	Oceanography	Reproduction	Kingsford 1989
		Feeding	
		Ontogenetic	
100–1000 km	Habitats	Recruitment	Holbrook et al.1990
	Oceanography	Reproduction	Cowen 1985
		Feeding	
		Ontogenetic	
		Geographic range	
		Physiology	
1000 km +	Habitats	Recruitment	Nil
	Oceanography	Geographic range	
		Reproduction	
		Feeding	
		Ontogeny	
		Physiology	

TEMPORAL

Scale	Mechanism Environmental	Fish Biology	Examples
Within a day	Time of day	Reproduction	Hobson & Chess 1978
	Tide	Feeding	Bray 1981 Jones 1983
		Predation	Kingsford & MacDiarmid 1988
Within a month	Lunar cycle	Reproduction	Ochi 1989
	Habitat destruction	Feeding	Tzioumis & Kingsford 1995
		Predation	
Within a year	Season	Reproduction	Schmitt & Holbrook 1985
	Habitat destruction	Feeding	Matthews 1990
		Ontogeny	
Among years	Habitat destruction	Recruitment	Ebeling et al. 1980
	Environmental change	Feeding	Ebeling & Laur 1988
		Predation	Love et al. 1991
		Ontogeny	Holbrook & Schmitt 1998
		Interactions	
> 10 years	Habitat destruction	Recruitment	Ebeling et al. 1985
	Environmental change	Interactions	Tegner & Dayton 1987
	Oceanographic events	Feeding	Choat et al. 1988
		Predation	

soft-bottomed habitats). Some relevant literature on soft-bottom environments includes: by-catch issues (Andrew & Pepperell 1992), methodology (Rozas & Minello 1997), seagrass beds (Pollard 1984; Bell & Westoby 1986a, b, c; Larkum et al.1989; Ferrell & Bell 1991; Worthington et al. 1992a, b), seagrass beds and reefs (Burchmore et al. 1985; Howard 1989), trawl surveys in deeper water (Tilzey 1994; Francis 1995; McClatchie et al. 1996) and methods of capturing fishes, including widely used techniques such as trawling (Kailola et al. 1993; English et al. 1994; Sainsbury 1996).

6.2 Methods and considerations for assessing reef fish assemblages

6.2.1 Sampling designs

Large- and small-scale spatial and temporal variation in the abundance of reef fishes is common and should be considered in all sampling designs (Table 6.1). Many biological problems will need to address more than one source of variation in the same sampling design. For this reason, nested or mixed-model sampling designs are common. For example, sites (200–500 m apart) may be nested in locations (tens of kilometres apart), and days of sampling may be nested within seasons. Repli-

cation in space and time will enable an investigator to make more general statements about the results of mensurative or manipulative sampling designs (see Chapter 3).

Spatial variation is great at scales ranging from less than a metre to thousands of kilometres on coastal reefs. It is also common for great differences in fish faunas to be found between mainland and offshore island locations at the same latitude in temperate regions (Choat & Ayling 1987; Kingsford 1989) and tropical reefs (e.g. Russ 1984a, b; Williams 1991). Offshore islands (e.g. Poor Knights, New Zealand; Solitary Islands, New South Wales) often have diverse assemblages of fishes of special interest, and in some cases may have endemic species (e.g. *Odax cyanoallix*, Three Kings Islands, northern New Zealand: Choat & Ayling 1987). It is important to consider patchiness of the environment in which you are working for descriptive and experimental investigations. There is generally great variation in the abundance of fishes with respect to habitat type and depth at virtually all latitudes (e.g. Russ 1984b; Jones 1988a; Syms 1995; but see also Holbrook et al. 1990). Some habitats are only found at certain depths, thus the relative importance of habitat and depth in influencing the distribution patterns of fishes can only be separated using manipulative experiments (see 6.3). More precise estimates of abundance can generally be obtained by stratifying sampling according to habitat type and/or depth. If you are comparing different sampling methods, try to keep replicate sampling units (e.g. belt transects) within a particular habitat. Ideally, the whole procedure of preliminary sampling should be repeated in more than one habitat and at replicate locations. The size of patches of habitat, their complexity (e.g. *Macrocystis* forest) and the logistics of sampling them (e.g. bottom time) will have a major influence on the type of method you choose. At the sampling-design stage, consider structured or stratified designs that involve sampling in different habitat types. Where sample units traverse a number of habitats, precision can drop dramatically (e.g. Lincoln-Smith et al. 1991), increasing the difficulty of detecting certain effect sizes (see Chapters 2 and 3). For many biological questions on the distribution of fishes, and processes influencing them, it is relevant to collect concurrent data on change (or lack of change) in habitats.

Temporal variation – the persistence of spatial patterns on ecological time scales is likely to vary according to variation in recruitment, change in habitats, physical forces (e.g. cyclones, ENSO), food resources, ontogenetic changes in habitat requirements, reproductive behaviour, variation in mortality rates through predation, interspecific interactions and the degree of fidelity fishes have to reefs (Table 6.1). For examples of the latter, some territorial fishes may show little variation in abundance for long periods of time (e.g. Pomacentridae: Holbrook et al. 1994). Fishes such as territorial pomacentrids and tripterygiids (Thompson 1986) have high fidelity to small areas of reef within a habitat type for all of their lives. In contrast, some types of fishes may move hundreds of metres and even kilometres within reef systems (e.g. planktivores feeding: Bray 1981; Kingsford & MacDiarmid 1988; reproduction: Johannes 1978) and among environments (e.g. estuarine to reef environments: Vieira 1991; Gillanders & Kingsford 1996). Some species have a loose association with reefs, and variation in their numbers may indicate excursions over the continental shelf (e.g. monacanthid, *Parika scaber*: L. Paul pers. comm.) or excursions over reefs by fishes of open-water habit (e.g. carangids including *Trachurus novaezelandiae*: Kingsford 1989). The behaviour and reef fidelity of fishes should be considered, therefore, for studies measuring mortality and other variables relevant to the population dynamics of species.

6.2.2 Methods

It is crucial to carefully consider the nature of your sampling problem before proceeding with the choice of method and allocation of your sampling units in your sampling design.

Many different methods are used to study reef fishes. This can be confusing when choosing a method, but a focus on the biological problem rapidly narrows the approaches that are likely to be appropriate. It is rare for investigators to use more than a couple of methods to address biological questions (Table 6.2).

The methods used for investigating reef-fish assemblages can be divided into three categories: destructive, non-destructive and non-destructive/ destructive methods (see Table 6.3, p. 138). Although destructive methods have mainly been used for non-quantitative collections of fishes, poison (e.g. rotenone) and explosives have been used in quantitative assessments (Russell et al. 1978; Williams & Hatcher 1983; Howard 1989). Other applications may include the assessment of different fishing methods in reef environments, and the spearing of fishes for experimental and age/growth studies.

I have given greatest emphasis to methods that can be used non-destructively. Non-destructive methods don't run into problems of potential serial depletion as a result of removing fishes from an area. The historical use of destructive methods, however, may be useful in identifying the fish species that have suffered most from fishing (e.g. the relative abundance of fishes caught in spearfishing competitions).

In general, it is very difficult to get accurate estimates of the abundance of all fishes using the same sampling method. Invariably, estimates of some or all of the species are strongly compromised if only one method is used. Multi-species estimates can be improved by counting different groups using different procedures (Lincoln-Smith 1988a). The vast majority of investigations of fishes in temperate waters, however, have focused on a single species (e.g. Moran & Sale 1977), a particular group (e.g. two species of embiotocids: Holbrook & Schmitt 1996), large demersal fishes (e.g. Cole et al. 1990, 1994) or a trophic group (e.g. planktivores: Kingsford 1989). Sampling methods can be optimised for one species, but precision-related compromises are invariably made when a number of species are targeted. Some techniques are clearly best suited for specific tasks such as quantitative studies (e.g. belt transects), behavioural work (e.g. following individuals) and collections for the treatment of live individuals (e.g. tagging fishes at baited stations, anaesthetics). Ultimately the choice of method must depend on the ecological problem, study species, local conditions and a knowledge from the literature of the advantages and disadvantages of each method.

Regardless of the method, your data sheets should include information on date, time of day, tide, habitat (e.g. kelp forest), location, site, size and number of sampling units. If multiple observers are used in the study, record the name of each diver. You may also want to consider adding abbreviations that are used in your database alongside the names of each species or group of fishes.

6.2.3 Non-destructive/destructive methods

Some types of gear are primarily considered as destructive methods, but can have non-destructive applications; for example, applications for tag-and-release or esti-

mates of abundance of species (cryptic species, deep-water species).

(a) Hand lines and set lines
Fishing for the purpose of tag-and-release, using barbless hooks and fishing in shallow water (to avoid having to vent swim bladders). In some cases you may consider hooking and tagging fishes under water. Replicate long-lines or replicate fishers using handlines for defined periods of time can be used for measures of catch per unit effort. Adequate pilot studies would be required to assess the influence of time of day, time spent fishing (catch per unit effort = CPUE) and variation among days before broad-scale sampling designs could be considered; for example, in replicate habitats, reefs, islands and latitudes. Fishes with swim bladders are unlikely to survive retrieval from great depths.

Table 6.2 Examples of applications of methods based on a specific problem.
(D = descriptive or mensurative experiment, E = manipulative experiment.)

Problem	Method design	Sources
Impact of fishing on rocky reefs (D)	Gill nets	Hickford & Schiel 1996
Broadscale distribution patterns (tens of metres to tens of kilometres) of small labrids (D)	Belt transects Habitat stratified	Jones 1984a
Total numbers of fish inside and outside MPAs (D)	Belt transects Habitat stratified Optimal sampling	McCormick & Choat 1987
Patterns of abundance of highly mobile herbivorous fishes in temperate and tropical waters (D)	Time counts – shallow water only	Meekan & Choat 1997
Impact of fishes on plankton (D)	Belt transects Behaviour of fishes Plankton abundance Oceanography	Kingsford & MacDiarmid 1988
Periodicity of spawning of damselfish (D)	Mapped areas Belt transects Daily sampling	Tzioumis & Kingsford 1995
Seasonal movements of sparids within a 120 x 120 km area (D)	Tag and recapture	Crossland 1982b
Reproductive success of small tripterygiid fishes (D and E)	Mapped territories of individual fishes Behavioural observations	Thompson 1986
Influence of topography on the survivorship of small tripterygiid fishes (D and E)	Belt transects of recruits Manipulation of topography	Connell & Jones 1991
Determinants of territory size in a pomacentrid fish (E)	Smorgasbord and fish-removal experiments	Norman & Jones 1984 Jones & Norman 1986

Advantages
✔ Light gear.
✔ Cheap and widely available.
✔ Experienced fishers rather than scuba divers can be used in some studies.
✔ Can be used beyond safe diving depths (Ralston 1982; Ralston et al. 1986).

Disadvantages
✗ Fish may not be caught under some conditions (e.g. dominating presence of other species, timing of reproduction, etc.).
✗ High mortality may be experienced from damaged swim bladders, internal and external damage by the hooks, and predation while hooked.
✗ CPUE can vary greatly with fishers, so it is advisable to calibrate individual differences and/or use the same personnel at all times.

Table 6.3 Methods for studying reef-fish assemblages and typical spatial scale of application.
Platform of application: B = From a boat; CO = Continuous recording in absence of diver; SC = Using scuba; SN = Snorkeller; SU = Submarine; TO = Towed diver.
It should be noted that some of these techniques may have may risks to health (e.g. explosives, poisons and anaesthetics) and appropriate safety precautions should be taken.
Spatial scale: S = Small(< 10 m), M = Medium (tens–hundreds of metres; L = Large (kilometres– hundreds of kilometres). Where a range of spatial scales is given for a method, it generally refers to nested designs; for example, replicate transects are 5 m long, multiple transects are deployed at sites about 100 x 100 m in area, multiple sites are sampled at locations that encompass a kilometre or more of reef, and locations are separated by over tens or more kilometres of coastline.

	Platform of application						Spatial scales
	B	CO	SC	SN	SU	TO	
DESTRUCTIVE							
Poison	✔		✔	✔	✔		S–L
Explosives	✔						S–L
NON-DESTRUCTIVE / DESTRUCTIVE							
Hand and set lines	✔		✔	✔			M–L
Traps	✔			✔	✔		M–L
Gill nets	✔						M–L
Barrier nets			✔	✔			S–M
Baited stations			✔	✔	✔		S–L
Spear			✔	✔			S–L
NON-DESTRUCTIVE							
Belt transects			✔	✔	✔		S–L
Strip transects			✔	✔	✔		S–L
Line transects			✔	✔	✔		S–M
Rapid visual counts			✔	✔	✔		M–L
Stationary visual method			✔	✔			S–L
Cine/video	✔	✔	✔	✔	✔	✔	S–M
Still photography	✔		✔	✔	✔		S
Manta tows			✔	✔		✔	M–L
Diver-propulsion vehicle			✔	✔			M–L
Presence/absence	✔		✔	✔	✔	✔	S–L
Spot mapping			✔	✔	✔		S–M
Behaviour			✔	✔	✔		S–M
Anaesthetics			✔	✔	✔		S–M

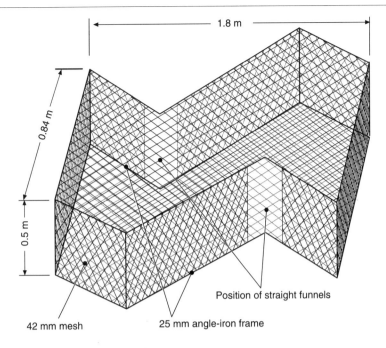

1.8 m

0.84 m

0.5 m

Position of straight funnels

42 mm mesh

25 mm angle-iron frame

Figure 6.2 Diagram of an Antellian or Z trap. Funnels are positioned in the V of each limb of the Z shape.

ADAPTED FROM DAVIES 1989

(b) *Traps*

Traps and their advantages and disadvantages have been studied in detail by researchers working on crustacea (Miller 1983, 1990). Traps have been used for studies on fishes in tropical waters (Miller & Hunte 1987; Koslow et al. 1988; Arena et al. 1994) and, to a lesser extent, in temperate waters (Crossland 1976). Traps vary in shape (e.g. square, round, Z-shape: Cappo & Brown 1994), and the Antellian trap (Fig. 6.2) is one of the most extensively tested for large reef fishes. Traps may be baited, or unbaited, and different species of fishes may respond to the presence of the bait, structural attributes of the cage, and presence of any occupants. Traps can be used for studies of distribution and abundance as well as tag-and-release studies.

Advantages

✔ Fishes can be tagged and released from boats or while diving.

✔ Arrays of traps can be left unattended.

✔ Cryptic fishes and/or fishes found in deep water can be collected.

✔ Can be robust and inexpensive.

✔ Fishes can be collected over a broad range of depths and potentially in different habitats (especially with concurrent observations by divers).

✔ Fish are generally alive at capture and can be brought to the surface in stages to avoid damage to swim bladders.

Disadvantages

✘ High variances and low catches are common.

✘ Many replicate cages are generally required for studies, and the equipment is bulky to transport.

✘ The catch of different species can vary with soak time and mesh size. Small sizes (12.5 mm) generally have lower catches than large sizes (42 mm) (Santurtun 1995).

✘ Residents in the trap may complicate things: some fish are attracted to their own kind or other species in the cage (Cappo & Brown 1994).

✘ Currents can alter the bait halo around the trap.

✘ Bait pickers may vary with location, habitat and time.

✘ It is often hard to convert catches to number of fish per unit area.

✘ Bait can be expensive.

✘ Traps often require conditioning before fish will go near them (e.g. fish dislike fresh zinc galvanising, and bubbles emerging from wood that is not water-logged can frighten fish).

✘ It may be difficult to target a wide range of species because there may be great variation among species in terms of the most appropriate soak time, etc.

✘ Comprehensive pilot studies are required to determine appropriate soak times, etc.

✘ Fish can be damaged if investigators are not careful.

(c) *Gill nets*

A wide range of nets are used in fisheries applications (e.g. Kailola et al. 1993). Nets may be used for destructive sampling or collecting fishes alive for tagging, measuring size, etc. (Moreno & Jara 1984). Monofila-

139

ment gill nets may be used on reefs, and sets of nets can be replicated at multiple locations for fixed periods of time (e.g. three hours: Howard 1989); most of the fishes collected using this method will be dead. Nets used on reefs are weighted so that they rest on the substratum. Floats enable the net to sit as a curtain of mesh. Nets may be set in random directions or in particular habitats or depth strata; for example, perpendicular to the low-tide mark so that fishes moving along the reef will encounter the net. The mesh size depends on target species, and in some cases a mixture of meshes is used in the net (e.g. 25, 63, 102 mm: Howard 1989). Time-of-day effects may be experienced, such as the peak catches at crepuscular periods found by Howard (1989).

Advantages
✔ Multiple locations can be fished at once.
✔ Samples of fishes are obtained that may be used for studies of morphometrics, ageing and diet.
✔ Can be used in water beyond safe diving depths.
✔ Some fish are hardy enough to survive gill netting for tag-and-release studies (e.g. *Arripis trutta*: Bradstock pers. comm.).

Disadvantages
✗ Gill nets can have a great impact on local fishes, and the by-catch of non-target crustaceans (e.g. lobsters and cetaceans) can be unacceptable.
✗ If the markers of the net are lost, or bad weather prevents retrieval, the net can continue to fish and may pose a hazard to divers.
✗ Serial depletion of fishes that have high reef fidelity (e.g. Cheilodactylidae) is likely to be high; sampling in time, therefore, is likely to produce a non-stationary time series with a negative slope (*sensu* Chatfield 1980; Chapter 3) that simply relates to the destructive method.
✗ Nets may snag in areas of complex topography and strong currents, especially when deployed with no knowledge of the habitats in which the net has finally rested.

(d) *Barrier nets*
Divers herd fishes into a diver-deployed net. Plan how you are going to herd the fishes (i.e. proximity of shelter sites, small valleys in the reef topography, etc.). Some fishes (e.g. cheilodactylids) are easily herded and will simply swim against the net rather than trying to swim over it. Mesh sizes will depend on the target species, but small mesh will reduce the chances of accidental gilling. Moreover, fishes seem to have a knack of squeezing through what appear to be impossibly small holes. Fishes can be collected by bunching up the net or scooping them with a dip net. Juvenile fishes may be collected in very small barrier nets (e.g. 2 mm mesh,

0.5 x 10 m) particularly in parts of the reef where there is a loose substratum, such as a boulder field.

Advantages
✔ Collect live fishes for tagging, measuring, etc.
✔ Fishes can be taken for destructive samples.
✔ Little damage to the reef.
✔ Nets can usually be transported without taking a lot of space.
✔ Divers can position the nets for maximum effect.

Disadvantages
✗ Can be cumbersome to use under water.
✗ Usually not suitable for quantitative studies.

(e) *Baited stations*
This is simply an extension of the stationary visual census technique (6.2.4e) to include bait to attract fishes to the observation area in which fishes are counted during a fixed time period (e.g. 10 minutes). The technique was designed to detect cryptic species and those that avoid divers.

Advantages
✔ Good where visibility is poor and fishes are cryptic.
✔ Non-destructive.
✔ Can be used with video methodology to further reduce diver-negative effects (6.2.4f).
✔ No structures that may influence the behaviour of fishes (cf. traps, 6.2.3b).
✔ Relatively easy to do multiple replicates.
✔ Useful for tag and recaptures, either by hand net or more specialised equipment such as diver-controlled electrofishing (Davis & Anderson 1989).

Disadvantages
✗ There have been no comparisons of this method with others.
✗ All of the bait-related biases listed for traps (p. 138).
✗ Davis and Anderson (1989) found that baited stations did not provide estimates of density as accurate and precise as those from visual transects.

6.2.4 Non-destructive methods

The information presented in this section is largely derived from the following references: Russell et al. 1978; Sale & Sharp 1983; DeMartini & Roberts 1982; GBRMPA 1985a, b; Bortone et al. 1986; McCormick & Choat 1987; Davis & Anderson 1989; Mapstone & Ayling 1993; Cappo & Brown 1994; and my own experience. Because non-destructive methods such as counts of fishes in transects and rapid visual counts have been particularly well used in biological assessments, more detailed coverage is given in section 6.3. Most of the non-destructive approaches to research indicated here have been done on scuba, though some of the methods may be applied from submarines and remotely operated vehicles (ROVs) for access to deeper reefs (e.g. Ralston et al. 1986). Studies using ROVs, however, have encountered many problems with snagging of the umbilical cord (Okamoto 1989).

(a) *Belt transects*

Diver counts fishes within a defined area (Fig. 6.3, p. 142). The area may be delineated before the count, or a tape measure is laid out and the fishes are identified within a given distance of the tape (e.g. 2.5 m each side). See 6.2.6 for a discussion on size of sample units and replication. Width of the sample unit can be estimated by body lengths and/or a known distance on the tape. The diver swims in a zig-zag path; for comparative purposes it is useful to record the time taken to complete transects. Transects are usually placed randomly or haphazardly within sample strata (e.g. different habitats, which will often give greatest statistical power and range of options with analyses. Although some hypotheses require the use of permanent transects (Wilkins & Myers 1992), this approach will violate the assumptions of many statistical tests because of non-independent replication in time. It may be appropriate, however, to use repeated measures.

Advantages
✔ With replication a precise estimate of density can be obtained.
✔ The sizes of fishes can also be estimated.
✔ May combine counts of fishes with those of large and discrete invertebrates, or a description of benthic cover by using a second diver.

Disadvantages
✗ Time-consuming.
✗ Biases include observer error (e.g. estimating belt boundary), how transect size relates to target population, recounts of fishes, how results change according to diver swimming speed, etc., and variation in estimates of abundance can be great when multiple divers are used.
✗ Generally requires good visibility.

Results: Number per unit area or volume (e.g. Holbrook et al. 1990); biomass if there are estimates of size and a knowledge of fish weights.

(b) *Strip transects*

Counts fish within a defined area. The area may be delineated before the count, or a tape measure is laid out and the fishes are identified within lanes, sometimes at different distances from the tape (Fig. 6.3). McCormick and Choat (1987) used multiple-strip transects to record the positions of cheilodactylids in a mapped area. The

use of a diver in each lane reduced the chances of multiple counts of fishes within the study area. In some cases fishes are counted in two lanes (one each side of a tape), but this does increase the chances of recounts. The distinction between belt and strip transects is sometimes blurred, and investigators have sometimes used these terms synonymously. Strip transects are a useful way of determining the influence of transect width on estimates of abundance of fishes. See 6.2.6 for a discussion on size of sample units and replication.

Advantages
✔ With replication a precise estimate of density can be obtained.
✔ Pilot studies on the influence of transect width are possible.
✔ The sizes of fish can also be estimated.
✔ May combine counts of fishes with those of large and discrete invertebrates, or a description of benthic cover, using a second diver.

Disadvantages
✗ Time-consuming.
✗ Biases include observer error (e.g. estimating lane boundaries), how transect size relates to target population, recounts of fishes, and other disadvantages as for belt transects.

Results: Number per unit area; correction factors are sometimes used (e.g. Willan et al. 1979; Sale & Sharp 1983).

(c) *Line transects*

Diver counts fishes within a defined area. The area may be delineated before the count, or a tape measure is laid out and the fish identified and position estimated within a given distance of the tape (Fig. 6.3). This method is commonly used in terrestrial studies (Thresher & Gunn 1986), but its application to highly mobile and sometimes secretive fishes is problematical. It is unlikely to be very useful in sampling designs that encompass broad spatial scales, but may be useful for pilot studies to determine appropriate width of transect. See 6.2.6 for a discussion on size of sample units and replication.

Advantages
✔ With replication a precise estimate of density can be obtained.
✔ The sizes of fishes can also be estimated.
✔ Corrections can be made based on a knowledge of estimates of abundance at different distances from the tape.

Disadvantages
✗ Time-consuming.
✗ Biases include observer error (e.g. estimating

distances from the transect line) and recounts.

✗ Difficult to make accurate estimates of the position of fast-moving fishes.

✗ Requires good visibility.

Results: Number per unit area; correction factors are sometimes used (e.g. Willan et al. 1979; Sale & Sharp 1983).

(d) *Rapid visual counts*

RVCs have been used in coral reef (e.g. Williams 1982) and rocky reef environments (e.g. Brandon et al. 1986). This type of count is also called 'timed counts' or 'interval counts' (Cappo & Brown 1994). Diver swims freely over an area (and potentially multiple habitats) and records fishes. Duration of swim is defined before the dive; intervals of 3–45 minutes have been used. See comparison of methods, 6.2.6.

Advantages

✔ Fast method of obtaining the relative abundance of a large number of species.

✔ No tapes required, thus greater ease of counting in topographically and complex areas with a diverse range of species.

Disadvantages

✗ Difficulty in delineating area of count (e.g. influence of currents, visibility, etc.).

✗ Where the technique is poorly applied (e.g. no replication), it may disregard variation in spatial distribution of species;

✗ Requires good visibility.

Results: Rank abundance; arbitrary categories (e.g. rare, abundant); number of individuals per unit time; log abundance per unit time; frequency of occurrence (see 6.2.8).

(e) *Stationary visual technique*

Diver takes up a position some distance from quadrat (e.g. 2 x 2 m), and all fishes that are in or pass through a cube (all or part of the water column) during a set time interval (e.g. 10 minutes) are counted; during the last minute the area is closely searched for secretive or small fishes. Bohnsack and Bannerot (1986) described a stationary visual census technique that they argued was suitable for quantitatively assessing community structure of reef fishes. Censuses of reef fishes were made within a 7.5 m radius by a diver (for five minutes; Fig. 6.3) at randomly selected stationary points (recommended range of visibilities, 8 m+). The study area should be marked out in a way that does not influence the behaviour of the fishes. The technique can be useful for monitoring experimental treatments (e.g. Andrew & Jones 1990) that are too small for transects,

and measures of abundance are augmented with behavioural data. Furthermore, it can be modified so that baits are left in the demarcated area for a fixed period of time. This modification may be particularly good for cryptic predators (see 'baited stations' above).

Advantages

✔ Area of observation easy to delineate.

✔ The method can provide information on frequency of occurrence, fish size, abundance, behaviour and composition of the assemblage.

✔ Large numbers of samples may be obtained (Bohnsack & Bannerot 1986).

✔ Useful in studies where investigators are concerned with use of the substratum (e.g. Choat & Bellwood 1985) or experimental manipulations (e.g. Jones & Andrew 1993).

✔ May be modified to use baits (see p. 140).

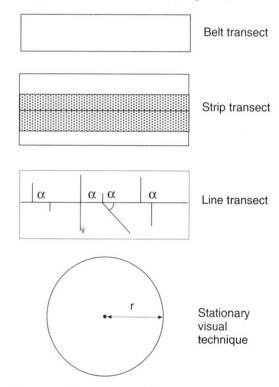

Figure 6.3 Different transect techniques and the stationary visual technique. Fish are counted in a single lane in 'belt transects' (see Table 6.4 for typical dimensions) and more than one lane for strip transects. Two lanes have been used most commonly, but multiple lanes at different distances from the centre line are sometimes used to estimates bias in counts as the width of the transect increases. The position of individual animals is recorded in line transects, but this is often hard to do for fishes. The stationary visual technique involves a diver counting fish in an area determined by a fixed radius from the middle. Bohnsack and Bannerot (1986) used an radius of 7.5 m; other investigators have positioned themselves on the outside of a smaller count area.

Disadvantages

✗ Biased towards fish that move about.

✗ Covers a small area over the duration of a dive.

✗ May underestimate the abundance of rarer species.

✗ The presence of a diver may influence counts (cf. video methods).

✗ The maximum size of the sample unit will be strongly influenced by visibility.

Results: Number and/or behavioural events per unit area and time.

(f) *Cine/video methods*

Modern video systems have all but replaced the need for cine cameras. Cameras may be used by free-swimming divers or from fixed mounts. Applications include the following: swimming with a hand-held underwater camera along a transect; stationary techniques where the diver pans the camera through a circular area and exposure time may be controlled (e.g. two minutes per revolution); mounting the camera on a well-weighted tripod so that a given area can be studied for a pre-determined period of time.

The latter may be particularly useful for behavioural work where long-playing tapes are used, or for time-lapse photography over long periods of time. With all of these techniques, fishes are counted and behaviour recorded on replays back at the laboratory. If estimates of size are required, then stereo-video measurements are required (e.g. Klimley & Brown 1983). Video techniques can be particularly useful for a description of habitats in which fishes are found (see Chapter 4).

Advantages

✔ The diver does not have to be competent at fish identification.

✔ The film can be replayed any number of times for counts to be viewed by many observers.

✔ Provides a permanent record.

✔ Sampling may be done for extended periods of time using time-lapse.

✔ Provides records of abundance and behaviour.

✔ Stereovideography can be used to obtain estimates of size (see 6.2.9).

Disadvantages

✗ Requires expensive and specialised equipment.

✗ Although the tolerance of modern Hi-8 video recorders to a wide range of light conditions is great, there may be problems with focus and exposure.

✗ Quality of the results may vary according to habitat, light conditions and visibility (see Larson & DeMartini 1984) – some of these problems may be overcome if cameras are used from fixed positions.

✗ Davis and Anderson (1989) found that video transects did not provide estimates of density as accurate and precise as those from visual transects.

✗ Analyses are time-consuming and are additional to the time spent in the water.

Results: Number per unit area, biomass if there are estimates of length and length-weight relationships; behavioural categories.

(g) *Still photography*

Still photography has also been used as a quantitative method by taking exposures at each of the four compass co-ordinates and repeating the procedure at a number of sites (Bortone et al. 1986). Photography is probably most useful for supplementing quantitative data from another technique to provide information for taxonomy (e.g. Francis & Randall 1993), descriptions of colour morphs, categories of behaviour, etc. Stereo-photography, however, has been used to estimate the size of reef fish. In this case care should be taken to get large sample sizes (van Rooij & Vidler 1996). Still photography can also be useful for descriptions of habitat type (see Chapter 4).

Advantages

✔ Records types of fishes, coloration, categories of behaviour and identification marks for behavioural work.

✔ Enables estimation of size of nests of demersal eggs such as those of pomacentrids and tripterygiids (tracing onto acetate sheets should also be considered for this) and habitats in which fishes are resident.

✔ Estimates of size with stereophotography; good to augment other methods.

✔ May be particularly useful in studies that use submarines.

Disadvantages

✗ As for 6.2.4f.

✗ Generally unsuitable for quantitative measures of fish abundance.

✗ Developing large numbers of photographs can be expensive, although this is often trivial compared to the cost of the entire diving operation.

(h) *Manta tows*

This method is to tow and stop at two-minute intervals for 10 seconds so that the diver can record observations (see also 4.3.1c).

Advantages

✔ Covers a large area quickly.

✔ Useful for identifying areas for more detailed investigation (e.g. Williams 1982).

✔ Supplements detailed information from strip transects or rapid visual counts.

Disadvantages

✗ Observer has little control over area towed or speed of tow and may have problems remembering data.

✗ Unlikely to have time to estimate the sizes of fishes.

✗ Difficult to keep consistent search image.

Results: Number per unit area; presence/absence.

(i) *Diver-propulsion vehicle (DPV)*
Method as for manta tows (see also 4.3.2).

Advantages

✔ Cover a large area.

✔ May control direction and speed of tow.

✔ As for manta tow.

Disadvantages

✗ Cost.

✗ Battery charge time (approximately three hours).

✗ Effect of motor noise on fishes is unknown.

✗ Problems remembering and transcribing data.

✗ Unlikely to have time to estimate sizes of fishes.

Results: Number per unit time.

(j) *Presence/absence*
Record all species encountered during a dive.

Advantage

✔ Rapid.

Disadvantages

✗ No estimate of abundance or size.

✗ Very limited application.

✗ Number of species recorded may vary according to the knowledge of different observers.

Results: Presence/absence

(k) *Spot mapping*
Diver counts all fishes within a defined area; the position of fishes is often plotted on a map of the study area. In some cases plotting the positions of fishes may be facilitated by marking out the area with strip transects (see 6.2.4b).

Advantages

✔ Absolute abundances of species obtained (usually within a small, well-defined area).

✔ Enables bias of other techniques to be assessed, where a record of the distribution pattern for diver neutral species is recorded on a map (e.g. Scorpaenidae: Larson 1980a,b)

✔ Sizes of fishes can also be estimated.

✔ Description of the spacing of individuals and territorial behaviour (i.e. combine with behavioural work, see 6.2.4l).

Disadvantages

✗ Time- and cost-intensive to map areas.

✗ Recounts of fishes.

✗ Generally small spatial scales.

Results: Number per total area; map of distribution, distribution by biomass could also be plotted; utilisation of habitats and spatial descriptions of interspecific and intraspecific interactions.

(l) *Behaviour*
Diver follows fish for defined periods of time. Behaviours are categorised and type of substratum over which fishes are found can be recorded. Categories of behaviour used by Jones (1981) are given in 6.4.5 and an example of a data sheet in Figure 6.4. The position of fishes and important events (e.g. spawning, feeding, conspecific and interspecific interactions) can be recorded on a map (see 6.4.5), as well as the substratum on which the behaviour takes place. Behavioural studies may be combined with tagging so individuals are easier to identify. Some kinds of tags may enable movements of fishes to be tracked from the surface (Kaseloo et al. 1992).

Advantages

✔ Obtain activity budgets for fishes of different sizes and/or sexes at different places and time.

✔ Data on behaviour can augment data on abundance and give information on individuals.

✔ Behavioural data is important in studies where the objective is to describe habitat utilisation, reproductive behaviour (e.g. Thompson & Jones 1983), interaction between sexes, different sizes of fishes and species (e.g. Holbrook & Schmitt 1992) and the monitoring of experimental treatments (e.g. Andrew & Jones 1990).

✔ Although individual fish can sometimes be identified based on body markings (e.g. Connell & Jones 1991), fishes may have to be tagged for identification, especially if it is likely that they will move long distances; combine with mapping (see above).

Disadvantages

✗ Very energy-intensive and time-consuming.

✗ Generally small spatial scales.

(m) *Anaesthetics*
A diver anaesthetises all fishes within an area, or targeted individuals, with quinaldine, MS-222, benzocaine or clove oil (Munday & Wilson 1997). Fish recover quickly, which is useful when they are to be tagged/measured and released. It is useful to cover topographically rugose areas with a fine-mesh net to prevent fishes

from escaping. Nets that are weighted around the edge (e.g. with a chain) are useful for this.

Advantages
- ✔ Fishes can be targeted without killing all fishes within an area, as happens with ichthyocides (e.g. rotonone, or derris dust);
- ✔ Useful for calming fish for tagging and/or measuring.

Disadvantages
- ✗ Anaesthetics can only be used to target individuals or treat small areas (e.g. patch reefs).
- ✗ Difficult to use in areas with strong currents.
- ✗ There are health risks with some anaesthetics (e.g. quinaldine).

Figure 6.4 Typical data sheet for behavioural work. It is divided into 15-second time intervals so that different types of behaviour can be recorded (e.g. F = foraging, R = reproductive behaviour). Other details such as number of bites and the substratum from which they were taken can also be recorded. The final design of the data sheet will depend on the objectives of the study. A stopwatch can be attached to the top of the diver's slate to keep track of time. The position of fish could also be recorded on small-scale maps.

(n) *Tagging*

Many types of tags are available for use on fishes (Parker et al. 1990). Investigators must make choices based on the size of fishes, the period of time that the tag must persist, any need to identify individuals, behavioural changes that relate to tagging, the ability of fishes to sustain damage, and the ability to identify fishes without tags (i.e. from natural markings: Jones 1983; Thompson 1983; Connell & Jones 1991). The technique may be combined with exercises in validation of ageing where fishes are injected with tetracycline, strontium, or other chemical, and released in the field (see 6.2.10). Methods of tagging range from conventional dart and T-bar tags (examples see Fig. 5.7) to high-technology coded wire tags (CWTs) that enable the identification of individuals, from marks on the tags and the detection of magnetised tags in fishes with a magnetic detector (Buckley & Blankership 1990; Buckley et al. 1994; Beukers et al. 1995). CWTs and chemical tags are invisible. Chemical tags include antibiotics such as tetracycline (Tsukamoto et al. 1989), strontium (Gallahar & Kingsford 1996; Pollard et al. in press) and radioactive isotopes (Kato et al. 1991). Chemicals are introduced by injection or bathing for a fixed period of

Species						
Date and time			Site			
Location			Habitat			
Other			Observer			
Fish 1		SL	Sex			
min \ sec	15	15	15	15	Bites	Notes
1						
2						
3						
4						
5						
6						
7						
8						
9						
10						
Fish 2		SL	Sex			
min \ sec	15	15	15	15	Bites	Notes
1						
2						
3						
4						
5						
6						
7						
8						
9						
10						

time. The latter approach is useful for batch-tagging of large numbers of fishes. It is also possible to identify groups of fishes by analyses of natural variation in the chemistry of bone (e.g. Gillanders & Kingsford 1996). Small and large fishes may be tagged with microwire and other invisible tags, subcutaneous dye (e.g. tattoo ink, acrylic paint: Lotrich & Meredith 1974), fluorescent tags (external marker that often requires a torch to detect; Visible Implant Fluorescent Elastomer = VIE tag; Northwest Marine Technology Inc, unpublished; Dewey & Zigler 1996), fin clipping, heat or freeze branding (e.g. Jones 1987), liquid latex (Forrester 1990) and, if the species is robust enough, conventional T-bar tags, etc. Double-tagging can help calculate the frequency of tag loss. In some cases even parasites have been used as biological tags on fishes, but there are generally used where the spatial scale of the study is great (e.g. hundreds to thousands of kilometres: Buckley & Blankership 1990).

Advantages
✔ Identification of individuals or batches of fishes of a wide size range.
✔ Useful for studies on mortality (Hilborn & Walters 1992), movement and migrations, growth of individuals, spawning success of individuals (e.g. Jones 1983) and sex change of individuals.
✔ Assessments of the importance of nursery areas (e.g. Gillanders & Kingsford 1996); biological justifications for the size of marine reserves.

Disadvantages
✗ Energy-intensive work.
✗ Increase in mortality of fishes as a direct result of tagging.
✗ Increase in mortality of fishes as an indirect result of tagging (e.g. increased rates of predation).
✗ Sublethal effects (especially if chemicals are used) may change the behaviour of fishes (movement, social ranking, response of opposite sex during spawning).
✗ In some cases cleanerfishes may remove tags (pers. obs.); some techniques will only enable the identification of batches of fishes; if tags are relatively temporary (e.g. fin clips, brands), fast-growing fish may lose the tags first.
✗ Difficulty in recovering tagged individuals (time taken to recover and choice of search area).

6.2.5 General problems with visual techniques

Each method has its own advantages and disadvantages (see 6.2.4). Some problems, however, are common to all methods of visual assessment:

- Water visibility.
- Visibility of the fish.
- Error in size estimates.
- Varying visibility of fishes among habitats.
- Recounts.
- Experience of the observer.
- Variation in swimming speed and search pattern of the observer (where appropriate).

For many methods these problems can be minimised by giving detailed descriptions of the actual procedures. For example, fishes were counted in belt transects when minimum visibility was 6 m; approximate times to search for fishes in different habitat types given; description of search pattern (e.g. under boulders, in water column). Furthermore, training before the study begins will familiarise divers with details of the method (see Chapter 2).

6.2.6 Sample-unit size and number

The discussion in this section will focus on transects and RVC methods, but the principles also apply to other methods of sampling.

Regardless of the method used for estimates of abundance of reef fishes, consider the following:

- One or a few very large sample units are much harder to search than many small units, and the efficiency of searching may drop with increase in the sample unit size (Mapstone & Ayling 1993).
- Replication gives some indication of the variation in densities found within a sampling locality.
- Replication enables more rigorous statistical tests.
- Repeated counts along one transect line do not constitute independent replicates, moreover, space and time are confounded with this approach (Sanchez Jerez in press).
- It is easier to keep small sample units within a defined habitat;
- In large sample units there is increased possibility of losing count.
- Measures of species richness can be calculated for each replicate (or, if necessary, pooled) for comparisons of species diversity.

More detail on the issues of sampling accuracy and precision are given in Chapters 2 and 3.

Ideally, optimal sample-unit sizes should be determined in a pilot study and preferably at multiple sites and habitats for a more accurate 'guesstimate' of the mean densities and scales of patchiness of aggregations of fishes. The formal procedures described in Chapters 2 and 3, however, are appropriate only to a single species or closely related group of organisms. Accordingly, compromises must be made when a variety of species is counted.

Ultimately, transect size will depend on the question that is being asked as well as the behaviour and size of the study animals and the nature of the habitats and topography in which they are found. Although medium-to-large reef fishes (> 50 mm SL) have often been counted in 50 x 10 m transects in temperate waters of New Zealand, comparisons of transect lengths in New Zealand have often shown that 50 x 10 m transects generally do not give the precision of a larger number of smaller transects (Table 6.4). McCormick and Choat (1987) found that 20 x 5 m transects gave the most precise estimates of abundance for cheilodactylids in the Leigh marine reserve. Similarly, in counts of demersal reef fishes on temperate reefs of eastern Australia, I concluded that 25 x 5 m transects gave precise estimates of abundance for many species, useable mean values, reduced chances of missing fishes, and transects could be kept in different habitats (see Holbrook et al. 1994 and Fig. 3.1). Mapstone and Ayling

(1993) found that estimates of density of fishes were lower in very long transects, presumably as the concentration of the diver dimishes (e.g. 100 x 4 m transects). Kingsford (1989) concluded that for planktivorous fishes 25 x 10 m transects gave more precise estimates of abundance than those measuring 50 x 10 m. Furthermore, at depths of 20–25 m only three 50 x 10 m transects can be completed by a diver before getting into decompression time; this compares with five 25 x 10 m units. The information available, therefore, suggested that the use of 25 x 10 m units would result in more precise data in a general study on densities of fishes than 50 x 10 m transects.

The width of a transect may affect the accuracy of a count within the defined area. Sale and Sharp (1983) concluded that the error in estimates rises with increasing width of transect, and this was corroborated by Mapstone and Ayling (1993), who found estimates of density of large coral reef fishes were higher in narrow (i.e. 4 m cf. 12 m and 20 m) transects. When a diver is searching for benthic reef fishes, the width of the strips covered should not exceed 5 m (2.5 m each side of the diver, just over one body length), especially over heterogeneous habitats. If lanes are larger, individual fish feeding or resting near the substratum may easily be missed and searching a given area may become less efficient

Table 6.4 Examples of sample-unit sizes of transects that have been used to obtain estimates of abundance of reef fishes. In some cases the size of units is given as a three-dimensional measure, which is often appropriate because fishes are found throughout the water column.
NA = multiple counts were done in an area over a period of eight days, V = vary according to the requirements of precision in the study.

Type(s) of fish	Unit size (m)	n	Source	Country
Mixed medium–large	30 x 10	1	Russell 1977	New Zealand
Mixed medium–large	50 x 10	5	Ayling 1978	New Zealand
Cheilodactylidae	50 x 10	5	Leum & Choat 1980	New Zealand
Mixed small–medium	100 x 3	5	Sale & Sharp 1983	Australia
Labridae	50 x 10	5	Jones 1984a, b	New Zealand
Mixed medium–large	50 x 10	3	Schiel 1984	New Zealand
Whole assemblage	100 x 2	NA	Bortone et al. 1986	USA
Mixed medium–large	50 x 10	5	Choat & Ayling 1987	New Zealand
Cheilodactylidae	20 x 5	8	McCormick & Choat 1987	New Zealand
Mixed planktivores	25 x 12 x 30	5	Kingsford & MacDiarmid 1988	New Zealand
Fishes in *Macrocystis* forest	30 x 3 x 1.5	8	Holbrook et al. 1990	USA
Mixed small–large	60 x 2	4	Lincoln-Smith et al. 1991	Australia
Mixed medium–large	50 x 5	V	Mapstone & Ayling 1993	Australia
Mixed medium–large	25 x 5	10	Cole 1994	New Zealand
Mixed medium–large	25 x 5	6	Holbrook et al. 1994	Australia

for adequate coverage (cf. narrow belt transects). Moreover, size estimates may be very inaccurate at distances greater than 3 m.

Where transects of more than 5 m wide are used, the diver generally searches two narrower lanes (half of total width) down each side of the tape (i.e. strip transects); for example, in the survey of the Poor Knights Islands, 50 x 10 m transects were used (Schiel 1984). A diver swam along each side of a 50 m tape and counted fishes in 5 m lanes. Recounts may be a problem with this method. I counted benthic reef fishes at five localities within the Leigh marine reserve by use of five replicate 25 x 10 m strip transects and 50 x 5 m belt transects. If recounts were a major problem with the 25 x 5 m transects (two lanes), higher mean estimates of density would be expected, but this was not found (i.e. rank abundance varied with method). It was concluded, therefore, that counting fishes down each side of the tape did not significantly alter the accuracy of an assessment. The perceived problem can be avoided, however, by using belt transects. A logistical difficulty for a single diver with a long, narrow transect, however, is that they must rewind a long tape and therefore lose valuable bottom time. The biases of using two divers, one on each side of the tape, are unknown. Inconsistent counting abilities may be expected between divers, especially if training and calibration of individuals is not done. Planktivorous fishes swimming in the water column are easily seen at distances of up to 5 m from the tape (Kingsford 1989). For this reason these fishes may be counted in a wider lane than demersal fishes. There are, however, no formal validations of this. Cole et al. (1992) counted planktivorous fishes in a 5 m lane.

This section has focused on transect sizes that are most appropriate for general studys of medium-to-large fishes. Clearly, counts for small, cryptic fishes (e.g. tripterygiids: Thompson 1979; 10 x 1 m transects) or recently recruited fish will require units of a much smaller size.

The time intervals of RVCs that have been used for a single count vary from 75 minutes (Syms 1995) to three minutes (Kingsford 1989, Cole et al. 1992; Gillanders 1997a; Meekan & Choat 1997). The choice of time interval has been based on the physiological limits of divers (45 minutes: Williams 1982; 75 minutes: Syms 1995), the time when a species curve asymptotes (30 minutes: Russ 1984a, b) or the distance travelled during a count and the importance of replication (Kingsford 1989). Investigators generally give the approximate area covered during a count. It is important that this be known, especially if more than one observer is used during the study. Kingsford (unpub. data) found that four different divers travelled 42–56 m during a three-minute count at one location. Moreover, a single diver

tested the distance travelled at three locations; it ranged from 36 to 44 m (Kingsford 1989). Greatest variability was found in areas with strong currents, whereas replicate three-minute counts in areas with weak currents covered areas of similar size. The distance travelled during a timed swim may be calibrated by swimming along a tape.

The diver generally restricts the counts to a lane 10 m wide (Williams 1982; Russ 1984a, b). This width would need to be modified if cryptic fishes were being counted (e.g. 1.5 m: Syms 1995).

6.2.7 Comparisons of methods for estimates of abundance and species diversity

Transects and RVCs (see 6.2.4a–d) are the most commonly used in studies on distribution and abundance and to a lesser extent, the stationary visual technique. Methods f and g are useful for providing preliminary qualitative information (see 3.2.2). Although it is felt that video techniques are unnecessary for studies of abundance, they may have useful applications where submersibles are used (e.g. Ralston et al. 1986). In addition, a video may be left in position on the bottom to photograph fishes as they use the substratum (application of stationary count method: Harvey pers. comm.) and recordings can be carried out at preset times (including stationary counts).

General advantages and disadvantages of these techniques are given in 6.2.4. There is a tendency to use one of these methods because of tradition (e.g. strip transect 500 m^2) or an inherent suspicion of other techniques. Despite the fact that investigators continually use one technique for all types of fishes (e.g. Brandon et al. 1986), no single method will give accurate estimates of abundance for all types of fishes (Lincoln-Smith 1988a). Where one method is used, at best, reasonably accurate abundance measures will be obtained for a certain suite of fishes and totally inaccurate abundance measures for the rest. This is due to the different behaviours of fishes (e.g. cryptic, highly mobile, diver-negative, -positive, -neutral) and their colours (dark – cryptic; bright – easily seen). Moreover, fishes that are resting on the substratum require a totally different search image from those in the water column.

Some species or size-classes of fishes scare easily in the presence of a diver (Cole 1994). When a tape is unreeled, they may be observed, whereas on the return trip (during the actual count) few fish may be recorded. Fishes that are diver-negative may be counted as the tape is unreeled. Assessments of abundance may, however, be best achieved by using RVCs for these species, as Meekan and Choat (1997) found for butterfish (*Odax pullus*) and Gillanders (1997a) for a large labrid fish (*Achoerodus viridis*). Furthermore, many juvenile fishes hide quickly when a diver is first observed, so they, too, may be best counted by RVCs. RVCs were found to be particularly useful when counts of planktivorous reef fishes were made along the coast and offshore islands off Northland, New Zealand (Kingsford 1989); they gave a similar picture of fish abundances to strip transects at the same locations. Moreover, RVCs took half as long to complete, and the time-consuming problem of setting a transect line over rugged bottom types (especially vertical faces) was avoided. Similar problems are encountered in very shallow water. Cole (1992) also used this method at the Kermadec Islands.

Belt transects, with replication, may enable precise estimates of density to be obtained. Although for some schooling species it is virtually impossible to gain precise estimates of density, a number of reef fish species are diver-neutral and precise estimates of density can be obtained (e.g. Cheilodactylidae: McCormick & Choat 1987; Nototheniidae: Kingsford et al. 1989; surfperch, Embiotocidae: Holbrook et al. 1990). Furthermore, the tape provides a good reference for both direction and area searched when cryptic fishes are counted. The presence of a tape is also a useful reminder of how the lengths of fishes are magnified under water. Belt transects may also enable more than one group of organisms to be counted over the same line (e.g. cryptic fishes, fishes in the water column, mobile invertebrates: see 6.4.1; point counts of benthic organisms, Chapter 4). Davis and Anderson (1989) compared estimates of abundance of temperate reef fishes in California that were based on three methods: visual transects, video transects and time counts at bait stations. Visual transects provided the most accurate estimates. A tag-recapture method was also used to estimate numbers of fishes in the study area. Estimates of abundance were several times greater than densities previously reported for other methods. More of these types of comparisons need to be done, as relatively small numbers of some species of fishes were tagged by Davis and Anderson (e.g. *Paralabrax*), but a bias toward recounting tagged individuals more than untagged individuals was considered unlikely.

In any study, therefore, the decision whether to use RVCs, transect or the stationary visual technique will depend on the species or guild of fishes to be targeted, the time available at each location, ease of laying out a tape, the required level of precision, and the desirability of sampling other organisms with the same sample unit. Finally, the detailed quantitative information derived from these techniques may be supplemented with manta tows, DPV tows or casual observations (presence/absence data) of fishes outside the areas delineated in counts. This type of information may be particularly important for descriptions of biogeographic pattern (e.g. Kingsford et al. 1989). In some instances, destructive samples may need to be taken for the purposes of identification (e.g. Francis & Randall 1993; see also Chapter 10).

6.2.8 Information recorded in counts

Data can be recorded as any one or more of the following methods:

- Numbers.
- Log abundance.
- Rank abundance.
- Categories (e.g. rare, abundant).
- Frequency.
- Sizes (see 6.2.6).
- Microhabitats in which fishes are first observed.

For the same amount of energy, actual numbers or log abundances will give considerably more information than rank abundance or category methods. Actual numbers and log abundances will provide quantitative information on densities that can be used in powerful statistical tests. Furthermore, the data will enable species abundances to be ranked. Log abundances are generally used where it is accepted there is a large error in the count e.g. schooling fishes or a large number of species. For example, Williams (1982) counted 146 species on the Great Barrier Reef in log_5 abundance groups (1 = 1, 2 = 2–5, 3 = 6–25, 4 = 26–125, 5 = 126–625, 6 = 626–3250, 7 = 3250+). In contrast, Russ (1984a, b) used log_3 for 51 species of herbivorous fishes in the same area (1 = 1, 2 = 2–3, 3 = 4–9, 4 = 10–27, 5 = 28–81, 6 = 82–243, 7 = 244–729, 8 = 729+). Kingsford (1989) also used log_3 to estimate abundances of planktivorous reef fishes along the coast of Northland, New Zealand.

Where densities of fishes and the number of species are relatively low, it would be unnecessary to use log abundances; actual counts would suffice (e.g. Gillanders & Kingsford 1993; Gillanders 1997a; Meekan & Choat 1997). Where large numbers of species or extremely high densities of fishes are encountered, however, log abundances will cater for exponential error in density estimates. In a study by Kingsford (1989), actual counts gave a similar picture of the abundances of planktivorous reef fishes along the coast of Northland to those gained from log counts. Thus it made no difference how the data were recorded in this situation. Where log abundances are used in statistical analyses, the average value of the category should be used, and not the lowest end of the range, or the raw log category. Care should be taken in interpreting variance structures when log categories are used, because in some cases variance can be zero for a set of replicates that have the same log category.

Some major problems have been found with rank-abundance techniques (see DeMartini & Roberts 1982). Species are scored according to their order of encounter in a free swim of a specific time interval (e.g. 10 minutes). Where 12 intervals are used, a species will score 12 if it is observed in the first time interval, 11 in the second, 10 in the third, and so on; a total score is derived after the last interval (DeMartini & Roberts 1982). Different methods have been proposed by Jones and Thompson (1978) and Kimmel (1985), but both of these overemphasise the importance of widespread, albeit rare, fishes, while underemphasising patchy, although abundant, species (DeMartini & Roberts 1982).

The use of categories is a very crude approach. The method, however, may be of some use when the diver's ability to record data is restricted while being towed by a manta board, DPV or where data are collected as incidental (Francis et al. 1987). If data are collected in this way, it is hard to know whether observed differences in abundance accurately reflect distribution patterns or differences in the area searched or search image.

Frequency data have sometimes been collected. For example, Syms (1995) analysed the relative frequencies of blennioid fishes and did not attempt to standardise the counts to densities.

It is also possible to record the microhabitat in which fishes are found while counts are made. Syms (1995) used 13 classes of habitat to relate to the habitat on which blennioid fishes were found during time counts.

6.2.9 Target species, size and sex

In many cases, estimates of size will be as important as measures of abundance. For exploited species in particular, data should be available to enable adults to be differentiated from juveniles. It is important that divers are briefed on the length that is to be estimated from the snout, whether standard (to the end of the spine, before tail), fork (to the V of the tail) or total (to the tip of the tail). The use of total lengths should generally be avoided, as fishes are sometimes observed with the end of the tail missing. I investigated the ability of divers to estimate the standard lengths of fishes using model fishes. It was found that abilities varied widely among divers on their first attempt. However, experienced divers who knew their personal bias had an 80% chance of placing large reef fishes in the correct 50 mm size-class. Despite this, in detailed studies of small fishes (under 30 cm) of a single species, an experienced diver may consistently estimate their lengths to within 1 cm of the correct value. Accuracy can be particularly high

for divers who have been regularly collecting fishes of a wide size range for other purposes (e.g. descriptions of diet) and measuring them. In most cases, however, it is spurious to present size-frequency information on large target species in anything less than 50 mm size-classes. If the sizes of a large number of medium and large species are estimated (e.g. serranids, cheilodactylids, sparids), 100 mm size-classes would be most realistic (Bell et al. 1985a; E. Harvey pers. comm.). A good training exercise is to take objects underwater to check the ability of a diver to estimate size before a study (English et al. 1994). Differences between divers can be great, and recalibration is critical (Mapstone & Ayling 1993). Stereo-video technology can provide very accurate estimates of the sizes of reef fishes, as long as the orientation of the subject to the camera is less than 75° (Harvey & Shortis 1996). The camera may be used while free swimming or from fixed positions (Klimley & Brown 1983; Harvey & Shortis 1996). For impact studies it is often relevant to include estimates of length for exploited species (Hancock 1994) such as sparids, cheilodactylids, latrids, monacanthids, zeids, and large labrids and odacids (see Chapter 10).

For some biological problems it may be necessary to get estimates of biomass. This is particularly true in a fisheries context, because most of the catch statistics are in units of weight. Estimates of biomass can be obtained if length/weight relationships are known for individual species. Data on the sizes of fishes can then be translated into information on weight (e.g. Russell 1977). If biomass estimates are to be obtained, the influence of minimum size of fishes counted should be considered. Bellwood and Alcala (1988) found that the minimum size of fishes counted had a greater influence on estimates of abundance than on estimates of biomass.

The sex of many fishes can be determined underwater. Differences between sexes include size, di- or polychromatism, dimorphisms, behaviour and nest guarding (Thresher 1984; Fig. 6.1). Differences in colour may be permanent or temporary. Differences in size and colour are particularly common for labrids (Doak 1972; Hutchins & Swainston 1986; Barrett 1995; Fig. 6.1; Plates 29, 31) and the different stages of life history of these protogynous fish can be identified to levels that include female, transitional (changing sex), superfemales and males (Russell 1988). Although not always as obvious as for the Labroidei, dichromatism is found in other fishes and may intensify during spawning (e.g. some pomacentrids, *Parma microlepis*: Tzioumis & Kingsford 1995; callanthiids: Francis 1988). Dichromatism in some fishes is only obvious during the spawning season (e.g. *Chromis dispilus*: Kingsford 1985; Tripterygiidae: Francis 1988). The shape of fishes may provide information on sex, as do special morphological features such as nasal horns of cheilodactylids (McCormick 1989; Schroeder et al. 1994; Fig. 6.5). Nest-guarding behaviour is also useful for distinguishing sexes (e.g. tripterygiids: Thompson 1986). The ability to identify sexes can greatly increase the range of biological problems that can be addressed underwater.

6.2.10 Age and growth

The age and growth of fishes is often of great relevance in studies on reef fishes. Questions relating to age since spawning, age to settlement, age since settlement, maximum age (= stored years of recruitment), age to maturity, growth rates of different cohorts, and growth at different times and places all require some knowledge of age. With adequate validation the age of fishes can be determined by using increments in skeletal components such as the earbones, otoliths (e.g. snapper: Francis et al. 1992), cleithra, opercula (Hostetter & Munroe 1993), vertebrae (mainly for elasmobranchs: e.g. Cailliet et al. 1986), spines (e.g. catfish: Davis 1977), scales (e.g. labrids: Jones 1980) and even soft rays (e.g. pectoral fin rays of scombrids: Prince et al. 1995). An advantage with structures such as spines and scales is that you do not need to kill the fish. In general, scales are not recommended because they tend to underestimate the age of old fishes (e.g. Paul 1976; Beamish & McFarlane 1983), but this could be checked in a comparison of methods and validation studies. Fishes can be aged using daily and annual increments in otoliths. Comprehensive reviews of the methods used for the ageing of fishes are found in Bagenal (1978), Smith (1992) and Stevenson and Campana (1992). Before counting increments, all prepared otoliths should be randomised to avoid any bias caused by prior knowledge of fish size, and repeat counts by more than one experienced observer are recommended.

In my experience, the most reliable method for getting information on the age of reef fishes using annual increments is to prepare a transverse section of the otolith (e.g. Fig. 6.6). Thus thin sections of otoliths should always be included in any pilot study on the best method for estimating age. Otoliths are extracted from fishes, cleaned well and treated immediately, or stored dry for later examination. A slice can be taken from the otolith (often already embedded in resin), in the vicinity of the primordium, using a low-speed diamond saw (e.g. Buehler, Isomet saw); the slice is then mounted on a slide and viewed under a microscope, or a grinder can be used to grind toward the primordium. Otoliths (usually sagittae) can be immersed in a bed of resin (e.g. Spurr's fixative), ground toward the primordium,

mounted on a slide (ground side down) then ground toward the primordium from the other side (Fig. 6.6). Large amounts of calcium carbonate can be removed from otoliths using coarse wet-and-dry, grit paper (120 grit), but approaching the primordium you should change to a finer grit (e.g. 1200). Lapping film can be

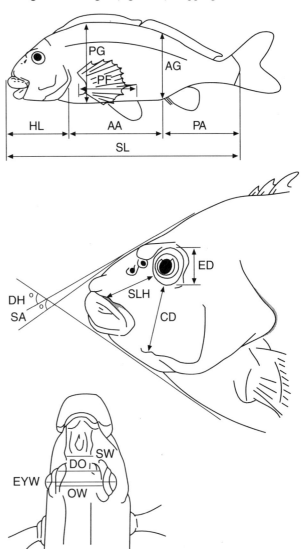

Figure 6.5 Morphological differences between sexes of the red moki / striped morwong (*Cheilodactylus spectabilis*, Cheilodactylidae). The diagram shows the morphometric measurements that were taken by McCormick: SL = standard length; AG = anal girth; PG = pectoral girth; PA = posterior anal length; AA = anterior anal length; PF = pectoral fin length; HL = head length; SA = snout acuteness; DH = dorsal head angle; CD = cheek depth; SLH = snout length; ED = eye diameter; SW = snout width; PO = preorbital width; EYW =eye diameter; OW = interorbital width. Females have small pectorals, small snout angle and a short mid-body. Males have large pectorals, large snout angle and a long mid-body.

ADAPTED FROM McCORMICK (1989) WITH PERMISSION

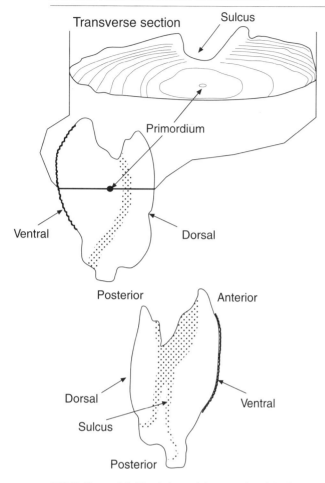

ABOVE: **Figure 6.6** Morphology of the sagittal otolith of a labrid fish. Dorsal and ventral views of the otolith (earbones that are used to age fish) is shown, with a transverse section that shows annual increments deposited sequentially from the primordium of the otolith. Transverse sections are commonly used to gain the best resolution of annuli in otoliths (see Fig. 6.7). The groove in the transverse section (= 'sulcus') is shown as a shaded area in the ventral view.

used for polishing and may be done in a series (e.g. 9 μm–3 μm abrasive film). Some of the equipment required for grinding is expensive and is often designed for jewellers or geologists. There are cheap alternatives, however; for example, a washing-machine engine, a metal disk to mount the grinding paper, and the base from an old microscope to enable the preparation to be moved to the grinding disk using the focus control (design of D. Ferrell). Annuli can generally be viewed at magnifications of less than 200 x, and their visibility can be improved by a range of techniques; for example, immersion in stains and reflective media (e.g. immersion oil), and by briefly burning.

Sagittae or lapilli are used to age small fishes using daily increments. The otoliths can be extracted from small fishes soon after they hatch from eggs. This can generally be done by squashing larvae on a slide with a coverslip, then removing the cover slip to tease away excess tissue and further expose the otoliths, which can then be viewed at 1000 x in immersion oil. For large fishes (e.g. newly settled juveniles), the lapilli and/or sagittae can be removed with fine forceps or tungsten needles (Kingsford & Milicich 1987; Jenkins & May 1994). These otoliths can generally be viewed at 200–1000 x without grinding. In some fish, however, the visibility of increments can be enhanced by leaving otoliths in immersion oil for a month or so, or some polishing may be required with lapping film.

In all studies on ageing, validation that the increments correspond to days or years is essential (Geffen 1992; Hilborn & Walters 1992). Validation should be

BELOW: **Figure 6.7** Transverse sections of the sagittal otoliths of *Achoerodus viridis* (Labridae). Annual marks are shown with dots. **A:** age 2, 220 mm SL. **B:** age 3, 300 mm SL. **C:** age 9, 444 mm SL. **D:** age 19, 527 mm SL. V = ventral edge; D = dorsal edge; scale bars 0.5 mm

FROM GILLANDERS (1995a), PUBLISHED WITH PERMISSION OF *MARINE AND FRESHWATER RESEARCH*

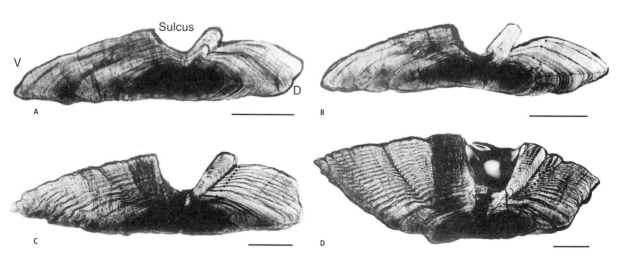

done at more than one location, especially if the information is to be used to justify a method that will be used over a broad geographic range. Forms of validation include sampling cohorts of a known age (this becomes difficult after 1–1.5 years), sampling fishes of a known age that are kept in tanks for short (daily increments) or long periods (annuli), and marking the otoliths with antibiotics such as tetracycline (especially adults and juveniles) or, in some cases, alizarine complexone (larvae). Stress markers (e.g. Victor 1982) may be used, but in my experience are less equivocal than a fluorescent mark. Fishes are then released into the field or kept in tanks and recovered x days or years after the treatment (larvae and/or juveniles: Kingsford & Milicich 1987, Gallahar & Kingsford 1992; adults: Fowler 1990; Ferrell et al. 1992; Francis et al. 1992). You may even want to double-tag fish with antibiotics, which is a good way of reducing edge effects, particularly where increments are closely spaced. Where daily increments are sampled, the only way of determining the time when the first increment is deposited (before hatching, at hatching, yolk absorption, metamorphosis to a juvenile: Kingsford & Milicich 1987) is by raising larvae so that the date of hatching is known. It is common for there to be a protein 'plug' at the centre of the otolith (including the primordium), so increments may start at 10 µm or more away from the primordium (e.g. snapper, *Pagrus auratus*: Kingsford & Atkinson 1994). Even when annuli have been validated, it is sometimes difficult to discern which increment is the first annual band. Although bands often form in winter when growth slows (e.g. Ferrell et al. 1992), extended periods of recruitment (e.g. four months or more) can make the interpretation of early increments difficult. It is beneficial therefore, to intensively sample the 0+ year class (i.e. every month or so, at different locations) so that the sizes of the otoliths early in life can be examined and compared with those of adults. This will help interpretations of the first annual ring even if you know from tetracycline experiments that increments are deposited annually. It is worth doing tetracycline experiments at more than one location, because the clarity of otoliths can vary on spatial scales ranging from 50 to 1000+ kilometres.

Another means of validation is 'marginal increment analysis', which has been used to validate daily (Jenkins 1987) and annual increments (e.g. Jones 1980; Gillanders 1995a). This technique involves measuring the distance from the last increment to the edge of the otolith along a predetermined axis. For example, if an annulus is laid down at a certain time of the year, soon thereafter the marginal increment would be narrow. As time since the last increment increases, so does the width of the marginal increments, until the next annual band is

formed. You should use fish of the same cohort for this procedure, as increment spacing tends to drop dramatically with age. The technique is difficult for daily rings because of edge-of-the-otolith effects. In general, marginal-increment analysis and 'back calculation' using the increment widths of other annuli are poor substitutes for marking otoliths, combined with detailed sampling of the 0+ age group, and/or having fish of a known age.

Many reef fishes are surprisingly old. For example, snapper *Pagrus auratus* and other sparids may reach 40–60 years old (Kalish 1990; Buxton & Clarke 1992) and blue moki (*Latridopsis ciliaris*, Francis 1981) 21 years, while even small reef fishes such as the pomacentrid *Parma microlepis* (SL max 160 mm) can reach ages of 37 years (Tzioumis & Kingsford unpub. data). Small blennioid fishes (e.g. *Forsterygion varium*) may only live two to three years, while serranids (e.g. 30 years for *Paralabrax clathratus*: Love et al. 1996) and scorpaenids (80 years, *Sebastes diplopora*: Bennet et al. 1982) live much longer. Labrids such as *Semicossyphus pulcher* (Cowen 1990) and *Achoerodus viridis* (Gillanders 1995a) live to maximum of ages of 21 and 35 years respectively. Choat and Axe (1996) found the age maximum of 10 species of subtropical surgeonfish (Acanthuridae) varied between 30 and 45 years.

When considering the dynamics of populations of reef fish, therefore, a knowledge of age is important because the population may be the result of the so-called 'storage' (*sensu* Warner & Hughes 1988) of many years of recruitment. A knowledge of the reproductive state of fishes, preferably by histological investigations (e.g. Pankhurst & Conroy 1987; Nakamura et al. 1989), can enable age at maturity to be calculated. There are still many fishes in temperate waters for which there is no information on age and growth.

Growth models that use the von Bertalanffy growth equation are generally good descriptors of the growth of reef fishes (Ricker 1987), but other models may be more appropriate to some species or particular stages of life history (reviews: Kaufman 1981; Schnute 1981). Fishes generally grow quickly over the first 5–10 years, and growth in length drops off dramatically after that time. When the age composition of fishes on a reef is relatively young, it is possible to make an age/length table where the probability of estimating age correctly based on length is very high (Meekan & Fortier 1996).

6.2.11 Foraging and diet

A knowledge of feeding biology is often important for understanding differences in patterns of distribution and abundance, and behaviour of the fishes (Jones 1984b;

Levin 1994; Holbrook & Schmitt 1998), and for assessing the impact of foraging on species of prey (plankton: Bray 1981; epiphytal invertebrates: Choat & Kingett 1982; large reef dwelling invertebrates: Cowen 1983; soft-bottom molluscs: Jones et al. 1992). Foraging behaviour may be described by following individuals (6.2.4l) and recording activity within a defined area. A behavioural data sheet of the type shown in Figure 6.4 could be used. Number of bites can be recorded, as well as the substratum from which prey were taken. It may be possible to identify large prey (e.g. Sweatman 1984). The feeding activity of planktivores is generally clear, particularly for fishes that target individual plankters. Planktivores will often feed freely even when a diver is close (e.g. de Boer 1978; Kingsford & MacDiarmid 1988; Glasby & Kingsford 1994). Although bite rates have generally been counted by following individual fish, video may enable this to be done for multiple fishes in an aggregation.

The most appropriate methodology for assessing diet will depend on the ecological question that is to be addressed, how fishes treat their prey (e.g. swallowed whole and slowly digested versus well masticated and rapidly digested) and the nature of the prey (e.g. soft- or hard-bodied). In general, however, more than one measure is required to gain an accurate picture of the diet (review: Hyslop 1980) and can include numerical abundance of prey, percentage occurrence (i.e. percentage of fishes that have a particular prey type represented in the gut contents), proportional representation and prey volume in millilitres.

Numerical abundance is only a suitable index where prey have been swallowed whole, as with many planktivores and large predatory fishes (e.g. Kingsford 1993). Subsampling may be required if many small organisms have been eaten. Proportional representation is a measure of percentage cover, and the methods used are similar to those of the substratum (see 4.3.3). Methods include using an ocular grid, a dish with dots marked on graph paper that is stuck on the bottom of the dish (this can be a problem if the gut contents are very dark and make it difficult to see dots) or the transect method (e.g. Choat & Clements 1992). The transect method requires a small belt of acetate sheet marked with a transect line and 10 intersections. Replicate transects are placed at random positions in the dish of gut contents until the required number of points (determined in a pilot study: see section 3.3) have been recorded. Measures of the volume of gut contents (using a measuring cylinder) may be useful, and in some cases it may be easy to separate types of prey so their volumes can be measured separately. For many of these procedures, pilot studies may be useful to determine number of points, in the case of percentage cover, and

check accuracy and precision of subsampling techniques. The categorisation of prey will depend on the question being asked. In some cases it may be necessary to identify prey to species level (e.g. the impact of fishes on a particular species: Cowen 1983) or classification to broad groups may be adequate (e.g. for a benthic carnivore: mollusc fragments, polychaetes, crabs, etc.). Hyslop (1980) described an Index of Relative Importance (IRI) for biological questions that relate to the importance of certain types of prey to a species of fish.

$$IRI = (\% \, N + \% \, V) \times \% \, F \, \ldots \ldots (6.1)$$

where: $\% \, N$ = percentage of prey items in the diet, $\% \, V$ = percentage volume, $\% \, F$ = frequency of occurrence

The parts of the gut that are examined will depend on the ecological question and the morphology of the gut. Many fishes (e.g. planktivore *Chromis dispilus*: Kingsford & MacDiarmid 1988) have an obvious stomach, while other types (e.g. labrids: Gillanders 1995b) have no stomach. The morphologies of fish guts are diverse, especially for herbivores (Horn 1992; Clements 1996). Investigations of the degree of digestion in different sections of the gut, and perhaps studies on the microbiota, would require sampling of the entire length of the gut (e.g. Rimmer & Wiebe 1987; Anderson 1991; Clements et al. 1994). In some cases, measurements of structures that are related to feeding (e.g. size and musculature associated with the pharyngeal apparatus) may help in the interpretation of dietary differences among fish of different sizes and taxa (e.g. Wainwright 1988). This should not, however, be done at the expense of ecological considerations for differences in diet (e.g. Coates 1980).

In any comparisons of diet (e.g. among species and locations) several aspects of the sampling design should be considered. Size- and age-related variations are often great (Werner & Gillium 1984), and temporal differences in diet are common; for example, many herbivorous fishes are primarily carnivorous as small juveniles (e.g. Bell et al. 1980). Planktivores (e.g. Kingsford & MacDiarmid 1988) and benthic carnivores often show great changes in the size range and composition of prey (Jones 1984c, 1988a; Cowen 1986; Gillanders 1995b). In some cases the timing of feeding may also change; for example, newly settled *Pempheris adspersa* feed as diurnal planktivores and later in life as nocturnal planktivores (MacDiarmid 1981).

Temporal variation in diet may range from within days to among seasons and years. Some fishes show great variation in feeding with time of day; for example, Choat and Clements (1993) found that butterfish

Odax pullus fed at high rates and feeding activity reduced to zero during the afternoon once the gut was full, digestion of the day's takings happening at night. Seasonal changes in the availability of food may also have an influence on diet; for example, Kingsford (1980) and MacDiarmid (1981) found great variation in the diet of adult planktivores with time of year. *Chromis dispilus* and *Scorpis violaceus* generally consumed appendicularians, calanoid copepods, and a mixture of other plankton. The *Scorpis* also consumed floating fragments of algae. In spring, the total quantity of food increased greatly (by up to 40% of volume for *Chromis*) and the diet changed to 95% or more pelagic fish eggs.

Variation in diet among habitats and locations has also been demonstrated for reef fishes (Jones 1983; Cowen 1986) and some size-related differences in diet may be a consequence of different size-classes foraging in different habitats or environments (e.g. estuaries and reefs). Spatial variation, combined with variation in diet with size/age of fishes and temporal differences, should be considered in sampling designs.

6.2.12 Reproductive biology and condition

Many demographic studies of fishes require information on reproduction (e.g. Hilborn & Walters 1992) to determine one or a combination of the following: sex (especially when fishes cannot be sexed by external characteristics); age to maturity; determination of sexuality (i.e. sex-changing or gonochoristic species); the timing (phenology: Miller 1979) and nature of spawning (e.g. serial spawners); fecundity (e.g. for the egg-production method: Zeldis 1993); and the development of gonads and how it relates to the somatic condition of animals. Some information on reproduction can be obtained using non-destructive techniques, particularly where the sexes can be identified from external characteristics (see 6.2.9). The spawning season and timing of spawning are best defined by observations of the fishes (e.g. Jones & Thompson 1980; Tzioumis & Kingsford 1995). Clearly, spawning behaviour and factors determining mate choice and guarding success can only be determined from behavioural observations (e.g. Thompson 1986; Warner 1990; Sikkel 1995). Observations of demersal spawning fish can also enable eggs to be collected for description (e.g. Kingsford 1985), and for the design of experiments with eggs or larvae. The behaviour of fishes can also be compared with plasma levels of hormones without destroying fishes, if blood can be extracted from fishes *in situ* (Pankhurst 1995).

A great deal of information on reproduction can be obtained from fish by using destructive methods. Although the sex of many fishes can be determined by the shape and form of gonads, this is not always the case, particularly when they change sex. Most reef fishes have bi-lobed gonads (Fig. 6.8), and differences between sexes are most clear during the reproductive season. The gonads of females are generally cigar-shaped, and large ova that are ready to be spawned give the gonad a grainy appearance (ova generally have a diameter of greater than 0.5 mm). In contrast, the gonads of males appear milky white and are more angular in cross-section. If fishes are 'running ripe', eggs or milt can be squeezed from the urinogenital opening and identification of gender can be successful without destroying them. Field staging of oocytes from the gonads of fish specimens is possible by categorising oocytes that are visible through the ovarian wall, or by slitting the ovary open for a more direct examination.

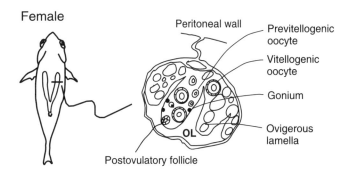

Female

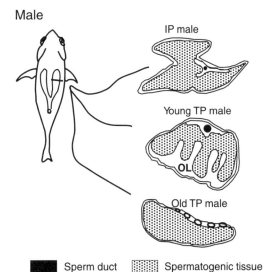

Male

Sperm duct ▓▓▓ Spermatogenic tissue

Figure 6.8 Schematic transverse sections of the gonads of female and male labrid fish; the position of sections from complete gonads are shown. **OL** = ovarian lumen. The three male sections show typical views at different stages of sex change. IP male shows a single sperm duct ; young TP male shows the remnant structures of a female, indicating prior function as a female; old TP male has multiple sperm ducts and the gonad is dominated by spermatogenic tissue.

MODIFIED FROM NAKAMURA ET AL. (1989)

This method, however, requires considerable experience and the technique should be validated (West, G. 1990).

The size of gonads is used as an indication of the timing of spawning, and these data are often presented as gonad indices (see Formula 6.2). It is important that gonad indices be compared for organisms of a similar size range. Gonad indices assume a linear relationship between gonad weight and body weight, and this is often not the case. For example, small male *Chromis dispilus* (90–110 mm SL and 40–80 g) can have gonads that are as large, if not larger (0.1–3.8 g) than those of large, nest-guarding males (125–160 mm SL, 100–142 g, gonad: 0.8–3 g) midway through the spawning season (January; Kingsford unpub. data). In most cases it is probably best to plot the weight of gonads (g) rather than an index, because actual gonad weights are accessible to readers without being confounded with body weight.

$$GI = GW/BW \times 100 \ldots\ldots (6.2)$$

where GI = gonad index, GW = gonad weight, BW = body weight

A great deal of valuable information on sex can be obtained from histological sections. For example, although sex change can be inferred from size/sex age/sex and sex ratios (the most unreliable method), histology is required for an unequivocal assessment (e.g. Nakamura et al. 1989). Gonad weights or indices are commonly used as an indication of an increase in the production of gametes, through changes in the proportional representation of gametogenic tissue may not be represented properly by changes in the size of gonads (Pankhurst & Conroy 1987). Age to maturity, age versus sex, and sex ratios can be determined unequivocally with information on histology and age.

Histological sections are classified according to the proportional representation of the different stages of gametogenesis. Classifications of fish maturity may include immature, mature (different to immature fish in that brown bodies are present in gonads), ripening, ripe and spent. Excellent photographs showing different stages of oogenesis and spermatogenesis are given in Pankhurst and Conroy (1987), and for evidence of prior function in the case of sequential hermaphrodites, see Nakamura et al. (1989). Sampling for histological purposes is often done on a monthly basis, but it should be noted that the later stages of vitillogenesis can take place over a very short period of time. For example, Hoffman and Grau (1989) found that female *Halichoeres* can hydrate eggs quickly over a period of hours so that spawning coincides with particular phases of the tide.

Typical treatment of gonads for histology is to preserve them in fixative (e.g. Bouin's for 24–48 hours). The specimens are then blotted dry and transferred to 70% ethanol. After fixation, the preserved gonads are embedded in paraffin and sectioned transversely at 5–10 microns (e.g. Gillanders 1995a). Pilot sections of different parts of the gonad (e.g. proximal, medial and distal transverse sections) should be done to check that different types of gametogenic tissue are not found in localised areas of the gonad (e.g. some sparids). Multiple fields of view and slides should be examined for measurements of the size of stages of gametogenesis.

Condition factors are sometimes used to compare fishes from different locations and times, and perhaps relating to impacts such as sewage outfalls (e.g. Nowak 1997). Condition factors will vary greatly according to the time of year that fishes are sampled and how this is related to the timing of reproduction. For example, the condition of *Chromis dispilus* increases before the reproductive season; using Formula 6.3 with the modification of *b* from Formula 6.4. Total condition drops off at the end of the reproductive season. Clearly, this could just relate to variation in the size of gonads, but when somatic condition alone is compared (i.e. deleting gonad weight from total weight), the fish are in poorer condition than at the beginning of the reproductive season at any other time of the year (Kingsford unpub. data). Tzioumis and Kingsford (unpub. data) found that *Parma microlepis* had large quantities of fat on the viscera while gonads were in the 'resting' stage, but this fat disappeared over the reproductive season when gonad size was large.

Condition factors that are commonly used are presented by Bagenal (1978) as follows:

$$\text{Fulton's } K = 100 \ w/L^3 \ldots\ldots (6.3)$$

where w = total weight of fish, L = total length

In the formula it is assumed that growth is isometric. If fishes are compared where the range of sizes is large, then the value 3 in Formula 6.1 should be replaced with *b* from Formula 6.4.

$$\text{Log}w = \log a + b\log L \ldots\ldots (6.4)$$

where $\log a$ = intercept of the line on the y-axis, b = slope.

6.3 Experimental methods

In situ manipulative experiments have been successfully in studies on reef fishes. Experiments are also done on reef fishes in aquaria, but it is difficult to argue that the results relate to field conditions. Tank experiments can, however, be of some value for generating hypotheses that can subsequently be tested in the field. Approaches that are used for experiments, and examples of the ecological questions that have been addressed, are described.

In some cases the ethics of field experiments may be questioned, particularly those that require the removal of fishes (e.g. by fishing) or cause injuries that may result in increased rates of mortality (e.g. by tagging), or other aspects that are relevant to population dynamics such as a reduction in spawning output. If this is the case, the value of the experiment has to be considered carefully. Experiments are the only valid method of investigating processes that influence fishes or other organisms. Thus the loss of tens to hundreds of fishes that are considered valuable may be justifiable if the information is crucial for an understanding of biology for management, and if the numbers of fishes are a small proportion of the population. An additional perspective is that the disturbance caused by many experiments (i.e. to other organisms or the fish themselves) is generally trivial compared to the scale of natural disturbances.

The methods used to do field experiments on reef fishes include:

- Exclusion using cages.
- Inclusion using cages.
- Exclusion of all or selected fishes by spearing.
- Translocation to treatments.
- Manipulation of the substratum (by adding or removing).
- Manipulation of organisms (by adding or removing).
- Manipulation of regimes of fishing (MPAs, access to different fishing methods and recreational and commercial groups).
- Tag, release and recapture (chemical and physical tags); for some types of tags, see Figure 5.7.

Before embarking upon a costly experiment, consider the general comments on the utility of experiments in this book and the advice on analytical aspects of sampling design, in Chapters 2 and 3.

Although fishes are often mobile (which poses difficulties not encountered for sedentary or sessile organisms: see Chapters 5 and 7), they can be excluded from areas using cages (e.g. Choat & Kingett 1982) of different sizes, or by spearing (e.g. Cowen 1983; Jones 1983). Furthermore, aspects of the substratum (e.g. rock or coral cover) or associated organisms (e.g. algae) can be manipulated, regimes of exploitation can be altered through protection (i.e. MPAs); or similarly, the type of fishing gear and level of fishing can be altered in areas. Fishes may be caged with or without other organisms to assess mortality (e.g. predator exclusion: Doherty & Sale 1985). In some cases fishes may be collected and allocated to different treatments such as experimental reefs (e.g. Jones 1987), but attention should be given to problems of immigration or emigration among treatments. Finally, in some experiments fishes are treated with chemicals or different types of tags. For example, it is common to tag fishes with tetracycline and release them into the field for a known period of time before they are collected and the otoliths examined for fluorescent marks (see 6.2.10).

Although it is unlikely that caged fishes behave normally, caging should not be ignored as an option. Small cages ($1 m^3$) were used by Bray (1981) to relocate juvenile *Chromis* to demonstrate that juveniles fed best on the incurrent sides of kelp forests. In general, however, caging fishes is not appropriate unless the cages are exceptionally large. For example, a very large cage (about 10 x 10 m and divided into half) was used by Carr (pers. comm.) to study the influence of kelp bass (*Paralabrax clathratus*) on the mortality and behaviour of recruit surfperch in California. One half of the cage included predatory *Paralabrax*, and the other half was a control. Replicates were repeats of the treatments in the same cage, but the half used for each treatment was alternated. Ideally there should have been multiple cages for each treatment, but logistics and expense precluded this. Half-cages always raise questions of independence, and some evidence should be provided that the close proximity of treatments has no behavioural or experimental effect. Connell (1994) used large cages (4 x 4 m and 2 m high) cages to exclude large predatory reef fishes in tropical waters (especially serranids and lutjanids) from pairs of damselfish (*Acanthochromis polyacanthus*) that were guarding their brood of juveniles. The large cages were necessary to encompass the maximum distance that juveniles were likely to range from the shelter sites of the parent fish. Based on *a priori* power analysis (see Chapter 3) for an estimated effect size of 0.14, Connell calculated that he would need 12 replicate cages for each treatment. Cage controls were included in the design (part cages that large predators could still enter) to test for cage artefacts (see Chapter 2), and there was also a treatment for high and low densities of juvenile *A. polyacanthus* in broods.

This experiment required a very large area, very calm conditions (rare in temperate waters) and a lot of bottom time. It was a good experiment in that it was specifically testing for a biological effect with an estimated effect size, and had open controls and cage controls, adequate replication and independent replicates. Finally, behavioural observations were done to check some of the assumptions of the full cages and the controls. Also note the use of cages in Steele's (1997) multifactorial study of gobies, where he assessed the effects of resident conspecifics, potential interspecific competitors, predators and reef location on patterns of recruitment.

An experimental approach has been used to understand interactions between fish and other organisms. Fishes have been manipulated to determine the influence of predation by fishes on large and small benthic invertebrates. The influence of predation is potentially a major determinate of patterns of abundance. Andrew and Choat (1982) experimentally examined the influence of predation and conspecific adults on the abundance of juvenile sea urchins (*Evechinus chloroticus*). Juvenile urchins were placed in full cages to exclude predatory reef fishes, in the presence and absence of adults. It was possible that the presence of adults could influence the mortality rates of juveniles. It was concluded that the exclusion of predatory fishes did influence rates of mortality of juvenile urchins, with greatest rates of mortality found in open controls. Andrew and Choat concluded that although mortality was higher in the presence of fishes, successful recruitment of urchins was sufficient to sustain populations of adults. The cages used were 2 x 1 m and all treatments were randomly placed at one location in the Leigh marine reserve, northern New Zealand. The conclusions, therefore, have to be restricted to that location because it is possible that different processes operate at other locations. Cowen (1983) hypothesised that the feeding activities of sheephead (a large labrid reef fish, *Semicossyphus pulcher*) could have a measurable effect on the behaviour and numbers of urchins (*Strongylocentrotus franciscanus*). He measured urchin abundance and recorded the distance that urchins moved from shelter at two sites. Sheephead were cleared from one site and the other site was used as a control. It was concluded that sheephead influenced both the abundance and the behaviour of urchins. Urchins did not move far from shelter in the presence of reef fishes. Although Cowen's conclusion may be correct, the sampling design was pseudoreplicated in that there was one experimental and one control site (see Chapter 3).

The possibility that fishes influence the distribution and population dynamics of small epiphytal invertebrates (e.g. amphipods) has also been investigated experimentally. Although epiphytal invertebrates gain some protection in algae (Coull & Wells 1983), and the degree of shelter may vary among different species of algae (Edgar 1983a; Taylor & Cole 1994), it was still considered possible that fishes could determine patterns of abundance of these invertebrates, although there was little direct evidence (Choat 1982; Jones 1988a). Choat and Kingett (1982) used 2 x 2 m cages to exclude predatory snapper (*P. auratus*) and mullids (*Upeneichthys lineatus*). Some cages only had roofs that allowed mullids in but did not enable snapper to feed (because they orientate their bodies vertically to feed). All treatments and controls were arranged randomly at one location. It was concluded that fishes had no influence on seasonal patterns in the abundance of invertebrates. The possibility that a significant proportion of invertebrates migrate from the substratum each night (and potentially into and out of cages of different treatments) was not considered likely. Migrations of epiphytal animals into the water column are considerable (e.g. Alldredge & King 1977; Jacoby & Greenwood 1988), and need to be considered in experiments of this type. This type of problem is also found with macroalgae (e.g. the laminarian *Ecklonia*), where up to 20% of the amphipod fauna of an individual plant may migrate per day (N. K. Gallahar pers. comm.).

In some experimental studies, investigators have concluded that competition, and in some cases food resources, are important for determining abundance patterns of reef fishes, particularly along the coast of California (Hixon 1980; Larson 1980c; Holbrook & Schmitt 1989; Schmitt & Holbrook 1990; see also 6.4.7). In contrast, Thompson and Jones (1983) did an experimental study (fish removals) on interactions between a wrasse and a blennioid fish. Although the species had an overlap in diet and behaved aggressively toward each other, the study concluded it was unlikely that the blenny had any influence on broad-scale patterns of distribution, although patterns of foraging within habitats were influenced.

Topographic complexity can have a strong influence on the distribution and survivorship of fishes (review: Jones 1988a; Hixon & Beets 1989). Connell and Jones (1991) collected descriptive information showing that although recruit blennioid fishes were found in habitats of a wide range of topographic complexity, adults were rare in areas of low topographic complexity. Because these fishes are strongly territorial from the time of settlement, it was suggested that differential mortality was responsible for the pattern.

Topographic complexity was manipulated by altering densities of boulders in each treatment. The treatments were as follows: (1) add boulders to areas of low complexity and compare with (2) controls; (3) remove boulders in areas of high complexity and compare with (4) controls; and (5) a control for the disturbance of boulders (picked up and put back in place). Recruits were found in average densities of 2.1 per 2 m^2 in all treatments at the beginning of the experiment, but no recruits were found in treatments 2 and 3 at the end of the experiment. The use of stone patch reefs is analogous to the use of coral patch reefs on tropical reefs (reviews in Sale 1991). Topography can influence the number of territorial fishes in an area. Larson (1980a, b) used experimental removal of rock fishes to argue that the maximum number of territorial *Sebastes* in an area was determined by the number of caves (used for shelter) and the presence of territorial males.

Topography may also influence the distribution (e.g. Hixon 1981) and reproductive success of fishes. For example, Thompson (1986) found that the spawning success of *Forsterygion varium* was positively correlated with the size of males and the presence of large boulders adjacent to their small territories. The experimental manipulation of numbers of rocks in territories changed the reproductive success of males. The addition of small rocks into the territories of previously unsuccessful males resulted in a significant increase in spawning success; the converse was also true.

Evidence for the influence of habitat types on the distribution of reef fishes has come from descriptive (e.g. Ebeling et al. 1980; Choat & Ayling 1987) and experimental studies. For example, the experimental removal of canopy-forming algae and tufting algae (Levin 1993) generally has a major influence on the distribution and behaviour of reef fishes such as labrids and surfperch (Jones 1984a; Bodkin 1988; Carr 1991, 1994).

The interrelationships between fishes and algae can be complex, and experiments have been important for elucidating processes. Not only do algae influence the distribution patterns of fishes, but herbivory by fishes may influence the composition of algae by selective cropping. Furthermore, the chemical composition of the algae may determine foraging preferences (e.g. experiments of Hay & Fenical 1988; Duffy & Hay 1990). Although herbivorous fishes are thought to have an important role in determining the standing crop of algae in tropical waters (e.g. Scott & Russ 1987; Hixon & Brostoff 1996), their contribution to the dynamics of subtidal habitats in temperate environments is poorly understood (Horn 1989). Some impacts of herbivorous fishes have been detected from descriptive and experimental studies in temperate waters. In Australia, Andrew and Jones (1990) demonstrated that grazing by reproductive female *Odax cyanomelas* may contribute to the mosaic distribution patterns of kelp (*Ecklonia radiata*) by removing small patches. Grazing by territorial damselfish in temperate and subtropical waters of the northern (Hixon & Brostoff 1996) and southern hemispheres (Norman & Jones 1984; Jones & Norman 1986; Jones & Andrew 1990) has been demonstrated to alter the composition of primarily tufting algae. The evidence

to date, however, suggests that the impact of fishes is small compared to that of grazing invertebrates (e.g. urchins) for altering the nature of seascapes (*sensu* Jones & Andrew 1993).

In some cases, individual fishes or an age/size-class of fishes are removed to study aspects of reproductive behaviours (e.g. 6.4.5). For example, Jones and Thompson (1980) described the timing of spawning of small and large spotties (*Notolabrus celidotus*) at two sites and found that large females spawned earlier than small females. When large females were experimentally removed from one site, small females spawned early in the season, rather than late as at the control site. Jones and Thompson concluded that large females were responsible for the social inhibition of small females. Similar to the Cowen (1983) example above, this study can only be treated as suggestive because the design was pseudoreplicated (see Chapter 3). Experiments of this type have been done at a variety of latitudes to determine the processes influencing sex change (e.g. Fricke & Fricke 1977; Warner 1988), and other experimental methods have been used to study the sexuality of fishes (e.g. Warner 1990).

It is worth noting that recent reviews have argued for experiments that address more than one process at a time because a plurality of processes is the rule rather than the exception (e.g. Jones 1991). The challenge is to determine the relative contribution of each process to variation in numbers or other biological variables.

A variety of hardware has been used for subtidal experiments, but caging material may include plastic mesh (from garden centres), stainless mesh, netting or even plastic milk crates. Ayling (1981) used perspex sheeting, attached low to the substratum to exclude monacanthids (which feed while orientating vertically) from areas of substratum. Substratum type can have a major influence on the means or ease of attachment (see Chapter 2). Check that materials are not susceptible to corrosion. See section 2.11 for suggestions on the attachment of cages.

6.4 Examples of studies

6.4.1 One-off survey of an offshore island

Example: Survey of fishes of the Auckland Islands (Kingsford et al. 1989).

Purpose: Obtain baseline information on fishes associated with rocky reef environments.

Situation: Extremely limited background information suggested that slow-moving notothenids were present.

The amount of fieldwork that could be completed was determined by logistical constraints of time and weather; two divers participated.

Approach: Sample as many localities as possible, encompassing a wide range of exposures (e.g. bays, points, on different sides of the island). The distribution patterns of reef fishes species generally vary with depth, so counts were made in shallow and deep water at each locality (as for crustaceans, 5.3.1).

Two techniques were used, one to cater for cryptic fishes and another for more mobile species; a single method was unlikely to record both accurately. At each of nine localities, fishes were counted in three replicate 25 x 10 m transects and three 25 x 5 m transects. Both divers swam the tape out: one counted large reef fishes and fishes in the water column within a 10 m lane and then returned along the transect, carefully searching for cryptic fishes within a 5 m lane; the second diver searched for crustaceans (see 5.3.1). Large differences were identified in the species composition of fishes at nine localities. Localities that were exposed and characterised by deep reefs and rich fish life were separated clearly, in analyses, from sheltered localities with few species of fish. Information was also gained on the densities and sizes of fishes, so comments could be made on rank distribution and size-frequency distributions of the fishes found.

Also see studies of the Kermadec Islands (Schiel et al. 1986 and Cole et al. 1992).

6.4.2 Temporal and spatial patterns of recruitment

Example: Recruitment patterns of spotties (Jones 1984a).

Purpose: Compare recruitment patterns of spotties (Plates 29, 30) among habitats and times in the Leigh marine reserve, northern New Zealand.

Situation: Detailed information was available on the distribution of habitat types in the marine reserve (Ayling 1978), so fish counts could be structured according to habitat. Considerable logistical support from the Leigh Marine Laboratory enabled regular monitoring of habitats over a long period. One diver counted the fishes.

Approach: The recruitment of spotties was examined in six habitat types and fishes were counted in four of the habitat types at two localities; the other two habitat types were examined at one locality only. Five replicate 50 x 10 m transects were used. Jones (1991) found that although large differences in the magnitude of recruitment were found among years, some habitats (e.g. shallow broken rock) consistently had high numbers of recruits. This emphasises the importance of habitat type, which can influence the species and size-frequency distribution of fishes present. Furthermore, important data can be obtained for relatively low cost by monitoring in local coastal environments, but note this study did not encompass a large area of coast. For reviews on recruitment, see Doherty and Williams (1988), Doherty (1991) and Jones (1991).

6.4.3 Total numbers of fish in a marine protected area

Example: Total count of red moki (Cheilodactylidae): McCormick & Choat 1987.

Purpose: Estimate the total numbers and size-frequency distribution of *Cheilodactulus spectabilis* (Plate 26) in the Leigh marine reserve and compare these with adjacent unprotected areas of coastline.

Situation: Detailed information was available on the distribution of habitat types in the reserve but not for adjacent areas of coastline. There was easy access to the area surveyed, and McCormick used stratified sampling to estimate total numbers with 95% confidence limits.

Approach: An outline of the procedures used to derive an estimate of total numbers is given in section 3.3. Stratified sampling gave estimates of the total numbers of moki with precise 95% confidence limits. The data indicated that there were more moki in the reserve than in the adjacent area. Furthermore, fish outside the reserve were, on average, smaller.

Once again, it would have been better if McCormick's stratified sampling could have been applied to this target species before the reserve was gazetted as well as after (see section 2.6). In this way natural variation in densities and size-frequency distributions could have been differentiated from that brought about by protection. Other concerns include: only one control area was used; there was no estimate of error on calculations for areas of habitats; and the study relied on a relatively old map for the proportional representation of habitats (see Fig. 4.8).

Other pertinent examples include surveys using RVCs by Kingsford (1989) and Meekan and Choat (1997). Each of these was carried out over a broad geographical area. RVCs were most appropriate in each of these situations because of a combination of the behaviour of fish, the time taken to complete counts, and complexity of the bottom type.

ABOVE: **1** Aerial view of most of the Cape Rodney to Okakari Point Marine Reserve (Leigh marine reserve) northern New Zealand. The reserve is about 5 km long and the total exclusion zone extends 500 m seaward of the shore, including that of Goat Island. This is one of the most famous marine reserves in the world, and substantial knowledge on the ecology of temperate reefs has been gathered from the area.

BELOW: **2** A juvenile butterfish, *Odax pullus*, in shallow water swimming past stands of the fucoid algae *Carpophyllum maschalocarpum*, which is characteristic of reefs in northern New Zealand. Mayor Island, depth 3m. These algae reach 1.5 m long and float if they become detached during storms. Some fucoid algae (e.g. *Sargassum*) detach from reefs at predictable times of the year (usually spring).

Fucoid and laminarian kelp habitats in northern New Zealand.

TOP: **3** Shallow broken rock habitat at Leigh marine reserve, depth 3–4 m. The large seaweed is *Carpophyllum maschalo-carpum*. The silvery-grey fish are the planktivorous sweep *Scorpis aequipinnis*. The adults and especially juveniles of this species are generally most abundant in shallow water.

A. M. AYLING

ABOVE: **4** The laminarian kelp *Ecklonia radiata* on shallow broken rock, depth 3 m. The stipes of the kelp are short, which is typical for these plants in shallow water. Fucoid algae, *Carpophyllum* sp., can be seen in the background.

M. FRANCIS

LEFT: **5** *Ecklonia radiata* in a deep forest, depth 16–22m. The stipes of the kelp are long, which is typical for these plants in deep water. In Australia, this growth form is often found in relatively sheltered, turbid waters . In some studies it is necessary to study the kelp and the subcanopy flora and fauna.

M. FRANCIS

Urchin grazing and its influence upon two habitat types in New Zealand.

ABOVE: **6** Urchin-grazed barrens on large boulders at Whale Island, Bay of Plenty, depth 10–15 m. Urchins, *Evechinus chloroticus*, are sometimes observed grazing in aggregations at a habitat transition (e.g. tufting algae or kelp forest and barrens). Because urchins can change the structure of habitats, they are sometimes called 'habitat determiners'. Fish include the demoiselle, *Chromis dispilus*, a female spotty, *Notolabrus celidotus*, planktivores, *Scorpis* spp. and a red moki/striped morwong, *Cheilodactylus spectabilis*.

K. WESTERSKOV

RIGHT: **7** Urchins grazing on *Ecklonia radiata* at Karewa Island, depth 12–15 m. The rock on which the urchins are found is devoid of tufting or canopy algae such as kelp. Kelp forms a characteristic type of habitat, but grazing can alter the shape of forests and even remove some of them.

K. WESTERSKOV

Deep reef-type habitats in northern New Zealand.

LEFT: **8**   Leigh marine reserve, depth 18–20 m. Long-stiped *Ecklonia radiata* is present with sponges *Ancorina alata* (grey) and *Iophonopsis minor* (yellow).

A. M. AYLING

LEFT BOTTOM: **9**   At the sand interface, typical of reefs close to the mainland. Hahei marine reserve, depth 18 m. Kelp is sparse and massive (e.g. *Ancorina* sp.), and finger and encrusting sponges are common. It is common for areas of reef near sand to be covered by sediment or exposed during storms, which has a major impact on assemblages of sessile and mobile invertebrates.

K. WESTERSKOV

RIGHT: **10**   Shallow *Lessonia* (Laminariales) forest habitat at Karewa Island, Bay of Plenty, depth 0–10 m.

K. WESTERSKOV

BELOW: **11**   Deep sediment flats, also called 'sponge gardens' at Leigh marine reserve, depth 18–22 m. Finger sponges, including *Raspailia* sp. and *Rhaphoxya* spp., are common. Other sponges include massive forms (e.g. *Polymastia croceus*) and golfball sponges (e.g. *Tethya* sp.). Tufts of coralline turf are present and *Ecklonia* is rare. The flat rocky reef habitat is covered with a few centimetres of sediment. Storms may generate ripples (20–50 cm high) in loose sediment, and settling larvae and asexual products including sponge buds and fragments become attached on the exposed rock of the ripple troughs. The orientation of the fans of finger sponges indicates the predominate current and swell direction (i.e. perpendicular to the current).

C. BATTERSHILL

ABOVE: **12** A tunnel at The Labyrinth, Poor Knights Islands, New Zealand. The tunnel is rich in sessile invertebrates, and the silhouettes of kelp, *Ecklonia radiata*, can be seen. Scenic reefs such as these sometimes provide justification for gazetting as marine protected areas. They are also often characterised by very high species diversity, another justification for protection (see Table 1.1).

K. WESTERSKOV

RIGHT TOP: **13** Deep reef at the Poor Knights Islands, New Zealand, depth 35 m. Algae are rare at these depths, despite the water clarity, and the substratum is covered with sponges such as *Tethya fastigata* (golfball form) and *Iophon proximum*

(finger sponge). Fan gorgonians (*Primnoides* spp.) are also present. The demoiselle, *Chromis dispilus*, is abundant at all depths at this locality.

K. WESTERSKOV

RIGHT BOTTOM: **14** A close-up photograph of the substratum (0.7 x 0.3 m) in a canyon at Goat Island, northern New Zealand, depth 10 m. This great diversity of encrusting sponges is the type of problem faced in studies where an objective is to obtain estimates of percentage cover of encrusting organisms. Common species in such assemblages include *Tedania*, *Microciona* and *Pronax* spp.

LEIGH MARINE LABORATORY COLLECTION, K. WESTERSKOV

Urchin-grazed habitats in New South Wales, Australia.

LEFT TOP: **15** A habitat transition between a shallow algal fringe and urchin-grazed barrens, Cape Banks, Sydney, depth 2–3 m. Algae in the fringe are predominantly tufting coralline algae (e.g. *Amphiroa*) and reds (e.g. *Delissea*), but other algae, including short *Sargassum* spp. and occasional stunted kelp plants, *Ecklonia radiata*, are also found. Urchins hide in crevices during the day and graze the barrens at night.
M. KINGSFORD

ABOVE: **16** Barrens grazed by urchins, *Centrostephanus rodgersii*; and kelp forest, *Ecklonia radiata*. Montague Island, depth 10–15 m. This scene is relatively unusual in that the urchins are exposed during the day: they normally shelter in crevices at this time. This may in part be related to the high densities of urchins, sometimes more than eight per square metre.
M. KINGSFORD

LEFT: **17** Two species of urchin, *Centrostephanus rodgersii* (black) and the much rarer *Heliocidaris tuberculata* (red), on the edge of a crevice. Cape Banks, Sydney, depth 3–4 m. The position of kelp forests (*Ecklonia*) and tufting algae is strongly influenced by topography in central New South Wales. Kelp forest is usually on flat substrata three or more metres away from crevices and other types of shelter where *Centrostephanus* lurks, emerging to feed during the night. The limpets (including *Cellana* sp.) also graze algae, but it is the urchins that have the greatest influence in determining where algae are found.
M. KINGSFORD

18 Substantial urchin-grazed barrens consisting of tumbled boulders on bedrock at Cabbage Tree Island, Port Stephens, Australia, depth 12–15 m. These barrens would be speckled with feeding black urchins at night. The dark boulder in the distance is covered with tufting algae.

M. KINGSFORD

19 Giant kelp, *Macrocystis pyrifera*, at the Auckland Islands, New Zealand, depth 15 m. Giant kelp are found in temperate waters of the northern and southern hemispheres. Kelp habitat can be found over a broad depth range, 3–30 m. The three-dimensional nature of this habitat needs to be considered in sampling design.

M. KINGSFORD

OPPOSITE TOP LEFT: **20** Aggregation of abalone, *Haliotis iris*, on a rocky reef near Wellington, New Zealand, depth 3 m. The lack of algae in the area is a result of grazing by these gastropods.

K. WESTERSKOV

OPPOSITE TOP RIGHT: **21** A mesh enclosure used by D. Schiel to shelter the spat of abalone when they were released in the field. This gives the spat time to move to the undersides of boulders and greatly reduces the mortality of spat. The enclosure is about 1 m³ and floats. This is a similar design to that of some demersal plankton traps with a cod end on the top of the net.

S. MERCER

OPPOSITE: **22** Rock lobsters, *Jasus edwardsii*, in a cave, Cape Saunders, southern New Zealand, depth 15 m. Great diversity of encrusting organisms can be seen on the roof of the cave. It is possible that the abalone shell is evidence of predation by lobsters. Large aggregations of female lobsters are common near the sand–reef margin in late winter to spring. At this time females are brooding the eggs prior to hatching.

K. WESTERSKOV

RIGHT: **23** Rock lobster, *Jasus verreaux*, in an *Ecklonia radiata* forest in northern New Zealand. It is unusual to see non-crevice-bound lobsters during the day, but this female was in berry (brooding eggs). This species is common and fished in northern New Zealand and the east coast of Australia.

K. WESTERSKOV

ABOVE: **24** Octopus (*Octopus* sp.) are voracious predators of shellfish, crabs, lobsters and even sleeping fish. These animals are found over a wide range of depths on reefs and are generally cryptic, but can be lured to bait.

K. WESTERSKOV

A selection of reef-dwelling fishes of New Zealand and Australia that are good candidates for impact studies.

OPPOSITE: **25** Snapper (*Pagrus auratus*) are important predators on reefs, where they eat a wide range of invertebrates and fish. There is evidence from Leigh, New Zealand, that numbers of these fish increase on reefs within a marine protected area.

K. WESTERSKOV

OPPOSITE BELOW: **26** Blue cod, *Parapercis colias*, Castor Bay, northern New Zealand. These inquisitive fish are common on reefs in New Zealand and are targeted by recreational and commercial fishers. Data are available on patterns of abundance and the timing of reproduction at some localities.

M. BRADSTOCK

RIGHT: **27** A red moki/striped morwong, *Cheilodactylus spectabilis*, in urchin-grazed barrens, northern New Zealand. *C. spectabilis* are particularly well studied (see Fig. 6.5). There is good evidence that protected areas have strong positive effects on abundance and size of these fish.

M. FRANCIS

BELOW: **28** White-eared damselfish, *Parma microlepis*, Cape Banks, Sydney, Australia. This species is sexually dimorphic; males are generally black and females brown. They are excellent subject material for studies because they are territorial and spend most of their lives in an area of 100 m² or less. This means that estimates of variation in abundance are unlikely to be confounded by movements in and out of a study area.

M. GALETTO

29/30 Initial- (above) and terminal- (right) phase spotties, *Notolabrus celidotus*, Leigh, northern New Zealand. Most initial-phase fish function as females and, if they live long enough,

change sex into a male (see Fig. 6.1). This species is one of the most thoroughly studied reef fish in temperate waters.
M. FRANCIS

31 An initial phase labrid, *Notolabrus gymnogenis* (scarlet wrasse) swimming over barrens, Cape Banks, Sydney, Australia. This protogynous species is one of the most common labrids in New South Wales. Small juveniles are most commonly found in algal-rich habitats.
M. KINGSFORD

32 A terminal-phase *Notolabrus gymnogenis* (scarlet wrasse), Bare Island, Sydney, Australia. This fish probably functioned as a female before changing sex into a male. There are some males, however, that never function as females; they probably steal fertilisations from adult fish as initial-phase males.
M. KINGSFORD

OPPOSITE TOP: **33** Aggregation of planktivores, Poor Knights Islands, northern New Zealand, depth 35 m. In waters above a sponge-encrusted rock are demoiselle, *Chromis dispilus*, pink maomao, *Caprodon longimanus* and the splendid perch, *Callanthias australis*. The latter two species are also found in Australia. Counting planktivorous fish is challenging in large aggregations; the best approach is often to count some groups of ten and then estimate how many groups of each species are in the aggregation.
M. KINGSFORD

OPPOSITE BOTTOM: **34** A large aggregation of demoiselles, *Chromis dispilus*, Poor Knights Islands, northern New Zealand, depth 25 m. These fish have been found to feed on different sides of archways and other promontories, with most adult fish feeding on the incurrent side. They may swim more than a kilometre to suitable shelter to sleep at night. Aggregations of feeding planktivorous fish can cause large reductions in the number of some plankton.
M. FRANCIS

Southern temperate and subantarctic shore profiles in southern New Zealand.

ABOVE: **35** An exposed rocky shore, Puddingstone, Otago. Bull kelp, *Durvillea antarctica*, is typical of the lower intertidal at latitudes south of about 36ºS, and also commonly found drifting in coastal waters. A conspicuous band of pale barnacles can be seen immediately above the kelp.

K. WESTERSKOV

BELOW: **36** Intertidal in Perseverance Harbour, Campbell Island, one of the subantarctic islands of New Zealand. Bull kelp can be seen in the lower intertidal. The rock has been grazed of algae by gastropods above the kelp, followed by a conspicuous band of *Porphyra columbina* and lichen before a carpet of mosses at the top of the profile. The narrow vertical range of the intertidal (about 2 m) suggests relatively sheltered waters.

K. WESTERSKOV

Temperate shore profiles in northern New Zealand and New South Wales, Australia.

RIGHT: **37** Great Barrier Island, northern New Zealand. In the lower intertidal is a mosaic of tubeworms, *Pomatomus caeruleus*, and different types of algae, including green *Ulva*, light brown *Pachymenia* and dark red-brown *Xiphophora*. A band of olive-green *Gigartina alveata* can be seen below a broader band of barnacles, *Chamaesipho brunnea* and *C. columna*.

M. FRANCIS

BELOW: **38** A gently sloping intertidal platform, Jervis Bay, New South Wales. A mosaic of bare rock, grazed by gastropods (including *Austrocochlea constricta*), the encrusting alga, *Hildenbrandia prototypus,* and Jupiter's necklace alga, *Homosira banksii,* can be seen on the lower right. The dark brown algae at the lower intertidal is *Sargassum* spp., along with some light coralline algae. The anglers are targeting luderick/parore, *Girella tricuspidata,* using green algae as bait, and bream, *Acanthopagrus australis,* with commercial baits and limpets.

M. KINGSFORD

39 An intertidal pool on a gently sloping rock platform near Lottin Point, New Zealand. The pool is fringed by algae, *Hormosira banksii*, and clearly shows why rock pools generally should be treated separately in studies of the intertidal. The rock flats adjacent to the pool have been grazed of algae by gastropods.

K. WESTERSKOV

40 A broad intertidal rock platform, Kaikoura, New Zealand. The reef is largely covered by ephemeral green algae, *Enteromorpha* sp., and *Porphyra* spp., The locations of large limpets (*Cellana* sp.) are conspicuous as areas of bare rock. It can be seen that algal spores also settle on the shells of the limpets.

K. WESTERSKOV

OPPOSITE TOP: **41** A close-up of a large limpet *Cellana* sp. (4 cm long), in the intertidal, Jervis Bay, New South Wales. A small *Patelloidea* sp. can be seen grazing on the back of the *Cellana*. The barnacles are primarily *Chamaesipho tasmanica*. Large limpets and other gastropods (e.g. *Turbo* sp.) are commonly targeted by users of reefs, and estimates of the size and abundance of these grazers are appropriate to many impact studies.

M. KINGSFORD

OPPOSITE BOTTOM: **42** Closely packed green-lipped mussels, *Perna canaliculus*, on the rope of a mussel farm, Malborough Sounds, New Zealand. This species is sought after for food and is common in the lower intertidal of shores ranging in exposure from exposed to sheltered.

M. BRADSTOCK

44 Trampling can result in changes to the cover of algae of the intertidal. This is the result of a trampling experiment on *Hormosira banskii* at Kaikoura, southern New Zealand. This was a treatment for heavy use with 100 passes along the same track. The resultant loss of algae is shown after one tidal cycle.

D. SCHIEL

45 Aerial infrared photograph of the intertidal from 50 m above the substratum, near Melbourne, Australia. Algae in the intertidal can be seen as a red colour. With adequate 'ground truthing', this can be a powerful method of measuring changes in the cover of intertidal algae. It has also been used to study cover of subtidal algae that surfaces (e.g. *Macrocystis*), but we advise caution where strong currents may sweep algae below the surface, as IR film will not detect algae more than a few centimetres beneath the surface.

M. KEOUGH

43 Treatments of an intertidal experiment by Peter Jernakoff, Cape Banks, New South Wales, Australia. All grazers have been removed from the rock, resulting in a uniform cover of ephemeral green algal species (e.g. *Enteromorpha* and *Ulva*). Clearances and enclosures were used to examine the effect of grazers on the carpet of algae.

R. CREESE

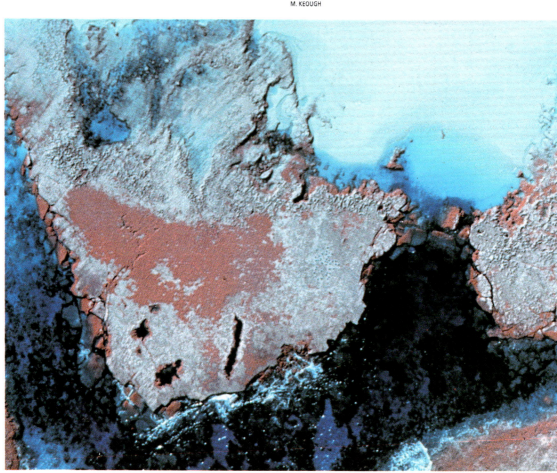

Mangroves (*Avicennia marina* var. *australasica*) in the upper reaches of the Whangateau estuary, northern New Zealand.

RIGHT: **46** Pneumatophores from the tree and neighbours can be seen. Small white tags mark the position of seedlings. The algae are *Hormosira banksii*.

R. CREESE

BELOW: **47** Seedlings and the pneumatophores from large plants at high tide . Barnacles are on some pneumatophores, and gastropods (*Amphibola crenata*) are abundant among them.

M. FRANCIS

ABOVE: **48** A mangrove tree encrusted with oysters, *Crassostrea glomerata*, at high tide, Whangapoua estuary, northern New Zealand. Oysters are also on the substratum, along with pneumotophores of the mangrove and an alga, *Hormosira banksii*. Oysters are targeted by recreational and commercial fishers and are good candidates for impact studies.

M. BRADSTOCK

LEFT: **49** The edge of a seagrass, *Zostera muelleri*, meadow, Tasman Bay, New Zealand. Seagrass is an important habitat on soft substrata in many sheltered bays and estuaries in New Zealand and Australia, and is often exposed at low tides. Plants can take a long time to recover from habitat destruction where the rhizome bases are destroyed.

M. BRADSTOCK

RIGHT: **50** The large starfish *Coscinasterias calamaria* on subtidal soft substrata and a bed of pipis (*Paphies australe*); Whangateau estuary, depth 3 m. Estimates of abundance and size of individual organisms such as starfish are easily obtained in belt transects and similar visual methods.

M. BRADSTOCK

Examples of aerial photography, northern New Zealand.

ABOVE: **51** Aerial view from 100 m of the shallow subtidal and intertidal near Goat Island beach, in the Leigh marine reserve. When waters are clear and the tide is low, subtidal habitats such as kelp forest and some topographic features can be observed and used for mapping (e.g. Fig. 4.8) and for records of temporal change of the position and size of habitats.

R. CREESE

RIGHT: **52** The entrance to the Whangateau estuary, from an altitude of 100 m. Fluorescein dye was being used to compare small scale patterns of oceanography with settlement patterns of bivalves. The dark outlines of habitats in soft sediment (e.g. shell beds) can be seen and, with 'ground truthing', could be used to map habitats for temporal change in the position, estimates of cover and for stratified sampling of organisms associated with different habitats.

R. CREESE

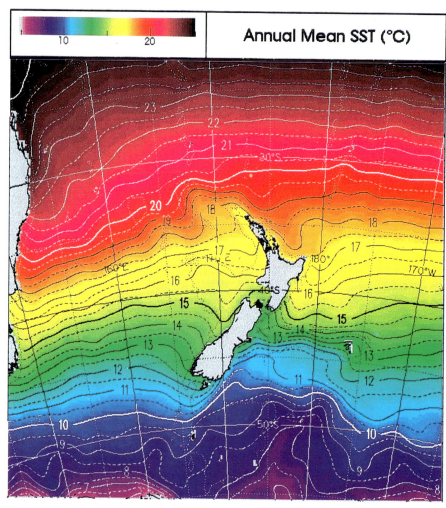

Annual Mean SST (°C)

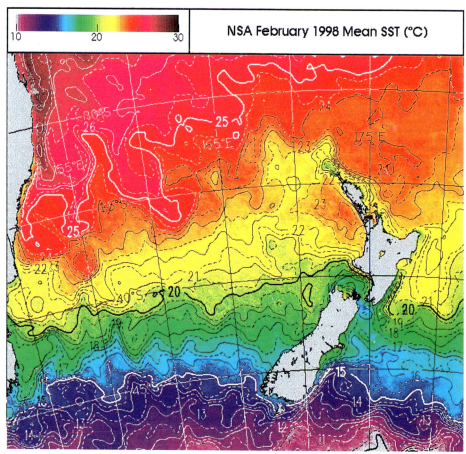

NSA February 1998 Mean SST (°C)

Satellite-derived analyses of sea-surface temperature (SST).

OPPOSITE TOP: **53** Mean SST in the waters surrounding New Zealand and the east coast of Australia. For the names of currents, see Figure 9.3. The 1000 m depth isopleth is indicated. Conspicuous differences in oceanography often correlate strongly with biogeographic boundaries of marine organisms (e.g. East Cape, New Zealand).

M. UDDSTROM AND N. OIEN FROM NIWA SST ARCHIVE

OPPOSITE BOTTOM: **54** SST in waters off the east coast of Australia and the Tasman Sea during February 1998. The East Australian Current is a strong south-flowing Western Boundary Current. It can reach speeds in excess of 3 knots and has a great influence on water temperature and plankton composition in coastal waters as far south as Cape Howe, 37° S, 150° E). Tropical fish and invertebrate larvae are transported to temperate waters during the austral summer and autumn. This current therefore influences the biogeography of many organisms, as does the south-flowing warm boundary Leeuwin Current on the west coast of Australia.

M. UDDSTROM AND N. OIEN FROM NIWA SST ARCHIVE

RIGHT: **55** SST in the waters off Northland, New Zealand, on 29 November (top) and 7–8 December 1994. These show how SST conditions can change very quickly on a scale of days to weeks. In the upper image there is marked upwelling in waters on the east coast, probably due to the effect of a northwest wind that transported surface waters inshore. In the lower image the upwelling conditions have ceased and warm offshore waters have been advected shoreward to cover most of the northeast coast of Northland. These dynamics can result from changes in wind direction and the extent of water-column stratification (Sharples 1997; Zeldis et al. 1998).

M. UDDSTROM AND N. OIEN FROM NIWA SST ARCHIVE

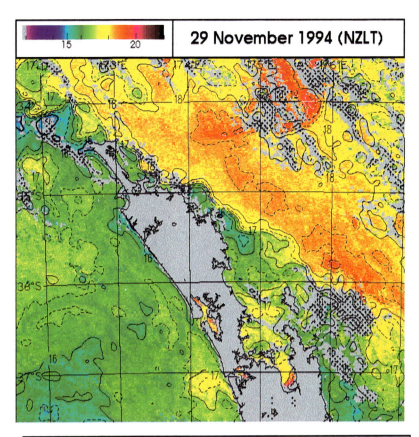

29 November 1994 (NZLT)

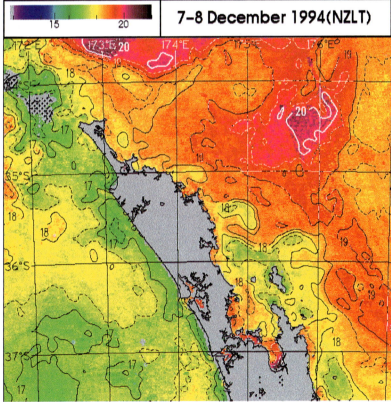

7–8 December 1994 (NZLT)

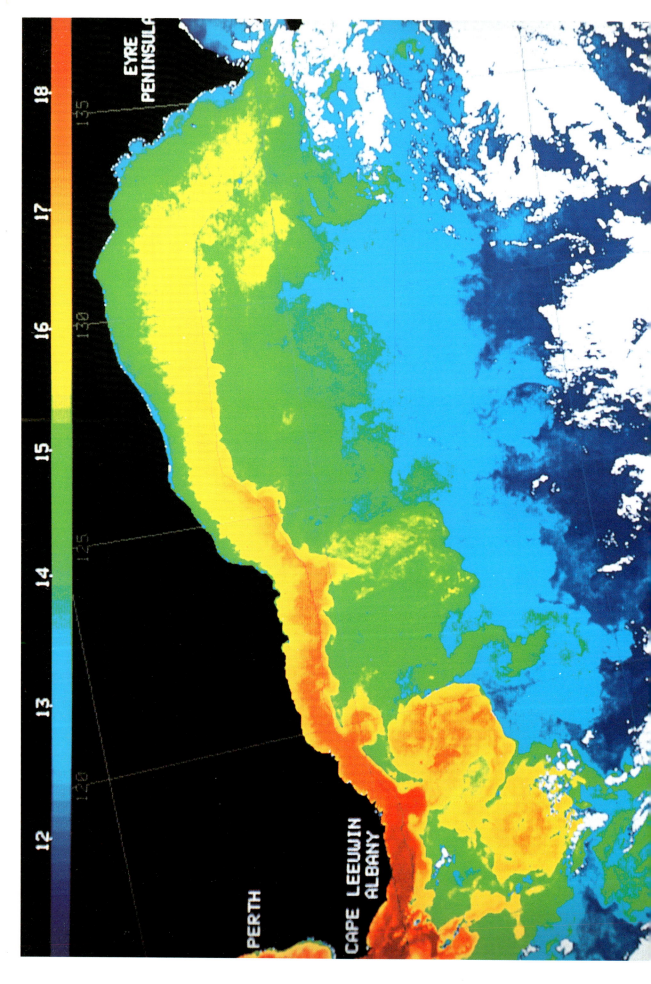

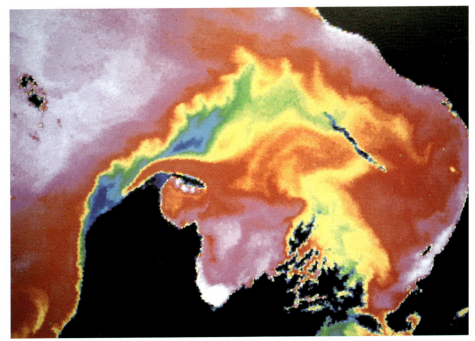

OPPOSITE: **56** NOAA AVHRR SST image of the warm Leeuwin Current, which has flowed from tropical waters of the Timor Sea down the west coast of Australia and into the Great Bight of Australia. Traces of warm water from this current can be observed as far east as Eyre Peninsula. There are strong correlations between the recruitment of many taxa (including fish, lobsters and molluscs) and the strength of the Leeuwin Current in Western Australia (Caputi et al. 1995).

IMAGE COURTESY CSIRO MARINE RESEARCH, HOBART

ABOVE: **57** Satellite image of SST near Farewell Spit, South Island, New Zealand. An area of upwelling, the Kahurangi Plume, can be seen advecting toward western Cook Strait and the Taranaki Bight. The cooler waters have a different composition of zooplankton compared with surrounding waters. The high plankton biomass attracts squid, and this area is the focus of a squid fishery.

M. UDDSTROM AND N. OIEN FROM NIWA SST ARCHIVE

BELOW: **58** An aerial view from 200 m of the plume of the Clarence River, near Kaikoura, southern New Zealand. The position and temporal changes in the extent of plumes, frontal margins and even plankton (e.g. *Munida gregaria*, see Chapter 9) can be plotted from the air, especially with the assistance of GPS fixes. The internal structure in this plume appears to be generated from the entrance, where water is being released in pulses, perhaps caused by the influence of swell and tide.

C. BATTERSHILL

OVERLEAF: **59** A SeaWiFs image of sea-surface chlorophyll off the coast of northern New Zealand, 1 January 1988. High-spatial/temporal resolution imagery, combined with 'ground truthing', is a powerful tool for studying processes influencing production. The image shows highly productive waters over the shelf, and filaments of chlorophyll-rich water being advected to waters well past the shelf break.

R. MURPHY FROM NIWA SS CHLOROPHYLL ARCHIVE/NASA GSFC

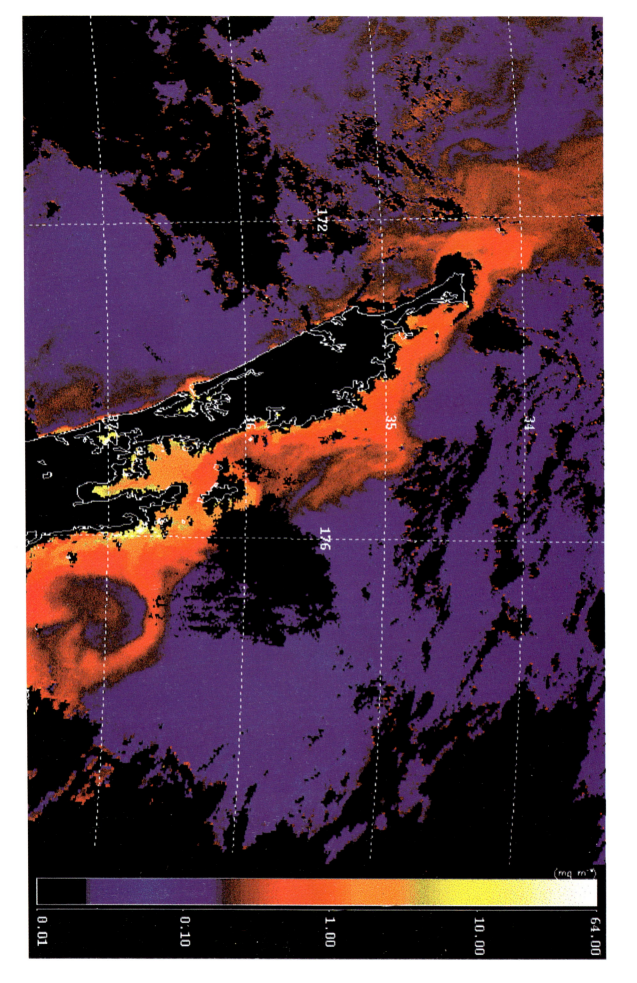

6.4.4 Regional studies for a biogeographic partitioning of marine protected areas

Example: Fish counts by M. Kingsford along the coast of New South Wales.

Purpose: Choose replicate marine reserves for fishes that encompass a broad geographic spread and are representative of major biogeographic regions (example based on a potential application of Kingsford data).

Situation: In a country or state that wants to have a broad geographic spread of MPAs (e.g. Walls & McAlpine 1993), a logical justification for the number and placement of reserves is that replicate MPAs should be situated in each of the major biogeographic regions (for other justifications, see Table 1.1). A set of data I have collected along the coast of temperate New South Wales shows how this could be done (Fig. 6.9).

Approach: A habitat-stratified study of demersal reef-associated fishes was done along 1000 km of coastline on temperate and subtropical reefs of New South Wales and northern Victoria. Belt transects were used (25 x 5, n = 6 replicates). A two-phase approach was used: a pilot study was done to optimise the number and size of transects (Fig. 3.1); and fishes were counted at multiple locations over a latitudinal range of 10°. A justification for the possible location of reserves based on these 'biogeographic' data is given.

In the pilot study, transect length was chosen based on estimates of abundance of fishes and a size that could fit within habitats. Transect width was approximately one diver length (2.5 m) each side of the tape and constituted a narrow belt that was easily searched for fishes.

Figure 6.9 Map of the coast of New South Wales. Areas of the mainland that vary greatly in the nature of subtidal habitats and the abundance patterns of fishes are indicated.

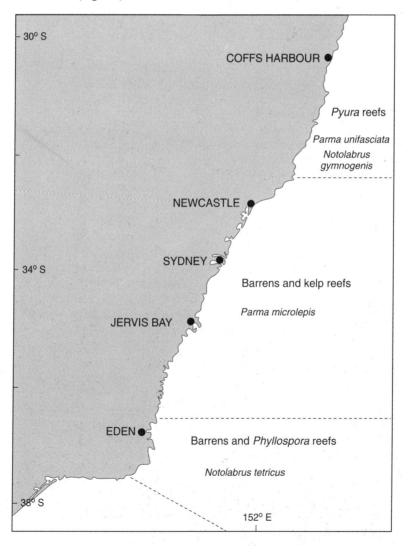

A sample size of six was chosen because it gave high precision for some target species (e.g. *Parma microlepis*, SE = 10–15% of the mean) and this number of replicates could be completed in a dive to depths of up to 20 m.

Belt transects were done at 24 locations. Where possible, sampling was also done at offshore islands. Sampling was stratified by habitat at each location (where more than one habitat was available). Different habitats included kelp (*Phyllospora*) stands south of 36° 30' S, kelp (*Ecklonia radiata*) and barrens at 32° 30'–36° 30' (Plates 15–18) and ascidian (*Pyura* spp.) dominated reefs north of 32° 30'. The latter two reef types are described in Underwood et al. (1991). Reef profiles were generally to 8–10 m in the north of 32° 30' and 10–25 m south of that latitude and at offshore islands.

Great differences in the distribution and abundance of reef fishes were found. Pomacentrid and labrid fishes were generally most abundant. At mainland locations, *P. microlepis* were most abundant in barrens habitats south of 32° 30' (Fig. 6.9). In contrast, *P. unifasciata* were most abundant north of this latitude, and few *P. microlepis* were seen. *P. polylepis* was also regularly observed at these latitudes (1–2 per 125 m²), whereas few of these fish were seen to the south.

Labrids showed similar great differences in abundance. *Notolabrus gymnogenis* were most abundant north of 33°, as was *P. unifasciata*. *Notolabrus tetricus* was only found south of Batemans Bay (36° 30') and was the most abundant labrid on the mainland to Gabo Island (37° 40'). The cheilodactylid *Cheilodactylus spectabilis* was similarly only abundant in the far south of New South Wales (3.8 ± 0.7 per 125 m² at Gabo Island). Individuals were very rare or absent north of Bateman's Bay (35° 45'). A broad range of tropical and subtropical pomacentrids (e.g. *Stegastes gascoynei, S. apicalis, Parma oligolepis, Amphiprion akyndnos* and *Dascyllus trimaculatus*), and labrids (e.g. *Thalassoma lunare, T. lutescens, T. janseni, T. hardwicki* and *Macropharyngodon choati*) were found at the Solitary Islands that were not observed on adjacent mainland sites.

Three biogeographic regions were identified on coastal reefs of mainland New South Wales (Fig. 6.9). Data on fishes, quantitative data on habitats (Underwood et al. 1991) and my records of habitat types where fishes were counted all corroborate. Managers would be advised to select a minimum of two replicate reserves within each of the biogeographic regions on the mainland. In New South Wales, as in other parts of the world (e.g. Choat & Ayling 1987), there were clear differences in the fish fauna found on the mainland and adjacent offshore islands. Offshore islands should be treated as a special case, and replicate islands at different latitudes (if they exist) could be protected (bioregionalisation, see also Edgar et al. 1997).

The issue of reserve size is not addressed in this example. Reserves with total protection range greatly in size from a few hundred metres (e.g. Jervis Bay and Solitary Islands in NSW) to 70 km in South Africa (C. Buxton pers. comm.). There is strong evidence that reserves such as the Leigh marine reserve have facilitated increases in the abundance and size of organisms such as rock lobsters and fishes (Cole et al. 1990; MacDiarmid & Breen 1993). However, more biological information is required on the influence of reserve size on the size-frequency, abundance and mobility of resident organisms. In general, however, it is very hard to police the margins of reserves, and the first 500 m or so at each end should virtually be treated as a buffer. Reserves that are less than 2 km long, therefore, are unlikely to fulfil their purpose.

6.4.5 Spawning-site choice by labrids

Example: Experimental study of spawning-site choice by female spotties *Notolabrus celidotus* and its influence on the mating system (Jones 1981).

Purpose: An experimental study that was designed to determine the importance of quality of site versus male quality on the choice of spawning sites by females.

Situation: Jones studied the mating system and behaviour of the protogynous labrid *N. celidotus* (Plates 28, 29) and found that most terminal-phase (TP) males guarded territories, but only certain deep sites were occupied consistently.

Approach: A broad area of reef was mapped. Jones marked the positions of territories of TP males and the positions of observed spawnings (Fig. 6.10). Two hundred and twenty-five hours of observations were made on individually recognised fish. The movements of fish were marked once per minute on slates etched with a scale drawing of the study area. Home ranges were calculated by the largest polygon method (Odum & Kuenzler 1955) for 30-minute periods of observations. For females, observed for several hours, there was rarely more than a 15% increase in estimated home range. The territory areas of TP males (Fig. 6.1) were estimated by drawing the boundaries through sites where border disputes between neighbouring males were observed.

Activities of fish were divided into three broad categories: maintenance, interactions and reproduction. Maintenance was subdivided into swimming and foraging. Interactions were subdivided into aggressive/submissive chases, aggressive/ submissive displacement, and border disputes between males (generally

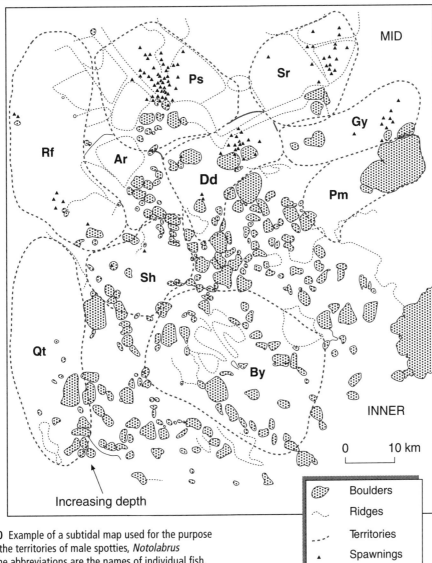

Figure 6.10 Example of a subtidal map used for the purpose of marking the territories of male spotties, *Notolabrus celidotus*. The abbreviations are the names of individual fish. Note how boulders and ridges have been indicated.

MODIFIED FROM JONES (1983), PUBLISHED WITH PERMISSION FROM *BEHAVIORAL ECOLOGY AND SOCIOBIOLOGY*

alternating chases). Reproductive activities were divided into courting and spawning.

Two experiments were done. In the first, the most successful TP male was removed and the subsequent redistribution of territories and spawning success was measured. The hypothesis of site selection would be supported if the site was rapidly taken over and the replacement male took over the ranking of the male that was removed. Alternatively, if the main area of spawning shifted to the territory of the male previously ranked second, then direct male selection would be favoured.

In the second experiment, all TP males were removed and the behaviour, movements, and spawning success of immigrant males was noted. A hypothesis of site selection would be supported if males orientated actively towards historically favourable territories and gained high spawning rates. Site selection would not be supported if the movements of fish and spawning within the study area were random.

Although the spawning success of TP males was not related to body size, age, coloration or territory size, testis weight:body weight ratios were positively correlated with spawning success. The experimental removal of the highest-ranked TP male (PS = name of individual fish) resulted in an adjacent male abandoning its own territory and taking over the territory. This male, which had not previously been observed to spawn, began doing so at the new site half an hour after the removal of PS. When all fish were removed, the first immigrant male occupied the territory of PS. This male was removed and a sequence of two more males also occupied the territory of PS. It was concluded that the spawning system was 'resource defence polygyny' because the defence of preferred spawning sites by TP males accounted for most spawning episodes. It was thought that females preferred territories that were in deep water and had boulders covered with kelp, in which they could shelter.

Work of this type is very time-consuming, which precluded its being done at a number of locations. Based on this detailed work, however, predictions could be made for other locations where it would be expected that spawning rates of males would be greatest where territories were in deep water (suggested to be related to low numbers of egg-eating fishes) and had shelter for females nearby. For a similar study, see Olla et al. (1981).

6.4.6 Do reef fish respond to disturbance?

Example: An experimental study to determine how *Atypichthys strigatus* (Pisces: Scorpididae) responds to disturbance of the substratum (Glasby & Kingsford 1994).

Purpose: Study the diet and behaviour of *A. strigatus* and experimentally determine how fish respond to disturbance.

Situation: No information was available on the ecology of mado *A. strigatus* in central New South Wales. The disturbance experiments were relatively easy to do, so they were done at multiple locations, which placed the authors in a better position to make general statements about the influence of disturbance. A pilot study and casual observations of *A. strigatus* responding to disturbance suggested that a quadrat size of 1 m² would be adequate for experiments. In addition, disturbances and controls of this size were easy to view in most conditions, and natural disturbances (e.g. by fish) were generally of this scale. Data on abundance indicated that fish were generally in aggregations and were abundant at all depths on rocky reefs. Moreover, dietary information indicated that there were size-related differences in the diet of *A. strigatus* and that large fish generally ate plankton, but some fish had consumed large numbers of benthic crustaceans and, in some cases, fish had eaten parasitic copepods, which suggested (and was later confirmed by subsequent behavioural studies) cleaning behaviour.

Approach: Two experiments were done. The first was only done in 'barrens habitat' (Plates 15–18). Sampling was done at six locations (a 'random' factor in analyses: see Chapter 3) that were separated by at least 300

m. At each location a maximum of 15 minutes was spent searching for an aggregation of *A. strigatus* and if no fish were found the experiment was done regardless. A 1 m² quadrat was put on the substratum and was disturbed (by brushing) for 10 seconds or left untouched (controls). Each quadrat (n = 5 for each treatment) was monitored for 10 minutes. The behaviour of fish at distances of greater than 2 m was not recorded, because it was not considered likely to be related to the benthic disturbances.

The second experiment was also done at six locations, but in kelp and barrens habitats. Quadrats were only monitored for five minutes, because the results of the first experiment indicated that most bites were recorded within this time.

Significantly more *A. strigatus* were found in disturbed areas than in undisturbed areas, and their feeding rate was greatest in the disturbed areas. Fish responded to disturbance in barrens and kelp habitat, but the greatest response was in the barrens. *A. strigatus* were found to respond to natural disturbances by large fishes (especially labrids). It was concluded that the behavioural flexibility of *A. strigatus* probably leads to increased feeding opportunities, and hence foraging success. In addition, it also raised questions about the accuracy of partitioning fish species according to one trophic group.

6.4.7 The influence of competition on depth-related patterns of abundance of reef fishes

Example: The use of field experiments to understand the abundance and dynamics of reef fishes (Schmitt & Holbrook 1986; Holbrook & Schmitt 1989, 1996).

Purpose: Explore the effect of interspecific competition on mean abundance of black surf perch and striped surfperch.

Situation: Two species of surfperch were found to co-occur on shallow reefs (less than 10 m) at Santa Cruz Island, California. Black surfperch were more common in deep water, while striped surfperch were most abundant in shallow water. Both species of fish are live-bearing micro-carnivores that feed on small crustaceans associated with algae (Love 1991). The availability of prey varied among types of algae; abundance of invertebrates was greatest in *Gelidium*, and both species of fishes preferred to feed on this alga. Based on these patterns of distribution, reciprocal removal/competition experiments were planned. A prerequisite for the experiments was a good knowledge of distribution of fishes, diet, and the distribution of the resource *Gelidium*.

Approach: All field work was done at six study sites within a 20 km stretch of coastline. There were two control sites and two sites each where densities of each surfperch were reduced by about 90% for a period of four years. Within each site, fish were counted in permanent 40 x 2 m belt transects at 3, 6 and 9 m depth contours. Counts were made eight times weekly along each transect, each season. All sites were sampled on the sample day (Schmitt & Holbrook 1986). Eight counts were made before and after each manipulation. The foraging behaviour of fishes and the availability of food was also sampled at each depth.

Overall abundances of fishes did not change in any area, partly because the experiment was not done in the recruitment season when these live-bearing fishes (Embiotocidae) release their young (see Love 1991). Black surfperch immediately moved into shallower water of the reef upon removal of striped surfperch, whereas striped surfperch did not shift habitats when black surfperch were removed. Furthermore, in the absence of competitors, each species greatly increased use of *Gelidium*, which is primarily found in shallow water. Neither species altered its use of deeper parts of the reef upon competitive release.

Subsequent experiments where the cover of *Gelidium* on reefs was reduced caused subsequent changes in the abundance of surfperch (Holbrook & Schmitt 1998). It was concluded that although interspecific competition can depress abundance of surfperch at different times and places, the availability of food resources explains much of the variation in abundance among reefs.

In years since the experiments there has been a great decrease (80–90%) in numbers of surfperch on reefs in the Californian Bight. Concurrent with these changes has been a 60–70% reduction in their food base (*Gelidium*), which appears to relate to physical forcing through changes in local oceanographic conditions and related production (greatly influenced by El Niño conditions). Holbrook and Schmitt (in press) concluded that their experiments provided a 'foundation' of knowledge that was needed to explore and understand (among-reef) spatial and temporal (decade and longer) patterns of variation. Moreover, the mechanistic understanding enabled them to interpret long-term environmental forcing of a resource-driven system. An ironic corollary to their work is that in the 1990s it would not be possible to do the experiments that were done in the 1980s because abundance of surfperch is now so low they would not have considered undertaking such a venture. Although the live-bearing life history of embiotocids is different from most reef fish, competitive interactions can still have a potent influence on the population dynamics of oviparous reef fish (e.g. Robertson 1996).

SELECTED READING

The references listed are examples of research that will lead the reader to a broader literature, also see the text of the chapter. * indicates a multi-authored book or collection of proceedings; references to volumes of a journal refer to a set of relevant papers.

Ageing: Beamish & McFarland 1983; (see also Growth); Chen et al. 1992; Paul 1992; Stevenson and Campana 1992; see 6.2.10.

Assemblages: Ebeling et al. 1990; Ebeling & Hixon 1991; Sale 1991*.

Artificial reefs: Bohnsack & Sutherland 1985; Pollard & Mathews 1985; Bohnsack 1989; *Bulletin of Marine Science* 1989 (44, 2), 1994 (55; 2, 3)*.

Behaviour: Johannes 1978; Jones 1983; Thresher 1984; Warner 1988; Schultz & Warner 1989.

Cleaner fish and parasites: Grutter 1997.

Competition: Jones 1988b, 1991; Holbrook & Schmitt 1996, 1998; Robertson 1996.

Creel surveys: Clarke & Buxton 1989; Kingsford et al. 1991; Jones & Robson 1991; Hancock 1994; West & Gordon 1994.

Diet: Hyslop 1980; Wallace 1981; Wallace & Ramsey 1983.

Distribution: Ebeling et al. 1980; Jones 1984c; Sano & Moyer 1985; Choat & Ayling 1987; Francis et al. 1987; Choat et al. 1988; Kingsford 1989; Kingsford et al. 1989; Holbrook et al. 1990; Lincoln-Smith et al. 1991; Love et al. 1991; Sale 1991*; Cole et al. 1992; Carr 1994; Syms 1995; Falcon et al. 1996; Francis 1996; Meekan & Choat 1997; Gillanders & Kingsford 1998.

Ecology: Larson 1980a, b; Jones 1988a; Sale 1991*; Holbrook et al. 1994, 1998; see also 6.3.

Elasmobranchs: McLaughlin & O'Gower 1971; Gordon 1993; Krogh 1994; Last & Stevens 1994; Pollard et al. 1996.; Smith & Pollard 1996.

Growth: Chen et al. 1992; Gillanders 1997b; Taylor & Willis 1998; see also 6.2.10 and Table 3.11.

Impact studies (e.g. MPAs, pollution-related): Cole et al. 1990; Buxton 1993; Carr & Reed 1993; Nowak 1996; Chapter 2.

Experimental studies: Holbrook & Schmitt 1998; see also 6.3.

Feeding: Hobson 1974, 1991; Hobson & Chess 1976; Bell 1979; Russell 1983; Cowen 1986; Jones 1988a; Kingsford & MacDiarmid 1988; Wainwright 1988; Sale 1991*; Choat 1982, 1991; Anderson 1991; Choat & Clements 1992, 1997; Horn 1992; Caceres et al. 1994; Gillanders 1995b, 1997b.

Fish attraction devices (FADs) (see also Artificial reefs): Rountree 1989; Hilborn & Medley 1989; Kingsford 1993; Hair et al. 1994; Druce & Kingsford 1995.

Fisheries: Gulland 1988; Hilborn & Walters 1992; Kailola et al. 1993; Tilzey 1994; Dalzell et al. 1996; *Ecological Applications* 1998, vol. 8, no. 1.

Fisheries methodology: Kailola et al. 1993; Sainsbury 1996.

Ichthyology: Lagler et al. 1977; Paxton & Eschmeyer 1994.

Larvae – fish: see Chapters 9 and 10.

Life history: Jones 1980; Thresher 1984; Kingsford 1988; Chapter 9.

Methods

General: Bagenal 1978*; Baker & Wolff 1987; Parker et al. 1990; Rozas & Minello 1997.

Fish counts: Thresher & Gunn 1986; Lincoln-Smith 1988a, b; St John et al. 1990; Harmelin-Vivien & Francour 1992; English et al. 1994

Movements: Gladfelter 1979; Johannes 1978; Stoddart & Johannes 1978*; Bray 1981; Bray et al. 1981; Burchett 1993; Howard 1989; Vieira 1991; Bell & Worthington 1993; Sakuri & Nakazono 1995; Gillanders & Kingsford 1996.

MPAs: Russ & Alcala 1989; Cole et al. 1990; Buxton 1993; Carr & Reed 1993; Cappo 1995; Smith & Pollard 1996; Edgar & Barrett 1997; MacPherson et al. 1997; Chapters 1 and 2.

Phenology: Johannes 1978; Miller 1979; Robertson 1991; Tzioumis & Kingsford 1995.

Population dynamics: Bagenal 1978*; Sale 1991*; Hilborn & Walters 1992; Hixon & Beets 1993; Pfister 1996; Hixon & Carr 1997.

Reproductive biology: Jones 1980; Olla et al. 1981; Thresher 1984; Thompson 1986; Warner 1988; Gladstone & Westoby 1988; Nakamura et al. 1989; Buxton & Garratt 1990; Cowen 1990; Darwall et al. 1992; Webb & Kingsford 1992; Pankhurst 1995; Tzioumis & Kingsford 1995; Gillanders 1995a; Andrew et al. 1996; Barrett 1995; Neira et al. 1997.

Seagrass fishes: Gray & Bell 1986; Bell et al. 1985b, 1988; Larkum et al. 1989; Ferrell & Bell 1991; Worthington et al. 1992a, b; Sanchez Jerez & Ramos Espla 1996.

Sharks: see Elasmobranchs

Surveys of fishers: see Creel surveys.

Taxonomy and Biology

Marine Fishes: see Chapter 10.

Estuarine and freshwater fishes: Australia: Merrick & Schmida 1984; McDowall 1996. New Zealand: McDowall 1990. North America: Page & Burr 1991. World: Sterba 1962.

Temperate versus tropical: Ebeling & Hixon 1991; Sale 1991*.

Tide pools: Beckley 1985; Polivka & Chotowski 1998.

SEVEN

ORGANISMS OF REEF AND SOFT SUBSTRATA INTERTIDAL ENVIRONMENTS

R. G. Creese and M. J. Kingsford

7.1 Introduction

Although they usually occupy only a narrow band at the intersection of land and sea, intertidal environments often contain a diverse array of organisms that interact in complex ways. They are subjected to a unique combination of both aerial and marine influences on a more-or-less regular, tidally defined basis, as well as irregular but often severe physical disturbance. A very few inhabitants of these environments are predominantly terrestrial species (e.g. the larvae of some insects and several spiders). Some predominantly intertidal species extend into subtidal habitats (e.g. some limpets, under-boulder species), and a few subtidal species extend into intertidal habitats (e.g. tidepool algae, sandy beach bivalves, fish feeding at high tide). Most species found in intertidal habitats, however, are restricted to this narrow band at the shoreline. Despite this, the intertidal environment is best considered as the terrestrial margin of the sea, and an understanding of the patterns and processes occurring here often benefits from knowledge of the adjacent subtidal or pelagic patterns.

Intertidal shores are usually the most accessible of marine environments both for members of the general public and for researchers (exceptions include exposed places). Although only accessible for a few hours around low tide (and often only during spring tides), intertidal shores are easier to sample than subtidal environments because of their more limited area and accessibility without boats or scuba, etc. Where baseline information about a particular coastline is required urgently, data can be gained quickly and inexpensively from intertidal habitats. These same features provide researchers with an excellent environment for assessing sampling procedures and identifying problems that may be encountered elsewhere.

Many of the field techniques and analytical protocols used in modern temperate reef ecology originated from intertidal studies. This, in turn, means that any biological assessment of an intertidal area can be carried out thoroughly and rigorously. In addition, manipulative field experiments, which are ideal for studying interactions among marine organisms, were pioneered in intertidal environments and have greatly expanded our understanding of the dynamics of marine systems (e.g. Peterson 1982; Underwood 1985; Underwood & Chapman 1995).

Accessibility, however, has its drawbacks. In terms of impact, intertidal reefs and sediment flats are particularly vulnerable to the activities of humans. The direct effects of humans include harvesting intertidal organisms such as sea urchins, gastropods, bivalves, crabs and flatfish for eating or use as bait. For example, a variety of intertidal organisms are taken from rocky reefs in southeastern Australia (Fairweather 1991a; Kingsford et al. 1991a; Keough et al. 1993), Chile (Moreno et al. 1984) and South Africa (Dye 1992), which can lead to changes in the ecology of these reefs (e.g. Castilla & Duran 1985; Castilla & Bustamante, 1989; Hockey & Bosman 1987). Similarly, there are many examples of the depletion of sediment-dwelling bivalves by harvesting from enclosed bays and harbours and open sandy beaches (e.g. Stace 1991), although the flow-on ecological impacts of this on the other biological assemblages have not been as well documented as for rocky shores. Another direct effect is the damage caused to intertidal species, particularly algae, by pedestrian trampling on rocky reefs (Brosnan & Crumine 1994; Brown 1996; Povey & Keough 1991). Many intertidal areas worldwide, especially in harbours, estuaries or sheltered bays, have been highly modified physically by port or road construction, flood mitigation or reclamation (e.g. Findlay & Kirk 1988; Zedler & Nordby 1986). Finally, many of the marine species that have been inadvertently transported to other geographic regions are intertidal (e.g. the New Zealand barnacle *Elminius modestus* and the Asian date mussel *Musculista senhousia*: see Chapter 1), and their introduction may result in unforeseen changes to native intertidal assemblages.

Indirect effects of human activity on intertidal environments include the following: decreased abundance and/or species diversity, reduced reproductive output or behavioural changes due to such pollutants as oils and dispersants (e.g. Battershill & Bergquist 1982; McGuinness 1990), sewage (Littler & Murray 1975; Roper 1990), urban runoff (Roper et al. 1988), pesticides (Pridmore et al. 1991), landfill leachate (Gowing 1994), antifouling paints (Stewart et al. 1992) or heavy metals and PCBs (Fairweather 1990b; Luoma *et al.* 1991). Filter-feeding molluscs on rocky reefs or sediment flats may pose a threat to public health through accumulating toxins and incubating diseases such as gasteroenteritis derived from faecal pollution.

All potential impacts, both direct and indirect, are likely to affect intertidal habitats even more than other marine environments (see Chapter 1). For the purposes of management and measurement of impacts, therefore, it is essential to have rigorous quantitative information on the intertidal environment. The objective of this chapter is to describe the sources of variation that should be considered in designing sampling protocols, and the apparatus, techniques and experiments most frequently used in intertidal studies. Both sampling procedures and experimental approaches are covered, the former more fully than the latter. Because the philosophy and many of the procedures used are applicable to many marine environments, there is some unavoidable overlap with other parts in this book. Wherever possible, however, this overlap has been minimised and reference made to other sections where greater details may be found.

7.2 Methods for assessing intertidal environments

7.2.1 Sources of variation

In any study of intertidal ecosystems, careful consideration of the nature of your sampling problem is needed before starting (see Chapter 2). Become as familiar as you can with the physical environments and sorts of organisms you might encounter. In particular, as has been emphasised in previous chapters, you must have a very clear and concise objective for any study and continually ask yourself, 'Is my design going to provide information that will allow me to meet my objective(s)?' This is especially important for monitoring studies set

up to assess the consequences of some future perturbation (which may be poorly defined, or may be quite specific but unpredictable, such as in the case of an oil spill). The likely magnitude and extent of such perturbations will be unknown (although predictions may be possible) and monitoring may continue for many years. Such studies need to be relatively simple to implement and robust in their ability to detect effects. Studies can be designed to detect effects of a magnitude (e.g. a 30% reduction in the abundance of a key species) that are unacceptable. Sometimes, however, the objective may be purely descriptive: to obtain a species list for an area or to delineate the limits to the vertical distribution of a particular species. Such descriptions may serve merely as pilot investigations for more detailed, mensurative studies, which may then generate hypotheses about processes that, in turn, can be investigated by manipulative field experiments, especially when done at different spatial or temporal scales.

Tides

The most basic variable to consider when working in intertidal environments is the tide. Unless the study specifically demands that observations (with a snorkel or scuba) be made at high tide (e.g. to study predation by itinerant fishes), sampling will be carried out at low tide. Low spring tides (which occur at full and new moons) will usually expose the greatest area of the intertidal and provide the best opportunity to sample comprehensibly. This is not always the case, however, and peculiarities of particular sorts of intertidal environments such as slow-draining marshes (Hubbard 1996) may require a different tidal state for the best sampling opportunities. Seasonal variations can also impose sampling constraints, especially if semidiurnal tides (i.e. two low-tide periods per day) are unequal. On the west coast of the USA, for example, this means that the best spring low tides occur at night in winter for northern areas but in summer for southern California. Weather also influences the rise and fall of the tide on a particular day. For instance, low-pressure weather systems raise the water level, sometimes completely overwhelming the tidal signal (e.g. on microtidal beaches in southwestern Australia: Hegge et al. 1996). Areas with substantial tidal ranges, however, are less affected. A good understanding of the local tidal and weather conditions is therefore essential when planning any intertidal study. In some countries, tide tables are available on disk as well as in printed form.

Spatial and temporal variation

Spatially, variation may occur along a coastline (for example, categories based on latitude or gross physical characteristics such as rock type, slope and exposure to wave action) or across it. Because of their limited spatial extent downshore and their generally two-dimensional structure, intertidal areas on rocky coasts often can be readily subdivided into different categories (e.g. boulder banks, vertical faces, sloping bedrock or flat platforms) and into more-or-less discrete habitats according to their biological assemblages and physical attributes (fucoid algae, barnacles, coralline turf, cobble patches, rock pools, etc.). This is more difficult on soft substrata, however, because the more three-dimensional nature of these environments means that many of the organisms are not apparent on the surface (see also Chapter 8). However, obvious surface features can still be used to delineate meaningful categories and habitat types. Especially in harbours or bays, different saltmarsh or mangrove vegetations can be used in higher tidal regions and seagrass beds, or accumulations of dead clam shells can be used in lower tidal areas. Even in seemingly homogeneous situations like sandy beaches, physical features such as morphodynamic type (reflective versus dissipative) can be used to categorise different types of beaches (McLachlan & Jamarillo 1995), and individual beaches can sometimes be subdivided visually into habitats according to such things as the line of accumulated beach wrack.

On the temporal scale, patterns of distribution and abundance of intertidal species commonly change on an inter-annual or seasonal basis as a result of growth, variations in the intensity of recruitment, predation, etc. (see also 7.2.4). On a shorter timescale, movement of mobile species may alter patterns of distribution or abundance on a daily, tidal or even hourly basis. Day/night differences in abundance and detectability can cause difficulties, especially if low tides occur during the day in some seasons but at night in others. Nested and partially hierarchical designs enable investigators to address hypotheses concerning multiple spatial and temporal scales in the same sampling design (see Chapter 3).

Taxonomy and life history

Equally as important as the selection of appropriate spatial and temporal scales is an appreciation of the classification, biology and life history of the organisms likely to be encountered. The majority of intertidal organisms spend time in the plankton as larvae or spores. Once settled, the juveniles of some species look dissimilar to the adults and may occur in different habitats. On the other hand, some things that look superficially similar may be distinct species (e.g. species of honeycomb barnacles, many red algae). It is true that basic patterns at an assemblage level can be described without knowledge of the names of the organisms (e.g. noting them only as taxa A, B, C, etc). This makes it difficult to do follow-up work, however, and expert advice on taxonomy should be sought wherever possible: it is a mistake to treat counts of organisms simply as data points! With an adequate knowledge of taxonomy, it is sometimes useful to pool species counts into functional groups (e.g. grazing gastropods) to aid interpretation (see Chapter 4). In addition, a basic understanding of such things as feeding biology, movement patterns or life cycles can assist the interpretation of many patterns and the processes influencing those patterns.

Examples of the life cycles of intertidal organisms are given in Figures 7.1 and 7.2 (see pp. 170–171); other examples are found elsewhere in this book. Most intertidal algae reproduce sexually by means of spores to produce new colonists, but many encrusting species may also spread vegetatively to cover new substrata. Seagrasses also employ both sexual reproduction and vegetative growth (Fig. 7.1a), but mangroves, another common flowering plant in the marine environment, disperse by means of large, fleshy propagules (Fig. 7.1b). Typically, intertidal animals produce very small dispersive larvae that may spend from days to months in the water column before settling on the shore to metamorphose into the adult form. Eggs may be shed directly into the water, or laid within benthic egg capsules or strings in which they undergo some of their larval development prior to emerging as free-swimming larvae (see intertidal gastropods in Fig. 7.2, p. 171).

7.2.2 Preliminary steps

It is worth emphasising that two preliminary steps are normally undertaken prior to any intertidal study. The first is reconnaissance, which might incorporate such things as literature searches of the area or organisms of interest, examination of maps and aerial photographs, or a familiarisation walk, flight, car or boat trip along the target stretch of coast (referred to as 'rapid visual assessment' throughout this book). During the latter, any qualitative notes, photographs or video footage may be particularly useful in terms of providing a 'feel' for the actual or potential study sites and the potential constraints, particularly logistical, that might be encountered.

Second, a more detailed pilot study should normally be undertaken once the specific areas have been selected or the broad outline of the sampling design decided upon (e.g. Underwood & Kennelly 1990a). This would incorporate quantitative sampling to obtain measures of variability within and between any spatial subdivisions of the area (e.g. different habitats) and to assess estimates of accuracy and precision using, for example, different numbers of different-sized or different-shaped

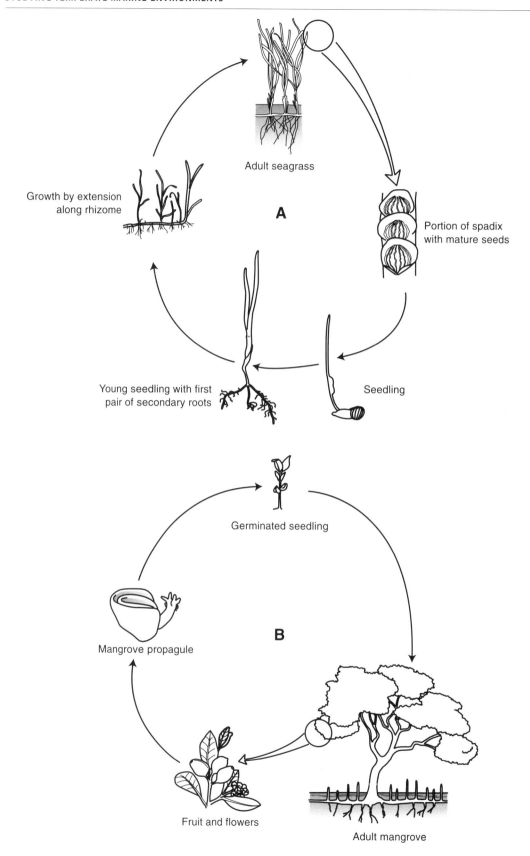

Adult seagrass

Growth by extension
along rhizome

A

Portion of spadix
with mature seeds

Young seedling with first
pair of secondary roots

Seedling

Germinated seedling

Mangrove propagule

B

Fruit and flowers

Adult mangrove

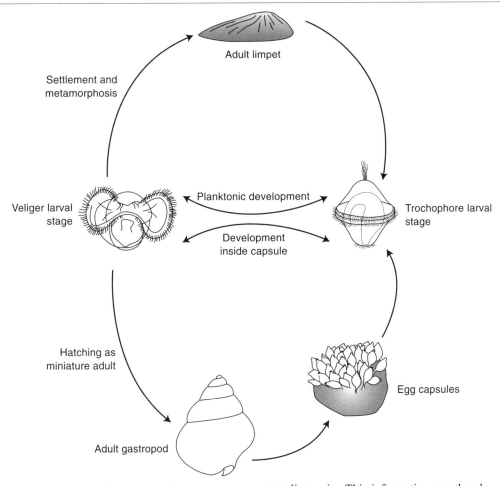

LEFT: Figure 7.1 Examples of the life-history features of two common marine angiosperms.

A: A seagrass of the genus *Zostera*. The flowers and seeds are very small and insignificant, but the latter do permit some dispersal from the parent plants. Most expansion within a seagrass bed, however, occurs vegetatively by means of subsurface rhizomes.

B: A mangrove of the genus *Avicennia*. Adult trees flower in summer, but flowers usually only occur on some branches in any particular year: branches that flower one year tend not to flower the next. The well-developed seed remains in its coat, the pericarp, until the seed drops from the tree. Seeds float for only a few hours before the pericarp is shed, after which the seed settles to the bottom and takes root in the mud. Young seedlings may rely on energy from the large cotyledons for several months after germination.

ABOVE: Figure 7.2 Common intertidal gastropods. The fertilised eggs of almost all gastropods go through a trocho-phore and then a veliger stage before metamorphosis into a miniature adult. Many species undergo their larval development in the plankton (i.e. in the water column) as shown in the upper part of the diagram for a limpet. Some species, however, lay benthic egg capsules in which most or all of the larval development takes place; this is shown in the lower part of the diagram for a whelk.

sampling units. This information can then be used to fine-tune the design of the main sampling so as to optimise effort and cost-efficiency. Sophisticated pilot studies can incorporate investigation of such things as observer biases (see section 7.4.3). Formal procedures are available to design and analyse such pilot surveys (Underwood & Kennelly 1990a; see also Chapter 3). Unfortunately, rigorous pilot studies are not always done, and when done, they are often limited to a single time or place. Because species abundances and spatial patterns can change dramatically over periods of years, seasons, weeks or even days, decisions made at any one time (e.g. one optimal quadrat size or number) may not apply at other times (Williamson 1992; McArdle & Pawley 1994; Thrush et al. 1996b). Ideally, pilot sampling should cover a range of time periods, but this is not always possible. Partial solutions to this problem of the unknown scale of temporal variability and its likely effect on sampling design can be obtained by repeating spatial sampling over short periods (e.g. days) and using pre-existing data (Ambrose et al. 1992; see also 7.4.3). In the latter case, pre-existing data were used to model likely temporal patterns and to simulate the effects of hypothetical perturbations. This procedure can also be done, on a hypothetical basis, in the absence of prior data (e.g. Fairweather 1993).

7.2.3 General methods

There are many different types of intertidal environments, and different techniques are needed to sample them. These are considered in detail in section 7.3. There are, however, two general approaches that are applicable, to varying degrees, to any intertidal situation. The first is remote sensing, a technology that is becoming more sophisticated and widely accessible, and which therefore has considerable potential. The second is 'zonation', a long-established philosophy for sampling intertidal shores that needs re-evaluation.

Remote sensing
Remote sensing merely means gathering information about an area or object without coming into contact with it, and in that sense human sight is probably the most useful piece of sampling equipment available. Our vision, however, is limited in the vertical plane to a metre or two

if we are at sea level. If you are interested in small spatial scales (0.25 m² or less), a hand-held camera or one mounted on a tripod provides an excellent means of recording patterns on the ground, and this technique is often used to sample individual quadrats (Brown 1996; see also 7.4.4). The recent intoduction of digital cameras holds the prospect of being able to record images in the field and then download them into a desktop computer for direct and very accurate analysis, pixel by pixel. Patches of intertidal habitat of the order of several square metres can still be monitored photographically with stepladders, mechanised devices such as cherrypickers, hot-air balloons or even remote-controlled minature aircraft (G. Branch pers. comm.). A reasonable visual overview of larger areas can sometimes be obtained by viewing an area from a hill or clifftop, but this is usually of limited value for mapping intertidal areas because of the parallax errors that are introduced by not being able to look down absolutely vertically. Images obtained from directly above the area of interest are needed, and these can be obtained from sensors mounted in, or under, such devices as balloons, kites, aircraft or satellites (Fig. 7.3). Two basic sorts of information can be obtained from satellites or aircraft: digital data from spectroradiometers and photographic images. The characteristics of these two sources are described fully in texts such as Kelly 1980, Dalby & Wolff 1987, Budd 1991; see also 8.2.1.

Intertidal areas, when photographed at low tide, are particularly amenable to remote sensing because the images are not affected by the sea state, a major problem with sensing near-shore habitats. Intertidal habitats are more suitable for remote sensing than even many terrestrial areas, and satellite data have successfully been used to map and monitor saltmarsh vegetation (e.g. Budd 1991). In these estuarine situations, there is usually little topography to confound the images, and vegetation such as *Spartina* and other saltmarsh plants, mangroves, seagrasses and mudflat algae such as *Ulva* and *Enteromorpha* often occur as monospecific, homogeneous patches with distinctive reflectance signatures. Similarly, aerial photography using infra-red film can provide quantitiative estimates of the cover of seagrass (Kirkman 1990) and some algae on rocky intertidal shores (e.g. *Hormosira banksii*: Plate 44; *Porphyra columbina*, M. Bradstock pers. comm.) as long as appropriate ground truthing is undertaken. Where algal flora is diverse and intermingled (e.g. at the low-tide mark), this can prove difficult, although kelp beds have been successfully mapped this way (see Chapter 4).

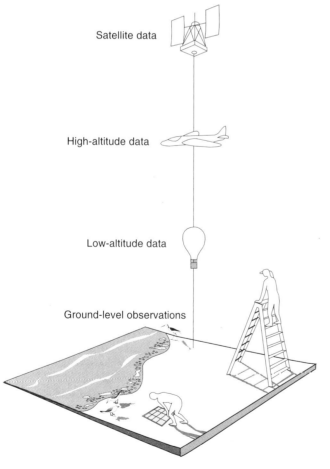

Satellite data

High-altitude data

Low-altitude data

Ground-level observations

Figure 7.3 Some of the techniques that can be used to obtain data from intertidal areas by remote sensing. These range from close-up photography on the shore itself to high-level images obtained from aircraft or satellites.

Aerial photography has two additional advantages. First, a record of past changes in intertidal habitats can be obtained if a time-series of images has been archived from the same area. Again, this is often the case with estuarine situations, and such historic records have been used to plot the progradation of mangrove forests in New Zealand (Young & Harvey 1996) and the demise and recovery of eelgrass in Chesapeake Bay, USA (Orth 1976). Film quality and photographic techniques are improving all the time, and it is now relatively easy and cheap to obtain high-resolution aerial photographs at scales of 1:4000. Many regional councils and other agencies now make regular aerial surveys of the coasts under their jurisdiction as part of their statutory obligations for maintaining coastal resource inventories. In New Zealand, photographs of this resolution are being used to map the positions and record the sizes (surface area covered) of individual mangrove trees in sparsely wooded mangrove forests, and monitor the spread of introduced *Spartina*. Second, the investigator has good control over the temporal interval of sampling, and data are less likely to be affected by cloud cover, a common problem with satellite images. This technique obviously has enormous potential for assessing changes in estuarine vegetation, and its use in other intertidal environments is bound to increase.

Vertical patterns
By definition, all intertidal habitats traverse the region of the shore between the low- and high-tide marks of extreme spring tides. This region therefore represents a gradient from almost fully marine to almost fully terrestrial. It has long been argued that along this gradient, physical factors such as temperature, desiccation, exposure to freshwater, etc., exert a gradation from mild at the seaward edge to extreme at the landward edge. Biological factors such as recruitment intensity, food availability, competition or predation may also vary along this gradient. According to numerous models developed over the past four decades, intertidal organisms on rocky shores, estuarine mudflats and sandy beaches are distributed across the shore in horizontal bands in response to these gradients of physical and biological pressures (Jones et al. 1990; Russell 1991; Dudgeon et al. 1995). This phenomenon, known as 'zonation', has become so ingrained in our concept of intertidal shores that it is often taken for granted and automatically incorporated into sampling protocols. This approach is not always helpful, and we would dissuade people from automatically searching for vertical pattern or uncritically using vertical zones for stratifying sampling effort. There are certainly situations where 'zonation' is visually apparent (e.g. vertical or steeply sloping rock faces; epifauna growing on the trunks of

mangrove trees), and in such cases it is obviously sensible to conduct sampling along downshore transects or stratify the sampling effort within the various vertical strata (depending on the question being addressed). In the absence of any prior information, running transects across a rocky shore as part of a pilot exercise is often desirable (see Creese & Ballantine 1986; also 7.4.1 and Black et al. 1979).

Universal schemes of 'zonation' have been proposed (e.g. Stephenson & Stephenson 1972, for rocky shores), and these are still accepted by some workers as a good basis for assessing intertidal coastal habitats (Russell 1991; McLachlan & Jamarillo 1995). There is no general agreement, however, on the recognised number of zones or where their boundaries occur for particular sorts of shores. The point we would like to emphasise is that the structure of intertidal assemblages is invariably patchy in space and time. As a result, it is probably only useful to recognise two zones on sandy beaches: an upper zone occupied by air-breathing animals, and a lower zone occupied by water-breathing animals (Brown 1983; Haynes & Quinn 1995). Even on rocky shores, where there has been such an emphasis on 'zonation' for so long, well-defined 'zones' containing predictable sets of species are rarely, if ever, found (Underwood 1994b). Certainly, variation across the shore may need to be taken into account as part of a sampling exercise, but this will depend on the questions being asked and the shore topography (e.g. low-relief rock platforms at Leigh, New Zealand, show little visual or quantitative evidence of non-random species distribution across the shore: see Creese 1988). Other sources of variation, such as along a shore or between micro-habitats, warrant at least as much attention, if not more.

7.2.4 Sampling rocky intertidal shores

Animals and plants on rocky shores generally live on the surface (epifauna or epiflora respectively) and are therefore visible, though often cryptic. These organisms can be sampled directly by placing sampling units on the substratum, and techniques for doing this are considered below and in 8.2.1. Organisms living within the substratum (infauna) can only be sampled by disturbing the substratum (invariably 'soft' sediments) by digging or coring; appropriate approaches and equipment for this are described in 8.2.1p. Although many methods are available, most published studies use only one or a small number of types of apparatus and applications based on the specific questions being addressed.

Transects and quadrats
Standard methods used on rocky intertidal shores include cross-shore profiles (using a line transect) for

mapping the widths and positions of habitats or species assemblages of interest; and quadrats, belt transects or a combination of these for quantifying abundance patterns. These are used in exactly the same way as on subtidal rock surfaces, with the same attendant advantages and disadvantages (see 4.3.1). The following discussion, therefore, relates to considerations that have particular relevance to intertidal shores.

Sampling units (quadrats) may be deployed randomly across the whole shore (simple random sampling) or allocated to subdivisions of the shore (strata) according to such categories as height above sea level, linear distance down a shore, or linear distance along a shore away from some feature (e.g. a potential source of disturbance such as a stormwater drain) or habitat (stratified random sampling). Alternatively, a belt transect can be run out over the entire vertical range of the shore, with quadats sampled along it either contiguously (e.g. Jones et al. 1980) or at intervals. As noted previously, the choice of method will vary according to the problem that is to be addressed: the profile of the reef (e.g. extended intertidal zone with a shallow versus steep gradient), the time available, amount of background information available, the ease of identifying different habitats, etc. Most often, however, some form of stratified random sampling provides the most effcctive strategy on intertidal shores. In addition, a combination of different techniques will often be needed. When monitoring an entire intertidal area, for example, larger, more scattered or more mobile species (e.g. starfish, abalone) may need to be counted in large belt transects rather than the small quadrats used to count barnacles, molluscan grazers or algal cover. Alternatively, a type of 'catch-per-unit-effort' measure could be employed here (i.e. the total number of individuals seen per person during the low-tide sampling period).

All these things need to be addressed during the 'reconnaissance' phase of the exercise. For example, if the extent of the reef was short, belt transects or sampling according to linear distance would be appropriate. If, however, habitats were clearly demarcated over a broad reef slope, stratified sampling according to habitat might be best. Moreover, in situations where the intertidal area has an extended platform of very similar habitat type, and a drop-off at the seaward edge, a combination of methods may be needed. In this case it would be most appropriate to devote a large proportion of the sampling effort to the drop-off and not distribute it cqually across the whole intertidal area. Regardless of the reef form, rock pools and boulder habitats should be sampled differently from open rock surfaces.

Unless the aim is specifically to describe a particular shore or area of beach (rarely the case), comparable sampling should be completed at a number of sites to reveal how consistent the distribution patterns of organisms are along a stretch of coastline (see section 3.2 for discussion of the allocation of sampling effort). An important component of all these sampling procedures is randomisation: sites should be randomly chosen if at all possible, replicate transects should be positioned randomly along the shore (randomly assigned distances can be read off from a base tape along the top of the shore), and quadrats should be positioned randomly within pre-selected strata such as habitat or height on the shore (e.g. by using random numbers to locate points on a grid laid over the stratum: see 3.3). Practical considerations will dictate the sampling devices to be used and how they are deployed. For instance, fibreglass survey tapes are a standard item for measuring distances and locating positions of sampling sites, but they are difficult to use in very windy situations (such as are often encountered on exposed coasts); simply pacing out distances can be just as effective, especially where the terrain is fairly flat and homogeneous and a person's pace can be readily 'standardised'. Lightweight quadrats made from PVC pipe are easy to use, but when working in an area washed by waves, a heavy metal quadrat may be needed to ensure that it stays in place.

The choice of a sample unit size should be made with a view to maximising precision, but again practical considerations may restrict what is finally used (see general considerations in 3.2.3). General surveys that have included quantitative assessments of organisms in the intertidal zone have often used 0.25 m^2 quadrats (e.g. Creese & Ballantine 1986; Schiel et al. 1986). Although an acceptable level of precision has been obtained for a wide selection of organisms in a variety of habitat types using quadrats of this size, optimal sample unit sizes have not often been assessed by formal procedures. When photoplots are to be used, rectangular quadrats (50 x 30 cm: Littler 1980; 50 x 75 cm: Ambrose et al. 1992) may more conveniently suit the shape of a camera image. In many situations, quadrats that are smaller than 0.25 m^2 may give more precise estimates of abundance. For example, Jernakoff (1983) studied temporal variability of algal abundance on an intertidal rocky shore: quadrats of three different sizes were compared (0.01, 0.02 and 0.04 m^2) with many 0.01 m^2 quadrats giving the most precise estimates of percentage coverage.

In most cases, as discussed above, random (or at least haphazard) positioning of quadrats is desirable, as this ensures an unbiased representation of the community under investigation (see Table 7.1). If sampling is carried out randomly on each of several occasions, the data could be analysed using ANOVA techniques with time as a random factor. Many monitoring studies, however, have used fixed quadrats that are repeatedly sampled through time (Ambrose et al. 1992; Lively et al. 1993). If the question being addressed concerns the fate of individual organisms, fixed plots are useful. Often, however, the main reason for using this approach is that natural intertidal communities are heterogeneous, requiring a large number of samples to be taken under a random-sampling regime to obtain adequate statistical power (ability to detect change in a population or assemblage; = effect size). Relatively small samples are needed for a fixed plot design, and these can be sampled more rapidly, with substantial savings in time and money (Table 7.1). Data collected from this sort of study, however, cannot be analysed using simple analyses of variance because the data from different times are not independent and there is no independent measure of residual variation to check the assumption of no interaction among times. These problems may be partially overcome by using multivariate or repeated measures analyses (e.g. Lively et al. 1993), but the validity of the interpretations made using these approaches is not universally accepted (see discussion in Underwood 1997). In some cases, a combination of fixed and random procedures may provide the best option. For example, Brown (1996) used fixed plots of 5 x 5 m within which five photoplots of 0.09 m^2 were randomly sampled, to document seasonal changes in intertidal algal abundances on shores in the Leigh marine reserve, New Zealand. No matter which procedure is selected, it is important to be aware of the limitations of that procedure when interpreting the results at the end of the study.

Measurements of abundance

The information recorded from each quadrat will vary according to the purpose of the investigation. Variables most often measured are density (number of individual organisms per unit area), sizes of non-clonal organisms, cover (proportion of area occupied by clonal or densely packed sessile organisms such as barnacles, mussels or algae), and frequency (percentage of quadrats in which a species occurs).

Direct counts and measures of intertidal organisms are relatively straightforward but can be time-consuming. We would recommend, however, that abundance scales (which have often been used for surveys of British shores: see Baker & Crothers 1987 for examples) only be used when a very broad description is required. In most cases, the extra effort needed to obtain an accurate count provides the researcher with much greater ability to determine differences between places or times.

Direct measures of cover can also be made in some situations. For example, where encrusting or turfing algae form discrete patches (such as in the territories of some large patellid limpets: see Fig. 7.4, p. 176), tracing the outline of the entire patch enables the exact area to be measured. The use of image-analysis software is invaluable in these situations. In most cases, however, patches of sessile organisms are too large or too indistinct to measure directly, and some form of estimation is necessary. Cover can be estimated directly in the field, but it is often convenient in intertidal studies to use photography to record the contents of a small area such as a quadrat. This minimises time spent in the field and also provides a permanent record of each and every quadrat sampled. A variety of estimation techniques can then be used either in the field or on captured images (Table 7.2, p. 177). The effectiveness of the final technique chosen will depend on the details (e.g. the number of points assessed per sample unit for the 'point' methods) and the particular intertidal situation being sampled. For instance, Dethier et al. (1993)

Table 7.1 Advantages and disadvantages of fixed versus random (including stratified) quadrats.

	Fixed plots	Random sampling
Advantages	✔ Small sample sizes adequate for monitoring dynamics. ✔ Logistically easier and more time-efficient. ✔ Changes within plots readily detectable. ✔ Can follow known individuals or colonies through time. ✔ Particular assemblages can be targeted at quadrat level.	✔ Gives unbiased representation of whole assemblage (or stratum). ✔ Comparisons among sites, in time, can be statistically examined. ✔ Particular assemblages can be targeted at stratum level in a stratified design.
Disadvantages	✘ Not representative of whole assemblage unless multiple plots originally chosen randomly. ✘ Comparisons among sites, in time, cannot be statistically examined. ✘ Uncertainty about interpretation of repeated-measures analyses.	✘ Large sample sizes necessary if heterogeneity large. ✘ Sampling may be complex and time-consuming. ✘ May be difficult to achieve sufficient power in statistical tests. ✘ Cannot track individuals through time.

compared point-contact estimates of cover with visual estimates obtained in the field for a wave-exposed shore in Washington, USA. They found that visual estimates (obtained by dividing their quadrats into 25 smaller sub-units, determining by eye the amount of space occupied by each species within each sub-unit, and then summing these counts to get a percentage cover) allowed better detection of rare species, took half as long to carry out and had fewer biases (e.g. between observers).

Another consideration is the fact that it may be more difficult to correctly identify species in photographic records versus direct field observations (e.g. Foster et al. 1991), especially if photographs are taken when it is raining or overcast. This problem is exacerbated if dealing with a multilayered assemblage such as foliose algae lying over encrusting species. This again highlights the need to do a thorough pilot sampling exercise: there is rarely any excuse for not doing this for easily accessible intertidal communities.

Figure 7.4 Many intertidal organisms are distributed on the shore in small patches. An extreme example of this occurs for some encrusting and turfing algae that may occur as discrete 'gardens', forming the feeding territories of some large patellid limpets. These algal gardens are important features of shores in Southern Africa (Branch & Branch 1981) and at the Kermadec Islands (New Zealand's northernmost territory), where *Patella kermadecensis* defends patches of turfing algae such as *Gelidium* sp. (shown below).

ROGER GRACE

Dealing with three-dimensional structure

The previous section dealt with relatively flat surfaces, but there is obviously some three-dimensional structure to most shores. In particular, two features that often warrant special consideration on rocky shores are pools and areas of cobbles or boulders. The presence and numbers of these on any particular shore will vary depending to the geology of the area. Regardless, the organisms that are found in these situations may differ considerably from those on adjacent rock surfaces, and this may pose a dilemma for any sampling of an entire shore. The alternatives are:

- To sample whatever falls within the randomly chosen quadrat, whether it is a pool, clump of cobbles or bedrock surface. If this is done, large variances in densities of organisms may occur (e.g. for anemones, which are most abundant in rock pools, or for barnacles, which usually occur on bedrock), and these are interpreted as spatial variation caused by the presence of the pool or cobbles.
- To arbitrarily move the randomly chosen quadrat to one side so as to avoid the pool or cobble patch (i.e. these features are ignored).
- To structure sampling so that pools and/or cobbles are treated separately from adjacent rock surfaces.

Unless these features are extremely uncommon on the shore of interest, the third option is usually recommended. By excluding these features from the standard

Table 7.2 Different techniques for assessing cover of sessile intertidal organisms. Some methods are dealt with in more detail in 4.2.4.

Method	Techniques	Advantages	Disadvantages
Visual estimation	Within single large plot.	✔ Very rapid; suitable for very simple situations.	✘ Inaccurate; cannot be used to assess change.
	Within small subdivisions of a quadrat, then summed.	✔ Rapid; picks up rare species.	✘ Estimates may be biased and vary among observers.
Linear point-intercept	Randomly chosen points along transect line.	✔ Rapid; suitable for single-species assemblages (e.g. seagrass).	✘ Inaccurate for multi-species situations, especially if few intersections are used.
	Point quadrat bar.	✔ Rapid; good for large foliose algae.	✘ As above.
Area point-contact	Random-point quadrat (e.g. random dots on perspex, randomly selected intersection points on grid).	✔ Can be used either in the field or on photographic images; unbiased.	✘ May be time-consuming.
	Fixed-point quadrat (e.g. regular arrays of dots or all intersection points on a grid).	✔ Can be used either in the field or on photographic images.	✘ May bias estimates if organisms are arranged in regular pattern.
Direct measures	Tracing outlines of patches, then digitising areas.	✔ Most accurate; most applicable to fixed plots.	✘ Most time-consuming.

sampling, more precise estimates can be obtained from the rock face; this is an example of the sort of habitat stratification discussed earlier. The abundances of organisms occurring in these special habitats need to be assessed carefully. In particular, further stratification often will be required. Because rock pools are three-dimensional, for example, estimates of abundances should consider the diameters and depths of pools and the possibly separate assemblages on the sides and bottoms of pools (e.g. Underwood & Skilleter 1996). Similarly, the upper and lower surfaces of cobbles or boulders should be treated separately, and other variables such as their sizes (and possibly also their shapes), what they are resting on, and their positions within the cobble field should also be taken into account (e.g. McGuinness & Underwood 1986; Sousa 1985; Swarbrick, Schroeter & Connell unpub. data).

At a smaller scale, a more detailed understanding of the distribution patterns of some rocky intertidal species may require sampling that takes account of such features as crevices (e.g. the effects of whelk predation on barnacle abundances; Fairweather 1988; Lively & Raimondi 1987) or small cracks and depressions in the rock surface (e.g. the effects of small-scale topographic complexity on dispersal patterns of some high-shore gastropods: Chapman & Underwood 1994).

Seasonal components

It is well known that some algae on rocky intertidal shores in Australasia exhibit consistent cycles of seasonal abundance, whereas others show no constant seasonal cycle (Jernakoff 1983). The recruitment patterns of sessile intertidal invertebrates (e.g. Caffey 1985; Gaines & Bertness 1993) and mobile grazers such as limpets (e.g. Creese 1981) also vary seasonally and among years. In New Zealand, the encrusting brown alga *Pseudolithoderma* sp. forms new colonies more frequently during late winter and spring, but existing patches may decrease in size or disappear at any time of year (Williamson & Creese 1996). Conversely, another brown alga, *Xiphophora gladiata*, has an autumn peak in density and biomass (Gillanders & Brown 1994). This is relevant to the sampling of rocky shores in a number of ways: the detailed patterns measured on any reef on one occasion are likely to change, often dramatically, over time; if seasonal changes cannot be monitored, a finished document should warn of probable changes to distribution or abundance patterns suggested by the existing literature; and if a shore is affected by a major disturbance (such as a large storm or an oil spill), the species that subsequently recolonise the substratum will, in part, be determined by the time of year that the incident occurred (Jernakoff 1983). Despite the fact that seasonal variations in abundances will often be found, large-scale differences among localities may still be detectable (e.g. on wave-exposed versus sheltered shores: Creese & Ballantine 1986; Vadas et al. 1990).

7.2.5 Sampling surface-dwelling organisms on intertidal sediment shores

Plants

Some intertidal soft-sediment habitats contain virtually no epiflora (e.g. wave-exposed sandy beaches); others contain relatively few species (e.g. sand flats in the lower reaches of harbours); while others may be dominated by them (e.g. mudflats in the upper reaches of estuaries). It is in the latter situation where some sampling variations may be needed because such places are often dominated by dense vegetation. This complex vegetative component and the more extensive areas often involved (e.g. where vertical and horizontal components are not obviously distinguished because of the meandering of tidal channels) usually mean that distribution patterns are much less obvious by eye at ground level than they are on rocky shores. *A priori* recognition of assemblages to be used in stratified random designs is therefore more difficult.

In temperate latitudes lower than about 37° S, mangroves are often a feature of upstream mudflats; above this latitude, saltmarsh plants such as *Salicornia*, *Spartina* and *Arthrocnemum* dominate. Seagrasses may be found on either tropical or temperate coasts, subtidally as well as intertidally, and on rocky shores as well as the more usual sediment shores (Larkum et al. 1989). All these plants are angiosperms and therefore possess many characteristics of terrestrial vegetation. Consequently, the techniques widely used to analyse terrestrial grassland communities (belt transects, quadrats, aerial mapping, photographic recording, etc) are applicable to studies of saltmarshes or seagrass beds (e.g. Dalby & Wolff 1987; Zedler & Nordby 1986; Kirkman 1990).

As with rocky shores, permanent transects or quadrats have been widely used to document the dynamics of saltmarsh plant assemblages (Polderman 1980) or seagrass meadows (Kirkman 1985), but these have been largely descriptive studies, and the attendant analytical problems with the non-independence of consecutive

temporal samples yet to be examined adequately.

The structure of mangrove forests can be assessed using similar tools to those used by terrestrial forest ecologists (diameter at breast height, basal area, mean stand diameter, tree height, etc: an excellent summary of these techniques is given by Cintron and Novelli 1984). These techniques rely on characteristics of individual trees rather than on assemblages of organisms. Densities of large trees can be assessed using large plots (of the order of 0.1 ha), or more commonly by plotless techniques such as point sampling or the point-centred quarter method (see Cintron & Novelli 1984). Densities of seedlings and pneumatophores (common in temperate mangrove species) are usually estimated by counting them in small quadrats (e.g. Clarke & Allaway 1993; Osunkoya & Creese 1997). Placing these quadrats in a truly random way is often difficult because of the dense nature of many mangrove forests, and it is often best to pace off random distances from randomly chosen points along a transect line (which will rarely be straight!). Photoplots are also difficult to use in this situation, because of the generally low light levels under the forest canopy and the 'upright' nature and often high densities of pneumatophores and seedlings. In addition to estimating abundances of these elements, it is often instructive to measure their heights and the epibiota living on them.

Both macro- and micro-algae occur in soft-sediment intertidal communities. They may be present among mangroves, seagrasses or saltmarsh, where they often attach to the stems, trunks, pneumatophores, prop roots, etc., of these maritime angiosperms. In tropical areas, these algae (together with cyanobacteria, other microbes and a diverse epifauna) form a complex and ecologically important component of mangrove communities. These have received detailed study (e.g. Farnsworth & Ellison 1996), but their distribution and abundance patterns in temperate mangrove forests have been rarely examined (Anderson & Underwood 1994). In southeastern Australia and New Zealand, an estuarine form of the fucoid alga *Hormosira banksii* is an occasional feature of the floor of mangrove forests and may influence the recruitment of young plants (Clarke & Myerscough 1993). An assessment of its abundance is best obtained using random quadrats as outlined above.

On open intertidal sand- and mudflats, microalgae are ubiquitous, abundant and often highly productive, especially at low tide on warm days. In most studies, however, it is unlikely that they will be comprehensively sampled (there are numerous unresolved taxonomic problems, quite apart from time limitations), but even if major groups are recognised (usually by subsampling within standard quadrats, followed by laboratory analysis), useful ecological interpretations may be possible.

For example, cyanobacteria are generally rare on many mudflats but may be abundant in polluted areas (Wolff 1987). Generally, estimates of microalgae are made by measuring concentrations of chlorophyll *a* (see example in Page et al. 1992). Further details of methods for the study of soft-sediment microalgae and microbes are given by Round and Hickman (1984) and Dye (1983) respectively.

Macroalgae also occur in sediment-flat habitats, but their distribution is usually patchy in time and space, with fast-growing green algae such as *Ulva* and *Enteromorpha* particularly common. The proliferation of these is often linked to organic enrichment and may therefore provide evidence of organic pollution (e.g. Fairweather 1990b) Their ephemeral nature and opportunistic life-history characteristics, however, make detailed sampling difficult. Snow (1995), for example, took photographs of *Ulva* in both random and fixed quadrats within fixed, gridded plots at daily, weekly, fortnightly and monthly intervals. There was an obvious seasonal pattern in abundance, but differences between sites were impossible to determine because of the large variability at all spatial and temporal levels, leading to significant interaction terms in the analyses. In situations like this, the best approach may be sampling by remote sensing and measuring cover at a scale of hectares rather than abundance at a scale of square metres.

Epifauna

The epifauna of intertidal sediment flats is dominated by echinoderms (which are usually subtidal but may move into intertidal areas at high tide, where they may sometimes be stranded on the outgoing tide; Fig. 7.5), predatory or deposit-feeding gastropods (which are found on the surface at both high and low tides), and crabs (which live in burrows but feed on the surface at either low or high tide). Gastropods and echinoderms are readily sampled using quadrats or transects. Kaly (1988) used five randomly placed quadrats of 0.1 m^2 to sample gastropods among mangrove forests in New South Wales. Because many of these species are small, very clumped and difficult to distinguish, it was necessary to carefully pick all individuals from the sediment surface using forceps and place them into a tray to enable accurate counts and identification. Ten species were encountered, which fell into two loosely defined groups: those that tended to be found low on the shore, and those that tended to be found high on the shore. Apart from this, there were no clear distributional patterns that were consistent among different shores or among sites at the same shore.

Techniques for measuring densities of intertidal crabs are not as straightforward, and seagrass meadows, salt

Figure 7.5 Echinoderms are often found buried at the lowest levels of sandy shores and extending into the shallow subtidal. Species encountered in New Zealand may include comb starfish (*Astropecten polyacanthus*), heart urchins (*Echinocardium cordatum*) and sand dollars (*Fellaster zelandiae*).

R. CREESE

marshes and mangroves are notoriously difficult habitats in which to determine absolute abundances of burrowing species. Many workers have relied on estimating crab numbers by simply counting numbers of crab holes within quadrats, but the accuracy of such estimates has rarely been checked using independent techniques. Warren (1990), working in a temperate mangrove forest in New South Wales, counted holes and numbers of crabs (*Heloecius cordiformis*) emerging from burrows during a 20-minute period. In this case, there was a good correlation between burrow density and crab numbers. McKillup and Butler (1979), however, found that density of burrows of the crab *Helograpsus haswellianus* was a poor indicator of crab density – there were invariably more holes than crabs, because individual crabs had multiple entrances to their burrows. Before using densities of holes as an estimator of crab abundances, therefore, it is important to identify the crabs likely to be encountered, and know something about their burrowing behaviour and, if possible, the relationship between absolute and apparent abundances. Even when this is done, uncertainty will usually remain: visual counts of crabs assess only those actively emerging from their burrows, and digging crabs out of a predetermined area is also difficult and unreliable. An alternative is to use baited traps, but this will only give relative measures of abundance because the area from which the trapped crabs came will not be known.

Many invertebrates inhabit seagrass beds, and it is often difficult to make a clear distinction between epifauna and infauna because many species move between the substratum and the seagrass blades, depending on state of the tide, time of day or stage in the invertebrates' life cycle. Itinerant visitors including fish and crustacea are also important in these habitats, especially where seagrass beds extend subtidally (Bell et al. 1995b; Ferrell & Bell 1991; McNeill & Bell 1992; Worthington et al. 1992a, b; Sanchez-Jerez & Ramos Espla 1996). The methods used to investigate this fauna include push nets, trawls, seines, drop nets and traps, and are described in detail by Heck and Wilson (1990). In intertidal areas, core sampling (as described below) is the most widely used technique for collecting samples from the underground component of seagrasses (Lewis & Stoner 1981). Quadrat sampling, *in situ* photography, or the cutting and removal of blades can be used to document organisms encrusting on or living within the above-ground component. Consideration may need to be given to differences in the morphology of seagrasses with depth, latitude or other variables (e.g. West 1990). An alternative, or augmentative, approach to examining the ways in which invertebrates and fish use seagrasses is to mimic the blades using artificial surfaces such as polythene tubing or plastic strips (see Bell et al. 1995b).

In many temperate mangrove forests, marine bivalves, barnacles and gastropods are commonly found living on the trunks and pneumatophores of mangrove trees. At the seaward edge of mangrove forests, encrusting and mobile epifauna may occur more than 3 metres above the surface of the mud. These assemblages are not as rich as those on tropical mangroves (Sasekumar 1984; Farnsworth & Ellison 1996), but they should be taken into consideration when assessing the fauna of these environments. Because they occur in very sheltered habitats and on surfaces that are vertical or nearly so, clear vertical distribution patterns can often be expected (Sasekumar 1984). The nature of tree trunks lends itself to the use of narrow, vertical belt transects divided into contiguous quadrats. In an eastern Australian mangrove forest, for example, Coates and McKillup (1995) used contiguous 3 x 3 cm quadrats to document the distributional limits of two co-occurring barnacle species. They also used a novel technique for relating the levels on tree trunks to tidal heights. A vertical, closely spaced series of empty capped vials, each with a single hole drilled in its side, was attached to each of three tree trunks. The vials that filled with water and those that remained empty over two complete tidal cycles were recorded and the data were then compared with the published tidal predictions for the area. This would be a useful technique for determining both relative and absolute tidal heights when comparing distribution or abundance patterns of epifauna on trees situated in different parts of a mangrove forest.

7.2.6 Sampling infauna on intertidal sediment shores

Sampling apparatus

Sedimentary shores range from fine muds in sheltered areas in the upper reaches of estuaries, to coarse sands on exposed surf beaches. Many other physical, chemical and biological variables often vary along this estuarine gradient (see Table 7.3). The assemblages of infaunal organisms that live within these sediments are determined, at least in part, by the grain-size distribution (e.g. Rhoads & Young 1970). Methods for determining grain-size composition, other characteristics of sediments themselves, and several methods for sampling benthic infauna are described in Chapter 8.

Two methods (really variants of the same technique) that have more relevance to intertidal areas are considered here. They involve inserting a device into the sediment and taking what is effectively a three-dimensional quadrat. The first employs a box quadrat, a square metal frame (aluminium or stainless steel are preferred) with

Table 7.3 Gradients found in estuaries. Salinity is given as parts per thousand and may vary widely in some parts of an estuary because of vertical stratification and/or rainfall.

Position	Sediment type	Sediment sorting	Currents	Salinity	Dominant organisms
River	Gravels	Well sorted	Strong; riverine	Nil	Freshwater species
Upper reaches	Fine muds	Poorly sorted	Negligible	< 18; variable with depth	Plants, deposit feeders
Mid reaches	Sandy muds	Poorly sorted	Moderate; tidal	18–35	Deposit and filter feeders
Harbour mouth	Coarse sands	Well sorted	Strong; tidal	18–35	Filter feeders

a given height (10, 20 and 30 cm are commonly used) pushed or driven into the sediment until the top is flush with the sediment surface. The contents can then be dug out with fingers, spade or trowel (Wolff 1987). This technique is well suited to large, sturdy organisms such as bivalves, and Hooker (1995) used it effectively to sample the infaunal bivalve *Paphies australis*, which often occurs in dense beds in both intertidal and subtidal habitats in New Zealand (Fig. 7.6). Care needs to be taken when soft-bodied organisms are involved, as the spade may damage a large proportion of them.

Figure 7.6 In sheltered bays and harbours, intertidal sand flats are often dominated by dense beds of filter-feeding bivalves. Shown here is a layer of dead pipi shells (*Paphies australis*) overlying a bed of living animals in Whangateau Harbour, New Zealand.

R. CREESE

The second method uses cylindrical corers, which are commonly made of PVC or stainless steel with the bottom end open, the lower edge sharpened to achieve better penetration (especially necessary where the roots of seagrasses or mangroves are dense) and a rod through the top for use as a handle. These can be constructed cheaply and easily, but they are also sold commercially in some parts of the world as 'clam guns'. In small corers or in situations where the sediment is moist, fine and cohesive, the top end may also be open. In larger corers, coarser sediments or very wet sediments, the top end is closed either permanently or temporarily with a rubber bung. It is essential to have a small hole near the top to allow air to escape during the sampling process. For ease of handling, corers are usually 25–80 cm long, and it is useful to have a scale engraved on the outside so that depth of penetration can be accurately determined. These can be used subtidally by divers

(see 8.2.1), but find their greatest application in intertidal situations because they are rapid, efficient and do not have the sharp corners of a box quadrat, which are difficult to excavate. Both core diameter and depth to which cores are taken will depend on the nature of the survey and sediments to be sampled, and should be formally investigated during the pilot study (an example of this for a seagrass study is given in Lewis & Stoner 1981).

Corers should be at least several times the maximum size of the organisms sampled and are commonly 0.01–0.03 m^2 in cross-sectional area (Wolff 1987), although larger ones have been used in some studies (e.g. Grange & Anderson (1976) designed one of 0.1 m^2 for sampling in the Manukau Harbour, New Zealand). They should penetrate deep enough to catch all the organisms of interest, but are most commonly used to depths of 200–250 mm (Wolff 1987). Dexter (1984) sampled to 100 mm in her study of sandy beach fauna in New South Wales, having previously determined that there were no differences in species composition or abundances between cores taken at 100 and 200 mm on a variety of beach types. On estuarine mudflats, however,

bivalves and polychaetes may occur down to 50 cm and a deeper core is necessary. For example, Page et al. (1992) used 10 cm diameter cores to a depth of 55 cm to collect three bivalves from a southern California salt marsh.

The contents are usually placed in a sieve to separate the organisms from the sediments. The mesh size of sieves will depend on the questions being asked (James et al. 1995; see also Chapter 8 for a discussion of this issue), but sieves with meshes of 0.5, 1.0 or 1.5 mm are commonly used in intertidal surveys. The latter size, or even larger, is often used for coarse beach sediments where the median grain size is greater than 1 mm (McLachlan 1983). In general, however, if the whole infaunal community is to be sampled, the smaller mesh size is needed to ensure that all the small polychaetes and crustaceans are adequately sampled. In many saltmarsh or mangrove habitats, sieving is impractical because samples are often collected a long way from the nearest source of seawater, the sticky mud does not pass easily through a sieve, and a sieve rapidly clogs with fine mangrove roots and other organic matter. In these situations, the sample may be spread out on a polythene sheet or sorting tray and hand sorted in the field (e.g. Sasekumar 1984) or back at the laboratory. This procedure will obviously miss the smaller organisms, but if these need to be counted, smaller subsamples could be taken.

On open beaches or when large samples are being collected, sieving material with a standard sieve (i.e. a

Figure 7.7 Sandy beaches also support dense beds of bivalves from time to time. Shown here is an excavated area of 0.25 m^2 from Ocean Beach in northeastern New Zealand, which witnessed a large influx of tuatua (*Paphies subtriangulata*) in the winter of 1992.

TARA ANDERSON

square metal or wooden frame with the mesh attached to it) is also impractical because material may be lost over the edges of the frame while sieving in the surf. Placing the entire sample in a mesh bag that can be tied off at the top, then rinsing it in the water, is a rapid and efficient alternative in these cases (e.g. Dugan & Hubbard 1997). The material retained in the mesh bag is then manually sorted to separate the organisms from the remaining sediment and organic matter. Alternatively, if only large, sturdy bivalves are to be sampled (e.g. the surf clam *Paphies subtriangulata* on New Zealand beaches; see Fig. 7.7), 'finger dredging' within a quadrat may be a viable alternative to sieving (James & Fairweather 1995; see also 7.4.6).

Sampling procedures and designs

The same basic considerations as discussed for rocky shores (see 7.2.4) apply to sampling sedimentary infauna, but two features of the substratum lead to special difficulties. First, sediment flats are often much more extensive than intertidal rocky shores, and it is therefore difficult to achieve adequate dispersal of sample units throughout the whole area. Unless very large numbers of replicates are taken (which is usually impractical), sampling using randomly placed quadrats will only provide an accurate description if there is very little spatial heterogeneity in the abundances of the organisms being sampled. On rocky shores this problem can be overcome by stratifying the area into subunits using visual criteria and then randomly sampling within these sub-units. Many soft-sediment areas appear homogeneous on the surface, however, which makes it impossible to stratify the area to be sampled into meaningful habitats except on a very coarse scale (e.g. adjacent to a channel versus far away from a channel). Second, the three-dimensional substratum provides a further opportunity for spatial variability. Abundances are usually expressed per unit area, not volume, so any variability in depth distributions is ignored. In effect, however, the sample is integrated across a predetermined depth into the sediment. The assumption that organisms are randomly distributed within the sampled volume is seldom true, but this is rarely taken into account unless a specific question about the vertical distribution of organisms in the sediment is being addressed.

These considerations should be taken into account wherever possible, because large variations in numbers of organisms and other variables on small spatial scales are common on sediment shores (e.g. Thrush et al. 1989; Morrisey et al. 1992b), even in areas that outwardly appear physically homogeneous. Such techniques as grid sampling and spatial autocorrelation analysis are often used in these situations (see example in Fig. 7.8),

but there may be interpretational problems (e.g. based on the scale of the grid) associated with these (see detailed discussion in 8.2.3). In general, we agree with Morrisey et al. (this volume, Chapter 8) that the use of nested or hierarchical designs with sampling units arranged at a variety of spatial scales is the best way to proceed when investigating the infauna of soft sediments, particularly when dealing with large expanses of relatively uniform substratum such as an estuarine mudbank or sandflat.

One situation where an alternative approach has been successfully employed is on sandy beaches. Here, there is an obvious gradient in tidal height running across what is usually a relatively uniform, sloping surface, and some form of transect sampling can be effectively employed. It is usually instructive to classify the beaches under investigation according to their morphodynamic type, using a measure such as Dean's parameter (Brown & McLachlan 1990). This is a dimensionless index that incorporates sediment type (expressed as sediment fall velocity) and wave energy (expressed as wave height and period: Short & Wright 1983). Reflective beaches

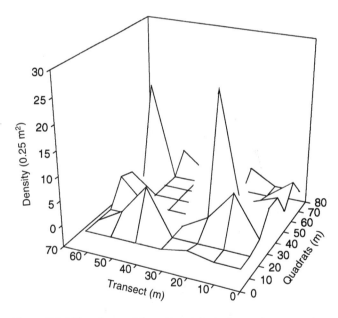

Figure 7.8 Although some infaunal invertebrates from soft-sediment shores occur in extensive beds (e.g. Figs 7.6, 7.7), many others aggregate in small clumps. One such species is the carnivorous gastropod *Cominella adspersa*, which is clumped at a scale of tens of meters in low intertidal and shallow subtidal habitats. This three-dimensional mesh plot shows a typically patchy distribution, and was generated by taking single samples (0.25 m²) at all intersection points of a 10 x 10 m grid extending 80 m along a shore and 90 m down a shore. This technique has also been used to good effect in studies of the invertebrate infauna of the Manukau Harbour (see Thrush et al. 1996 and earlier publications).

(Dean's parameter < 1) have small waves, narrow surf zones, steep slopes, coarse sediments (gravel or cobbles) and generally few macrofauna, while dissipative beaches (Dean's parameter > 5) have large waves, wide surf zones, gentle slopes, fine sandy sediments and generally abundant macrofauna. There are several intermediate varieties with characteristics between these extremes.

On all beach types, the sediment is very dynamic and moves around considerably in response to waves and tides. In some situations, this may result in a 'spur and groove' arrangement that might profitably be sampled using a stratified design. The infauna of sandy beaches (especially the crustaceans) is also dynamic, and distributional patterns can change from hour to hour or day to day (Brown & McLachlan 1990). Sampling therefore needs to be done as quickly as possible, especially if comparisons among several beaches are to be made. Vertical distributions of organisms can be enumerated by taking replicate cores at predetermined distance intervals down a cross-shore transect (e.g. Dexter 1984; Haynes & Quinn 1995), but this often gives imprecise estimates of abundance because many of the infauna are aggregated into very narrow vertical bands. To overcome this, a modified vertical transect technique was developed for temperate beaches in southern California by Dugan and Hubbard (1997), in which cores at closely-spaced intervals (0.25–1 m, depending on the intertidal width) are rapidly taken down a transect line and combined into a single mesh bag for sieving. This effectively integrates the abundances across the beach, and numbers are expressed per vertical metre of beach to allow comparisons between beaches of different widths. Depending on the nature of the survey, this procedure can be modified by subdividing the transect into smaller intervals. Replication is achieved by doing several transects at randomly chosen distances along the beach, but care should be taken not to sample too close to a previous transect, because the digging and trampling associated with sampling can cause organisms such as mole crabs to move out of the immediate area (Dugan & Hubbard 1997). These authors recommend that a buffer zone of 2–5 m be left between transects. When trying to provide a quantitative description of the whole community on a beach, it is further recommended that the total area sampled should not be less than 4 m^2, especially on dissipative beaches. Sampling less than this may seriously underestimate species richness (Jaramillo et al. 1995).

7.3 Experimental methods

Rocky intertidal shores have been widely used to investigate ecological processes in the marine environment. It is beyond the scope of this chapter to fully review the range of experimental techniques used and the problems associated with their analysis and interpretation; readers are referred to the excellent coverage given by Underwood (1985, 1986, 1994c). Rather, we wish to emphasise three main points here. First, realistic interpretations of both patterns and the processes that determine them are often best obtained by generating testable hypotheses and then using manipulative field experiments to directly test these hypotheses. Second, doing such manipulations is relatively easy on rocky shores, given a bit of imagination and innovation. Replication can usually be adequately achieved, and pseudo-replication can be avoided with careful planning (see Underwood 1986; also Chapters 2 and 3). Third, artefacts generated as a result of these experimental procedures are always likely to be present; these need to be adequately controlled for wherever possible, and certainly taken into account when interpreting the results in a wider context. The latter two considerations are particularly problematical on soft-sediment shores where the substratum is more mobile and the effects of altered water currents may have severe consequences. In these environments, procedures such as inclusion of predators using cages are not usually recommended (see Hall et al. 1990; Wilson 1991). Predator exclosure experiments are also difficult, especially in situations where there is even a moderate current flow (Hooker 1995), but they can be done in areas of low current such as tidal lagoons (e.g. Peterson & Black 1993; see also section 8.3).

There are many different types of manipulative experiments that can be done on rocky intertidal shores, and researchers have used an enormous number of experimental devices to test various hypotheses (pot scourers, copper strips, plastic mesh, PVC pipe and fibreglass moulds of organisms, to name but a few). Here, we note some of the more commonly used types of experiment (summarised in Table 7.4), and briefly mention a few examples of how they have been used. Further details of experimental procedures are given for one or two of the case studies described in section 7.4. Analyses of variance techniques are particularly appropriate for analysing manipulative experiments, and these are discussed in Chapter 3.

Planktonic dispersal, larval settlement and recruitment of intertidal organisms and the effects of these on population dynamics and the organisation of marine assemblages (commonly known as 'supply-side' ecology) have received considerable attention in the past

Table 7.4 Examples of experimental manipulations used on intertidal shores. Note that many of the procedures listed can be used in a variety of contexts. The most profitable experiments may be multifactorial and incorporate several techniques in the one design. Single-factor models rarely explain complete biological systems. All of these procedures require comparisons with controls.

Ecological process	Procedures used
Settlement/recruitment	Removal of organisms to provide bare space; provision of artificial settling surfaces.
Assemblage development	Removal of organisms; provision of artificial surfaces; periodic simulated disturbances.
Competition	Removal of putative competitors; barriers to keep organisms in; augmentation of densities by transplantation but without artificial barriers.
Predation/grazing	Removal of predators or grazers; barriers to keep predators or grazers in or out; transplantation of predators but without artificial barriers.
Limits to vertical distribution	Amelioration of physical conditions.
Movement	Tagging of organisms; tethering of organisms; translocation of organisms.
Environmental impact	Mimicking the likely impact at a small scale.

15 years (see Underwood & Fairweather 1989; Gaines & Bertness 1993). While it is sometimes possible to monitor the settlement of intertidal species (e.g. by counting barnacle cyprids on the shore on a daily basis: Wethey 1984), most have very small or cryptic juveniles that are only apparent to the human eye some time after settlement. This 'recruitment' can be experimentally investigated in a number of ways. The intensity of recruitment to the shore can be done by simply clearing patches of bare rock (e.g. with a scraper or propane torch) to provide clean settlement surfaces (Lively et al. 1993). The role of adult conspecifics (or any other species for that matter) as either attractors or deterrents to recruitment can be studied in the same way, by removing them from patches on the shore and monitoring recruitment to plots with (controls) and without the adults. Another common technique for monitoring levels of recruitment (e.g. at different levels on the shore or in different microhabitats) is to attach artificial surfaces made from a variety of substances. These have the advantage that they can be removed and taken to the laboratory for careful enumeration under a microscope. They can then be cleared to provide a clean surface again, returned to the same place on the shore to monitor subsequent recruitment or community development (e.g. species succession), or moved to a new location to test different hypotheses (e.g. about survival of juveniles under different conditions). Because commonly used materials like fibrolite and perspex are artificial, levels of recruitment to them may not be a realistic reflection of what occurs naturally on a shore. It is therefore best to use natural rock and to set inserts flush with the rock surface wherever possible. For example, Caffey (1985) examined barnacle recruitment in New South Wales by using rock inserts obtained from

a variety of different shores. Where artificial surfaces are used, appropiate checks should be made to determine how well they mimic natural substrata. Williamson and Creese (1996) used fibrolite plates to monitor recruitment (new plates positioned on the shore at regular intervals) and growth (old plates returned to the shore after data collection) of an encrusting alga to shores in northeastern New Zealand (Fig. 7.9, p. 186). Recruitment and growth of the alga on natural limpet shells were used to check for possible artefacts associated with the use of artificial plates. The temporal development of an assemblage can be further investigated experimentally by carrying out periodic disturbances (such as turning over boulders: McGuinness & Underwood 1986) and monitoring the response of the target organisms.

To monitor the performance (usually measured in terms of growth, survival or reproductive output) of a species in the absence of a competitor or predator, it is usually necessary to reduce the numbers of these species or eliminate them altogether. The easiest way to do this is to manually remove them from large areas. For example, Fairweather (1988) removed predators (whelks) from within pools and crevices, and was able to demonstrate that barnacles only survived in the absence of these. In the presence of whelks, densities of barnacles decreased with time over increasingly greater distances from these features, creating the 'halo' effects often observed around pools and crevices. Molluscan grazers that shelter in crevices can also produce this effect on algal distributions (see Fig. 7.9). Grazers can also be removed from large areas such as isolated boulders to investigate their effects on algal abundance (e.g. Jernakoff 1983). At a smaller scale, physical barriers can be used to either include or exclude target organisms, and Jernakoff (1983, 1985) used artificial turf

185

ABOVE: **Figure 7.9** The encrusting alga *Pseudolithoderma* sp. occurs in large, rapidly growing patches on shores in north-eastern New Zealand. Its spores only travel short distances, so that new patches tend to form close to existing patches. This was shown experimentally by attaching fibrolite plates to the rock surface at various distances from the edges of patches. New growth (the dark colour) only occurred close to existing patches, while plates further away remained uncolonised. Note also the cleared strip in the *Pseudolithoderma* patch adjacent to a crevice. This is caused by the grazing activities of chitons (*Chiton pelliserpentis*), which shelter in the crevices during the day and emerge at night to forage short distances away. A similar 'halo effect' can occur when carnivorous whelks sheltering in crevices kill the nearby barnacles (e.g. Fairweather 1988). R. CREESE

BELOW: **Figure 7.10** Plastic mesh cages nailed to the soft mudstone shore of a reef at the Leigh marine reserve, northeastern New Zealand, were used to include combinations of grazing gastropods to assess inter- and intra-specific competitive interactions among them. R. CREESE

nailed to the rock to exclude limpets and littorines from some of his experimental plots.

Competitive interactions among grazing limpets have often been demonstrated by using fences or cages constructed of plastic or stainless-steel mesh (Fig. 7.10). These enable specific numbers and combinations of species to be maintained within a fixed area to investigate both interspecific and intraspecific interactions (Creese & Underwood 1982; Fletcher & Creese 1985; Marshall & Keough 1994). Such structures may be time-consuming to construct and affix to the shore, but are usually very effective at keeping organisms both in and out of the experimental plots. Another commonly used technique involves painting antifouling paint onto the rock (Dethier 1994; Schiel unpub.), but this is effective only for keeping grazers out rather than in. Copper strips nailed to the rock may be even more effective (Johnson 1992), but are obviously more difficult to install. As in all these types of experiments, it may be necessary to fully investigate potential artifacts arising from the use of a particular technique. Johnson (1992), for example, warns that partial copper barriers (used as a control for exclusion artifacts) may not operate in the expected manner.

As well as the above-mentioned biological factors, physical factors that may influence the distributional characteristics of intertidal species can be investigated experimentally. For example, various devices can be used to ameliorate conditions in areas normally outside a species' distribution. It was noted by Worthington and Fairweather (1989) that large individuals of a grazing turbinid snail were most abundant around the edges of coralline turf. To test whether this was providing cover for the snails, they nailed plastic pot scourers (to mimic turf) within patches cleared of the turf, and demonstrated

that large snails remained around these artificial shelters as predicted. In another experiment in New South Wales, Underwood (1980) provided shade (in the form of mesh roofs and cages) above a low-shore algal band to demonstrate that algae were limited in their vertical distribution not by desiccation effects but by grazers.

Another method of experimentally investigating limits to distribution (i.e. ability to survive and grow in places other than where they are normally found) involves transplanting organisms. This can be done with anything from mangrove seedlings (Osunkoya & Creese 1997) to barnacles (Coates & McKillup 1995). This technique can also be successfully used to investigate such things as dispersal and movements of intertidal snails (Chapman & Underwood 1994). Plants or animals are usually tagged as part of these experiments so that they can be subsequently identified (see Fig. 7.11). All these procedures involve disturbance of the organisms and, as with all experimental procedures where the manipulated organisms are the subject of subsequent tests, it is important to establish that the disturbance has had no measurable effect on the organism's response. In addressing this issue for a high-shore littorine in New South Wales, Chapman (1986) advocated a wide range of control procedures to check the effects of tagging, physical handling, translocation, the densities of conspecifics in the area to which snails are moved, and other possible confounding effects. These are often time-consuming and may be considered tedious but they are nevertheless a vital pilot component of such experiments. As a result of such a pilot experiment, it may be unnecessary to include all possible controls in subsequent experiments.

The above examples all come from hard substrata where it is relatively easy to attach structures or to relocate tagged and transplanted organisms. Such procedures are more difficult on soft substrata, especially as virtually all experimental manipulations will disturb the sediment in ways that are nearly impossible to control for. Examples of successful experimental manipulations in this environment include infaunal invertebrates

Figure 7.11 Affixing plastic tags (numbered or colour-coded) to the shells of gastropods (in this case the New Zealand abalone, *Haliotis iris*) is one way of uniquely marking individuals for studies of growth, survivorship or movement.

R. CREESE

placed in tubs of sediment or in mesh fences set into the sediment, to investigate the effects of such things as deposit and filter-feeding bivalves on the standing crop of epibenthic microalgae in estuaries (Page et al. 1992), and interspecific competition among bivalves (Peterson & Andre 1980; Peterson 1982). Further details of these sorts of experiments are given in section 8.3.

A final example concerns using experiments, which may need to be on a relatively large scale to be realistic, to assess the likely impact of an anthropogenic disturbance on infaunal biota. Regulatory authorities often are unwilling to permit this, but field experiments are often the most definitive tool available for this sort of assessment because they most closely mimic what would occur in a natural situation. Their use should be encouraged wherever possible. For example, Pridmore et al. (1991) introduced a chemical (technical chlordane) into a plot on sandbanks in the Manukau Harbour, northern New Zealand, and by comparison with a control plot, showed that juvenile bivalves responded by moving out of the contaminated areas. Similar *in situ* procedures can be used to investigate the possible effects of oil spills and use of dispersants on both soft sediments (McGuinness 1990) and rocky shores (Battershill & Bergquist 1982).

7.4 Examples of descriptive and manipulative sampling

The following are examples of particular studies that have used the techniques discussed in the previous sections. The first five examples are from rocky shores and are arranged in complexity from simple descriptive studies of poorly known areas, to sophisticated experimental investigations of intertidal patterns. These are followed by examples from three different sorts of soft-sediment shore.

7.4.1 Survey of the rocky reefs on an exposed offshore island

Example: Survey of the Poor Knights Islands, New Zealand (Creese & Ballantine 1986; Creese in prep.).

Purpose: To obtain baseline information on the structure of the biotic assemblages in the rocky intertidal zone, in terms of the benthic organisms present.

Situation: The waters around these islands were declared a marine reserve in 1980, but there had been no quantitative descriptions of rocky-shore habitats. The only background information came from a single descrip-

tive account obtained several decades previously. The islands were visited on a number of occasions from a launch, but the rugged, steep nature of the terrain and the unpredictable weather meant that the number of shores that could be visited and the amount of time that could be spent on any one shore were limited.

Approach: A two-phase approach was used. Initially, effort was concentrated in one bay. The distribution patterns of sessile organisms and grazers were described down a reef profile. Two transects were examined and, because the shore was relatively short, sampling was structured according to linear distance. Densities and percentage covers (estimated by eye from 25 grid squares within the quadrat) were measured in two replicate 0.25 m² quadrats placed at 1 m intervals on either side of a transect line down the shore. This provided a description of the cross-shore sequence of the major species assemblages, identified a new species of chiton that was unique to the islands (Creese & O'Neill 1987), and found other species that were rarely encountered at other localities in northeastern New Zealand. Results from this pilot survey were then used for the second phase, which entailed a broader investigation of the densities of molluscan grazers at seven localities around the islands. These seven shores were sampled at three levels: approximately upper, mid, and lower, corresponding to biological habitats determined in the first phase. Each level was sampled at two sites for each locality to provide a hierarchical design (closely approximating the example given in 3.5.6, which provides the analytical rationale for this design; see also Chapter 3). The distribution patterns of molluscan grazers down the shore varied among localities, emphasising the importance of not assuming that results from one sampling site are representative of a broad stretch of coastline. The information gained during these studies provided a further rationale for protecting the marine habitats at the Poor Knights. Similar sorts of descriptive surveys that can be done quickly on remote offshore islands that are difficult to get to have also been used at the Kermadec Islands, another New Zealand marine reserve (Schiel et al. 1986).

7.4.2 Temporal and spatial patterns of algal abundance on a rocky reef

Example: Variations in algal diversity and abundance on an intertidal rocky shore in New South Wales (Jernakoff 1983, 1985).

Purpose: To describe the seasonal abundances of algae commonly found at mid-shore levels of an intertidal barnacle zone.

Situation: The study was restricted to the barnacle zone because previous work had demonstrated that some species of algae occurred only in that area. The shore was accessible by road, so the cost of sampling was minimal for the long-term investigation. Many previous studies had been done in the area, and these were used to choose the size and number of sampling units to be used.

Approach: Algal cover was estimated at two sites (150 m apart) of similar tidal height, wave exposure and aspect. The barnacle zone was 6 m in vertical extent, and this was divided into three equal segments. Percentage cover was estimated by randomly placed quadrats (10 x 10 cm) within a horizontal distance of 5 m at each of the three levels. Cover was estimated at monthly intervals for 33 months to determine patterns of algal seasonality. Jernakoff found that temporal increases in diversity were caused by increased abundance of some species and the appearance of annual algae. In addition, clearances were made at different times of the year, which demonstrated that different species settled depending on when the clearances were done, and this subsequently altered any successional algal sequences. Experiments were also set up on short- and long-term bases to investigate the effects of grazers (excluded by artificial turf) and the crevices between barnacles (by filling in some areas with fibreglass resin) on these patterns. Only the former had a significant effect. This example illustrates the usefulness of simultaneously combining several different experimental manipulations with the monitoring of natural populations.

7.4.3 Long-term monitoring of a large section of coastline

Example: Establishing an inventory of the shoreline assemblages in Santa Barbara County, California, USA (Ambrose et al. 1992).

Purpose: To provide information about the nature of the rocky intertidal community, to assess the temporal dynamics of key species, and to provide information that could help assess damage caused by any future oil spill.

Situation: Considerable background information existed on the shores in the area (allowing an appropriate choice of sample sizes and numbers) and the species on them (e.g. Littler 1980; internal reports for Channel Islands National Park). The area is adjacent to extensive offshore oil-drilling activities, and oil spills had occurred in the past.

Approach: The main part of the sampling desig[n] [in]volved taking photographs from five fixed quadrats from each of eight assemblages of sessile organisms (= habitat types) on eight shores twice a year, and determining percentage cover of the major space occupiers in each quadrat. Supplementary procedures involved estimating cover of surf grass using point-intercepts on a transect line, and counting large mobile organisms (limpets, starfish and abalone) in larger fixed plots. The reason for providing this example, however, is not the details of the final design, but the lead-up work that preceded it. Before the final design was decided upon, the following pilot measures were undertaken:

- Developing five criteria for selecting key species or species assemblages (ecological importance, special relevance to the Santa Barbara area, availability of background information, special human interest/ value, practicality), and evaluating possibilities for monitoring using these criteria.

- Extensive reconnaissance of the entire coastline in the area using video footage to select 13 initial sites, which were then reduced to eight after further qualitative and quantitative sampling and application of a six-point set of criteria.

- Statistical analysis (using repeated-measures ANOVA) of a previous four-year data set to determine the extent of annual variability in the organisms likely to be encountered.

- Modelling this existing data set by introducing a perturbation causing an average 50% reduction in species numbers (arbitrarily set) to determine the number of replicates needed to detect an effect of this size.

- Testing the reliability of field counts by comparing scores from the same plots sampled on three consecutive days by each of three experienced scientists, and comparing scores from the scientists and three inexperienced samplers for one of those days.

These pilot procedures meant that the researchers (and the agency funding them) had considerable confidence that the proposed design would meet the specified objectives.

7.4.4 Determining the likely impact of an activity

Example: Predicting the effects on intertidal algae of human visitation to rocky shores in Victoria, (Povey & Keough 1991) and northeastern New Zealand (Brown 1996).

Purpose: To determine the effects of various levels of trampling on the cover of two common rocky shore algae, *Hormosira banksii* and articulated coralline mats (*Corallina* sp), by experimentally simulating the activities of pedestrians, and to monitor recovery after cessation of trampling.

Situation: Information was available in both cases on the expected annual variation in cover of these species in the respective study areas. Human visitation to shores in both areas was at high levels, but the actual shores chosen for the experimental manipulations were rarely visited. In the New Zealand study, the location was in the Leigh marine reserve, and the actual numbers of visitors walking over parts of the intertidal areas had been documented.

Approach: Both studies used a modified BACI design (see Chapter 2). Brown used three trampling regimes (high, 150 steps per day; medium, 50 steps; low, 10 steps) as well as a control (0 steps) in small plots (2 x 0.5 m). There were three replicate sets of these four treatments. Five fixed quadrats (0.09m^2) were randomly positioned within each of the 12 plots, and these were photographed once prior to trampling, immediately following experimental trampling (two months later) and at two-monthly intervals for one year. Povey and Keough's design was similar, but incorporated only two trampling regimes (high, 25 steps per day; low, 2 steps) and larger recording quadrats (60 x 40 cm) that were always sampled from the same place in the plot. Their cover estimates were made in the field using 100 regularly-spaced intersection points on a gridded quadrat, rather than photographically. Both studies found short- and long-term reductions in the cover of *Hormosira*, even at low rates of trampling, but the cover of coralline turf was generally unaffected. By measuring the height of the turf within the same experimental framework, however, Brown (1996) was able to show that the turf became compressed at mid and high trampling rates, remaining this way for at least nine months. Povey & Keough noted, but did not measure, the same sort of effect, but they did measure an increase in the abundances of molluscan grazers where the *Hormosira* had been reduced. Because the results (in terms of the nature of the impacts) initially documented in Victoria were later repeated in New Zealand using similar procedures, this pair of studies illustrates the very real impacts humans can have on intertidal communities, and suggests that better management of these impacts (especially in a marine reserve situation) may be required.

It is worth noting that the documented impact of trampling on coralline turf (i.e. long-term compression) also seriously reduces the abundances of many of the small macrofauna that live within this turf (Brown 1996). The investigation of very small macro-inverte-brates and meiofauna is a topic that has not been dealt with in this chapter. Although a good microscope and considerable taxonomic expertise are required, their study may be useful for detecting changes to intertidal communities as most meiofaunal species have short generation times, potentially allowing impacts to be detected rapidly. However, differences in abundances of intertidal meiofauna, of many orders of magnitude, are well documented both in space and time for these animals (e.g. Coull & Wells 1983), and these natural variations would have to be differentiated from any change caused by humans, by setting realistic effect sizes (the amount of change that would cause concern among coastal managers).

7.4.5 Determining the influence of a physical variable (rockpool size) on species assemblages

Example: Describing the influence of pool size (diameter) on the assemblages of invertebrates and algae in intertidal rock pools in New South Wales (Underwood & Skilleter 1996).

Purpose: To use artificially created rock pools to test the hypothesis that there should be more species in pools of larger diameter when depth, period of colonisation and time of formation of pools are held constant.

Situation: Despite the fact that they occur on rocky shores worldwide, there had been few detailed studies of rock-pool assemblages because of inherently large spatial and temporal variability and the difficulty of obtaining suitable replicate pools. Because pools can be considered as islands of one habitat surrounded by a different habitat (Plate 39), the authors considered them to be useful systems to test more general predictions arising from models of patch dynamics. The problem of confounding variables found among natural tidepools was overcome by constructing 360 artificial pools, 15 at each of 24 sites (six each in low/exposed, low/sheltered, mid/exposed and mid/sheltered situations). These were of different diameters and depths (with replicates) and had been constructed three to four years prior to the study to allow colonisation to be completed.

Approach: An inverted periscope was used to count mobile animals and estimate percentage cover of sessile organisms in 4 x 4 cm quadrats in up to four different depth strata within the pool. The number of strata varied according to size of pool (diameter and total depth). Differences in the biota were examined among diameters for 5 cm deep pools (three diameters tested: 15, 30 and 50 cm), 15 cm deep pools (two diameters tested: 30 and 50 cm) and 30 cm deep pools (two diameters tested: 30 and 50 cm). Within each of these three sets, the effects of diameter for each depth stratum were sampled and analysed separately because different strata within the same pool would not have been independent samples in terms of the subsequent statistical tests (i.e. a different pool was used for each depth stratum). Pools were sampled on three separate occasions and 12 variables were analysed, yielding 324 separate univariate, partially hierarchical analyses of variance. In addition, a multivariate analysis (MDS) was done to examine the overall effects of pool diameter on the structure of the assemblages. The conclusion from all this careful analysis was that there were very few consistently significant effects of pool diameter on the abundance of most taxa found in the pools. This was different to what had been expected under previously published models of the effects of patch size. The point to note from this example is that variables that have been claimed to influence patterns of species assemblages can be controlled in replicate sets of entities (in this case intertidal pools) in orthogonal arrays of those factors. This then enables hypotheses about different features of that entity to be rigorously tested. It is not always easy to arrange, and considerable thought needs to go into the experimental design, but the rocky intertidal does allow opportunities to undertake this sort of innovative ecological assessment.

7.4.6 Sampling a single species on a sandy beach

Example: Assessing the distribution and abundance of a sandy-shore bivalve in New South Wales (James & Fairweather 1995).

Purpose: To determine the most accurate and efficient methods to sample Australian pipi (*Donax deltoides*) by comparing two rapid measures: finger dredging and a knife technique.

Situation: Information on abundance patterns of the sandy beach fauna of temperate eastern Australia is scarce, with only a few published studies (e.g. Dexter 1984; Haynes & Quinn 1995). Bivalves on sandy beaches are usually sampled by sieving sand from quadrats or cores, but consideration of other methods was desired because of factors such as the large amount of time and labour required for sieving, the difficulty of sieving coarse sediments, and the need for rapid sampling in the swash zone or in situations where bivalves can burrow or emigrate laterally in response to sampling disturbance.

Approach: Finger dredging involved using a box

quadrat, but leaving some of it above the surface of the sand, filling it with water to create a temporary pool, and then raking one's fingers through the saturated sand to a depth of 10 cm to force the bivalves to the surface, where they could be manually collected. In the first test, the proportion of pipis recovered by this technique was determined by comparison with standard digging-and-sieving techniques. A depth comparison was also incorporated. Less than 2% of pipis were missed by finger dredging, and the authors concluded that this technique was as accurate as sieving for quantifying abundances of animals larger than 5 mm, but more time-efficient (twice as rapid). In a second test, two quadrat sizes (0.1 and 0.2 m^2) were compared; both gave similar estimates of abundance (determined by ANOVA), but sampling efficiency (time taken to process samples) was better in the smaller quadrat in 75% of cases. A final technique was trialled in which a knife was systematically run through the sand to locate pipis and provide an index of abundance (number of hits per 2 m run). There was only a poor correlation between knife hits and density estimates from finger dredging, and many small pipi were missed. The technique was useful, however, as a rapid way of determining the upper limit of distribution on a beach. These techniques would not be suitable for all sandy-beach studies, or even for all studies on beach-dwelling bivalves, but the study does illustrate the importance of doing careful pilot work to determine appropriate techniques to address particular questions. A similar approach, using a stratified sampling protocol, was used to examine surf clam biomass on New Zealand beaches (Haddon et al. 1996).

7.4.7 Investigating the small-scale distribution patterns of an estuarine plant

Example: Assessing the spatial pattern of distribution and abundance of seedlings of the mangrove *Avicennia marina* var. *australasica* in northeastern New Zealand (Osunkoya & Creese 1997).

Purpose: To document where seedlings occur within mangrove forests and to investigate the possible influence of several factors (e.g. gaps in the canopy) on seedling establishment.

Situation: A previous study (Kaly, Creese & Jones unpub. data) had shown that the structure of *Avicennia* forests differed markedly throughout its geographic range in northern New Zealand. Eight sites covering three regions (north, mid and south) were selected from the original set of 24 sites. Sampling of forests was done at all eight of these sites, and experimental manipulations of seedlings at five.

Approach: In the initial assessment stage, two replicate 5 m wide belt transects were run through each forest from the seaward to the landward edge, and all trees were counted and measured. Seedlings were counted in two 0.25 m^2 quadrats randomly placed (by pacing out) within each 5 m segment of the transect. This approach described the structure of the natural mangrove forest at each site. Young seedlings were then individually tagged in 10 m sections of one of the transects; 10 underneath-canopy trees and 10 in light gaps (Plate 46). Their growth and survival was monitored four times over 18 months and analysed using a three-factor analysis of variance (canopy condition, intertidal level and forest). Significant effects from these analyses suggested that survival and growth were best in light gaps. There were also differences among sites (as expected) and among tidal positions (better growth but poorer survival at lower levels). Experiments in which seedlings from a single small area within each forest were dug up and transplanted were used to further examine aspects of early growth and survival in light gaps. This approach of combining quantitative surveys, demographic analysis of plants in natural conditions and experimental manipulations to test specific hypotheses provided useful insights on how mangrove stands might establish. This information can then be used to help develop protocols for the possible restoration of damaged estuarine areas (Staines 1997).

7.4.8 Sampling the infauna on intertidal sand flats

Example: Assessing the spatial distributions of infaunal invertebrates on sediment flats in the Manukau Harbour, New Zealand (Thrush et al. 1989), and the influence of these on temporal patterns (Thrush et al. 1996b).

Purpose: To establish a monitoring programme to provide an inventory of infaunal species within the harbour and to establish a framework of distribution patterns against which process-oriented studies could be conducted.

Situation: The Manukau Harbour is a large, shallow inlet, 40% of which is composed of intertidal sand flats used extensively for recreation and food gathering. Data on macrofaunal populations have been collected bi-monthly, from 1987 to 1996, from six sites (each 9000 m^2) spread around the harbour. The accumulated data set can be used to describe scales of spatial patchiness and the influence of this patchiness on interpreting time-series data.

Approach: Each site was divided into 12 equal sectors and initially three cores (13 cm diameter by 15 cm deep) were randomly collected from within each sector. Sediments were sieved through a 0.5 mm mesh and all retained organisms removed, preserved and then counted. Data were analysed by spatial autocorrelation using 5 m distance intervals. Most species at most sites showed significant spatial autocorrelation (i.e. were clumped at some scale; similar to the example shown in Fig. 7.8). For subsequent sampling, only 12 samples were collected (one per sector) following an optimisation exercise balancing effort against precision. Two techniques (one using ranges of abundance, the other using estimates of between and within group variances from one-way ANOVAs) were then utilised to generate ratios of spatial to temporal variation. Finally, a simulation exercise (repeated 100 times) was undertaken in which subsets of four samples were randomly chosen from the initial 12 to illustrate the changes in temporal patterns (e.g. the timing and magnitudes of peaks in abundance) that might be detected with decreased precision on each sampling occasion. Most populations exhibited density variations in time similar to or less than those in space. The accuracy with which a reduced sampling effort (four cores) mimicked the full patterns (generated from 12 samples) depended on the ratio of spatial to temporal variation; those populations with small spatial compared to temporal variation were adequately interpreted in up to 44% of the simulations, but populations with similar scales gave close correspondence in less than 30% of the simulations. These analyses illustrate the critical role that relative variability (i.e. spatial versus temporal) may play in the detection of ecologically important change, and further emphasises the need to carefully evaluate the numbers of replicate samples to be collected in any monitoring survey. Unfortunately, Thrush et al.'s study also stresses that there are no general guidelines that can be used when sampling such habitats. This is a very real possibility in any intertidal sampling exercise: each species is likely to vary differently in different places, and sampling procedures need to be as robust as possible to account for this.

SELECTED READING

The references listed are examples of research that will lead the reader to a broader literature, also see the text of the chapter.
* indicates a multi-authored book or collection of proceedings; references to volumes of a journal refer to a set of relevant papers.

General descriptions
Morton & Miller 1973; Bradstock 1989 (New Zealand shores); Bennett 1987 (Australian shores); Branch et al. 1994; (South African shores); Brown & McLachlan 1990; McLachlan & Jaramillo 1995 (sandy beaches).

Intertidal ecology
General: Rafaelli & Hawkins 1996; Underwood & Chapman 1995.
Competition: Connell 1983; Underwood 1986; Wilson 1991.
Disturbance: Pickett & White 1985; Sousa 1985; McGuinness 1987; Petraitis et al. 1989.
Estuaries: Jones 1983; Day et al. 1989; Grimes & Kingsford 1996; Attrill 1997.
Growth: Petraitis 1995.
Keystone species: Foster 1990; Paine 1994; Menge et al. 1994; Navarrete & Menge 1996.
Predation: Quammen 1984; Fairweather 1990a; Petraitis 1990; Menge 1991; Wilson 1991.
Plant–herbivore interactions: Lubchenco & Gaines 1981; Creese & Underwood 1982; Gaines & Lubchenco 1982; Hawkins & Hartnoll 1983; Bustamente et al. 1995; Jernakoff et al. 1996.

Settlement and recruitment: Butman 1987; Gaines & Rough-garden 1987; Raimondi 1990.
Succession: Luckens 1976; Jernakoff 1983; see also Disturbance above.
Use of experiments in intertidal ecology: Paine 1994; Underwood 1993a, 1997.

Groups of intertidal organisms
Epiphytal invertebrates: Edgar & Moore 1986; see also Chapter 10.
Gastropods: Underwood 1979; Creese 1988; McQuaid 1996; see also Chapters 4, 6, 10.
Bivalves: McArdle & Blackwell 1989; Haddon et al. 1996.
Barnacles: Raimondi 1990; Anderson 1994.
Mangroves: Chapman 1979; Snedaker & Snedaker 1984; Crisp et al. 1990.
Salt marsh: Pomeroy & Wiegert 1987; Packham & Willis 1997.
Sea grass: Phillips & McRoy 1980; Larkum et al. 1989; Clayton & King 1990; Jernakoff et al. 1996; Kuo et al. 1996.

Fisheries: Dugan & Davis 1993; Botsford et al. 1997.

Impact assessment: Siegfried 1994; Botsford et al. 1997; Castilla & Fernandez 1998*; see also Chapter 2.

Methodology: Price et al. 1980; Rafelli & Hawkins 1996.

Other related topics: see Chapters 4, 5 and 8.

Taxonomy: see Chapter 10.

EIGHT

SUBTIDAL ASSEMBLAGES OF SOFT SUBSTRATA

D. J. Morrisey, S. J. Turner and A. B. MacDiarmid

8.1 Introduction

Most of the Earth's surface is covered by the sea, and the largest portion of the seabed below it consists of soft substrata, such as muds, sands and gravels. These habitats are subjected to numerous impacts from humans, especially in the area between the coastline and the edge of the continental shelf. Sands and gravels, for example, are mined for use in the construction industry, as exemplified by recent applications to mine sand off the coasts of New South Wales and northeastern New Zealand. Other areas are dredged for navigation or obliterated by construction of ports, marinas or structures such as airport runways. Contaminants introduced into the marine environment by human activity, such as heavy metals, organochlorines and polycyclic aromatic hydrocarbons, tend to adsorb to the surface of fine particles of sediments and are thereby deposited in association with them. Areas of the seabed where fine sediments accumulate, particularly in sheltered coastal areas that receive runoff from rivers, therefore act as sinks for these contaminants. The same sediments may subsequently act as sources of contaminants to the water column if they are disturbed by, for example, dredging. Needless to say, all of these impacts can have severe ecological consequences on the animals and plants living on or in the sediments, or on adjacent ecosystems.

Subtidal soft-substratum environments are influenced by many of the same physical and chemical forces that act on soft substrata in intertidal areas. These forces include currents, which erode and deposit sediments, determining their depth and particle-size composition; and the physico-chemical processes that control the rate of diffusion of oxygen and nutrients into the sediments. Sedimentary processes in both intertidal and subtidal environments are responsible, at least in part, for the supply of organic matter to the organisms living in sediments. The microbial processes of decay by which this material is made available to higher trophic levels are also similar in the two environments. In some cases, subtidal habitats may be continuations of adjacent intertidal ones, such as mud- and sandflats extending below the low-tide area into muddy and sandy seabed. Sediments are generally finer in subtidal areas than in adjacent intertidal areas, illustrating one of the differences between intertidal and subtidal soft substrata, namely the relative strength of water movement over them. Intertidal areas are subject to the ebb and flow of the tide across their surface, and to wind-driven waves. Subtidal areas are also subject to waves in shallow areas and to currents such as those driven by tides, but these tend to be weaker than those experienced by intertidal areas. As a consequence of the slower water movement offshore, finer sediments are able to settle there. The deeper the water, the slower the currents and the finer the sediments. There are, of course, many exceptions to this pattern, such as areas around the entrances to estuaries and other embayments, where tidal currents are concentrated, and near wave-exposed shores. In these areas, sediments are coarser or may be absent altogether, leaving the *in situ* rock exposed.

Subtidal soft substrata range from coarse sands in areas where water currents are strong, to fine muds. The range of particle sizes in any given sediment may be broad ('poorly sorted') or narrow ('well sorted'). Sandy areas are often patterned with ripples with wavelengths ranging from a few centimetres to several metres ('megaripples'). Sands and muds may contain gravel derived from underlying older deposits and from the shells of molluscs living within the sediment or brought into the area by currents. The surface of the sediment may be dotted with the burrows and tubes of polychaetes and other types of worms, crustaceans (crabs, amphipods, burrowing shrimps), octopus, and the holes made by the feeding siphons of bivalves living within the sediment. Dense populations of surface-living bivalves may almost cover the surface of the sediment in some places. In the case of mussels, which are sedentary, assemblages of organisms usually found on hard substrata often develop on the mussels' shells. The bodies of soft corals (alcyonarians) and burrowing anemones and the feeding structures of brittle-stars, sea-cucumbers and other animals may also project above the surface of the sediment, but are able to withdraw rapidly if disturbed.

The majority of animals in soft substrata, however, live below the surface. To identify and count them, and to identify patterns in their distribution, it is necessary to remove them from the sediment, either on the spot or, more often, by taking a sample of sediment and sorting it back on land. This unavoidable need to disturb the animals' habitat in order to observe and count them represents an important difference in methods of study and experimentation between soft and hard substrata.

The nature of the sediment has a major influence on the organisms living within it, partly as a result of the movement of oxygen and nutrients into the sediment. Such movement is more rapid in coarser sediments, but these tend to be less rich in organic detritus, which provides food for many sediment-living animals. In intertidal areas there may be considerable primary production by microalgae occurring on or within the sediment, particularly at low tide. In subtidal areas, the reduced availability of light, especially in turbid areas such as estuaries, greatly reduces the amount of *in situ* primary production. Inputs of dead plant material from nearby intertidal areas (seagrasses, mangroves and saltmarshes, for example; Plates 46–50) and, particularly in deeper areas, from the plankton that rains down from above may, however, be considerable. The large algal assemblages found on subtidal rocky reefs are not usually present in soft-substratum areas because of the turbid water and the lack of hard substrata to attach to, but there may be large amounts of drifting algae present.

Where the overlying water is shallow enough for sufficient light to penetrate, beds of seagrasses may occur. These often serve as habitat for different assemblages of animals to those found in nearby unvegetated areas. They also represent important nursery areas for juveniles of many species of fish and invertebrates. Sampling of seagrass beds is not discussed specifically in this chapter, but relevant sources of information are cited in Selected Reading (p. 226).

Organisms living in subtidal soft substrata do not face the same problems of desiccation and extremes of temperature encountered by their intertidal counterparts. Such problems are not, in any case, as severe for intertidal organisms living in soft substrata as for those on rocky shores, because the sediment acts as a buffer, damping rates of change in temperature, salinity, etc. between periods when the substratum is exposed and when it is covered. Conversely, as mentioned earlier, processes such as the supply of nutrients and oxygen are also slowed down.

Methods for sampling subtidal soft substrata and for studying their ecology are similar to those of intertidal sediments (see Chapter 7), but the intervening layer of water has an influence on the methods of study, just as it does on the environments themselves. Some of the techniques discussed are relevant to both environments but have been covered in this chapter to avoid repetition. Much of the discussion of the philosophy and methods of experimentation in Chapter 7 is also relevant to the present chapter, and the discussion of methods for extracting animals from sediments and for describing the sediments themselves apply to both chapters.

8.2 Studying soft-substratum environments

8.2.1 Methods

A diverse array of sampling methods is currently available to identify, map and monitor soft-sediment subtidal habitats and their associated biological assemblages. These range from techniques providing broad overviews of entire systems, identifying the main types of habitat and assemblage in an area and their spatial distribution, to very detailed quantitative studies providing fine-scale information about the distribution and abundance of individual organisms. The precise methods selected will depend very closely on the objectives of the study, but there are, perhaps, more constraints on the methods that can be used than in intertidal areas. These constraints include the depth of the water, the sea conditions and the cost and availability of ships and equipment such as grabs or side-scan sonar.

No one method is suitable or appropriate for the complete characterisation of habitats or biological assemblages at all sites, and many biological problems will require a combination of methods. The choice in any particular situation will depend on a number of factors. In particular, the quantity and quality of the information needed to address the aims of the sampling programme, the resources (both logistical and economic considerations) available for collecting and processing the data, as well as the physical nature of the site under study (e.g. water depth, sea conditions, substratum type) all need to be considered. While it is desirable to use a

	Sampling method[1]	Spatial scale of samples[2]	Usual purpose[3]	Destructive/non-destructive	Diver operated/remote	Continuous/discrete sampling strategy[4]	Cost[5]
(a)	Existing information	Various (historical)	Habitat/assemblage/species	Non	N/A	Various	Low
(b)	Satellite imagery/aerial photography	Broad	Habitat	Non	Remote	Continuous	High
(c)	Echo-sounding	Broad	Habitat	Non	Remote	Continuous	High
(d)	Side-scan sonar and swath-mapping systems	Broad	Habitat	Non	Remote	Continuous	High
(e)	Manta tows	Medium	Habitat/assemblage/species	Non	Diver	Continuous	Low
(f)	Diver-propulsion vehicles	Medium	Habitat/assemblage/species	Non	Diver	Continuous	Moderate
(g)	Benthic trawls and dredges	Medium	Assemblage/species	Destructive	Remote	Discrete	Moderate
(h)	Remotely operated TV and video cameras	Medium	Habitat/assemblage/species	Non	Remote	Continuous	High
(i)	Diver-operated video	Medium	Habitat/assemblage/species	Non	Diver	Continuous	Moderate
(j)	Diver depth profiles	Medium	Habitat/assemblage/species	Non	Diver	Continuous	Low
(k)	Diver counts	Small	Assemblage/species	Non	Diver	Discrete	Low
(l)	Spot checks	Small	Habitat/assemblage/species	Non	Both	Discrete	Low
(m)	Remotely operated still photography	Small	Assemblage/species	Non	Remote	Discrete	Moderate
(n)	Diver-operated still photography	Small	Assemblage/species	Non	Diver	Discrete	Moderate
(o)	Benthic grabs and remotely operated corers	Small	Assemblage/species	Destructive	Remote	Discrete	Moderate
(p)	Diver-operated hand cores	Small	Assemblage/species	Destructive	Diver	Discrete	Low
(q)	Airlift sampling	Small	Assemblage/species	Destructive	Diver	Discrete	Moderate

range of techniques for sampling different components of a subtidal assemblage, this may not always be possible. In such cases, compromises between the ideal method and what is actually possible may be unavoidable. For example, the ideal method of sampling for a particular study may be to have divers collect precisely placed core samples, but, given the depth of water and the available funding, grab-sampling may be the only feasible method. The aims of the programme therefore need to be carefully defined at the outset. It is particularly important to be aware of the limitations of the sampling and processing techniques available, and that they may not adequately sample some habitats or assemblages.

There are a number of reviews in which the equipment used for sampling subtidal soft-sediment assemblages is described in detail (e.g. Holme & McIntyre 1984; Baker & Wolff 1987; English et al. 1994). In this section, we briefly review the main methods currently used in identifying, mapping and monitoring subtidal habitats and biological assemblages, and evaluate the advantages and limitations of each method. The main features of the different sampling methods discussed are summarised in Table 8.1. References describing methods for sampling seagrass beds are listed at the end of this chapter.

OPPOSITE: **Table 8.1** Summary of methods appropriate to determining the distribution of subtidal habitats and assemblages (See text for further details).

1. The different sampling methods can be used quantitatively, semi-quantitatively or qualitatively. This in turn will determine the level of post-sample-collection processing, costs etc.
2. Sampling methods may provide a broad-scale (hundreds of metres to kilometres) overview of subtidal habitats and assemblages through to detailed information at small scales (millimetres to metres).
3. Sampling methods may be used to provide information for broad-scale habitat mapping, estimates of the distribution and abundance of assemblages and assemblage structure and composition, and/or estimates of the distribution and abundance of species (as well as the taxonomic identity of species).
4. Sampling methods may produce discrete samples (e.g. cores, grabs, still photographs) or continuous information (e.g. side-scan sonographs, video film) on subtidal habitats and assemblages.
5. The high cost of some methods reflects requirement for access to appropriate equipment and a suitable vessel to handle the gear; low-cost methods generally employ divers using relatively simple equipment and working from a small boat. These costs reflect the cost of collecting the samples, and do not include sample processing costs, etc.

(a) *Existing information*

An important first step in obtaining data on soft-bottom subtidal habitats and assemblages may be a review of the available information. Much large-scale information on the distribution of subtidal habitats and assemblages can be inferred, for example, from sedimentological and bathymetric charts. Old charts, aerial photographs, etc., may also provide a useful historical perspective. Although these exclude much of the small-scale diversity of the marine environment, they can provide useful information to assist with the design of a sampling programme.

More detailed information on the distribution of habitats and assemblages and occurrences and abundances of species, usually from small areas, may be found in a wide variety of unpublished and published literature (e.g. scientific papers, environmental impact assessments, reports and theses). A major limitation, however, is that different objectives and the use of a wide range of sampling techniques means that spatial and temporal comparisons of the information collected can become very difficult. It is, therefore, important to define carefully the objectives of a particular study and to keep these in focus throughout (see Chapter 2).

Advantages
✔ May give an indication of species likely to be encountered and an idea of relative abundance.
✔ Can be used for planning sampling programmes.
✔ May provide useful historical perspective.
✔ Cheap and relatively straightforward to undertake.

Disadvantages
✘ Rarely at the scale required.
✘ By definition, early studies are not contemporary.
✘ The mixture of methodologies makes studies difficult to compare.
✘ Data sources and methodologies are often poorly documented.
✘ Often include a sample bias.

Example: Powell (1937) undertook a study of the subtidal soft-bottom communities in the Waitemata and Manukau Harbours, Auckland, New Zealand. From his qualitative samples, collected using either a small naturalist's dredge or a small conical dredge, he produced maps showing the distribution of the main benthic communities in the harbours. In 1993, the Auckland Museum (Hayward et al. 1994a, 1995) undertook a study to document the distribution and composition of the soft-bottom communities in the Waitemata Harbour. Part of this project included a resurvey of Powell's study sites to assess changes in the communities over the past 60 years.

(b) *Satellite imagery and aerial photography*

Remote-sensing techniques involve the acquisition of imagery of the properties of the Earth's surface from a distance, mainly from aircraft or satellites (Dalby & Wolff 1987). They are efficient and relatively cost-effective, and provide a synoptic overview of the entire system that is otherwise unobtainable from ground level.

Two broad categories may be recognised:

Digital remote sensing – where information on environmental features and conditions is based on the transmission of digital data from scanning spectro-radiometers recording the intensity of reflected radiation over specific spectral bands, and expressed in a form directly ready for computer processing or as pre-processed analogue images (Dalby & Wolff 1987).

Aerial photography – where information is expressed as a photographic image.

Satellite imagery has both advantages and disadvantages compared with aerial photography (Orth et al. 1991; Dobson et al. 1995).

- Satellite data generally have greater spectral resolution than aerial photography, but less spatial resolution.
- Satellite imagery is more appropriate and less costly for rapid repeated observations over broad areas.
- Satellite data are already in a digital format that can be directly processed by computer, whereas aerial photographs must be digitised or scanned into computer if the information is to be analysed quantitatively.
- Aerial photography is more suitable for smaller-scale, high-resolution studies.
- A major advantage of aerial photography over satellite imagery is flexibility of timing. Because of the fixed orbital paths of satellites, satellite images may not always be acquired under optimum conditions (e.g. cloud cover).

To date, aerial photography has been used most widely for biological studies of coastal areas. However, digital systems (e.g. LANDSAT multi-spectral scanners, LANDSAT thematic mappers and SPOT high-resolution visible scanners) have provided useful information on subtidal habitats along large areas of coastline (e.g. Dalby & Wolff 1987; Kirkman 1990; Dobson et al. 1995). As digital remote sensing systems become more widely used and the ground resolution of the systems available improves, they are likely to be increasingly used in coastal-survey work (Dalby & Wolff 1987).

Both digital remote sensing and aerial photography need to be undertaken in conjunction with comprehensive field observations (ground truthing) to confirm the identification of recognised features and to identify unknown features on the images.

In suitable conditions, aerial photography of the near-shore environment can provide useful information on subtidal topography and the distribution of different habitats and assemblages (e.g. mussel and oyster beds, seagrass beds, microalgal mats at the sediment surface, in addition to coral reefs, macroalgae, and rocky reefs) down to depths of about 10–15 m (Orth et al. 1991; Dobson et al. 1995; Pasqualini & Pergent-Martini 1996). It requires suitable weather (avoiding low sun angles, haze, cloud and mist) and sea/tide conditions, including low turbidity (i.e. avoiding phytoplankton blooms or immediately following periods of heavy rainfall or persistent strong winds), a low tide and no wind or wave activity. This broad-scale information in very shallow water may be difficult to obtain by ship-based mapping methods.

In most cases conventional aircraft (preferably high wing) are used for acquiring the photography, although hot-air balloons and helicopters have also been used. The sensor can range from a simple 35 mm hand-held camera, a permanently mounted mapping camera or a video camera. A variety of different types of film (e.g. black and white, infra-red, colour daylight and false colour) and film-filter combinations have been used, depending on the information required. For example, colour photography is better for water penetration and for distinguishing different species of seagrasses (Kirkman 1990).

The scale of the photograph is important and depends on the objectives of the work and the detail to be shown (Dalby & Wolff, 1987). This in turn will determine the flight height and the number of photographs to be taken. Scale is a compromise between resolution of the different ground signatures, coverage of study area, inclusion of land features sufficient for horizontal control, and cost (Dobson et al. 1995). Scales of 1:5000 are regarded as generally acceptable for biological mapping in most areas; scales of 1:10,000 to 1:25,000 are suitable for more general studies of larger areas (Dalby & Wolff 1987). Sixty percent end-lap (overlap of adjacent photographs along a flight-line) and 20% side-lap (overlap of photographs along parallel flight-lines) enables stereo-photo interpretation and ensures complete coverage (Orth et al. 1991; Dobson et al. 1995)

Depending on the accuracy of the mapping, it may be necessary to have precise ground control for scale and orientation using points whose positions are known accurately and which are clearly visible in the photograph. In relatively featureless areas such as sandflats it is necessary to deploy scale-objects of known size and position in the field. These should be of contrasting colour to the background, for example, strips of plastic sheeting showing a white cross on a black background (white on its own is not very visible against sand). The photographs can either be incorporated onto scale maps of the coast or used to provide background information.

Advantages
✔ Can very quickly survey long stretches of coastline (tens to hundreds of kilometres).
✔ A lot of permanent, detailed information can be collected that enables future interpretation.
✔ Photographs may be less costly and more efficient than labour-intensive field surveys.
✔ Historical photographs can provide information on systems from earlier years.
✔ Sampling is non-destructive and non-intrusive.

Disadvantages
✘ Restricted to shallow water (less than 10–15 m).
✘ Depends on flat seas, low tides, good underwater visibility and optimum atmospheric conditions.
✘ Ground truthing for habitat descriptions is necessary. This is particularly so when mapping different types of vegetation or substrata, to ensure that variation in the photograph corresponds reliably to variation in the features being mapped.
✘ Depending on the accuracy requirements, precise ground control for scale and orientation may be necessary.
✘ Post-survey interpretation of photographic data requires some experience, is labour-intensive and therefore relatively expensive.
✘ Commercial aerial photography is expensive (but improvised apparatus can be mounted on an aeroplane).

Example: Larkum and West (1990) investigated long-term changes of seagrass meadows in Botany Bay, New South Wales, using historical aerial photography (1930–1985) in conjunction with field observations (1970–1987) and *in situ* sediment coring.

(c) *Echo-sounding*
Echo-sounders measure the two-way travel time of a pulse of acoustic energy transmitted from the vessel to the seabed and reflected back to the ship, and from this determine water depth using the velocity of sound in water (Hooper 1979). Echo-sounders thereby provide rapid, continuous and accurate depth measurements of the seabed immediately beneath the ship.

Swath-sounding systems use echoes of sound projected sideways as well as downwards (as with a conventional echo-sounder). The sideways-projected sound is focused into almost pencil-like beams at discrete angles to the vertical, so that with simple geometry it is possible to calculate the depth where each beam strikes the seabed. Instead of a single line of echo-soundings, the system produces parallel lines of echo-soundings from each beam, giving a swath of depth measurements (swath-bathymetry).

Recent advances include the development of hydro-acoustic processors (e.g. RoxAnn) that, when connected to a standard echo-sounder, can discriminate between seabed material types and provide detailed texture and environmental maps.

Advantages
✔ Can quickly cover large areas (tens to hundreds of kilometres); echo-sounders can operate at vessel speeds of up to 5 knots.
✔ Relatively easy to install and operate.
✔ Calibration is straight-forward, particularly in shallow (less than 30 m) water.
✔ Provides an instant profile of the seabed along the sounded line.
✔ Can easily interface with Differential Global Positioning System (DGPS), thereby enabling accurate positioning of sounded lines.
✔ With correct selection of operating frequencies, can detect differences in seabed material and can therefore assist in ground-truthing side-scan data.
✔ Sampling is non-destructive and non-intrusive.

Disadvantages
✘ Provides depths only along sounded line and can thus miss seabed features between the lines.
✘ Poor results in certain habitats (e.g. fluid mud).
✘ Depending on vessel size, soundings can be adversely affected by rough sea conditions.

(d) *Side-scan sonar and swath-mapping systems*
The principle of operation of a side-scan sonar is essentially similar to an echo-sounder, the fundamental difference being that the side-scan sonar transmits pulses of acoustic energy sideways rather than vertically downwards. The pulse of energy is transmitted in the form of a fan-like beam, narrow in the horizontal plane and wide in the vertical plane, so that scanning can be achieved from directly under the vessel to a selectable range on either side of the vessel track (D'Olier 1979). The swath width is about double the water depth, so it provides the greatest coverage in deeper water.

A recorder displays an image of the varying intensity of sound that is reflected back from the seabed. This builds up a 'side-scan sonograph' or 'swath image' that resembles a strip aerial photograph of the seabed's 'texture'. Variations in the signal returns are interpreted to provide information on the lateral change in sediment type (e.g. mud, sand, shell) and morphological features (e.g. protrusions or depressions such as scour-marks, sand dunes, trawl marks).

Side-scan images alone provide no information on depth. However, side-scan data can be draped over existing bathymetric data, or data on depth can be simultaneously collected by the ship's echo-sounder.

The new technology of swath-mapping produces intricate maps of seabed 'texture' (i.e. the extent of mud, sand, rock, etc.) and its shape (as precisely located contours) for a wide strip or swath of seabed on either side of the vessel (Lewis 1994). Swath-mapping systems project highly focused multiple-sound beams to intersect the seabed perpendicular to the ship's track and at a precise angle to the vertical. It is only when both swath-bathymetry and swath-imagery can be compared for precisely the same area of seabed that it is possible to interpret the geological processes, biological habitats, and oceanographic effects on the seabed. This has been facilitated by the recent development of 'dual sonars' that will simultaneously produce intricate three-dimensional maps of the seabed incorporating information on both depth and seabed texture for a wide strip of seabed on either side of the ship. This coincided with the availability of satellite-based Global Positioning Systems (GPS), which accurately positions the vessel and its data anywhere on Earth 24 hours a day.

Advantages
✔ Can quickly cover large areas (tens to hundreds of kilometres) for gross habitat determination.
✔ Excellent for identifying different subtidal habitats (e.g. mud, sand, gravels, boulders) and, depending on the system used, it can identify different assemblages (e.g. beds of horse mussels).
✔ Can easily interface with Differential Global Positioning System (DGPS), enabling accurate positioning.
✔ Most efficient in deeper water because larger areas are covered.
✔ There are no problems associated with low visibility.
✔ Relatively easily deployed.
✔ Sampling is non-destructive and non-intrusive.

Disadvantages
✘ The narrow swath width in shallow water requires many parallel tracks.
✘ Shallow coastal waters with complex emergent reefs are difficult to survey.
✘ Some form of ground-truthing using other techniques (see below) is necessary.
✘ The post-survey interpretation is relatively labour intensive and therefore expensive.
✘ Relatively costly.

Examples: In 1993, New Zealand and French scientists undertook a study to swath-map an 86,000 km^2 area of the continental shelf of New Zealand from north-east of the East Cape to Kaikoura, to produce detailed maps of the continental shelf region (Lewis 1994). Cole et al. (1995) used side-scan sonar to map dense populations of the venerid bivalve *Tawera spissa* in 18–20 m of water at a site near Tauranga, New Zealand.

(e) *Manta tows*
A snorkeller or diver is towed on a 'manta board' behind a small boat at a constant speed (up to 1.5 knots) (Moran et al. 1989). Manta boards are wooden or plastic boards (approximately 40 x 60 x 2 cm in size) provided with hand-grips at the leading and rear edges of the board for the snorkeller or diver to hold on to (English et al. 1994). The diver either records a brief description of the habitat or abundance of a particular organism at set intervals on data sheets affixed to the board or using an underwater tape-recorder (e.g. Gamble 1984), or surfaces periodically to be debriefed.

In general, manta towing is regarded as a useful tool for rapid assessment of qualitative differences among sites and for estimating broad-scale changes in distributions and abundances of species or assemblages (e.g. Lassig & Engelhardt 1994). Up to 1 km of sea-bed can be covered in approximately 30 minutes (Hiscock 1987). Manta towing has, however, a number of inherent and largely unmeasured limitations that may render counts biased and imprecise (Fernandes 1990; Fernandes et al. 1990). These include 'perception bias' (which occurs when potentially visible target organisms are missed by observers) and 'availability bias' (which occurs when target organisms are unobservable because of viewing conditions). To ensure consistency among observers, some form of prior training must be undertaken.

Advantages
✔ Compared with other sampling methods involving divers, relatively large areas (tens to hundreds of kilometres) can be covered in a short time with little observer fatigue.
✔ Inexpensive, as no costly equipment is required.
✔ Simple to perform, but some form of training is necessary to reduce variability among observers and boat drivers.
✔ Can be undertaken under difficult field conditions (e.g. wind speeds up to 30–40 knots: Moran et al. 1989) and is useful where the costs of boat-time and personnel are high.
✔ Sampling is non-destructive.

Disadvantages
✘ Data can be biased and imprecise.
✘ Divers can become saturated with information, particularly if many variables are being recorded.
✘ Diver has no control over speed or direction, and is unable to stop and more closely inspect areas of

interest without releasing the manta board.

✘ Limited to diveable depths and good underwater visibility.

✘ May become difficult and dangerous in areas of foul ground.

✘ Many profiles required to make detailed habitat maps.

✘ Limited to providing categoric information (e.g. percentage cover) about the distribution of habitats and abundances of assemblages.

✘ On scuba, dive profiles are not standard (i.e. compatible with decompression tables) and can be very dangerous even at shallow depths.

Example: As a first step in a five-year study to follow changes in seagrass community structure in southern Western Australia, Kirkman (1985) surveyed the entire seagrass bed in the study area by underwater towing.

(f) *Diver-propulsion vehicles*

This method uses a diver-propulsion vehicle (DPV) to tow the diver but otherwise is very similar to that of a manta survey (see above). DPVs overcome some of the problems of manta boards. In particular, they have the advantage of that the diver has direct control over speed, stopping/starting and positioning, thereby allowing the operator to stop and make detailed observations whenever necessary.

If a calibrated current meter is attached, it is possible to obtain a measure of the distance travelled, and a compass can be used to determine direction. Alternatively, it is possible to use a boat to lay a lead-line rope between two points fixed by DGPS and then use the DPV to survey the length of the rope (e.g. Murdoch & Grange 1989).

Advantages

✔ Compared with other sampling methods involving divers, relatively large areas (tens to hundreds of kilometres) can be traversed quickly (within the limitations of depth and time imposed by diving).

✔ Because the operator has direct control over speed, stopping/starting and positioning, detailed observations can be made whenever and wherever necessary.

✔ DPVs are generally cheap to hire.

✔ Accurate positioning is possible.

✔ Sampling is non-destructive.

Disadvantages

✘ Limited to diveable depths and good underwater visibility.

✘ Many profiles are required to make detailed maps.

✘ The initial outlay is high because DPVs are expensive to buy.

✘ The power sources have a limited operational life.

✘ As with manta boards, dive profiles can be dangerous.

(g) *Benthic trawls and dredges*

Qualitative/semi-quantitative sampling of the macrofauna may be carried out by means of dredges or trawls. There are many kinds of trawls (e.g. Agassiz trawl, beam trawl, otter trawl), sledges (e.g. epibenthic sled) and dredges (e.g. naturalist's dredge, anchor dredge, hydraulic dredge), the choice depending on the size of the operating vessel and the equipment available on board to handle the gear. Trawls are large, wide-mouthed nets designed to be dragged along the seafloor to catch epibenthic fauna. Dredges essentially consist of a heavy metal frame, to which is attached a mesh bag to retain the organisms. The penetrating power of dredges is generally limited and they typically sample only the upper layer of sediment, collecting slow-moving or sedentary organisms and shallow-burrowing infauna larger than the mesh size. In the case of the hydraulic dredge, jets of water are shot into the sediment ahead of the dredge to fluidise the sediment and enable collection of deeply buried infauna.

Gear efficiency tends to vary depending on biological, environmental and operating conditions such as the size of the organisms, sediment characteristics, sea conditions, depth and towing speed. It is possible to standardise the speed and duration of a tow, so obtaining an estimate of density, but most trawls and dredges sample only a portion of animals on the surface and shallow-burrowing fauna, so results underestimate real abundances. In some sediments (e.g. mud, loose gravel) clogging of the net occurs, and thus the method needs to be calibrated using some other direct-count method.

Advantages

✔ Large areas (tens to hundreds of kilometres) can be covered.

✔ Not limited to diveable depths.

✔ Permanent sample collected (useful for taxonomic identifications).

✔ Provide useful preliminary data on types and distributions of assemblages and sediments, as a basis for planning the main sampling programme.

Disadvantages

✘ Only qualitative/semi-quantitative unless any biases involved in sampling are quantifiable and consistent.

✘ Efficiency of sampling varies with type of sediment.

✘ Generally selective for certain groups of organisms and do not provide adequate samples of larger, widely distributed or more mobile species. This may not matter, depending on the objectives of the study.

✗ Mesh size determines lower size limit of organisms collected.

✗ Long tows obscure small-scale spatial variation. The scale of tows should, therefore, be compatible with the aims of the study.

✗ Post-survey sorting of samples can be labour-intensive.

✗ May cause considerable damage to benthic habitats and to specimens.

Examples: Grange (1979) sampled the subtidal soft-bottom benthic communities in Manukau Harbour using a small naturalist's dredge covered with 1 mm mesh and fitted with large cutting blades to facilitate good penetration into the sediment. At each site, the dredge was towed for four minutes to collect comparable samples. Hayward et al. (1994b) used a two-litre capacity bucket dredge, hand-hauled from a 4-metre aluminium dinghy powered by a 5 hp outboard engine or by oars, to collect samples of the benthic foraminiferal associations in sediment samples from Stewart Island, New Zealand.

(h) *Remotely operated television and video cameras*
Television was first used underwater in the 1940s, but only with the recent development of compact video cameras and recorders has the full potential of this methodology been realised (e.g. George et al. 1985). Owing to recent advances in video resolution and ease of use, videos are becoming a standard sampling tool for subtidal habitat and biological assemblage survey and analysis.

Television and video cameras in suitable pressure housings may operate at any depths when mounted on a remotely operated vehicle (ROV), suspended frame or benthic sled. The cameras can be used in conjunction with external lighting units where light levels are too low for the use of ambient light. However, bright lights can cause considerable backscatter from particles in the water, which can seriously affect picture quality. The equipment is generally powered and operated via connecting cables to the supporting vessel. This linkage to the ship allows for instantaneous and continuous viewing and recording.

One of the primary considerations in the use of both remotely operated underwater video and still photography (see below) is the deployment of the camera to ensure that the subject is in the field of view and in focus and, once this is achieved, the remote operation of the camera. If the camera is lowered vertically from the ship or attached to a towed structure, some means of obtaining a constant distance from the object must be achieved so that the pictures remain in focus, and the distance and camera angle are standardised, so that

problems of distance and scale do not confound interpretation of the photographs or film (Holme 1985).

Television or video cameras are particularly well suited for habitats below safe diving range with good underwater visibility, and can be used alone to construct maps of habitat distribution or to ground-truth side-scan sonar surveys. Large areas can be traversed because there are few restrictions on time and depth of sampling. The size of the sample footprint will be determined by underwater visibility and the camera's angle of view. The resolution depends on water clarity, the amount of natural light and/or the use of artificial light, and the distance of the camera from the substratum (Berkelmans 1992). Video is generally more adaptable than still photography in terms of both light and colour, thus it is possible to obtain good results without an underwater lighting system. The incorporation of silicon-intensified target low-light-level cameras allows operation without lights and so avoids backscatter from suspended material in turbid waters (e.g. Potts et al. 1987).

The primary advantages of employing underwater videos or cameras (either remote or diver-operated) are their ability to acquire data rapidly and produce a permanent record of habitat type and abundance of the larger fauna and flora (e.g. sponges, corals, horse mussels). One of the major benefits of underwater photography is that film and photographs can be analysed at a later date in the laboratory, under controlled conditions. They can also be referred to at a later stage if the data are lost or erroneous, or if more detailed data interpretation needs to be carried out. The data are in a form that is amenable to electronic image-processing and analysis. Using these techniques, film and photographs can be analysed to provide accurate estimates of abundance, size and shape. For example, size estimation is possible using constant-distance-from-target, visible scale-bar, laser-dot scaling or stereo-video methods. An additional advantage is that film and photographs can provide information which cannot readily be obtained by more conventional sampling-techniques (e.g. behavioural observations, the appearance and position of individual organisms in their natural habitat, as well as spatial relationships between groups of organisms and interspecific and intraspecific associations). Relatively long-term changes can be recorded using time-lapse video.

Both film and photographs suffer from the disadvantage of not adequately sampling smaller cryptic organisms (Done 1981). Biases in estimates of abundance of more mobile species may also occur, especially if lights are used (Spanier et al. 1994; MacDiarmid et al. unpub. data). The resolution of video photography is not as good as *in situ* observations, and is generally not as good as that of quality still photography, and the relatively coarse resolution precludes the use of underwater videos, in particular, to discriminate fine taxonomic detail (Berkelmans 1992). Community structure, therefore, has to be described in terms of higher taxonomic or morphological variables.

Advantages

✔ Fairly large areas (tens to hundreds of kilometres) can be traversed quickly, making the technique useful for detecting rare, conspicuous organisms.

✔ Accurate positioning is possible.

✔ With real-time video there are no time or depth restrictions other than those imposed by the facilities available on the boat to handle the equipment.

✔ Provides instantaneous and continuous viewing and can convey valuable information in terms of activity and sequence.

✔ A detailed permanent record of the habitat type and the larger fauna and flora is produced.

✔ Sampling is non-destructive.

✔ The data are amenable to electronic image processing.

✔ Video film is cheaper than still film and more samples can be taken between film changes and video provides more images per unit of time.

✔ Ideal for conveying information to a lay audience.

✔ The expense of data collection is nominal once the equipment has been purchased.

Disadvantages

✘ Depends on good underwater visibility.

✘ Many profiles are required to make detailed maps.

✘ With towed sledges there is the danger of damage or loss of the gear through collisions with objects on the sea floor.

✘ Post-survey interpretation is relatively labour-intensive and therefore expensive.

✘ Suitable only for documenting larger-scale differences and for larger flora and fauna.

✘ Lacks the resolution of still photography and is less accurate in terms of the number of species documented, in particular smaller species.

✘ Inability to discriminate taxonomic detail.

✘ Bias in abundance estimates of mobile species may occur, especially if lights are used.

✘ The initial outlay to purchase equipment is relatively high.

✘ With standard Hi8 video filming, time is limited by the tape length and battery life.

(i) *Diver-operated video*

Until recently, underwater video has lagged behind still photography as a means of obtaining qualitative and quantitative data by divers, as early systems were bulky and expensive. Modern compact video camcorders

(video cameras with built-in video recorders) have minimal bulk and are easy to use and a variety of hand-held video units are available. An independent or surface-powered light source can also be easily incorporated.

The use of videos (and diver-operated cameras: see below) maximises data acquisition with limited diver time, thereby enabling the time spent underwater by divers to be used most effectively. Compared with other sampling methods involving divers, relatively large areas can be traversed, within the limitations of depth and time imposed by diving, because all data extraction and interpretation is post-survey. This has the additional advantage of avoiding potential observation biases caused by physiological and environmental stresses on a diver. The fact that a permanent record is obtained means that highly trained (and expensive) taxonomic specialists are not required in the field.

Advantages

✔ Fairly large (tens to hundreds of metres) distances can be surveyed, while video can also be used in place of still photography to collect small-scale (less than 1 m) information.

✔ Accurate positioning is possible.

✔ A detailed permanent record of the habitat and associated larger fauna and flora is produced.

✔ Estimates of size are possible using the methods described in (h).

✔ Easily deployed.

✔ Sampling is non-destructive.

✔ The expense of data collection is nominal once the equipment is purchased.

✔ Ideal for conveying information to a lay audience.

Disadvantages

✘ Limited to diveable depths and good underwater visibility.

✘ Post-survey interpretation is labour-intensive.

✘ Usually only larger flora and fauna can be surveyed because of limits of resolution.

✘ Inability to discriminate taxonomic detail.

✘ Lights and diver activity may bias estimates of abundance of some mobile species.

✘ Many profiles required to produce detailed maps.

(j) *Diver depth profiles*

A diver equipped with scuba swims from a well-defined point, usually along a measured tape, generally in a direction perpendicular to the shoreline, and makes notes of habitat type and boundary changes. There are no problems with varied topography of the bottom because the tape can easily be laid across different topographic features. To provide sufficient information for descriptions of habitats and assemblages, multiple profiles should be done at each location. This technique has been widely used to describe subtidal habitats (see Chapter 4) and is ideally suited to habitat descriptions near shore where other techniques are difficult to use.

Advantages
✔ The diver is free to modify speed and to record detailed observations.
✔ Can provide quantitative data on species distributions and abundances if data are recorded on the number and size of transitions along the tape measure (see 'Transects', below).
✔ Requires no specialised equipment.
✔ Accurate positioning is possible (see 'DPV', above).
✔ Sampling is non-destructive.
✔ Can be combined with other methods, such as video-recording.
✔ Efficient in terms of the information gained per unit of time spent underwater.
✔ Can record information from varied bottom topography.

Disadvantages
✘ Compared with techniques such as underwater video, only small areas (tens of metres) can be covered in each dive.
✘ Restricted to diving depths and good visibility.
✘ Many depth profiles are required to produce detailed maps.

Example: Ayling (1978) used a combination of aerial photography and diver-surveyed transects to map the distribution of major subtidal habitats (e.g. 'sediment-covered rock flats', 'mixed rock and sand', 'sponge garden', 'sand and gravel') in the Leigh marine reserve, New Zealand (see Fig. 4.8, p. 96). Approximately 100 transects, ranging in length from 50 to 450 m, were undertaken. Transects were surveyed using two divers, one swimming a compass course and noting depth, the other laying out a 50 m tape and recording habitat types. These maps of the abundance and distribution of habitat types were used to locate study areas and survey quadrats.

(k) *Diver counts*
In deeper water (more than 30 m), remote sampling is generally the standard practice. In some shallow-water situations (e.g. strong currents, high turbidity), remote sampling may also be safer, quicker and/or cheaper than diving. Despite these qualifications, diving is generally unrivalled as a means of directly collecting underwater samples. There are, however, major constraints to the effective field work that can be undertaken by divers in the subtidal environment. These include viewing limitations (underwater visibility), problems of

mobility and stability (currents, bottom surge), problems of communication, time restrictions (air supply, decompression limits, water temperature) and human limitations (nitrogen narcosis, fatigue, experience, physical abilities, attention span). In addition, the presence of divers may affect the numbers and/or behaviour patterns of some species (e.g. fish).

A variety of quantitative study methods have been applied using divers to map and census subtidal habitats and assemblages. These include:

Quadrats: Grids of a pre-determined size are placed over the bottom and estimates of one or more variables are made. These commonly include the percentage cover of the various substratum types or taxonomic groups/species within the quadrat, the numbers of species (or other taxonomic/morphological group) per quadrat, or the numbers and/or sizes of individuals of each or of selected species per quadrat. Careful attention needs to be given to quadrat size in relation to the scale of patchiness within a community (see Chapters 2 and 3).

Transects (e.g. broad-belt transects, line or chain intersect transects): Weighted lines or tapes are placed over the substratum, in a direction chosen from the surface or by means of depth profiles or underwater compass bearings. The linear 'area' and identity of different substratum types (e.g. sand, seagrass, etc.) occurring beneath the line are recorded, or the numbers and sizes of specific organisms are recorded within a fixed distance (e.g. 50 cm) adjacent to the transect or at specific locations along the lines. A hand-held cross-bar or T-bar of the required transect width can be moved along the tape by the diver to provide an accurate measure of the width of the transect. Transects may be used to quantify distribution, abundance and diversity. Transect sampling is very efficient in terms of the information gained per unit time spent underwater. There are no problems with varied bottom topography because the tape can easily be laid across different topographic features.

It is also possible to use a combination of both transects and quadrats; for example, when undertaking quadrat counts in sections along a transect.

A number of studies have evaluated line transect and quadrat methods (e.g. Bouchon 1981; Weinberg 1981; Dodge et al. 1982; Ohlhorst et al. 1988). There appears to be no uniform consensus as to which method is most appropriate for the majority of community studies. The transect method is generally quicker and simpler to use in the field. Furthermore, a much smaller area of the bottom can be covered by quadrats compared to transect techniques, in terms of the results obtained for equal time.

Permanently marked (fixed) quadrats or transects can be used to monitor the occurrence, growth, mortality and recruitment of organisms through time (Mundy 1990; English et al. 1994; Nelson 1996). Photography (see above) is an appropriate method for recording any changes.

Timed counts: The numbers of target organisms are recorded per unit time rather than area. The difficulty of the method, however, is that estimates of density cannot be readily derived from such data. Nevertheless, the general advantages of rapid visual techniques make them ideal for locating suitable habitat or high densities of patchily distributed animals for detailed studies.

These methods are useful for rare or fast-moving species, such as crabs or octopus. For these species, an area-based measurement of density is unsuitable because of the large areas that need to be searched. Sampling based on timed counts may be stratified by type of habitat.

The major advantage of all these techniques is that, without taking samples to be processed in the laboratory, counts made by divers can provide a substantial body of quantitative information on the distribution of species that have conspicuous features at the sediment surface (e.g. horse mussels, worm tubes, scallops).

Advantages
✔ Positive habitat or taxonomic group/species identification and size measurements can be made during the survey, thereby reducing post-survey data extraction.
✔ Provides detailed information on size, abundance, distribution and population structure of species.
✔ Allows sampling of species from micro-habitats that are otherwise difficult to sample.
✔ Relatively cheap because no specialised expensive equipment is necessary.
✔ Sampling is non-destructive.

Disadvantages
✗ Only small areas (metres to tens of metres) can be covered in each dive.
✗ Restricted to diveable depths and good visibility.
✗ Bias may occur with the ability of individuals to locate target species and in abundance estimation of mobile species.
✗ Lots of transects or quadrats need to be undertaken to produce detailed habitat maps.
✗ Although generally regarded as being of limited use for infaunal assemblages, these methods may be appropriate to document above surface structures (e.g. seagrass, horse mussels, etc.).
✗ Requires personnel with taxonomic skills in the field.

Example: In a study to document changes in seagrass community structure in southern Western Australia, Kirkman (1985) set up two 300 m long transects. Each transect was permanently marked every 10 m by a wooden stake. Each time the transect was traversed, one diver laid out 20 m of tape between three stakes, and the other swam along the tape describing the seagrass 50 cm each side of the tape. After each interval was described, the tape was moved on to the next and the procedure repeated along the length of the transect.

(l) *Spot checks*
Spot checks can be made to assess rapidly the habitat immediately below the boat by looking over the side with a face mask or using a glass-bottomed box in clear shallow water, by a brief dive, or with remote-sampling equipment such as video cameras. The technique may be used alone to map large areas, albeit crudely if station density is low, or used to ground-truth other techniques such as aerial photography and side-scan sonar.

Advantages
✔ Each spot check is quick to carry out.
✔ No specialised or expensive equipment is required unless remote-sampling equipment is used.
✔ Can be used to ground-truth other wide-area survey techniques.
✔ Sampling is non-destructive.

Disadvantages
✗ Restricted to diveable depths unless remote-sampling techniques are used.
✗ Can be dangerous in repetitive dive situations – usually an absolute maximum of four spot dives per day per diver.
✗ High station density required to create detailed maps.

Example: Lee Long et al. (1993) mapped seagrass habitat between Cape York and Hervey Bay, Queensland (10° 40' – 25° 15' S). They employed a survey pattern that included dives at intervals of 250–400 m along transects aligned parallel to the shore and 5–20 km apart. Between transects, dives were made at least every nautical mile (at varied depths and distances from shore) to check the continuity of cover of seagrass. In shallow clear water, the distribution of benthic vegetation was recorded from the surface. Divers investigated areas of at least 5 m^2 at selected sites. They recorded type of sediment, estimated seagrass cover and collected samples of seagrass for later identification.

(m) *Remotely operated still photography*
Underwater photography can be used to provide qualitative or quantitative records of habitats and organisms.

Over the past 40 years, cameras have been used increasingly as a means of recording subtidal information (e.g. George et al. 1985). There has been a substantial improvement in the quality of pictures as a result of developments in camera and lens design, improvements in the speed and quality of films, and in the development of compact electronic flash units (e.g. Holme 1985; Shreeves 1991).

Many purpose-built underwater cameras are available, as well as a range of waterproof housings that enable everyday cameras to be used underwater (Buehr & Picton 1985; Shreeves 1991). In suitable housings, cameras can be made to operate at any depth on a ROV, suspended frame or benthic sled. Remotely operated cameras are simplest and cheapest to operate automatically with bottom-contact exposure triggering, or time- or distance-lapse automatic exposure. The system is most flexible when operated from the surface in tandem with a video camera to provide real-time images of the bottom habitats and assemblages.

Cameras can be used with or without lighting. Electronic flashes are most widely employed as a light source, and are best housed in a separate underwater case to allow lateral separation from the camera, thereby minimising the illumination of suspended particles in the field of view (Holme 1985). The underwater housing should provide space for sufficient battery power for a large number of flashes. A major problem is achieving complete and even illumination of the camera's field of view.

Digital still cameras have recently been developed that are able to operate in very low light conditions. The images are captured to computer disk at a quality approaching that of conventional cameras and with almost instantaneous image availability.

An ideal system to provide both qualitative and quantitative analysis involves the use of a pair of calibrated stereocameras capable of providing highly accurate photogrammetric measurements. Stereophotography (see Chapter 6) offers many advantages during analysis of photographs that a single photograph does not, including improved resolution of detail and an excellent perception of depth; species are easier to separate and identify; increased ability to view under canopy-forming species and to assess the three-dimensional spatial relationships of the organisms in the photographs; as well as facilitating more accurate measurement of the true dimensions and shapes of benthic organisms (e.g. Done 1981; Christie et al. 1985; Hiscock 1987).

Advantages

✔ No depth restrictions (other than those imposed by the facilities available to handle the equipment).

✔ Higher-quality images than video cameras, enabling better species identification.

✔ Detailed analysis is possible in the laboratory.

✔ Permanent, detailed records are produced and can now be stored on compact disk.

✔ Estimation of size is possible using the methods described in (h).

✔ Samples are undisturbed.

Disadvantages

✘ Relatively small sample areas (centimetres to less than a metre) are covered by individual photographs, so more samples need to be taken compared to other photographic techniques.

✘ Dependent on good underwater visibility.

✘ Deep-water cameras can be expensive.

✘ There is no choice of subject matter if the camera is not controlled from the surface.

✘ Potential bias in estimating abundance of mobile species.

✘ Post-survey data extraction is labour-intensive.

✘ With towed sledges there is the danger of damage or loss of the gear through collisions with objects on the sea floor.

Example: Goshima and Fujiwara (1994) used an underwater camera to provide information on the distribution and abundance of cultured scallop *Patinopecten yessoensis* in extensive sublittoral beds. A standard 35 mm film camera with a motor wind, mounted in an underwater case and with an electronic flash connected, was mounted on the top of a stainless-steel frame to photograph 1 m^2 of seabed. The camera was fired by a rigid foot switch activated on contact with the seabed. A TV monitor connected to an underwater camera attached to the frame allowed operation of the still camera to be monitored. Persons with experience in observing scallops under natural conditions counted the numbers of individuals in the photographs.

(n) *Diver-operated still photography*
The use of still photography maximises data acquisition with limited diver time, thereby enabling time spent underwater to be used most effectively.

Advantages

✔ Cameras for shallow-water work are widely available and easy to use.

✔ Divers are able to select the subject.

✔ Permanent, detailed records are produced.

✔ Images are of better quality and resolution than video (including macro-images) and aid taxonomic identification.

✔ Estimation of size is possible using the methods described in (h).

✔ Detailed analysis is possible in the laboratory.
✔ Relatively rapid, therefore useful where diving operations are restricted by time limits.
✔ Sampling is non-destructive.

Disadvantages

✗ Small sample areas (centimetres to a metre) are covered by individual photographs.
✗ Restricted to diveable depths and visibility of at least 1 m.
✗ Post-survey data extraction is labour-intensive.
✗ The presence of divers may lead to bias in estimates of abundance of mobile species.

(o) *Benthic grabs and remotely operated corers*

The grab is an efficient tool for quantitative sampling of surface sediments and the associated infauna, and for slow-moving or sedentary epibenthic species. Grabs are lowered on a rope or wire from a stationary boat to obtain a sample of sediment of a given surface area (usually 0.1–0.2 m^2) and down to a depth in the sediment determined by the particular instrument in use (English et al. 1994). Most grabs penetrate to about 10–15 cm depending on the weight of the grab and the coarseness of the sediment.

A great variety of grabs have been designed and used, each with their own inherent advantages and disadvantages (e.g. Petersen grab, Campbell grab, Day grab, van Veen grab and Smith-McIntyre grab). The latter is considered the most suitable for use from a small boat in the open sea (English et al. 1994). The grab is mounted in a pyramid-shaped stabilising frame. Trigger plates on either side of the frame ensure that the grab is resting flat on the bottom before it is activated. Springs assist penetration of the buckets into the sediment.

The Campbell grab is similar to the Petersen grab but is larger (0.55 m^2) and heavier (410 kg), and is generally regarded as unsuitable for use from small boats.

A variety of box-samplers and corers have been developed to provide a means of obtaining deep and relatively undisturbed samples. The Reineck box-corer weighs 750 kg and samples an area of 0.96 m^2 to a depth of 45 cm. Because of their weight and large size, box-corers are difficult to work, requiring a large vessel and calm conditions for safe deployment.

Craib corers are considered to give the best results for remote sampling of meiofauna in subtidal sediments (McIntyre & Warwick 1984). They consist essentially of a frame-mounted core tube (5–7 cm diameter), which is forced into the bottom by weights. The core tube is pushed very gently into the sediment, ensuring minimal disturbance of the surface sediments where the

Figure 8.1 A full Hunter grab (above) and a Smith-McIntyre grab being deployed (below).

ABOVE: **Figure 8.2** A small dredge equipped with teeth to lift animals from the sediment.

BELOW: **Figure 8.3** A small toothed dredge and two small grabs.

meiofauna are usually concentrated. Craib corers can be used from small boats but may not work efficiently in bad weather or on uneven seabeds.

Advantages

✔ Collects a quantitative sample of the surface sediments and slow-moving and sedentary organisms.

✔ Produces permanent samples that can be reworked for smaller organisms at a later stage.

✔ The smaller grabs can be easily handled from small boats.

✔ Not limited by depth of water (other than practical limitations such as those imposed by the facilities available on the boat to handle the grab).

Disadvantages

✘ Grabs with different bite profiles sample different sediment depths unequally.

✘ The small sample-footprint (0.1–0.2 m²) requires many replicates to be taken.

✘ The pressure wave produced by the descent of the grab may affect quantitative sampling of the surface sediments and organisms living at the sediment/water interface.

✘ Grab penetration and sample volume depend on substratum type. This may be overcome to a degree by adjusting the amount of weight added to the grab. Grabs that are able to penetrate to greater depths into the sediment often require greater force to retrieve them.

✘ Current speed can influence the angle at which the grab hits the seabed and thereby affect efficiency: the stronger the currents, the heavier the grab must be to work effectively.

✘ Larger, widely distributed or more mobile organisms are not adequately sampled.

✘ Premature triggering of grab may occur during descent, caused by momentary slackening of the cable; for example, through rolling of the ship.

✘ The grab may not completely close if stones or shells become wedged in the jaws, causing loss of sample.

✘ Rough sea conditions or inexperienced operators may influence the size of the samples taken, as the grab may be lifted off the bottom before the jaws have closed on a full sample.

✘ Labour-intensive post-survey sorting and identification.

✘ Larger, heavier grabs (e.g. Campbell grab) require special handling gear and are generally unsuitable for use from small boats.

✘ Sampling is destructive.

Examples: Probert and Anderson (1986) estimated biomass, faunal density and composition of the soft-bottom macrofauna on the continental shelf and upper slope (23–938 m depth) off the west coast of the South Island, New Zealand, from four replicate benthic samples collected at several sites using a box-corer (0.066 m²). Luckens (1991) used an orange-peel grab to collect samples to document the distribution, growth rate and incidence of octopod and gastropod predation on the bivalve *Tawera bollonsi* at the Auckland Islands. Long et al. (1994) used a modified orange-peel grab and corer to collect data on seagrass biomass in a study area of 37.649 km² in Moreton Bay, Queensland. They identified seven discrete seagrass strata from aerial colour photographs of the study area. Sites were allocated to the strata in proportion to their areas and to estimated variances in the above-ground biomass of seagrass.

(p) *Diver-operated hand corers*
Cylindrical corers can be made of various materials (e.g. perspex, PVC, stainless steel) and of varying sizes (generally 10–15 cm diameter and 10–20 cm deep depending on the nature of the samples required for macrofauna, and 1–2.5 cm diameter, 10–30 cm deep for meiofauna). These can be used for rapid collection of relatively small samples of sediment and associated infauna. The corer is pushed vertically into the sediment with the top open. When the required depth of sediment is reached, the top is closed (e.g. with a bung or flap-valved lid), the corer is pulled out and the open end sealed to prevent sediment escaping from the tube. The bottom end of the corer is usually sharpened to facilitate penetration into the sediment. Larger corers may need to be hammered into the sediment.

Advantages

✔ Produces quantitative samples, and equal-sized samples can be collected at each site.

✔ Yields detailed, fine-scale data on occurrence, abundances and size distributions of species.

✔ Inexpensive. Corers are relatively cheap to make.

✔ Easy and rapid to deploy.

✔ Upper sediment layers are much less disturbed than if grabs are used.

✔ Corers generally penetrate deeper than grabs.

✔ Sampling sites can be positioned accurately.

Disadvantages

✘ Sample small areas of sediment, therefore many replicates are usually required.

✘ Labour-intensive sample processing.

✘ Do not work effectively on coarse sediments: cores are most useful in mud or muddy sand.

✘ Limited to diveable depths.

Example: Morrisey et al. (1992a) used diver-operated plastic core tubes (10 cm diameter, 10 cm deep) to

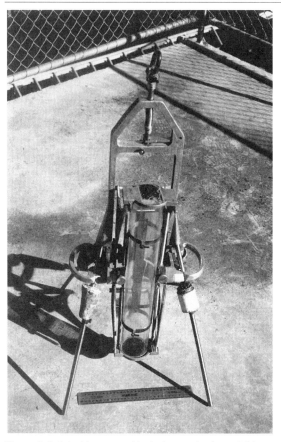

Figure 8.4 A Jenkins corer, with closing mechanisms visible at the ends of the core tube.

Figure 8.5 A suction sampler with a fine mesh cod end and a mesh tray.

examine spatial variation in the soft-sediment benthos in Botany Bay, New South Wales.

(q) *Suction or airlift samplers*

Airlift suction-samplers employ suction to draw sediment and the associated fauna up into a tube leading to a self-sieving mesh bag (e.g. Gamble 1984). The sampling area may be defined by pushing a cylinder or quadrat into the sediment and excavating the contents. Diver-operated suction-samplers provide the most accurate samples of the soft-bottom assemblages, since contiguous samples of uniform size and depth may be consistently obtained (Hartley & Dicks 1987). If it is required to sample a larger surface area than a core would provide, to sample deeper into the sediment than a grab would penetrate, or to sample at accurately placed locations, a diver-operated suction-sampler may be the most appropriate sampling technique (Hiscock 1987).

Suction-samplers range in size from small models powered by compressed-air cylinders, with modest 'lift' ability, to large surface-supplied models capable of excavating gravel and small boulders (e.g. Gamble 1984).

Advantages

✔ An accurate method of quantitative sampling to whatever depth in the sediment is required.
✔ Yields detailed fine-scale data on species occurrences, abundances and sizes.
✔ Sample sites can be accurately positioned.
✔ Equal-sized samples can be collected at each site.
✔ Small or thin patches of sediment can be sampled.
✔ For larger samples, suction-sampling may be quicker than coring or grab-sampling.

Disadvantages

✗ The sample footprint is relatively small, with many replicates usually required.
✗ Although remotely controlled suction-samplers could theoretically be used in deeper water, this technique is generally limited to diveable depth.
✗ Cumbersome equipment.
✗ If the suction-sampler is being operated from compressed-air cylinders, the number of samples that can be collected from one cylinder is often small.
✗ Creates a cloud of suspended sediment, thereby limiting underwater visibility.

✗ Post-survey sorting and counting is labour-intensive and thus expensive.

✗ Tendency to damage soft-bodied organisms.

✗ Few samples can be collected in a given time.

Examples: In a study of the benthic infauna associated with an estuarine outfall in Tauranga Harbour, New Zealand, divers collected three randomly positioned suction-dredge samples of benthic macroinvertebrates from each site along two transects (Roper 1990). Each sample was taken by sucking out the sediment inside a 0.1 m² quadrat hammered approximately 30 cm into the seabed. The collecting bags attached to the suction dredge had a mesh size of 1 mm.

Kennelly and Underwood (1985) trialled a simple suction device as a means of quantitatively sampling

Figure 8.6 Diver-operated hand corers (above) and sieves of a variety of mesh sizes (below) for separating infaunal animals from sediment. Circular sieves have the advantage that animals cannot become trapped in corners.

small invertebrates from natural substrata, including silt-covered hard substrata. The sampling device consisted of a water pump connected to an air-driven drill running off a scuba tank. The pump pulled water through two 6 mm diameter copper pipes leading through a sealed sample container. The inhalant attachment was a plastic tube with a 23 x 6 mm aperture, run back and forth over the sampling area while the suction pump was running. A vacuum-cleaner end may be used to control the size of the sample unit.

8.2.2 Size and number of sampling units

The selection of the size and number of sampling units should be decided by the methods described in Chapters 2 and 3, with choices based on the accuracy and precision needed to address the objective of the study. Again, if large enough samples cannot be obtained because of practical restrictions, it may be necessary to use different methods. For example, if the number of cores divers can collect in the available dive time is not sufficient to provide adequately precise estimates of the abundances of animals, it may be preferable to collect grab samples, which, although they cannot be placed so carefully and accurately, can be collected in larger numbers.

8.2.3 Sampling design

General considerations for sampling are discussed in Chapters 2 and 3. Designs that have commonly been used in sampling of subtidal soft bottoms include random sampling, stratified random sampling, distribution of samples on a grid pattern and sampling along transects. Examples are also discussed by Andrew and Mapstone (1987) and Hartley and Dicks (1987).

The choice of sampling method and allocation of sampling effort depends on the purpose of the study. One of the major points in deciding on a design is to be sure that samples really represent what you assume they do. For example, if you want to compare the seabed around a sewage outfall with another part of the coast where there is no outfall, does your set of samples from the outfall represent the area in general or just a small patch of seabed near the outfall? If the latter, then any effects of the outfall may be confounded with differences among patches, as described in Chapter 2. Or, if you were sampling on a grid pattern with the aim of mapping the types of sediment on the bed, one sample of area 0.1 m^2 is unlikely to represent adequately, say, a 1 km^2 sector of the grid. The charts of distribution of different types of subtidal sediments produced by the New Zealand Oceanographic Institute (now NIWA) are a good illustration of this (Godfriaux, 1973). The

samples on which the maps were based were very widely spaced, representing areas of 20–100 km^2 (Godfriaux 1973). While this may be acceptable in cases where only a broad indication of the nature of the seabed is required, it may be completely inadequate for studies at finer spatial scales (e.g. mapping the distribution of habitats of a species with very specific requirements). If information is required at a finer scale, as in many inshore studies (see section 8.3), the maps should not be relied on too closely.

As in all ecological studies, pilot studies provide an important source of information on scales and sizes of variation likely to be encountered in sampling of subtidal sediments and, consequently, on the allocation of sampling effort to different scales of replication and on practical difficulties likely to be encountered during sampling. As Green (1979) pointed out, 'Those who skip this step because they do not have enough time usually end up losing time.' On the other hand, McArdle and Pawley (1994) advise a conservative approach in using estimates of variance obtained from pilot studies. Such estimates can be misleading if, for example, variance changes through time and there is a lag between the pilot sampling and the main sampling. Nevertheless, pilot studies provide useful information, which is always better than a complete guess, but the use of that information should be conservative. It should also be made with due consideration of other sources of information, such as methods used in previous similar studies, providing that these were based on sound principles. This is definitely *not* the same as copying a design of sampling from a previous study simply because 'that's the way it's always been done'. This attitude seems to be a common failing in studies of subtidal habitats and should be avoided. It is important to think carefully about the objectives of your own study and to design the sampling specifically to address those objectives.

A design that has been used several times for the study of impacts of oil rigs and other localised sources of disturbance in offshore areas is described by Olsgard and Gray (1995). This consists of a series of transects radiating out from the source in a cross or star pattern. Sampling stations are usually closer together nearer the centre of the pattern. The radiating transects are designed to take into account predominant directions of currents and to achieve greater concentration of sampling nearer the source of disturbance. The more distant points of the several transects serve as controls.

Random distribution of samples within an area can be achieved by dividing the area into a grid, identifying points on the grid by Cartesian co-ordinates and using tables of random numbers to select a subset of points to be sampled. Random sampling is appropriate in areas where the habitat is fairly homogeneous and a representative sample of that area is required, and where there is no prior information on the distribution of organisms. An example would be an area known from sediment charts to be homogeneous and which, therefore, might be expected to have a reasonably homogeneous benthic fauna. Random sampling would be an appropriate method for assessing the types and abundances of animals. In areas where the habitat is likely or known to be variable, random distribution of samples will often lead, by chance, to the undersampling of some types of habitat. This is particularly likely if the sample size is small. Furthermore, random sampling of heterogeneous areas is not very efficient because the variation in the numbers of animals (or whatever is being measured) among samples is likely to be large. As a consequence of this, statistical tests of differences among different areas will not be very powerful.

When the heterogeneity of the area to be sampled is known (e.g. from a pilot study or from previous work in the same or similar areas), stratified random sampling can be used. The area is divided up into sub-areas, or strata, on the basis of variation in the type of sediment, assemblages of animals (e.g. mussel beds or patches containing tubes of polychaetes), depth of water, etc. Sampling effort is then divided among these strata. Numbers of samples to be collected in each stratum may be apportioned on the basis of the relative areas of each stratum or on the relative variability of the measured variable in each stratum (as determined by pilot-sampling). Stratified random sampling thus ensures that each stratum is sampled with adequate intensity. Individual samples are located randomly within each stratum. Differences in the measured variables can be compared among strata within an area and, perhaps more usefully, when stratified random sampling is done in two or more places, differences among places can be compared for each stratum. In this way, variation among strata is separated from variation within each stratum and the test for differences among places becomes more statistically powerful (see Chapter 3).

In subtidal habitats, it is often impossible to assess the homogeneity of an area visually because of limited visibility and the time required to survey the area. Even when such an assessment is possible, variations (e.g. in numbers of animals) may be surprisingly large within areas that appear to be homogeneous to a human observer. For example, Morrisey et al. (1992a) found that spatial variation in the abundance of several kinds of animals occurred at scales ranging from metres to kilometres in a study of apparently very similar sandy sediments in Botany Bay, New South Wales It is, therefore, generally a good idea to use a nested or hierarchical design when sampling such areas (see Chapter 3). Within the area to be sampled, two or more sites are

chosen, and at each of these replicate sub-sites are chosen where replicate samples are collected. This spatial hierarchy of sites and sub-sites, with replicates at each level, enables the total spatial variation in the measured variable to be divided up among different spatial scales. The number of levels of subdivision depends on the size of the largest scale and on the size of sample units.

For example, if the overall aim of the study is to compare two places a kilometre apart, one might select replicate sites 100 m apart within each place. Within each of these sites, a number of replicate sub-sites could be selected, 10 m apart, and several replicate 1 m^2 quadrats taken at each sub-site. The data would then be analysed by, for example, a nested analysis of variance (see Chapter 3), which would apportion the total variation among sample units to that due to differences among places a kilometre apart, sites 100 m apart, sub-sites 10 m apart within each site, and among sample units within each sub-site. This makes the test for differences between the original two places of interest (i.e. those a kilometre apart) statistically more powerful. It also provides a more accurate biological picture, because differences between places (the difference of interest) are not obscured by differences within each place.

Tests of differences at larger scales are made more powerful by having more replicates at these scales (and therefore more degrees of freedom in the analyses) or by pooling lower terms in the analysis (see Chapter 3). Increasing replication at larger scales can dramatically increase the required sampling effort. Nested designs also provide useful information about patterns of variation within areas. They can be used in pilot studies to identify important scales of variation, and this information can then be used to design the main study (surveys or experiments), leaving out those scales of replication that prove to be unimportant.

Although nested designs provide some information about patterns of distribution of variables, they are not very efficient if the object of the study is to obtain a detailed description of patterns in a particular area. This is because the number of replicate points at each spatial scale in a nested design is limited by practical considerations (time, cost of collecting and sorting). While nested designs are appropriate for describing distributions over larger areas and for estimating scales of variation (e.g. for cost-benefit analyses), other methods are more useful for detailed descriptions of patterns. In such cases, grids and transects may come into their own, enabling detailed mapping of the distribution of animals and other variables.

Regular grids of sampling points have often been used to disperse sampling effort across the area of study (examples are given in Hartley & Dicks 1987). Hooker

(1995) used sampling grids as a means of collecting samples of clams (*Paphies australis*) at three subtidal sites in Whangateau Harbour, New Zealand. The use of grids ensured adequate dispersal of sample units (quadrats) throughout each site. Permanent grids were marked out by metal pegs in the sandy seabed at 5 m intervals along a series of transects at each site. Numbers of transects, and therefore the size of the grid, varied among sites according to the known size of the clam bed at each site. The sites were sampled repeatedly over a period of 17 months. The animals collected were measured to provide information on the size-frequency distributions of the populations and the rates of growth of individual clams.

One problem with grids and other methods using regular spacing of sample units is that there is a possibility that the spacing of the units will coincide with some environmental regularity and that variation in the measured variable will not, therefore, be estimated adequately. A good example would be the coincidence of the spacing of sample units with the spacing of megaripples on sandy bottoms (this is an example of 'autocorrelation', discussed below). It is not usually difficult to introduce some element of randomness into the design of sampling; in fact, it is often unavoidable because of inaccuracies in locating sampling points precisely. As mentioned, grid designs are often used to provide general descriptions, but one needs to be confident that samples collected at each point on a grid (or any other design) are representative of that place. This requires an estimate of variation, which can only be provided by replication of sample units. One or a few samples from the centre of a 1 km² square on a grid does not mean that the type of sediment or numbers of animals per unit area is the same across the whole square. It may be that variation within squares is at least as large as that among squares. Samples may be made more representative by combining grid designs with other arrangements, such as a nested design, to study patterns of distribution within the grid square in more detail.

One group of techniques for identifying patterns is analysis of spatial autocorrelation. These techniques compare the relative strength of correlations between pairs of samples at different distances along a transect or a grid (three-dimensional autocorrelation). This is done by collecting samples at, say, 1 m intervals and calculating a correlation coefficient for the variable of interest (e.g. numbers of snails) for all pairs of points 1 m apart along the transect (points 1 and 2, points 2 and 3, points 3 and 4, etc.). Next, one calculates the correlation coefficient for all pairs of points 2 m apart (points 1 and 3, points 2 and 4, etc.). This is done for all possible inter-sample distances (though as the distance

increases, the number of pairs of points decreases and the correlation becomes less reliable). The correlation coefficient (on the y-axis) is plotted against the distance between samples. The difference of each value from zero (i.e. no correlation) can be tested statistically and the graph (known as a 'correlogram') indicates the inter-sample distances at which there is autocorrelation. This indicates that samples spaced at these distances apart are likely to be more (in the case of positive autocorrelation) or less (in the case of negative autocorrelation) similar than would be expected for randomly located samples. In other words, patterns exist in the spatial distribution of the variables with the distances indicated by the correlogram. In such studies, detection of autocorrelation at different distances among samples becomes the object of the exercise. Descriptions of these and other methods of pattern analysis are given by Jumars et al. (1977), Sokal and Oden (1978), Legendre and Fortin (1989), Legendre (1993) and Legendre and Legendre (1994). Autocorrelation through time can also be studied using equivalent methods.

Thrush et al. (1989) used analysis of autocorrelation to identify patterns in the distributions of invertebrates in sandflats in Manukau Harbour, New Zealand. McArdle and Blackwell (1989) used a hierarchical sampling design to identify variation in the distribution of the bivalve *Austrovenus* (*Chione*) *stutchburyi* among and within sites in Ohiwa Harbour, New Zealand. Spatial autocorrelation and contour plots of numbers of individuals along transects down the shore were used to identify patterns in the distribution of the animals at one of the sites. Temporal autocorrelation was used to identify patterns of change in distribution through time. McArdle and Blackwell (1989) provide a useful and clear discussion of the use of sampling grids and autocorrelation in analyses of patterns of distribution of animals in sediments.

By comparing patterns of distribution among different variables, insights can be gained into the processes that determine their distributions. These can then be tested experimentally. In the case of large-scale patterns, however, experimental testing of derived models is often impossible for practical reasons. In such cases, correlations among patterns for different variables may be as close to demonstrating cause and effect as is feasible.

When there are obvious or suspected gradients in the distribution of the variables of interest, sampling along transects often provides an appropriate means of describing them. Again, it is important to have replicate sample-units at each point and replicate transects. Without replicate transects it is not possible to assert that differences among places are due to their different position along the gradient rather than that they are simply different places and would have differed even if sampled perpendicular to the gradient. To illustrate this, imagine that samples are collected along a supposed gradient of impact from an industrial outfall. Numbers of animals are significantly different between two points along the transect, one 10 m from the outfall and the other 20 m from it. Without replicate sampling points at each distance, however, we cannot determine whether this difference arises because they are at different points along the gradient or simply because they are any two points that happen to be 10 m apart. Again, the essence of the problem is to be sure that the samples are really representative of points 10 m and 20 m away from the outfall (or, in statistical jargon, that the populations of places 10 m and 20 m away have been adequately and randomly sampled).

Whatever methods are used to describe patterns of variation in distribution, small-scale variation can be as significant as that at large scales (e.g. Morrisey et al. 1992a) and must be taken into account when investigating the latter. Just as it is important to separate out the effects of small-scale spatial variation when identifying variation at larger scales of interest, it is similarly necessary to distinguish short-term variation in time from longer-term patterns of change. For example, if the objective of a study is to identify seasonal variation in the abundance of an animal, sampling only once in each season will not allow us to do this. Differences among times may not be seasonal but, rather, due to shorter-term variation within each season. The numbers of the animals might vary on a week-to-week basis, and if samples one season were collected during a week when abundance was large and in another season in a week when they were low, this variation from week to week would be mistaken for variation among seasons. As with spatial variation, this problem can be overcome by using a nested sampling design, as described by Morrisey et al. (1992b). By sampling on replicate days in each of two replicate weeks in each of two replicate months in each of two replicate seasons, they were able to identify variations in the abundance of several types of animals at scales from days to months. In such studies, it is necessary to have spatial as well as temporal replication so that variation among times can be distinguished from spatial variation within each time.

8.3 Sorting and analysis of samples

After collection, samples of sediment can be stored in plastic jars or bags. Labels should be written on waterproof paper and placed inside the container with the sample, but it is also worth labelling the outside of the container to enable each sample to be identified without having to sift through it to find the label.

8.3.1 Macrofauna

Infaunal animals are generally separated from the sediment in which they live by sieving. Traditionally, biologists studying animals in sediments have used sieves developed for use in geology and soil science, and this is reflected in the sizes of mesh and the shape and size of the sieve itself.

The geological sieves available commercially are usually circular, 30 cm in diameter with a mesh made of stainless steel and the sides made of brass or stainless steel. Because they are precision-made, they are quite expensive. It is also easy to obtain stainless-steel mesh of the appropriate aperture, again precision-made, and make up your own sieves. This is usually much cheaper. Circular sieves have the advantage that animals do not become trapped in the corners of the sieve. Nylon mesh of precise apertures is also available (designed for filtering) and can be used to make sieves of very small mesh size. Although not as long lasting, plastic mesh is less likely to damage the animals and may be appropriate if specimens are needed in good condition, for example, for adding to a reference collection. Circular plastic or aluminium sieves with a mesh size of 0.5 mm are also available commercially and are relatively cheap.

One legacy of the use of geological sieves is that the sizes of mesh available form a series differing from each other by a factor of two (8, 4, 2, 1, 0.5, 0.25, 0.125, 0.063 mm, etc.). Intermediate sizes are also available. The animals living in the sediments are grouped into size classes that were originally based on sieve sizes, for practical reasons. 'Microfauna' are those animals that pass through a 0.063 mm sieve, 'meiofauna' are those retained by a 0.063 mm sieve but passing through a 0.5 mm sieve, and 'macrofauna' are those retained by a 0.5 mm sieve.

The size of mesh used depends on the purpose of the study, but 0.5 mm is the most commonly used because it provides a generally acceptable compromise between the need to use as large a mesh as possible to allow most of the sediment through but as small as possible to retain smaller animals. Obviously some larger particles of sediment will usually be retained and small animals, including many juveniles, will pass through the sieve. Instead of using a different size of mesh depending on the nature of the sediment (e.g. using a larger mesh for coarse gravels), biologists have settled on the compromise of 0.5 mm (or, less often, 1 mm) mesh for general studies. If smaller organisms, such as juvenile stages of larger animals, are of particular interest, then other sizes can be used. If only larger animals, such as adult bivalves, are of interest, then a coarser sieve is suitable. Schlacher and Wooldridge

(1996a) found that body size of animals was a poor predictor of the efficiency with which they were retained on a mesh of a given size. They suggested that retention efficiency should be determined experimentally for each particular combination of habitat and assemblage.

The merits of different sizes of mesh are discussed by Eleftheriou and Holme (1984) and Hartley et al. (1987), with examples of various sizes used for specific studies. Pilot sampling, using a series of different mesh sizes, may help select the appropriate size (although this may change through time with the presence of juveniles in the population, etc.). In a study of estuarine macrofauna, Schlacher & Wooldridge (1996b) showed that 0.5 and 1.0 mm meshes gave inaccurate (underestimates) and imprecise (i.e. larger standard errors) estimates of abundance and biomass relative to 0.25 mm mesh. Of those species retained by all three sieves, the total number of individuals retained by the 1.0 mm mesh was only 8% of that retained by the 0.25 mm mesh. The equivalent value for total biomass was 49%. The authors pointed out that no single mesh size will perform optimally in every combination of habitat and assemblage. The results of this study emphasise the need for careful consideration of mesh size in relation to the objectives of the study, and the pitfalls of uncritically adopting methods from other studies.

James et al. (1995) have examined simultaneously the effects of mesh size and taxonomic resolution (in this case, identifying animals to species or family) on the detection of spatial patterns in the distribution of macrofauna in subtidal sediments at two sites off the coast of New South Wales. They concluded that these two factors had little effect when patterns were identified using non-parametric multivariate techniques. Differences in the assemblages of animals among sites were just as apparent if animals were only identified to the level of family, and when smaller individuals (which passed through the coarser sieve) were omitted. Both factors did have some effect on interpretation of patterns of distribution, however, when univariate methods of analysis were used. Samples sieved through a 0.5 mm mesh took significantly longer to sort than those sieved through a 1 mm mesh, and, not surprisingly, it also took longer to identify to species rather than family. The overall conclusion was, therefore, that the loss of information from using coarser mesh sizes and taxonomic resolution will generally be offset by using the time saved to collect larger numbers of samples. They also pointed out that if a choice is to be made between the level of taxonomic resolution and the size of mesh, it should be remembered that samples identified to coarse taxonomic levels can always be re-identified with greater resolution later, but resampling with a finer mesh cannot be done. When compromises such as these are being made in studies to detect human impacts, another point to bear in mind is that it may take so long to sort and identify samples using fine-mesh sieves and identification to the lowest possible level of taxonomic resolution that impacts may not be identified until after environmental damage has already occurred.

The actual sieving process is usually done in sea water (natural or made up from tap water with salt added) to avoid osmotic damage to the animals. Care should be taken to avoid abrading the animals on the sieve by agitating it too vigorously. For this reason, an up-and-down motion is better than side to side. Similarly, while a jet of water can help break up lumps of mud and other material, it must not be so strong that it damages the animals. Sticky muds and clays may be very difficult to sieve, and the prolonged sieving is likely to cause damage. Leaving such samples to soak in water, or freezing and then soaking in a solution of surfactant (such as Calgon), may help break up the mud (see Hartley et al. 1987, p. 132), but freezing can cause soft-bodied animals to rupture and is not advised. When animals are being extracted from coarse sediments, the sediment can be pre-sieved through a larger mesh before being put through the 0.5 mm mesh (or whatever size has been chosen for the study).

Another factor to consider is whether the samples should be preserved (with formalin, etc.) before or after sieving. Generally, samples are sieved before preservation and the material retained on the sieve is washed into a storage container before preservative is added. In some situations this may not be possible; for example when samples cannot be sieved for some time after collection and are likely to deteriorate (sediments become anoxic in closed containers, etc., and the animals begin to decompose) or when scavengers and predators in the samples are likely to eat other animals (crabs are notorious culprits). Whole samples of sediment can be placed in a container, preservative added (equivalent to about twice the volume of the sample) and the sample stirred or shaken to ensure that the preservative reaches the whole sample. In our experience, samples of mud and sand can be successfully preserved in this way as long as a sufficient volume and concentration of preservative is used to allow for water in the sample. One advantage of preserving samples before sieving is that animals such as worms cannot actively squeeze through the mesh. If sediments take a long time to sieve, this can be a particular problem and pre-preservation may be a solution. In a trial study with easily sieved fine sediments from Botany Bay, New South Wales (D. J. Morrisey & J. S. Stark, unpub. data) we did not, however, detect any differences in numbers of animals retained when samples were sieved either before and after preserving. Needless to say, whichever method is chosen, it should be used consistently throughout a study.

Suction samples, discussed earlier, generally discharge the sediment and water from the end of a hose, underwater. Mesh bags of known aperture can be tied over the end of the hose so that the sampler also acts as a sieve. The minimum size of mesh that can be used effectively, however, will depend on the nature of the sediment. If the mesh is too fine relative to the particle size, it will become clogged. The force of the water may also damage delicate animals and plants.

When the grains or pieces of detritus in a sediment are larger than the mesh of the sieve, sieving will not be much help in separating organisms and sediment. Pre-sieving through a larger mesh to remove larger particles may reduce the amount of sediment. The larger fraction of sediment should also be searched for larger animals that may also have been retained on the coarser sieve. Usually, however, it is not possible to remove all of the sediment and detritus by sieving, and the residual material must be sorted in a shallow tray of water, preferably under a low-power binocular microscope or magnifying lamp. If the sample is large and there is a lot of sediment and detritus left, sorting in this way can be inefficient as well as tedious. It can also be expensive, and in many studies of soft substrata the time required to sieve and sort samples is one of the major costs.

Staining of the samples prior to sorting helps to detect animals among the grains of sediment. The most commonly used stains are Rose Bengal and Biebrich's Scarlet, both of which stain animals (and some other organic matter such as detritus, clothes and skin) pink. The stains are added to the samples with the fixative solution and should be left for a few days to take effect. Very small amounts of stain are needed (e.g. a 0.1% solution in preservative), and overstaining may make subsequent identification of specimens difficult.

There are a number of other ways of separating animals from sediment, of which the simplest is elutriation. At its most basic ('decanting'), this consists of putting the sample into a beaker or similar tall, narrow container, filling it with water, stirring to suspend the sediment and animals and pouring the water out of the beaker through a sieve of the appropriate mesh size (usually 0.5 mm). By pausing slightly before pouring, the denser grains of sediment settle out, leaving the animals (and often some detritus) in suspension. The process is repeated until no more animals are seen to be washed out. The remaining sediment should be searched as described above to pick out any animals, particularly molluscs, that may have been left behind, but the process is much quicker than sorting the unelutriated sediment. Separation of sediment and fauna can be enhanced by using solutions of sugar, magnesium sulphate or zinc sulphate. The higher density of these solutions makes the animals float more readily. Sugar solutions are messy to work with and, in summer, are very popular with wasps. Use of other solutions for elutriating meiofauna is discussed below.

Other techniques exploit the behaviour of animals to separate them from the sediment. Some types of amphipods (e.g. corophiids) can be separated by slowly submerging the sieve containing sediment and animals in seawater. The amphipods tend to swim up into the water column and can be led off into a collecting sieve. Similarly, if the sediment is washed into one side of the sieve and the sieve left standing at an angle with the sediment lowermost, some amphipods will tend to climb up the mesh and can be collected (personal observation). Mobile animals can sometimes be extracted from coarse sediments by placing the sediment in a funnel and placing either seawater-ice on top of the sediment or shining a light on the sediment surface. Animals then crawl down through the sediment to escape the extreme conditions near the surface and fall down the neck of the funnel. Obviously these methods are only suitable for small samples.

Methods for separating organisms from sediments are discussed by Eleftheriou and Holme (1984, pp. 190–199) and Hartley et al. (1987). The effort required can also be reduced by subsampling the original sample. The sample must first be thoroughly homogenised, and this can be surprisingly difficult. The success of homogenisation should be checked by counting the animals in two or more subsamples, introducing another level of replication into the hierarchy of sampling.

8.3.2 Meiofauna

The problem of separating animals from grains of sediments that are of similar size is particularly great for meiofauna. The major taxa found in the meiofauna are turbellarians, nematodes, oligochaetes, harpacticoid copepods and ostracods. Many live in the spaces among the grains. Methods and reasons for sampling meiofauna are described in detail by McIntyre and Warwick (1984) and Bouwman (1987) and, because this group is fairly specialised, are only considered briefly here. Because of their small size, meiofauna are usually sampled with smaller cores than are used for macrofauna, but the same considerations for selecting the size and number of sampling units apply (see Chapter 3). Plastic syringes (1 or 2.5 cm diameter) with the end cut off are often used as convenient core-samplers. The plunger of the syringe provides a seal while the syringe is being withdrawn from the sediment, holding the sample in place inside, and can be used to extrude the sample.

As with macrofauna, meiofauna can be separated

from the sediment either before or after fixing, but soft-bodied forms such as turbellarians may be difficult or impossible to identify when dead. Fine sediments can be sieved through 0.063 mm mesh and the animals collected off the sieve. For coarse sediments, decanting with sea water can be used to separate live animals. Alternatively, the sediment can be place in a 0.063 mm sieve standing in a bowl of sea water and seawater ice placed on top to encourage the animals to migrate down through the mesh and into the water.

Preserved material can be sorted, after staining, by decanting or elutriation. Elutriation of meiofauna is often done with a colloidal silica polymer with the tradename Ludox (see McIntyre & Warwick 1984, p. 233; and Bouwman 1987, p. 150, for details). Separation of animals and sediment-grains within the Ludox can be enhanced by centrifuging the suspension for seven minutes at 6000 rpm. Ludox is toxic to humans.

8.3.3 Microfauna and microflora

The sampling and sorting of these groups are specialised processes and beyond the scope of this chapter. Methods are given by Round and Hickman (1984), Floodgate and Gareth Jones (1987), and Wolff (1987, pp. 88–94).

8.3.4 Sediment grain size

The description given below is for a general assessment of the nature of a sediment, such as might be used to provide complementary information to a biological study. More detailed and varied techniques are used for detailed, specific sedimentological studies, and these are described by Folk (1974) and Buller and McManus (1979).

Pretreatment
The manner of storage of samples depends on the variables that are to be measured. For analysis of grain size, samples can be stored wet at room temperature. If other variables such as the amount of organic matter or nutrients in the sediment are to be measured, the sample must be refrigerated or frozen to prevent microbial degradation. Freezing is not suitable if the samples are to be analysed for some chemical variables (see below). Drying is another way of preventing microbial degradation, but it is not recommended for samples that will be used for analysis of grain size.

Many marine sediments contain at least some organic debris and pieces of mollusc and other shells. These materials may have been derived from within or on the surface of the sediment itself (local populations of molluscs, local seagrass beds, etc.) or may have been brought to the area by water currents and incorporated into the sediment after the animals or plants had died. If a general description of the sediment is required, these materials can simply be considered as part of the structure of the sediment. Descriptions of the sediment for other purposes may require that this material be quantified and removed before analysing grain-size distribution. The amount of shell material present can be measured by drying the sample, weighing it, covering it with dilute hydrochloric acid until no further effervescence occurs, redrying and reweighing. The difference between the weights represents the shell fraction. This method will not differentiate between recent and fossil (including chalk, limestone, etc.) or subfossil material. Organic matter can be removed by pre-treatment with hydrogen peroxide to avoid including this material in the analysis of grain size. Methods of measuring the amount of organic material are described below.

Analysis of grain size
Traditionally, the distribution of grain sizes of the coarse fraction of sediments (larger than 0.063 mm) has been determined by sieving through a series of meshes of decreasing sizes and expressing the weight of material retained on each sieve as a percentage of the total weight. Obviously this method divides what is really a continuous distribution of sizes into discrete size-classes. The width of each size-class (i.e. the difference between successive sizes of mesh) depends on the purpose of the study. More generally, the methods of analysis and interpretation of the data should be decided on in advance on the basis of the purpose of the study. Some of the methods described below may, for example, produce information that is not appropriate for some of the questions asked. This is particularly likely when geological methods are adapted to answer biological questions about sediments.

If only a very general description of the sediment is required, it might be sieved first through a 2 mm mesh and then through a 0.063 mm mesh (the sieves are designed to be stacked on top of one another) and the material retained on each expressed as percentage of gravel (larger than 2 mm) and percentage of sand (smaller than 2 mm but larger than 0.063 mm). Material passing the 0.063 mm mesh is also retained and weighed, and constitutes the mud fraction. Sediment is usually sieved after drying to constant weight (at 60° C). The material larger than 0.063 mm in diameter can be divided into smaller size classes by using a range of mesh sizes, with successive meshes differing by a factor of two or by an intermediate factor (known as the 'Wentworth' series). It is important not to place too much material on the stack of sieves, otherwise the efficiency of sieving is reduced. For coarse sands, 100–250 g is sufficient; for fine sands 10–100 g is enough.

A continuous distribution of sizes for particles in the range 0.063–2 mm can also be obtained using a settling-column. This consists of a tall (about 2 m) glass column filled with water. A small pan at the bottom of the column is connected to a strain gauge. The sample of sediment is introduced at the top of the column and, as the grains sink, they collect on the pan at the bottom. The strain gauge records the accumulation of sediment on the pan through time at, for example, 0.5 second intervals and sends the information to a computer, where a program converts the rate of accumulation of sediment on the pan into a particle-size distribution for the sediment, based on the settling velocities of particles of different sizes. Because the settling velocities are calculated on the assumption that the grains are spherical, material such as shell fragments and other particles that are markedly non-spherical will distort the results. Grains larger than about 2 mm will sink so quickly that the method is not appropriate, and those smaller than 0.063 mm will take too long to sink down the column. These fractions of the sediment are, therefore, determined beforehand by sieving. The fine fraction can be analysed separately, as described below, and the results for the different fractions later combined to give an overall particle-size distribution. The use of settling columns is described by Buller and McManus (1979). Their main advantage is that the method is much quicker than the use of sieves, but it has been criticised by Smith (1992) on various grounds associated with the validity of assumptions underlying the relationship between particle size and settling velocity. Most of these considerations apply equally to the commonly used pipette analysis of fine sediments, described below.

If the mud fraction represents more than a few percent of the total weight, a more detailed description of the size distribution of particles smaller than 0.063 mm will probably be required. This fine material will form a hard cake if dried, making it difficult to sieve and, more importantly, altering the association among the finer particles (if the material has been dried, however, it can be redispersed by soaking in a 2 g/litre solution of Calgon – sodium hexametaphosphate). Prior to drying and sieving, therefore, the whole sediment is first wet-sieved in fresh water through a 0.063 mm mesh. The material retained on the mesh is dried and sieved as described above.

There are a number of ways of calculating the proportion of the total sediment represented by the material smaller than 0.063 mm. In the first, duplicate subsamples of the original sample of sediment may be taken. One subsample is then treated as described earlier in this paragraph and the fine fraction dried and weighed after being allowed to settle and the water taken off. The other subsample is wet-sieved through 0.063 mm and the material passing through the mesh is subjected to pipette analysis (described below). This method requires that the original sample be thoroughly homogenised so that the particle-size distributions of the subsamples are the same. In a second method, the washing water is shaken to thoroughly suspend the fine material and a known proportion by volume is removed, dried and weighed. The amount of fine material in the original sample can then be estimated by dividing the weight by the proportion of the washing water that was dried.

Description of the particle-size distribution of the fine material does not rely on sieving because, in wet-sieving, surface tension and electrostatic forces influence the passage of particles through the mesh. Instead, a method is used based on the relative time taken by particles of different diameters to fall a known distance through a column of water. The washing water from the 0.063 mm mesh is allowed to stand overnight or until the fine particles have settled out (this may be achieved more quickly by centrifuging). The overlying water is then carefully decanted or siphoned off, removing the salt present in the original sample, which would otherwise cause overestimation of the weight of sediment. The remaining slurry is washed into a measuring cylinder, fitted with a stopper, and the level of water made up to a known volume. The cylinder is inverted several times, suspending the particles, and a timer is started after the last inversion. Subsamples of known volume are drawn off by pipette at intervals of time corresponding to the settling velocities of particles following the Wentworth series (see above). The depth at which the subsample is drawn is clearly crucial, and the mouth of the pipette must be held at this depth while the subsample is taken. For example, a subsample taken after 58 seconds at a depth of 20 cm will contain only particles smaller than 0.063 mm because larger particles will have fallen further than 20 cm during this period. A subsample taken 8 hours 10 minutes after the last inversion at a depth of 10 cm will contain only particles smaller than 1.95 μm. Lists of the appropriate depths and times are given in references such as Buller and McManus (1979). Each subsample is then dried and weighed. The amount of sediment in each size fraction can then be estimated by successive subtractions of the weight of each subsample from that of the previous one and each expressed as a proportion of the weight of material less than 0.063 mm (i.e. the weight of the first subsample).

The pipette method is tedious and time-consuming. Since the finer grains take many hours to settle the required distance, it is important to start the sampling at an appropriate time of day. Other, more sophisti-cated and less labour-intensive methods of analysis for fine sediments are available, such as Coulter counters and laser particle-size analysers. These also provide a continuous size distribution, rather than the series of size classes provided by sieving and pipette-analysis, but obviously require specialised equipment. Pipette analysis is still a commonly used method because it requires simple equipment and is cheap to set up. When large numbers of samples are to be analysed, however, it may be cheaper to have them analysed commercially by an automated method rather than employ someone to do them by pipette analyses.

An important point to note is that all of the various techniques available for estimating distributions of particle sizes make assumptions about the shape, com-position and other characteristics of the grains in order to produce descriptions of 'typical' grains in each size class. An assumption common to several methods, for example, is that grains are spherical, which is obviously unlikely to be true. The particular assumptions vary among methods, so that results obtained by different methods may not be strictly comparable.

The results from particle-size analyses are usually summarised in the form of a number of descriptive statistics that characterise the sample. These include the median size of grain in the sample and various percen-tiles of the distribution (e.g. the size range that includes 90% of the grains, etc.), the skew of the distribution about the median and the 'peakedness' (kurtosis) of the distribution. Methods of analysing size-distribution data for sediments, for presenting the data, and for calculat-ing the various descriptive statistics, are given by Buller and McManus (1979) and Folk (1974). An alternative to trying to calculate a statistical descriptor of a hypothetical 'typical' grain in a sediment would be to use multivariate statistical techniques to characterise sediments based on a range of variables, including different size-classes of particles.

8.3.5 Other physical characteristics of sediments

The 'cohesiveness' of a sediment refers to its resist-ance to erosion by water flowing over it. This is an important indicator of rates of transport and deposition of sediments by currents, and of their stability for the animals living within them. Generally, larger particles requires stronger currents to erode and move them, and, conversely, as the speed of movement of water progres-sively slows, first large and then smaller particles will settle out. Fine particles may, however, pack together very closely, presenting a smooth, relatively erosion-resistant surface. Animals and plants can have a con-siderable influence on the erodibility of sediments by,

for example, incorporating sediment into faecal pellets or by secreting mucus that binds the sediment and in-creases its resistance to erosion (Meadows & Meadows 1991). The roots and emergent parts of plants such as seagrasses and mangroves and the tubes of worms and other animals also help to promote the deposition of sediment and to inhibit its re-suspension by reducing current speeds.

A simple and traditional method of characterising the erodibility of a sediment is by the use of a 'shear-vane meter'. This consists of a rod with several parallel vanes oriented along the axis of the rod at one end, and a torque meter at the other end. The rod is pushed into the sediment up to the top of the vanes, and the top of the instrument, with the torque meter, is gently turned until the sediment shears and the vanes and rod rotate within the sediment. The meter measures the force required to achieve this shearing. Because the resist-ance to movement of the sediment is measured over a depth of several centimetres, the shear-vane meter provides a relatively crude measure of erodibility. The upper layers of sediment, where most animals and plants live, are likely to differ from deeper layers because of the relative intensity of the various processes mentioned earlier. A more sensitive method of measuring the co-hesiveness of surface sediment has been developed by Paterson (1989). His 'cohesive strength meter' shoots a jet of water at the surface of the submerged sediment, and the velocity of the jet is increased until the parti-cles of the sediment begin to move.

The compaction, or penetrability, of a sediment can be measured by a 'penetrometer', which measures the force required to press a disc of standard surface area into the surface of the sediment. Like the shear-vane meter, this is a precision-made and calibrated instru-ment available from specialist suppliers of scientific equipment, and tends to be quite expensive.

An indication of the cohesiveness and compaction of a sediment can be obtained more cheaply, but prob-ably less accurately and with less replicability, by measuring the water content. Comparison of the change in weight on drying among sediments collected in the same way will give an indication of their relative degrees of consolidation. The disadvantage of this tech-nique for intertidal sediments is that the water content will change with time after the tide recedes, so meas-urements must be made at the same time after exposure for all samples. Alternatively, the rate of change in the water content over time for a site, or among several sites, provides information on the relative rates of drainage of the sediments, which may be useful information for a particular study. In subtidal sediments, samples often contain at least some of the overlying water, which is likely to become mixed with the surficial sediment during recovery of the sample. This will distort any measure of water content, so samples must be collected with as little overlying water as possible and handled very carefully, which is not always possible.

8.3.6 Chemical characteristics of sediments

The organic-carbon content of sediments provides a measure of the amount of food available to microbial decomposer organisms in the sediment and for animals that feed on detritus and its associated microbes. Many contaminants in sediments are also adsorbed onto the surface of organic particles, affecting their relative biological availability and, therefore, toxicity. Some human activities, such as aquaculture, sewage treatment and pulp milling, affect the rate of input of organic material to sediments, and measurement of this variable provides an indication of the extent of impact.

The simplest way to estimate the amount of organic matter present is to combust the predried and preweighed sediment at 400 or 450° C for 5–6 hours. The resultant loss of weight is assumed to be due to removal of organic material, principally carbon. Any shell or other material made of calcium carbonate will also decompose at high temperatures, affecting the result. Pre-treatment with dilute acid will remove carbonate before re-drying, weighing and combusting. Large animals are usually picked out of the sample by hand or sieved out before the amount of organic matter is measured.

More accurate estimates of the amounts of organic carbon in sediments are done by wet oxidation (the methods of Walkley and Black or El Wakeel and Riley: see Buchanan 1984). These require basic chemical reagents and equipment and, although routine, take more time than combustion. For many purposes, combustion is adequate. Automated CHN analysers provide accurate measurements of the amounts of carbon, hydrogen and nitrogen in sediments by combusting the sediment and analysing the gases evolved. The sample is dried and finely ground before analysis. This analysis is also routine and available commercially, and represents the most convenient, if not the cheapest, method available. The nitrogen in sediments is also most conveniently analysed using a CHN analyser.

Some sediments contain significant amounts of organic carbon, such as woody debris or coal, in forms that are not available as food to animals or microbes. If the purpose of analysing the amount of organic material is to estimate the amount of potential food or some other fraction of the organic material that does not include this refractory material, methods of analysis involving combustion or oxidation are not appropriate. Buchanan and Longbottom (1970) describe an alterna-

tive method based on spectrophotometric estimation of the amount of protein in the sediment.

8.3.7 Sampling of contaminants in sediments

The principles of sampling design for contaminants are generally the same as for any other variable (see Chapters 2 and 3). The abundance of contaminants can be surprisingly variable in time and space (e.g. Morrisey et al. 1994a, b), so that adequate spatial and temporal replication is as important in drawing conclusions about patterns of distribution of contaminants as for any other variable.

In most studies of contaminants, the major factor in deciding on the size of sampling unit is the amount of material required for the necessary analyses. This is particularly true for organic contaminants, which may be present in very small concentrations (parts per billion). In such cases, quite a large amount of sediment must be extracted (e.g. several hundred grams of wet sediment). In addition, extra material may be needed for separate extractions of different contaminants, for verification of the accuracy and precision of the extraction and analysis, and for archiving for future study or in case of accidents during the analysis.

A second major consideration, related to the amount of sediment required, is the depth to which the sediment should be sampled. Most contaminants deposited in association with sediments are adsorbed onto the surface of the particles. Consequently, the depth profile of a core of sediment contains, to some extent, a record of the history of deposition of sediments and associated contaminants in that place. This record is often obscured by the burrowing and feeding activities of the animals in the sediment, which mix the sediment vertically and horizontally ('bioturbation'). Nevertheless, the surface sediments are likely to be the most recently deposited. A sample of this sediment, therefore, will provide a measure of the concentration of recently deposited contaminants, while a deeper core will provide a time-averaged measure of deposition over a longer period. The surface sediments are also likely to be more mobile and the concentrations of contaminants in them more variable in the shorter term.

The purpose of the study will determine the appropriate depth of sampling. A relatively deep core, say 10 cm, might be more appropriate for providing a broad, time-averaged measure of contamination. Despite bioturbation, deep cores often preserve an adequate record of sedimentation to enable different layers within the core, and the contaminants associated with them, to be ascribed to a particular period of time. Dating of cores can be done by analysing the types of pollen present and linking changes in these with known changes in land use in the surrounding area, or by identifying classes of contaminants that can be linked to specific events, such as the products of accidents in chemical plants or atmospheric nuclear tests. The distributions of various radioisotopes through the depth profile of the core provide another means of establishing how long ago layers of sediment were deposited. If the purpose of the study is to determine the exposure of particular types of animals to contaminants in the sediments, then the depth in the sediment at which the animal lives and feeds, and the mechanism by which it is exposed to the contaminant (feeding, respiration, contact between the sediment and the body wall), will determine the appropriate depth of sampling. Exposure of surface-deposit-feeding bivalves to contaminants, for example, is likely to be via feeding on the surface layer of sediment. In this case, a core sample 10 cm deep may not provide an accurate estimate of the concentration of contaminant to which the animal is exposed, particularly if the concentration of the contaminant in the sediment varies through time.

The expected concentration of contaminants in the sediments to be sampled partly determines the degree of care needed to avoid inadvertent contamination. In the case of severely contaminated sediments, additional trace amounts from the sampling process (from sampling gear, containers, etc.) are likely to be of minor consequence. When sampling heavy metals, plastic or stainless-steel equipment (grabs, dredges, core-tubes) should be used. Galvanised equipment may produce flakes of zinc, for example, which will contaminate samples. Plastic rather than brass sieves should be used for sieving samples. Equipment used for sampling trace contaminants (e.g. organic compounds and some metals) should be suitably cleaned before use, as should sample containers. Samples for analysis of heavy metals can be transported in clean plastic bags and later transferred to acid-washed plastic containers. Low-density PTFE containers are commonly used. Solvent-cleaned glass containers must be used for organic contaminants. A convenient alternative to solvent cleaning is to heat the containers in a furnace at, say, 500°C for several hours to burn off any impurities on the glass.

The methods of storage and preparation of samples for analyses of contaminants depends on the nature of the analyses. If samples are processed soon after collection, refrigeration will usually be sufficient to prevent biodegradation. Samples for the analysis of metal contaminants can be air-dried or frozen for longer-term storage. Volatile organic contaminants, however, will be lost during air-drying.

For detailed information on preservation, storage, extraction and analysis of contaminated sediments, specialised texts should be consulted. Methods for heavy metals are described by Förstner (1979) and by NRCC (1988). The Guidance Manual on Sampling, Analysis, and Data Management for Contaminated Sites, produced by the Canadian Council of Ministers of the Environment (CCME 1993a, b), contains descriptions of standard methods for sampling, storing and analysing inorganic and organic contaminants, as does the US EPA's QA/QC Guidance for Sampling and Analysis of Sediments, Water, and Tissues for Dredged Material Evaluations (USEPA 1995). Another useful set of descriptions of standard methods for the sampling and analysis of organic and inorganic contaminants, and of other, complementary variables, is that of the US National Oceanic and Atmospheric Administration. These are contained in four volumes, of which the first (NOAA 1993) provides a summary and overview.

8.4 Experimental methods

Many of the experimental approaches used in studying the ecology of subtidal soft bottoms, and the techniques used, are the same as those used on intertidal sediments. Not surprisingly, given their greater accessibility, much of the earlier work on the ecology of marine sediments was done in intertidal areas. Detailed discussions of techniques used and the particular problems associated with experimental research in soft sediments are given in Chapter 7.

As with experiments on rocky shores, motile animals have to be contained by some form of enclosure or cage if their numbers are to be manipulated. Such experimental studies might include manipulations of the numbers of one or more species to identify effects of intraspecific or interspecific competition or predation. Skilleter and Underwood (1993), for example, used enclosures to examine the effects of intraspecific and interspecific interactions and the nature of the sediment on the growth rates of three species of snails living on sandflats on One Tree Island, in the southern Great Barrier Reef, Australia. The snails were confined within plastic tubs, roughly 0.4 x 0.3 m in area and 0.1 m high, with drainage holes (covered in 1 mm mesh) cut in the sides. The tops of the tubs were left open because the snails were not able to climb up the sides. Snails were added to the tubs at several combinations of densities of adults and juveniles either of the same species (to test the hypothesis that the presence of adults of the species *Rhinoclavis aspera* affects the rate of growth of juveniles of their own species) or of different species (to test the hypotheses that the presence of adults or juveniles of other species affect the rates of growth of juveniles of *R. aspera*, *R. fasciata* and *R. vertagus*). All of these combinations of species and sizes were done with each of several different types of sediment placed in the tubs, to determine whether the nature of the interactions varied with the type of sediment. This was based on the assumption that the availability of food in the sediment (the presumed mechanism of interaction among and within the species of snails) was correlated with the grain-size of the sediment.

Hansen and Skilleter (1994) used the same tubs to identify effects of different population densities of *Rhinoclavis aspera* on the microbial flora of the sediments. The biomass of microalgae was measured as the amount of acetone-extracted chlorophyll from the sediment, bacteria were counted after staining with the fluorescent dye, acridine orange, and the rate of bacterial production (i.e. the rate of synthesis of new bacterial material) was measured as the rate of uptake of radioactively labelled thymidine, which provides a measure of the rate of cell division of the bacteria.

Other studies on soft bottoms, such as that of Peterson and Black (1993) in Western Australia, have used cages to exclude predators. Peterson and Black confined equal numbers of two species of cockles in combinations of two different densities ($40/m^2$ and $160/m^2$) in each of three different types of cage. Cages with tops prevented access to the cockles by large mobile predators (fish, starfish, crabs, etc.), cages without tops allowed access by the predators but prevented the cockles from moving away, and cages with half-tops provided controls for any artifacts caused by the tops. The cages were $1\ m^2$ in area, and 5 cm in height above the surface of the sediment. They were made of 4.5 mm plastic mesh and extended 10 cm into the sediment to prevent cockles or predators burrowing underneath. The cockles were individually marked and measured so that growth rates could be determined. The experiment was run for three months. Comparison of growth and survivorship of cockles among treatments indicated that effects of large predators were greater at lower densities of cockles and that there was a negative effect of the tops of the cages on growth. The most likely predator was a starfish, and the variation in its rate of predation between large and small densities of cockles suggested that crowding of shells provided a structural barrier to its feeding activity.

Miller (1993) used small (12 cm diameter) tubs of sediment to examine the effects of a tube-building polychaete on settlement and immigration of other members of the sediment fauna. By comparing colonisation of tubs of sand containing live worms and their tubes, tubs containing tubes without worms and tubs containing imitation tubes with colonisation of tubs of bare sediment, he was able to determine whether the presence of the worms had an effect on colonisation and, if so, whether it was positive or negative. He was also able to test whether any effect was due to the presence of the worms themselves (e.g. disturbance caused by their feeding movement) or to the presence of the tubes.

Scanes (1993) used plastic tubs (35 x 60 x 20 cm deep) to move sediment among contaminated and uncontaminated subtidal sites in Lake Macquarie, New South Wales, to determine whether cockles (*Anadara trapezia*) accumulated contaminants from the sediments in which they live or from the overlying water. Cockles were collected from uncontaminated sites. Those placed in contaminated sediment that had been moved to uncontaminated locations did not accumulate contaminants. Cockles placed in uncontaminated sediments that had been moved to contaminated locations, however, did accumulate contaminants, indicating that the water-column is the major source of contaminants accumulated by this filter-feeding species. Without the manipulative experiments, it would not have been possible to separate the two potential sources of contamination (the sediments and the water). In this experiment, Scanes incorporated controls for potential artifacts caused by placing the sediment in the tubs and moving it around the lake. These involved tubs of sediment dug out of their native sites and handled in the same way as the translocated tubs, but then replaced at the native sites again. Further, undisturbed control sediments were also incorporated into the experimental design. Scanes also provides a practical example of the use of power analysis (see Chapter 3) to determine whether non-significant results were due to inadequate statistical power.

Morrisey et al. (1996) examined the effects of copper on the faunas of sandy sediments by burying blocks of plaster mixed with copper sulphate just below the surface of the sediment. Six blocks were buried in a circle of 1 m diameter. Samples taken before and 30 days after the blocks were buried showed that copper diffused out of the blocks into the surrounding sediment. There were two types of control treatment. First, areas of the seabed, 1 m in diameter, were marked out with pegs to provide a comparison between the copper-contaminated areas and undisturbed sediment. Second, to control for the presence of the plaster and for the disturbance caused by burying the blocks, blocks of plaster without copper sulphate were buried in a circle 1 m in diameter. There were four replicates of each of these three treatments, 5 m apart from one another and randomly interspersed within a site measuring roughly 15 x 15 m. This whole experimental set-up was replicated at two sites about 100 m apart. The depth of water at both sites was about 6 m. Core samples were collected from each treatment at intervals up to six months after the start of the experiment, to measure the concentration of copper and count the animals in the sediment. Concentrations of copper remained higher in the copper treatments than in the control throughout this period. Increased concentrations produced measurable effects on the macrofauna of the sediments, namely changes in the abundances through time of individuals of several species (Morrisey et al. 1996). Blocks of plaster without copper sulphate had no effect.

Field studies of the effects of contaminants on animals in sediments usually infer such effects from correlations between the concentrations of contaminants and the diversity and abundance of the animals. Reliance on correlative evidence, however, cannot eliminate the possibility that any effects observed are not due to the contaminant but to some other variable (perhaps not measured) that varies in the same way as the contaminants. Examples of such covariables would be other contaminants, the type of sediment, or patterns of water movement. By introducing copper experimentally into areas of sediment and comparing the response of the faunas of these areas with randomly interspersed controls, the experiment by Morrisey et al. (1996) was able to remove potentially confounding effects of other environmental variables. An earlier study of spatial variation in the distribution of animals in sediments in Botany Bay (Morrisey et al. 1992a, and see below) had demonstrated variation at scales of metres and tens of metres. To give generality to the results of the experiment, these scales were incorporated into the design by the replication of treatment within sites and the replication of sites. As it turned out, there was variation in the response of the fauna to copper at both of these scales, indicating that results obtained at one place (or time) cannot necessarily be assumed to apply to other places (or times).

For practical reasons, repeated sampling of the same plots is often unavoidable in studies such as that of Morrisey et al. (1996). It may, however, give rise to temporal non-independence (see Skilleter 1996 for a discussion). Another potential problem is that the disturbance caused by sampling may itself have an effect on the fauna (or whatever variables are being measured). This does not matter if effects of sampling and effects of experimental treatments are simply additive, because experimental treatments are compared to controls that have also received the disturbance from sampling. In other words, the effects of sampling on the controls cancel out those on the experimental treatments, leaving the net effect of the treatment of interest. It is possible, however, that effects of sampling and of treatments are not additive but interactive. For example, the disturbance caused by sampling might not affect animals exposed to small concentrations of an experimentally added contaminant, but might affect them at higher

concentrations because the animals are already subject to stress. In this case, subtracting changes seen in the controls from those seen at the larger concentration does not adequately remove the artefacts of sampling. Peterson and Black (1994) have discussed interactions between experimental artefacts and treatments in a more general context, while DeWitt et al. (1996) have considered the special case of *in situ* toxicity tests.

8.5 Examples of sampling designs

8.5.1 Stratified random design

A simple stratified random design was used to assess the potential significance of the impact of a long-line mussel farm near Auckland (Morrisey unpub. data). Under a worst-case scenario, it was predicted that organic waste, dead shells and live mussels (Plate 42) from the farm could accumulate on the seabed below and immediately around the farm. The question was how significant such an impact would be in the context of the general area of coast in which it was proposed to site the farm. The scant available information on the benthic fauna of the area did not suggest that any unusual organisms or assemblages occurred in the area of the proposed farm (natural beds of green-lipped mussels were fished out in the 1940s and 1950s). Information on the physical nature of the seabed (from unpublished theses and from charts produced by the New Zealand Oceanographic Institute, now NIWA) indicated that the proposed site lay within an area characterised by one of four types of sediment found off that part of the coast.

Ecological comparisons between the seabed at the proposed site and the general area around it were, therefore, based on the assumption that spatial variation in the fauna would broadly correlate with differences in the type of substratum. Differences among the different types of sediment shown on the chart were, in any case, fairly small, with mud or silt over the entire area but with differences in the relative amount of sand or shell/gravel present.

To test these predictions that the seabed and its fauna at the proposed site was generally representative of the surrounding area, samples were collected at the proposed site and from each of the sediment types in the general area, as identified from the charts. Sampling sites in studies such as this can be selected haphazardly to give representative distribution within each area. Within each of the areas, randomly located samples were taken with a Smith-McIntyre grab. The animals in each sample were identified and counted, and the grain-size distribution and organic content of the sediment determined.

This design was deemed an appropriate level of investigation relative to the small probability of impact, and the small predicted likelihood that if an impact did occur, important habitats or fauna would be lost (as predicted from the sediment charts and verified by sampling). The data collected indicated that the fauna and sediments in the area of the proposed farm were, in fact, very similar to those in all of the other areas sampled and that, therefore, in the unlikely event of a severe impact on the seabed below the farm, no ecologically unusual assemblages would be lost.

8.5.2 Nested designs

The importance of obtaining estimates of variance from a sample population in order to make valid comparisons among areas has been discussed earlier, along with the use of nested designs of sampling to separate variation at different spatial scales (see 8.2.3). Jones et al. (1990) used such a design to describe the spatial distribution of molluscs in sediments in One Tree Lagoon, Great Barrier Reef, and to identify scales of environmental variability to which the animals respond. The latter objective was intended to provide information (through correlation with scales of variation in the distribution of the molluscs) on factors likely to significantly affect the distribution of molluscs in the lagoon, as the basis for future experimental investigations of these factors (Jones et al. 1992).

In the absence of previous information on what might be important spatial scales, Jones et al. (1990) incorporated four scales into their study, the largest (10 km^2) covering the major types of habitat within the lagoon and the smallest being individual core samples. The lagoon was stratified into five zones, differing in the coarseness of their sediments and the depth of water. Within each zone, three random 'locations' (area 10 000 m^2, 100–100 m apart) were selected and within each of these, two random 'sites' (25 m^2) were sampled with four replicate core samples. Cores were collected using a diver-operated airlift (see 8.2.1q).

A pilot study was done to determine the optimal diameter and depth of cores (based on desired levels of precision, according to the method described by Andrew and Mapstone 1987) and mesh size of sieve. A further pilot study determined the number of samples that could be sieved in the time available and the optimal allocation of sampling effort among scales (see Chapter 3). This second pilot study included two zones, two locations, three sites and four replicate core samples at each site (i.e. minimal replication at each level), and showed that more variation was explained by the differences among zones than by smaller-scale differences. Consequently, in the final design most effort was put into the

larger scales (zones and location), giving better estimates of variation from the most important sources and maximising the power of the test for differences among zones (Jones et al. 1990).

The results of the main study were consistent with those of the pilot sampling in indicating that most variation in the distribution of molluscs occurred at the scale of zones (50–85% of the total variance), although differences were observed at all of the scales examined. Different species of molluscs were most abundant in different zones. Within each zone, however, there were significant differences in abundance among locations, indicating that zones did not represent homogeneous environments for the animals living in them, even though they might appear that way to casual human observation.

Morrisey et al. (1992a) also observed variation in the distribution of organisms at smaller spatial scales within apparently homogeneous larger areas, as described earlier. Using a similar nested design, they identified differences in the abundance of benthic fauna at five spatial scales, ranging from several km to less than 1 m (the distance between replicate core samples). Again, a large part of the total variance was related to differences at the largest spatial scale, but the residual (or 'error') variance in the analyses of variance used to compare scales was also large in the case of some of the taxa studied. This indicated that there was important spatial variation at the smallest scale examined (i.e. among cores spaced 1 m apart).

As part of the same overall study, Morrisey et al. (1994a) also examined spatial scales of variation in the distribution of heavy metals in the sediments of Botany Bay, New South Wales. Again, significant variation occurred at all of the scales examined, but the largest variation was associated with different scales for different metals. Copper, for example, varied most at scales of less than 1 m, whereas zinc varied most at scales of several kilometres. The finding that the most important scales of variation were not the same for different faunal taxa or metals, and that patterns of variation differed for animals and metals in the same sediments, is an interesting one.

In an equivalent study using nested temporal scales, Morrisey et al. (1992b, 1994b) identified variation in the distribution of faunal taxa and heavy metals at scales ranging from days to several months. Sampling (replicate cores at each of two plots, 10 m apart, at each of two sites, 100 m apart) was done on each of two consecutive days, in each of two consecutive weeks, in each of two consecutive months, in each of two consecutive seasons. Cores were collected by divers. The relative importance of different scales of temporal variation differed among taxa and among different metals.

Differences were not always consistent for the same variable among different sites or plots within locations (i.e. there were temporal-spatial interactions at various scales). These differences among variables in the relative importance of different sources of variation (temporal or spatial scales of sampling) suggest that appropriate sampling frequencies to detect temporal change will differ among variables (as for detection of spatial differences). Furthermore, differences in temporal or spatial variability will affect the amount of replication needed to achieve a given level of statistical power for different variables.

8.6 General summary

Living within soft sediments may shelter organisms from some of the environmental variability experienced by those organisms living above the sediment. Rates of exchange of essential materials, such as food and oxygen, are also reduced within sediments. Many natural and anthropogenic contaminants, such as heavy metals and some organic compounds, are closely bound to particles of sediment, particularly finer fractions. Consequently their impacts may be stronger in soft-sediment habitats.

The same basic considerations apply as in other habitats to the design of sampling programmes and experiments in soft sediments. There are, however, special difficulties associated with such work. These relate to the often unavoidable need to sample destructively and to use disruptive experimental manipulations. The former limits the range of sampling options available, while the latter increases the likelihood that experimental artifacts will obscure or confound conclusions. Careful planning is required to overcome them.

SELECTED READING

The references listed are examples of research that will lead the reader to a broader literature, also see the text of the chapter.
* indicates a multi-authored book or collection of proceedings; references to volumes of a journal refer to a set of relevant papers.

General sampling methods: Holme & McIntyre 1984*; George et al. 1985; Baker & Wolff 1987*; English et al. 1994; see also Chapter 2.

Sampling design: Andrew & Mapstone 1987; Hartley & Dicks 1987; Legendre & Legendre 1994; see also Chapters 2 and 3.

Treatment of biological samples: Eleftheriou & Holme 1984; McIntyre & Warwick 1984; Round & Hickman 1984; Bouwman 1987; Floodgate & Gareth Jones 1987; Hartley et al. 1987; Wolff 1987.

Analysis of sediment samples: Folk 1974; Buller & McManus 1979; Murdoch & MacKnight 1991*.

Sampling of contaminants: Förstner 1979; NRCC 1988; NOAA 1993; US EPA 1995.

Experimental methods: See Chapters 2, 3 and 7.

Seagrasses
Sampling and analysis of plant-related variables: Zieman 1974; Larkum et al. 1984; Phillips & McRoy 1990*; Mellors 1991; Inglis 1992; Ibarra-Obando & Boudouresque 1994; Inglis & Lincoln Smith 1995; Kuo et al. 1996*.
Sampling and analysis of fauna associated with seagrasses: Bell & Westoby 1986a, 1986c; Phillips & McRoy 1990*; Ferrell & Bell 1991; Conolly 1994a, 1994b; Inglis 1992; Edgar et al. 1994; Edgar & Shaw 1995; Kuo et al. 1996*.
Experimental studies: Bell & Westoby 1986a; Phillips & McRoy 1990*; Inglis 1992; Conolly 1994a; Kuo et al. 1996*.

NINE

PLANKTONIC ASSEMBLAGES

M. J. Kingsford and R. Murdoch

9.1 Introduction

The nature of planktonic assemblages
Planktonic assemblages are composed of organisms
that spend their entire lives as plankton (holoplankton)
and those that only spend part of their lives as plankton
(meroplankton). Plankton range in size from microbes
(under 0.05 mm) to jellyfish with gelatinous bells that
can be in excess of 1 m in diameter and have tentacles
that extend over 10 m in extreme cases (Fig. 9.1, p. 228;
Fenchel 1988). Plankton can be loosely grouped as
plants (= producers, called phytoplankton) and animals
(= consumers, called zooplankton) and microbes that
include bacteria and protists (Fenchel 1988). Many
protists are both producers and consumers, and may
account for a large proportion of primary production
at different times and places (Sherr & Sherr 1991). Protists can also cause the collapse of diatom blooms and
increase the 'sink' of material to the substratum (Garilov
1994).

Although plankton is generally most abundant in the
photic zone, it is found at all depths in the ocean. Some
plankton is found just below the surface miniscus
(neuston) while a few float on the ocean (pleuston, e.g.
Portuguese man-of-war). Many planktonic animals are
highly mobile (euphausids: Ritz 1994) and capable of
substantial vertical movements (Haney 1988) and, in
some cases, horizontal movements (e.g. medusae: Haney
1988; pre-settlement reef fish: Stobutzki & Bellwood
1994). The physical and chemical environment plays
a major role in structuring planktonic assemblages,
especially those of the lower trophic levels. The depths
of the surface mixed layer, upwelling events and
advective features such as fronts and eddies have a major
influence on primary production, and therefore subsequent levels of the food chain. In addition, structures
are found in the pelagic environment that may influence the distribution, feeding, dispersal, and potential
survivorship of planktonic organisms and other associated organisms (e.g. Edgar 1987; Ingolfsson 1995).
They include 'marine snow' (Alldredge & Silver 1988),
drifting algae (Kingsford 1995) and flotsam (Fig. 9.1).
The size of target organisms, their swimming speed and
the nature of the pelagic environment (i.e. presence/
absence of structures) in which they are found need to
be considered when sampling protocols and designs are
considered.

Early life history of marine organisms
Most marine organisms have a bipartite life cycle, where
early life is spent in the plankton, generally as a larval
form that bears little resemblance to juveniles and adults
(e.g. lobsters and fish; Figs 5.1, 6.1). Moreover, adults
are often highly fecund and individual organisms can
release thousands and even millions of eggs or, in the
case of algae, spores (Reed et al. 1988). In contrast to
familiar organisms on land, therefore, there is a decoup-
ling of reproductive output with recruitment back into
a population. This has made it particularly hard to
understand relationships between stock size, reproductive output, and subsequent recruitment back into local
populations (Hilborn & Walters 1992; Carr & Reed
1993).

Sampling planktonic organisms
The mobility of planktonic organisms, the often im-
mense spatial scales and highly dynamic nature of the
pelagic environment present many challenges for
quantitative sampling of plankton. Many studies have
demonstrated, however, that by carefully identifying a
biological problem and planning the sampling protocol and design (see Chapter 2) important biological
patterns and processes can be identified.

There have been great improvements in the design
and use of equipment that is used to sample planktonic
organisms and their physical environment (e.g. small
conductivity, temperature and depth devices, or CTDs,
data loggers, remote sensing, fluorometers and particle

counters). Some towed devices (e.g. batfish and aqua-shuttle profilers), enable real-time measurements of downwelling and upwelling, light, temperature, salinity, depth, fluorometry, size spectrum of particles and, in some cases, the identification of specific taxonomic groups (Stockwell & Sprules 1995). Remote sensing of surface chlorophyll is also possible but is not feasible for most pelagic organisms (cf. Jillett & Zeldis 1985). Nevertheless, towed nets are still the primary

means of collecting many plankters. The technology of nets has evolved greatly from simple cone nets to sophisticated combinations such as multiple opening and closing nets of the MOCNESS type (Wiebe et al. 1976). Flowmeters, better towing platforms, more attention to sampling designs and an awareness of standard problems relating to properties of plankton mesh and avoidance have all resulted in better-quality information. Plankton-mesh purse seines (Murphy & Clutter 1972; Kingsford & Choat 1985), light traps (Doherty 1987) and the use of fish attraction devices, or FADs (McCormick 1994a; Druce & Kingsford 1995) have enabled larger and faster members of the plankton and smallest members of the nekton to be sampled. Future development and refinement of acoustic and optical

Figure 9.1 A size hierarchy of planktonic organisms and structures that are found in the pelagic environment.
A = macroplankton; **B** = mesoplankton; **C** = microplankton; **D** = nanoplankton; **E** = picoplankton.

MODIFIED FROM FENCHEL 1988; KINGSFORD 1993

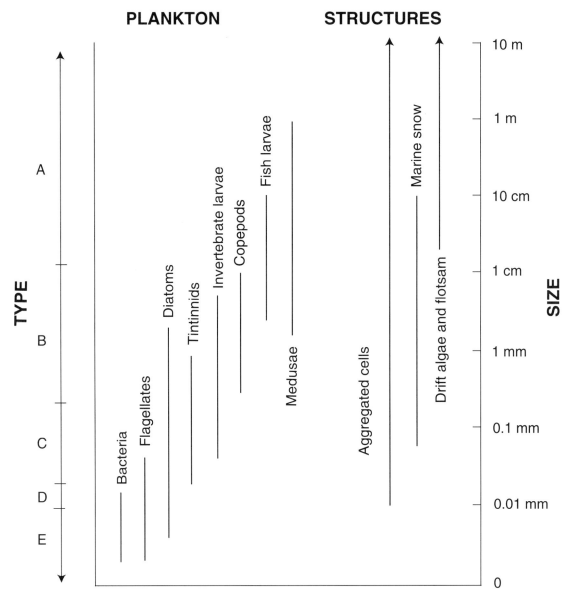

technology are likely to provide better quantitative estimates of biomass and distribution of these more mobile members of the plankton. The use of scuba diving and submarines (e.g. Hamner et al. 1975; Silver et al. 1978; Alldredge & Silver 1988) has enabled the collection of delicate structures and plankton that were undersampled using traditional methods but are now known to be important in pelagic environments.

Why study planktonic assemblages?

There are many reasons for studying planktonic assemblages. Blooms of phytoplankton and/or protists, for example, can be noxious to many pelagic and benthic marine organisms (Sundstroem et al. 1990), and this can be of particular concern for commercial enterprises that raise shellfish (e.g. mussels and oysters) or bony fish (Shumway & Cembella 1994). Toxins from protists not only kill some organisms but frequently accumulate in bivalve molluscs (e.g. paralytic shellfish poisoning: Sindermann 1996). Phytoplankton blooms, mediated through physical oceanographic events (e.g. facilitated by El Niño) can also have a great impact on the dynamics of hard- and soft-bottomed benthic assemblages. For example, during the 1982–83 El Niño event, unusually calm and cool conditions in northern New Zealand facilitated the growth of phytoplankton blooms. The settling of live and dead phytoplankton cells on the substratum clogged the gills of animals, and anoxia associated with the bloom decay killed benthic organisms such as *Echinocardium* (Taylor et al. 1985). In California, the composition of algae associated with rocky reefs can alter at these times, probably as a consequence of reduced nutrients associated with phytoplankton blooms (Foster & Schiel 1993).

Planktonic organisms also play a pivotal role in the cycling of carbon within the ocean. There are now a number of international initiatives (e.g. JGOFS – Joint Global-Ocean Flux Study: Anon. 1992) to determine the processes within the pelagic ecosystem that control the flux of carbon from the atmosphere to the sea floor. Knowledge of the factors that control phytoplankton productivity, and the trophic links from bacteria through to large zooplankton, is an especially important aspect of these studies. Carbon taken up by photosynthesis in the surface layers of the ocean is lost to the sea floor and deep ocean primarily through the sinking of dead planktonic organisms and faecal particles. Study of the factors governing this process is an essential part of understanding the geochemical cycles in the ocean, and the development of models to predict the fate of anthropogenic carbon dioxide and the influence of the ocean upon climate change.

The location of new fish stocks and the spawning grounds of known stocks have been identified using ichthyoplankton surveys (e.g. Murdoch & Chapman 1989; Murdoch et al. 1990). Measures of the production of fish eggs has been used to estimate stock size (Zeldis 1993), while stratified sampling (*sensu* Chapter 3) may be used to get estimates of stock size for commercially important plankters such as jellyfish (Sloan 1986).

The year-class strength of fishes and levels of recruitment to the populations of many marine organisms may depend on processes in the plankton. There have been a number of hypotheses presented, including the match of fish spawning with suitable conditions in the plankton (Match-Mismatch Hypothesis: Cushing 1975, 1990), the retention of larvae to prevent expatriation from populations (Member-Vagrant Hypothesis of Iles & Sinclair: Sinclair 1988), the presence of food-rich haloclines (Vertical Stability Hypothesis: Lasker 1975), abundance of predators (e.g. Moller 1984; Frank & Leggett 1982; Leggett & Deblois 1994) and a combination of the health of fish and vulnerability to predation (Purcell 1985; Bailey & Houde 1989; Paradis et al. 1996). Empirical methods (*sensu* Sissenwine 1984) of relating year-class strength (at recruitment) to some measure of variation in the physical environment that is a good correlate of oceanographic conditions have been used in many parts of the world. Advances in technology and a refinement of sampling methodology have greatly increased the chances of making them a predictive tool. For example, Caputi et al. (1996) have argued that as a result of the desert climate in Western Australia, where river runoff is minimal, the coastal oceanography is comparatively stable except for the dominating influence of the Leeuwin Current. From data sets that often extend for periods greater than five years, the strength of the current appears to have a major influence on the recruitment of rock lobsters, scallops, prawns and fishes. In temperate waters of northern New Zealand, it has been argued convincingly that there is an inverse relationship between seawater temperature and the successful recruitment of snapper (*Pagrus auratus*: Francis 1993). Zeldis et al. (1995) provide some insight to the biological basis for this relationship: they argue that high densities of salps (usually associated with warm water) essentially vacuum-clean the water column. As a result it could be hypothesised that food sources for the larvae may be compromised and result in high larval mortality and low recruitment.

In an anthropogenic context, it is common for sewage and other pollutants to be dumped into ocean waters (e.g. Fagan et al. 1992; Phillips & Rainbow 1993). Because of the dangers of eutrophication, investigators may want to measure the relative contribution of sewage to nutrient input and subsequent primary production so that it can be compared with natural levels of nutrient input (i.e. from upwelling, rivers, etc.). It is also possible that point sources of pollution can have detrimental effects on delicate larvae, and this can be assessed in the field and laboratory (e.g. Kingsford & Gray 1996; Kingsford et al. 1996). Estuaries are often subject to environmental degradation and are frequently argued to be important for the recruitment of coastal invertebrates and fishes in many parts of the world. The input or retention of potential settlers to these systems has often been assessed in studies of planktonic assemblages (e.g. invertebrates: Epifanio 1995; fishes: Blaber & Blaber 1980; Norcross & Shaw 1984; Roper 1986; Norcross 1991; Vieira 1991; Kingsford & Suthers 1996).

Approach

The application of information on planktonic assemblages is essential for many studies. Sampling the pelagic environment does not mean you use the closest plankton net that can be found in a field store. All of the considerations outlined in Chapter 2 are particularly relevant to the pelagic assemblages. In this chapter we give an overview of the methods that are used to make physical measurements in the pelagic environment, together with methods for qualitative and quantitative collections of planktonic organisms. The utility of experiments is emphasised and we give a brief guide to literature on the taxonomy of planktonic organisms. Examples of studies that give some perspective to the diversity of useful studies in the pelagic environment are also listed.

Planktonic assemblages are strongly affected by physical oceanography on scales ranging from entire ocean circulations (thousands of kilometres) to tens of metres (reviews: Denman & Powell 1984, Legendre & Demers 1984; Mann & Lazier 1991; Heath 1992; Grimes & Kingsford 1996). Horizontal oceanographic features vary greatly in shape from nearly circular potential retention areas to lines on the ocean (Alldredge & Hamner 1980; Kingsford 1990), but all of them have the potential to influence the distribution and survival of plankton. The vertical structure of the water column is also important, especially the depth of the mixed layer at the surface, since this influences the nutrient and light levels that control phytoplankton growth and assemblage composition. In general, therefore, direct or indirect physical measurements of the pelagic environment are an essential component of plankton sampling.

9.2 Methods and considerations for assessing plankton

9.2.1 Oceanography and considerations for sampling designs

In all studies of the pelagic environment it is crucial to consider variation in physical conditions and the abundance of organisms both in space and time (see general comments on sampling designs in Chapter 2). A hierarchy of spatial and temporal variation in abundance of planktonic organisms must be considered in sampling designs (Table 9.1).

Temporal variation in abundance is generally great, with changes in the behaviour, avoidance and distribution of planktonic organisms at different times of the day (e.g. Haney 1988; Heath 1992). Within seasons there are pulse events that may relate to upwelling, vertical-mixing events, storms, the intrusion of currents and the demography and life history of different taxa (e.g. *Euterpina*: Haq 1972; *Thalia democratica*: Heron 1972a, b; *Aurelia*: van der Veer & Oorthuysen 1985; benthic egg reserves: Marcus 1995; Madhupratap et al. 1996), and these events may take place over days, weeks and/or months (Plates 55, 57). Seasonal cycles in total production and abundance of individual taxa (e.g. *Penilia avirostris*, Fig 9.2, p. 232) are common (Dakin & Colefax 1940; Jillett 1971, 1976; Bradford, 1972; Kingsford 1980; MacDiarmid 1981; Colebrook 1982; Acuna & Anadon 1992; Neira & Potter 1992a, b; Kiorbe 1993; Liang & Uye 1996), and these are generally more pronounced with increase in latitude (Cushing 1975). Moreover, longer-term temporal variation may relate to changes in the southern oscillation and other long-term cycles and sporadic events (e.g. Colebrook et al. 1984; Glynn 1988).

Spatial variation in abundance is also high (Haney 1988). Great variation is found with depth in the water column, and plankton of a wide range of sizes are capable of migrating up and down in the water column (Forward 1988; Frost 1988; Leis 1991; Heath 1992). Although nocturnal (to surface at night), reversed (to surface at day) and crepuscular (to surface dawn and dusk) migrations have all been reported, nocturnal

Table 9.1 Examples of sources of spatial and temporal variation in abundance of zooplankton assemblages. Where names of spatial or temporal scales are given, these are according to the classification of Hauri et al. (1978). Mechanisms often refer to more than one 'type' for each spatial scale. 'Presence of conspecifics' – interactions that include reproduction, competitive exclusion and cannibalism.

SPATIAL VARIATION

Spatial scale	Type	Mechanisms
Millimetres – 1 m (Microscale)	Horizontal Vertical	• Behaviour (migrations, presence of conspecifics, other members of the assemblage, including predators and prey) • Oceanography (turbulence, transport) • Presence/absence of structures (e.g. drift algae, medusae)
1–999 m (Fine scale)	Horizontal Vertical	• Behaviour (migrations, presence of conspecifics, other members of the assemblage, including predators and prey) • Predation (e.g. planktivorous fishes) • Oceanography (turbulence, transport) • Presence/absence of structures (e.g. drift algae, medusae)
1–100 km (Coarse scale)	Horizontal Vertical	• Oceanography (fronts, eddies, river plumes, upwelling) • Distribution of adult spawners • Distribution of resting egg stages • Interactions with other members of the assemblage • Physiological tolerance to different water masses • Presence/absence of structures (e.g. drift algae, medusae)
100–1000 km (Mesoscale)	Horizontal	• Oceanography (major currents, gyres, pinched-off eddies, upwelling, riverine plumes) • Distribution of adult spawners • Distribution of resting egg stages • Interactions with other members of the assemblage • Physiological tolerance to different water masses
1000–3000 km (Macroscale)	Horizontal	• Oceanography (major circulations, currents) • Physiological tolerance to different water masses
3000 m+ (Megascale)	Horizontal	• Oceanography (major ocean circulations) • Physiological tolerance to different water masses

TEMPORAL VARIATION

Spatial scale	Type	Mechanisms
Within a day	Time of day Tide Stochastic (e.g. storms)	• Behaviour (migrations, interactions with conspecifics and other members of the assemblage) • Oceanography (transport – vertical and horizontal; convergence/divergence)
Within a month	Lunar Stochastic (e.g. storms)	• Behaviour (migrations, interactions with conspecifics and other members of the assemblage) • Reproduction, especially where generation times of organisms < 1 month. • Oceanography (transport, concentration) • Habitat changes (presence/absence of structures, e.g. drift algae, medusae)
Within a year	Seasonal Stochastic (e.g. storms)	• Oceanography (transport, concentration, degree of mixing and related changes in the presence of thermoclines) • Timing of reproduction, the hatching of benthic stages (where relevant), and recruitment • Habitat changes (presence/absence of structures, e.g. drift algae, medusae)
Among years	Environmental change	• Oceanography (changes in major global circulation patterns, e.g. ENSO; vertical mixing) • Changes in the abundance of other members of the assemblage (e.g. prey and predators) • Changes in number of young from benthic egg/polyp reserves • Habitat changes (presence/absence of structures, e.g. drift algae, medusae) • Changes in water quality (e.g. pollution)

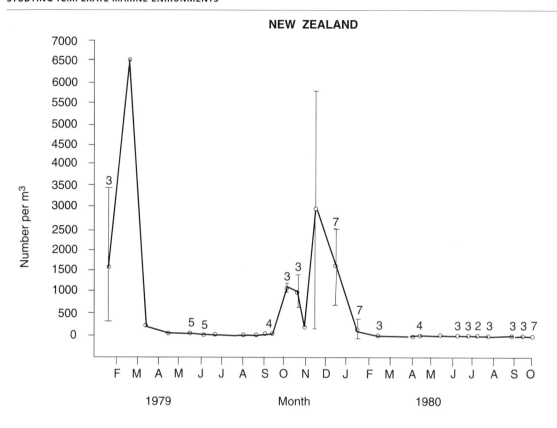

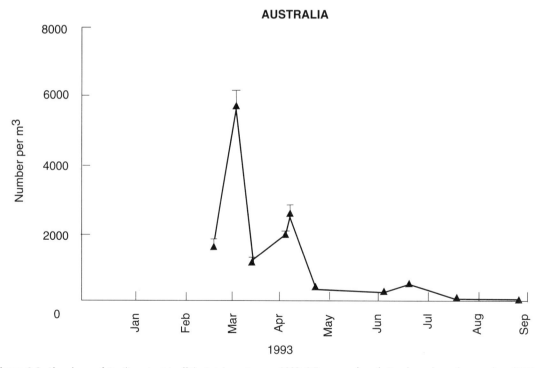

Figure 9.2 Abundance of *Penilia avirostris* off the Leigh marine reserve, northeastern New Zealand (Kingsford & MacDiarmid data; 0.28 mm mesh net) and Botany Bay, New South Wales (Chick 1993; 0.5 mm mesh net). Numbers above the error bars (95% CL) refer to number of replicate tows or n = 1. Great temporal variation in abundance is typical of many species of plankton.

migrations in shallow and very deep-water columns are common and have been reported worldwide (Haney 1988). Vertical distribution patterns may also relate to oceanography, such as the presence of thermoclines, but this is not always the case, especially in regions of rapidly changing oceanography (Gray 1996c).

Horizontal distributions also vary greatly. Although large zooplankters such as scyphomedusae are capable of horizontal movements over hundreds of metres and even kilometres (Hamner & Hauri 1981; Shanks & Graham 1988; Zavodnik 1987; Hamner et al. 1994), the horizontal movements of zooplankton are generally limited (Haney 1988). Great variation in abundance has been found on scales of metres (Steele 1976) to entire ocean circulations (thousands of kilometres: Davis 1991; Table 9.1). Interactions among plankters (e.g. Raymont 1983b; Fenchel 1988; Kiorbe 1993; Ianora et al. 1996) and physical oceanography (e.g. Mann & Lazier 1991; Lehodey et al. 1997), or a combination of these (Frank & Leggett 1982), have major roles at all spatial and temporal scales. The presence of structures (e.g. drifting objects: Kingsford 1993) also influences distribution patterns. Some variation may be attributed to predation by invertebrates (Purcell 1985) and fishes in marine (e.g. Hamner et al. 1988; Kingsford & Mac-Diarmid 1988) and freshwater systems (Lazzaro 1987; Mittelbach et al. 1995). Although plankton are not commonly considered in impact studies (Kingsford & Gray 1996), pollutants may affect their abundance (Karas et al. 1991) and health.

Physical oceanographic conditions can change quickly (Plate 55), and it is important therefore that most studies have concurrent information on the physical environment. In Australia and New Zealand, as in other parts of the world (e.g. Heath 1992), major oceanographic features include global circulations (e.g. the West Wind Drift) and mainstream currents such as the Leeuwin Current, East Australia Current, East Auckland Current and the Southland Current (Fig. 9.3, p. 234; Plates 53, 54, 56). Mainstream currents are generally characterised by different assemblages of plankton (e.g. Griffiths & Wadley 1986). Mainstream currents are often well defined, although velocities may vary considerably from the mean flow. Wind-forcing, tides and other processes (e.g. local upwelling) can influence the strength and direction of currents, may be complex in structure and may generate multiple eddies (Heath 1985).

Differences in water temperature often indicate different plankton assemblages. Frank and Leggett (1982, 1985), for example, found that cool water was rich in predators, which resulted in high mortality rates of fish larvae (caplin). Other coastal oceanographic features that influence the distribution and fate of

planklers include riverine plumes (Grimes & Kingsford 1996; Kingsford & Suthers 1996; Plate 58), 'Linear Oceanographic Features' (*sensu* Kingsford 1990) such as topographic fronts (Wolanski 1994) and internal waves (Shanks 1983; Kingsford & Choat 1986; Shanks & Wright 1987), areas of upwelling (Foster & Battaerd 1985), headland and oceanic eddies (Murdoch 1989; Wolanski 1994) and internal bores (Pineda 1991). The generation (e.g. internal waves) and shape (estuarine plumes: Kingsford & Suthers 1994, 1996) of many of these features can alter according to state of the tide. Tidal currents influence the transport, behaviour and fate of many zooplankters (Norcross 1991; Tankersley et al. 1995; DeVries et al. 1994).

Although the pelagic environment is vast, dynamic and three-dimensional, it is not unstructured. There is a great deal of structure that can provide a focus for sampling. Sampling designs that involve sampling over a grid of stations have often been used (e.g. Crossland 1980, 1981, 1982a). Designs of this type are appropriate for some ecological problems such as the egg-production method for measuring the size of fish stocks (Zeldis 1993) and identifying the presence of previously undescribed physical structure and how it influences plankton (Murdoch 1989). See 8.2.3 (p. 214) regarding the analysis of data from grids.

There are many alternative approaches: for example, sampling can be stratified according to the presence and absence of oceanographic features that manifest themselves at the surface (e.g. Kingsford 1990) or at depth (e.g. thermoclines: Frost 1988) as well as the presence and absence of structures such as drift algae (Kingsford 1992, 1993). Because planktonic assemblages vary greatly on a variety of spatial and temporal scales, nested and partially hierarchical sampling designs are suitable for many biological questions (see Chapter 3). It may be appropriate to sample at different timescales while following current-indicating buoys (Fortier and Leggett 1985; Heath & Rankine 1988, Murdoch & Quigley 1994), or to let experimental objects (e.g. Druce & Kingsford 1995) or mesocosms (see 9.6.2) drift for different periods of time. Sampling should be designed to address the question and not be constrained by traditional methodologies and sampling designs.

9.2.2 Sampling

There are comprehensive books on methodology for sampling the pelagic environment (e.g. UNESCO 1968; Steedman 1976; Parsons et al. 1993) and general accounts of sampling methodology in texts on plankton and specific taxa (e.g. Newell & Newell 1977; Heath 1992). These sources are useful, but in many of

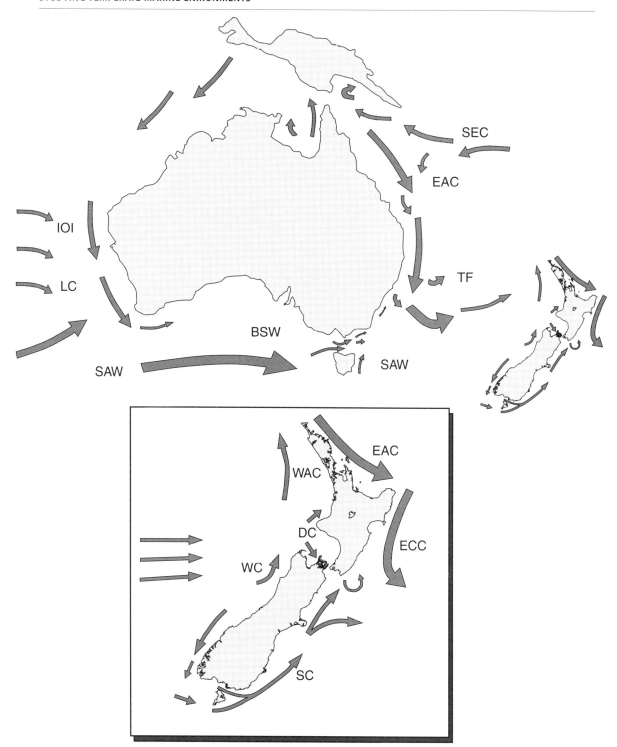

Figure 9.3 Major currents and average direction of flow off the coasts of Australia and New Zealand.
BSW = Bass Strait Waters; DC = Durville Current; EAC = East Australia Current, or East Auckland Current off New Zealand; ECC = East Coast Current; IOI = Indian Ocean Inflow; LC = Leeuwin Current; SAW = Sub-Antarctic Waters; SC = Southland Current; SEC = South Equatorial Current; TF = Tasman Front; WAC = West Auckland Current, WC = Westland Current.

these books the implication is that you should use a particular net if you are sampling ichthyoplankton or some other type of plankton. *We urge investigators to consider the biological problem first (see Chapter 2) because the questions/hypotheses and the spatial scale of the study should determine the type of gear that is used.* Moreover, you should consider modifying existing gear or developing a completely new type of sampling device if you find existing gear inappropriate. The later approach has proved particularly successful in recent years (e.g. plankton-mesh purse seines, light traps, use of scuba). New technology (e.g. particle counters, videography) is increasing the options for sampling planktonic assemblages faster than publications can go to press.

9.2.3 Water pumps

Water pumps allow continuous or depth-stratified sampling of organisms and water chemistry from discrete patches and depths. They are most frequently used to sample small organisms such as phytoplankton and microzooplankton, where only a small volume of water is required to adequately sample the populations. Large-volume pumping systems, however, have been used to sample larger zooplankton, including fish eggs (Checkley et al. 1997) and early-stage fish larvae (Herman et al. 1984; Murdoch & Chapman 1989). Jahn and Lavenberg (1986) used a pump to sample larval fish at different depths, just above the substratum. The pump was rated at 227 m^3/h, and waters were filtered with a 0.333 mm mesh net. Avoidance is likely to be a major problem with highly mobile larval fishes, as suction and visual cues may allow them to detect the device. The effect of suction can be minimised if the pump is towed so that the speed of pumping approximates the speed of intake.

The tube of the pump is lowered into the water and collections are made on board with nets of different mesh sizes and bottles for water samples. The concentration of samples, therefore, is generally done on deck. Appropriate mesh size will depend on the biological question. Decisions regarding a suitable diameter for the opening of the tube, and the pumping rate, would also have to be based on the mobility and size of the target organisms. The components of pumping systems are given in Sournia (1978). The release of dye may be used to determine passage times of water and associated organisms from the inlet to the collection area on deck, or estimated from the time to fill containers of a known volume (Murdoch & Chapman 1989).

The types of pumps that can be used are many, although centrifugal pumps are most commonly used (Cassie 1956; Sournia 1978). Careful consideration

should be given to the maximum depth to be sampled, the flow rate and diameter of the hosing necessary to sample the target organisms, and the system configuration when designing a pump sampling system (see Sourina 1978 and Miller & Judkins 1981 for design details). There are two main designs: one that has the pump on deck, and the other where a submersible pump is used, situated within the intake hose. In both systems, the hose intake is weighted and deployed on a hydrowire to required depths.

See 9.7.5 for a study that used pumps.

Advantages

✔ A wide size spectrum of zooplankters can be sampled.
✔ Ideal for studies on eggs.
✔ Nets of different mesh sizes can be used at the proximal end of the device and water samples for nutrients can be taken.
✔ Discrete sampling (i.e. accurate depth stratification of sampling).
✔ Continuous sampling can be done if the proximal end of the pump has more than one cod-end or the water is analysed for fluorescense.
✔ Can target small-scale spatial patchiness.
✔ Sampling in shallow water and near obstacles (Rutzler et al. 1980).
✔ Filtration rates can be altered to suit the fragility and mobility of the target organisms.
✔ Clogging problems can be seen on deck and alleviated by changing nets at outlets.
✔ Accurate measurement of volume filtered.
✔ The equipment required is relatively inexpensive and can be operated from a range of vessels.
✔ Pumps sample larger zooplankters more effectively than water bottles because they can sample larger volumes of water.

Disadvantages

✘ May require technical assistance in the field for reliable operations.
✘ Size of gear may cause problems for small boats.
✘ Avoidance by highly mobile plankters.
✘ Physical damage of organisms where pumping rates are too high.
✘ Require a knowledge of passage times up the tube from depth, so that species abundance and composition at each depth is not obscured through an error in methodology.
✘ Fouling of the tube.
✘ Efficiency of the pump will change with depth, the maximum sampling depth is restricted for pumps as a consequence of pump efficiency and limitations, and the limited hose length (for a given hose diameter) associated with frictional effects on water flow.

✗ Often more expensive and cumbersome than conventional net systems.

✗ Pumping systems are generally inappropriate as a means of collecting live organisms.

9.2.4 Water bottles

Water bottles are one of the oldest and most commonly used methods for collecting discrete water samples from specified depths (e.g. Niskin bottles). They are most frequently used for retrieving water samples for chemical analysis or to collect phytoplankton and microzooplankton. Bottles are generally inadequate for sampling larger planktonic organisms, primarily because of the small water volumes they sample and the avoidance behaviour of large planktonic organisms. There are a number of different designs, but the most commonly used consists of a plastic cylinder with opening/closing caps at either end. Bottles can vary in size, although one- to three-litre bottles are most routinely used. In its simplest form, the bottle is deployed on a wire or rope to the required sampling depth with the cylinder open, and closure for retrieval is triggered mechanically by a messenger (small metal object slide down the rope or wire). Several water bottles may also be deployed on a rosette system in conjunction with a CTD. In this system, closure of the bottles is electronically triggered in sequence from the deck at the depths required. Use of a CTD in conjunction with a rosette system has the advantage of enabling water samples to be collected from waters of a particular physical characteristic, or in relation to physical structures in the water column such as the thermocline. There are a wide range of designs commercially available, although it is also possible to construct bottles from readily available materials (see Sournia 1978).

Advantages

✔ Discrete sampling for nutrients, chemicals, chlorophyll *a*, particulates and potentially pollutants.

✔ Discrete sampling of microbes and other small plankters.

✔ Water temperature and salinity can be measured from water collected in the bottle.

✔ Multiple bottles can allow multiple depths and replicate samples to be taken on the same drop.

✔ Can provide live phytoplankton for primary production measurements, and live microzooplankton for experimentation.

✔ Can be used from a wide size range of vessels and platforms.

✔ Samples are relatively undisturbed and enable live collection of bacteria, phytoplankton and microzooplankton for experimental purposes.

Disadvantages

✗ Greater potential for contamination on deck.

✗ Trigger mechanisms are sometimes unreliable: you can only be certain that the bottle has closed properly once it is retrieved (labour-intensive).

✗ Unsuitable for sampling large zooplankton.

✗ Water bottles undersample some microzooplankton and larger zooplankton through avoidance and the collection of small volumes of water

9.2.5 Designing towed or dropped plankton nets

Nets have been successfully used to sample plankton of a wide variety of sizes. It is important, however, that the size, fragility and likely mobility of the target organisms are considered before a net is designed; again, the biological question should determine design. Other criteria for design may be based on the type of vessel or methodology that is to be used to deploy nets. For example, many large vessels cannot reduce speed and / or maintain a heading at less 2 knots, and therefore filtration efficiency is a major concern, particularly if the mesh size of the nets is small. If the nets are simply tethered to a reef or at the entrance to an estuary in strong tidal or wind-generated currents, efficiency should also be considered if the nets are going to sample large volumes of water.

The aim of a net is to retain target organisms that enter a net and collect them in the cod-end (Fig 9.4). The design of the net will have a great influence on what is retained. There are many different types of nets, but all have similar concerns (Figs 9.5, 9.6; Table 9.2).

Table 9.2 Concerns regarding sampling, construction of a plankton net and the method of towing.

Structure	Concern
Bridle	Avoidance
Mouth area	Avoidance, efficiency
Mesh size	Retention, extrusion, damage, efficiency, clogging
Mesh colour	Avoidance
Thickness or gauge of mesh	Abrasion and tearing of mesh, self-cleansing
Filtration efficiency	Clogging, calculation of volume filtered
Shape of net	Clogging, efficiency
Flowmeter and position in mouth of net	Accurate estimation of average flow into the net
Speed of tow	Avoidance, efficiency, extrusion, damage
Volume of water filtered	Efficiency, damage

A Rigging and sections of a plankton net

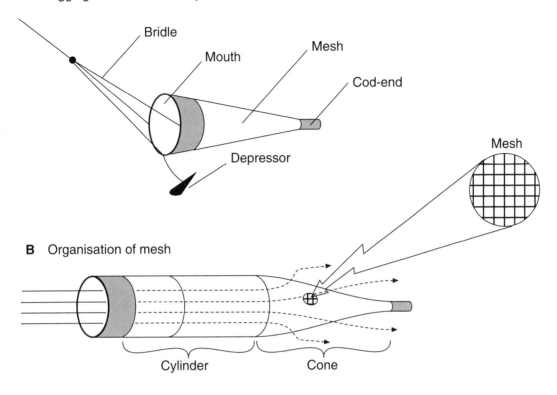

B Organisation of mesh

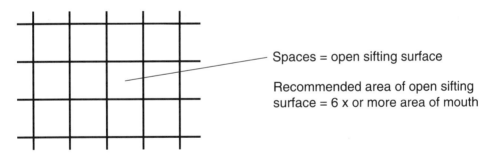

C Plankton mesh

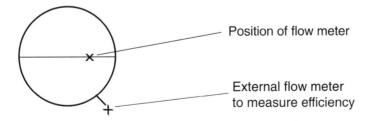

Spaces = open sifting surface

Recommended area of open sifting surface = 6 x or more area of mouth

D Net mouth and flow meter

Position of flow meter

External flow meter to measure efficiency

Figure 9.4 Attributes of a plankton net. **A:** Rigging and sections of a conventional cone plankton net; **B:** A cylinder/cone net showing the organisation of the mesh and typical pattern of flow through the net; **C:** Plankton mesh; **D:** Flow into a net varies within the area of the mouth and average flow can be measured at 0.4–0.5 of the radius. Filtration efficiency can be estimated by placing a flowmeter outside of the net and comparing counts to that of a flowmeter inside the net.

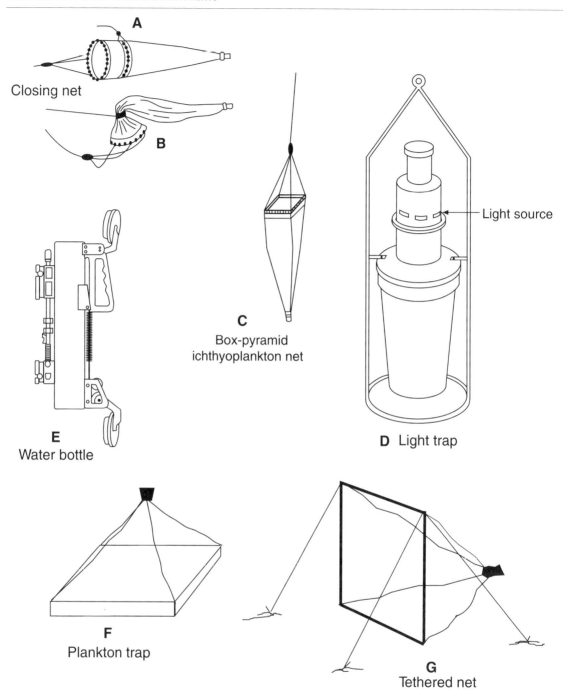

Closing net

A

B

C
Box-pyramid
ichthyoplankton net

E
Water bottle

Light source

D Light trap

F
Plankton trap

G
Tethered net

Figure 9.5 Equipment used to sample planktonic and small nektonic organisms. **A:** A net with a closing mechanism in the open position and **B** in the closed position: this method enables investigators to sample discrete depths. **C:** A square-framed net that can be used for a variety of applications including skimming the surface as a neuston tow (a Scripps depressor should be attached on the lower side of the frame for horizontal tows at any depth or oblique tows). **D:** A light trap used to catch a variety of large plankters including fishes, crustaceans and cephalopod larvae. **E:** A niskin bottle designed to be closed at a predetermined depth with a metal messenger (small weight that slides down the cable/rope to trigger the device). **F:** A demersal plankton net used to collect organisms that undergo migrations from the substratum (usually at night). These devices are usually weighted and tied down; the cod-end is shown in black and usually has a cone leading to it to reduce the chances of target organisms escaping from the sample jar. **G:** A tethered net used to sample plankton on reef crests with tidal movement and/or some swell (adapted from Doherty & McIlwain 1996).

Figure 9.6 Different types of nets and a particle counters.
Left: deployment of a MOCNESS net showing multiple sequential nets (first net open) and sensing systems; note the nets are black to minimise avoidance.

NIWA ARCHIVE

Above: an undulator called a 'Mini-batfish' that is towed behind a vessel and provides continuous data on temperature, salinity, pH and concentrations of the particles using an optical plankton counter (OPC).

IAIN SUTHERS

Left below: bongo nets: the larger nets have 61 cm diameter mouths and 350 mm mesh for sampling fish eggs and larvae; the smaller nets have 20 cm diameter mouths and 100 mm mesh for sampling microzooplankton.

NIWA ARCHIVE

Below: paired push nets 1 m wide x 0.6m high (500 mm mesh) designed to skim the neuston. The arms extend 2 m from the edge of the boat, and a flat metal floating device is attached to the frame immediately above the net (from Pham & Greenwood 1998).

TRI PHAM

239

The relative importance of some of these concerns will vary with the type of organism that is collected (Table 9.3, opposite). For example, the presence of a bridle that partly obscures the mouth of the net will increase avoidance by organisms with great mobility (e.g. fish larvae and euphausids: Munk 1988). A combination of mouth area and speed of tow will determine how easy a net is to avoid. In general, as mobility of the animal increases, the area of the mouth, mesh size and speed of tow should increase. The exception to this recommendation is the use of high-speed samplers (e.g. Gulf III) that are used to collect fast zooplankters (Heath 1992): they have a narrow mouth, coarse mesh, and are towed at high speeds (above 6 knots). Some high-speed devices are used to obtain continuous data on plankton such as fish eggs (e.g. Longhurst Hardy Plankton Recorder, LHPR, Fig. 9.7).

The mesh size and thickness of the mesh are important to consider in the design, because these will determine retention, the condition of the catch, and filtration efficiency. Problems of mesh distortion and weave displacement found in some old-style meshes (UNESCO 1968) are generally minimal for modern meshes.

Some properties, such as retention, loss and condition of the catch, will vary with changes in tow speed. Filtration efficiency is affected by the relationship between area of the mouth and open sifting surface (as in formula 9.1; see also Fig. 9.4) and is often expressed as an efficiency ratio (area of mouth : area of open sifting surface).

Open Area Ratio $R = a.b/A$ ...... (9.1)

where a = area of net mesh; b = porosity; A = area of mouth.

Where nets will filter large volumes of water, efficiency should be high (e.g. at least 1:6). The amount of open sifting surface per unit area of mesh changes greatly with mesh size (Table 9.4). The efficiency of the net also changes with the shape of the net. If large volumes of water are filtered, it is crucial to have a cylinder (round nets, Fig. 9.4) or box (square nets) section in the net to facilitate cleaning. Because these sections orientate parallel to the direction of flow, they tend to waft in the current and to some extent self-clean. The cone or pyramid section at the rear of nets (Fig. 9.4) remains taut in the current and is the first part of a net to start clogging.

Filtration efficiency can be measured by placing

Figure 9.7 A Longhurst-Hardy Plankton Recorder (LHPR) used for vertical distribution studies. The device collects a series of samples on a roll of gauze that is advanced at intervals and on which the plankton is collected. The system shown is fitted with a main 200 μm mesh net for fish eggs and small larvae, and a 53 μm mesh system mounted on top of the sampler for collecting food organisms. A Scripps depressor is attached to the lower side of the device.
STEVE COOMBS OF SPARTEL, DEVON, UK

Table 9.3 Properties of nets, tow speed and quantities required to sample different types of plankton. A wide range of values is given for measures such as speed of tow and volume sampled, because an investigator will choose values that are best for the relevant hypothesis, the organisms sampled (mobility and fragility), local conditions of abundance and speed of clogging. Size ranges for some of these organisms are given in Figure 9.1. Sources include UNESCO (1968) and Sournia (1978). See also 9.2.15 for advice on the sampling of specific groups. Full overlap of the bridle means that all wires of the bridle are orientated in front of the mouth.

Type of organism	Bridle overlap with mouth	Mesh (μm)	Net shape	Efficiency ratio	Speed of tow or vertical haul (knots)	Volume filtered (m³)
Phytoplankton	None–full	1–25	Cone Cylinder/cone	1:2+	< 0.5	< 10
Copepodites and Hydromedusae	None–full	50–100	Cone Cylinder/cone	1:2+	0.5-2	1–100
Invertebrate larvae	None–full	50-1000	Cone Cylinder/cone	1:2+	0.5-2	1–5000
Mesoplankton (meroplankton copepods etc.)	None–full	200-300	Cone Cylinder/cone	1:2.5+	0.5-4	1–5000
Fish larvae and euphausids	None–small	50–300	Cylinder/cone Box/pyramid	1:6+	1–4	50–1000
Scyphomedusae and ctenophores	None–small	Mixed box 1000+ pyramid < 300	Box/pyramid	1:6+	0.5-4	5–5000

flowmeters inside and outside the net. Efficiency drops as the inside flowmeter records lower rates of flow relative to that of the outside flowmeter (see formula 9.2). Alternatively, efficiency (F) can be calculated by towing the net over a known distance.

$$F = w/A.D \quad \ldots \ldots \quad (9.2)$$

where w = volume filtered (determined with a flowmeter); A = area of the mouth; D = distance over which the net was towed.

If large volumes of water are to be sampled, it would be judicious to have very efficient nets (we often design nets that are 1:10; e.g. Fig. 9.8) to allow for high concentration of plankters that could clog the mesh. It should be noted that mesh size simply refers to the maximum size of the mesh apertures. Effective mesh size will drop (especially in the cone section) as the net becomes clogged, and smaller organisms may collect. Sophisticated and expensive multiple opening and closing nets (e.g. MOCNESS, BIONESS: Wiebe et al. 1976; Heath 1992), can minimise loss of efficiency (e.g. where F < 0.85, UNESCO 1968) by closing one

net and opening another. Moreover, a CTD joined to the rig will provide concurrent physical information (e.g. temperature stratification of the water column). Some nets have mixed meshes (e.g. Williams et al. 1988). For example, nets that are designed to catch scyphomedusae have a coarse mesh (1–2 mm) box and

Table 9.4 Example of the relationship between mesh size and open sifting surface of plankton mesh for one type of nybolt mesh. Porosity of the mesh varies greatly with diameter of the mesh. The price of mesh per unit area increases rapidly at mesh sizes of less than 100 mm.

Mesh size (μm)	Porosity (% of open sifting surface per unit area)
27	27
100	38
200	44
280	53
500	47
1000	58

0.5 mm mesh net – porosity = 39% (10:1)

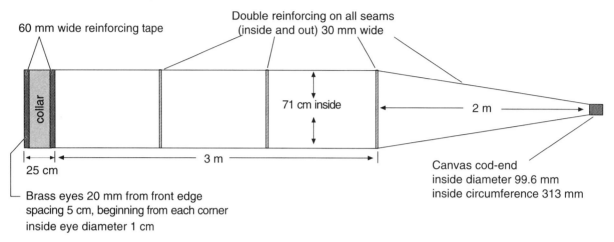

Square net – box section 28.52 m²
– pyramid 2.8 m²

Figure 9.8 Design of an efficient box-pyramid ichthyoplankton net suitable for collecting large invertebrate plankton and fish larvae. The square mouth also enables it to be used as a neuston net. The box section of the net facilitates self-cleaning as the net is towed. (For use as a tethered net, see Finn & Kingsford 1996.)

a fine mesh (< 0.5 mm) pyramid to give the animals a soft landing and minimise damage to specimens. Caution should be taken in using these nets for other applications (e.g. mesoplankton) because filtration will occur on the fine mesh until the pyramid is clogged (the rate of which may vary with location and time), then filtration will happen on the coarse mesh and greatly decrease the retention of mesoplankton.

Flowmeters are required to measure the volume of water filtered and, in some cases, the speed of the net. Tests in wind tunnels have demonstrated that average flow rate into a net can be measured with sufficient accuracy by placing a flowmeter 0.4–0.5 of a radius from the edge of the net (UNESCO 1968; Fig. 9.4). Flowmeters (e.g. General Oceanics 2030) are robust and with regular maintenance (e.g. changing water on the inside and periodic flushing with weak acid such as vinegar) will last for years. The flowmeter should be calibrated from time to time to check for accuracy, particularly if the propeller shaft has been bent and restraightened.

The properties of nets vary greatly with the type of organisms that are targeted. Phytoplankton are often sampled with fine-mesh nets (also see pumps, microbes, fluorescence). Because concentrations of phytoplankton are often high, small volumes of water are sampled and low towing speeds are used (see 9.2.14). Micro-

zooplankton and many invertebrate larvae are collected in fine-mesh nets. The filtration efficiency of these is often low because small volumes of water are sampled, tow speeds are slow and clogging is minimal. Mesoplankton is commonly caught in 280 and 333 mm mesh nets. Mesh of this size is ideal for catching adult copepods and other mesoplankton (e.g. Hobson & Chess 1976, 1978; Kingsford & MacDiarmid 1988). When relatively small volumes (e.g. < 30m³) are sampled at slow speeds, a cone net is adequate, but more efficient cylinder/cone-type nets are required when large volumes of water are sampled and a higher tow speed is used. We generally collect fish larvae, for example, with nets that have a 0.5 or 1 m² mouth and are towed at 2–4 knots (e.g. Kingsford & Choat 1989). Meshes of this size will undersample newly hatched larvae (Houde & Lovdal 1984), but they will sample a broad size range of preflexion and postflexion fish of most species (e.g. *Pagrus auratus* 3–14 mm SL: Kingsford & Atkinson 1994). Mixed-mesh nets are designed to give delicate organisms such as jellyfish a soft landing (see also discussion on mesh size above).

The number of zooplankters per unit volume (e.g. number per 1 m³) is calculated according to formula 9.3.

Catch per unit volume V = t x w   . . . . . .   (9.3)

where t = quantity of organisms in net; w = volume filtered as a proportion of V.

Concentration should be standardised to a common unit (e.g. number of organisms per 1, 10, 100 or 1000 m³; see section 9.4) that closely approximates the mean volume sampled.

In general, nets have the following advantages and disadvantages.

Advantages
- ✔ Practicality: robust, easy to construct and mend, can be used in a wide variety of ways (e.g. from ships, small boats, tethered or diver-controlled) and at a wide range of depths.
- ✔ Particularly good at collecting some types and size-classes of plankton, and the only practical method of collecting certain zooplankton.
- ✔ There is often comparative data on concentration from different parts of the world.
- ✔ Short turnover time from retrieval of the net, processing of samples and redeployment.
- ✔ Can be used in most conditions where it is possible to work on deck.

Disadvantages
- ✗ Avoidance by mobile organisms especially during the day (Murphy & Clutter 1972; Choat et al. 1993; Doherty 1987).
- ✗ Damage, destruction or extrusion of delicate organisms.
- ✗ Variable loss and retention according to conditions in the plankton.
- ✗ Clogging and loss of filtration efficiency (especially in some conditions, e.g. high concentrations of phytoplankton and marine snow).
- ✗ Difficult to sample multiple locations in a short period of time (i.e. space and time are confounded: cf. automated light traps) unless multiple nets are used at different locations.
- ✗ Unsuitable for sampling in shallow water with foul ground.
- ✗ Too much drag for some small vessels to tow.

It is always worth taking spare mesh and a sewing kit (with polyester cotton) on major cruises. Extra cod-ends and other spares should always be taken. It is often judicious to have a long line and float trailing the net for retrieval of the net should the towing line part.

9.2.6 Cod-ends and collecting buckets

The designs of plankton net cod-ends are many and varied, and generally depend on the materials available and the ultimate use of the plankton collected (Fig, 9.9). In its simplest form, the cod-end may be a pouch of mesh secured to the end of the net via a solid ring-and-clip mechanism. Delicate organisms, however, can be severely damaged using this design because of the high water flow through the cod-end mesh. More frequently, a tube (plastic, metal or glass) closed at the end, and of similar diameter to the end of the net, is used for col-

lecting the filtered sample. Often buckets are designed with clip mechanisms or screws for easier removal and attachment to the net (the male-female connections of plastic tubes are often suitable). Small windows of mesh glued to holes drilled in the sides of the bucket also enable the entire filtered sample to be washed into the buckets before removal from the net. Alternatively, buckets without meshed windows may be used if live,

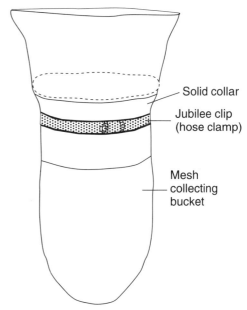

Figure 9.9 Cod-ends: made of mesh (above) and a solid cod-end made of plumbing tube and with drainage ports (below). Clips are shown on the solid cod-end, but a threaded attachment would suffice.

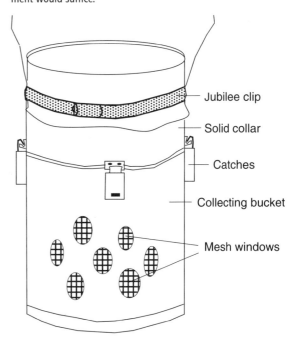

undamaged organisms are required, since this design prevents animals from being held against the mesh during towing. Standard collecting jars may provide plenty of spares. There are anecdotal accounts (I. Suthers pers. comm.) that larvae may begin to decompose in cod-ends if they are dead for 20 minutes or more in waters of above 25°C, but this is obviously a problem only in tropical waters. Rates of decomposition may, however, increase where larvae are also damaged (e.g. in fast tows of coarse mesh).

9.2.7 Methods for towing and tethering plankton nets

The method of deployment will depend on the biological question and background knowledge regarding the attributes of organisms (e.g. fragility), their distribution patterns (e.g. neustonic organisms) and the facilities available to the investigator. Depth-integrated tows, or hauls, sample the entire water column and include vertical hauls, drop nets, oblique and stepped oblique tows (Fig. 9.10). These methods are often used where there is no *a priori* knowledge of vertical-distribution patterns, plankters are well dispersed with depth, or a wide variety of plankters are sampled each of which has a different vertical-distribution pattern. If plankters

are only found in a very narrow depth strata (e.g. the neuston or in thermoclines) they will be undersampled or missed completely in depth-integrated tows (Kingsford 1988).

Sampling of the neuston (a few centimetres of the net protrudes out of the water so that the surface meniscus is skimmed; a square or rectangular net is required) and surface tows (within the top metre but not necessarily sampling the neuston effectively, especially with a round net) do not require use of a closing mechanism. The net must not be dragged in the wake of the vessel. This can be avoided by towing to the side (Kingsford & Choat 1986), towing in a circle or using a push net (Pham & Greenwood 1998). Microplankton may be sampled by sieving a small volume of water to the side of the boat.

If sampling is to be done at greater depths, a closing mechanism is required (Fig. 9.5). Although these are not always used (e.g. Leis 1986), they are generally necessary if the net is to fish a given depth stratum without contamination. The net must be lowered down to a specific depth in the closed position, opened to fish and retrieved in the closed position (e.g. Kimmerer 1984). Closing mechanisms should always be checked because contamination can be considerable if they are not working correctly. Nets may be towed or hauled vertically. The duration of tows should be determined in a pilot study or, if is not possible, the samples should be observed in the field to make a rough check that adequate quantities of plankton are being collected. The

Figure 9.10 Different ways of towing nets. The dotted lines indicate where a net is not fishing (i.e. the net is lowered cod-end first, or throttled with a closing mechanism).

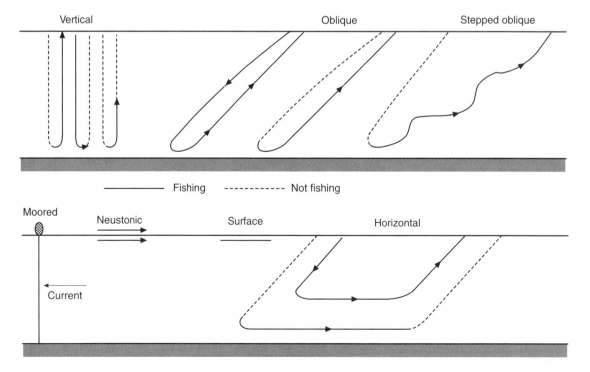

time taken to lower a net to designated stratum and retrieve it can be great. The whole procedure can be sped up considerably if nets of the MOCNESS type are used. These multiple nets (up to 10) allow samples to be taken at multiple depths during one deployment.

Determining the depth a net is at during a tow is best done by an electronic or acoustic depth detector whereby the position of a net can be tracked in real time aboard the tow vessel. This is not always possible (e.g. in areas that are not safe for large vessels), and triangulation may have to be used to estimate the depth of the net by calculating the angle of the tow rope from horizontal and a knowledge of the required depth to calculate the amount of cable or rope to let out.

Calculations are based on the following formula:

$$\text{Depth} = \sin(a) \times L \quad \ldots \ldots \quad (9.4)$$

where a = angle of tow rope from horizontal;
L = length of tow rope out.

You may have to recalculate if the angle changes because of idiosyncrasies of the sea or skipper.

Triangulation is unlikely to be accurate in very deep water columns because curvature the rope violates the assumption that the rope is straight (see Table 9.5).

Vertical hauls can be made by lowering the net down cod-end first, or by the WPII method, which uses a heavily weighted net rim, and fishes as it free-falls mouth-down through the water column. When the net reaches the target depth, it is hauled to the surface by a throttle rope around the middle of the net. Clearly with the latter technique there is no bridle to facilitate avoidance. Some consideration should be given to the shape of the net (see above) if large volumes are sampled.

Neuston nets: Because these nets have to skim the surface, they need to have a square mouth. Many neuston nets are wider than they are deep (e.g. 1 m wide, 0.3 m deep: Holdway & Maddock 1983; 1 x 0.36 m: Tully & O'Ceidigh 1989a, b), but many nets that have been used to sample the neuton are square (e.g. 0.4 x 0.4 m: Sameoto & Jaroszynski 1969); 1 x 1 m: (Kingsford & Choat 1989; Tricklebank et al. 1992).

Although neuston nets can be rigged and hung to the side of a vessel on a boom to avoid the wake, asymmetrical rigs can allow the net to fish in undisturbed waters at a greater distance from the vessel. If an asymmetrical rig is used, the net will require wings that float (e.g. Brown & Cheng 1981) or are angled in such a way to produce lift (Sameoto & Jaroszynski 1969). Another method of fishing in undisturbed water is by pushing a net rigged in a special frame on or near the bow of a small boat (Pham & Greenwood in press; Fig. 9.6).

Table 9.5 Example of estimated amount of cable (metres) required from the ocean surface to position a plankton net at four predetermined depths. The ratio of depth to length in these calculations ranges from 1:3 to 1:4. The net used in this example was rigged with a Scripps depressor (UNESCO 1968).

Depth required (m)	Degrees from horizontal		
	10	15	19
0	0	0	0
7	44	28	21
14	80	55	42
20	115	80	60

Sledge nets: Nets mounted on a sledge have been used for the collection of plankters close to the substratum. Because the net is lowered to the bottom, the device must have a closing mechanism; for example, a door (e.g. Rothisberg & Pearcey 1976) or a throttle around the net. If the sledge is very heavy, the net may not be strong enough to be throttled on retrieval. Because nets collect drifting algae, sponges and other organisms that are loose or attached, it is judicious to have a coarse inner mesh to protect the plankton mesh. If the sledge might encounter sharp objects such as mussels, coral, etc., a canvas sheet attached to the frame below the net should help protect the mesh. Nets of this type have been used to collect near-bottom plankton, including fish larvae.

Diver-controlled nets: Although drag often makes nets difficult to manoeuvre underwater, they have been used successfully by divers. Diver-controlled nets have been used to take spot samples (for identification and size) of aggregations of zooplankters such as a mysids (Ohtsuka et al. 1995) to obtain quantitative samples of fish larvae (Marliave 1986) and net plankton. In general, nets are easiest to control with two divers (see Hobson & Chess 1978). Diver control can be more useful for depth-stratified tows where the waters are too shallow for large vessels and where bottom topography is complex.

Tethered nets: Tethered nets have been used with great success in the tidal channels of estuaries (e.g. Roper 1986; Epifanio 1995) and reefs (Shenker et al. 1993; Finn & Kingsford 1996) and in the surf zone of reefs (e.g. Dufour et al. 1996). Although these nets have been primarily used to measure the input of invertebrates or pre-settlement fish, they have also been used to obtain estimates of plankton passing fixed points (e.g. Hobson & Chess 1978). In tidal channels, nets are generally tethered so that none of the rigging obstructs the mouth of the net, and they are rigged for neuston or surface sampling by the use of floats so that they move up and down with the tide. Nets of a variety of sizes have been used. For example, Roper (1986) used a round net (0.45 m² mouth, 0.45 mm mesh), Shenker et al. (1993) used a 2 x 2m square net of 2 mm mesh, and Finn and Kingsford (1996) used a net with a square mouth of 0.5 m² and a mesh of 0.5 mm. Nets are sometimes left for long periods of time or for short replicate collections. If long collecting times are used, consideration should be given to filtration efficiency (see 9.2.4). Nets can be tethered in surf zones using a robust rig. Dufour used stakes to pin a net with a 1 x 0.25 m mouth (mesh 0.5 mm) in Moorea, which has a very small tidal range. Doherty and McIlwain (1996) used nets that were very tall (2 m high, 0.5 wide) to allow for the relatively large tidal range of the Great Barrier Reef. The nets were staked to the reef and had running fore and back stays to the reef that did not obstruct the mouth (Fig. 9.5). Greatest catches in all of these studies were generally taken at night.

9.2.8 Plankton-mesh purse seine nets

Plankton-mesh purse seines can be used to encircle a body of surface water and therefore are ideal for sampling discrete targets. They can sample a substantial volume of water without any need to tow a net for a long period of time. They have been used for four main reasons: studying avoidance of conventional towed nets (Murphy & Clutter 1972), sampling natural and experimental drifting objects and adjacent waters without drift objects (Kingsford & Choat 1985; Druce & Kingsford 1995), sampling in and out of discrete oceanographic features such as topographic fronts (Kingsford et al. 1991) and estuarine plumes and fronts (Kingsford & Suthers 1994), and sampling small fishes in seagrass beds and other estuarine habitats (P. Sale pers. comm).

The mesh of these nets varies according to the purpose for which they were built. For example, the seine designed by Kingsford and Choat (1985) had a mesh size of 0.28 mm so that fishes of all sizes (i.e. recently hatched to pelagic juveniles) and mesoplankton could be caught. Coarser mesh sizes may be chosen if only large fishes or juveniles are the target of the study. When gear of this type is used, make sure you have fishing permits that entitle you to use undersized meshes.

It is a great advantage if a plankton-mesh sock or cone net is included in the design so that the catch can be wafted toward the cod-end, as for a conventional plankton net (Fig. 9.11). Purse seines should be designed so that the length of the net is about five times the depth.

A basic design and illustration of deployment is shown in Kingsford and Choat (1985) for a 2 x 10 m net. The design of a larger net used by Kingsford and Suthers (1994) and Kingsford (1995) is shown in Figure 9.11. Two lines of floats, about 100 mm long, were required to keep the net afloat, with sections of light chain as a weighted line. It is important that the net settles quickly to a curtain of mesh during deployment, so the weighted side of the net should not be too light. A loose line is attached to the bottom edge of the net for rigging the purse line. Large plastic or brass curtain rings are arranged on this line for the purse line. It is useful to have different colours of rope for the top and bottom lines and the purse line. A small sea anchor is required for deploying the net. If a custom-made canvas sea anchor is not available, an old plankton net will suffice.

Careful consideration should be given to direction of the wind before deployment of a purse seine net (Fig. 9.12, p. 248). It is important that after encircling the target the boat is left on the downwind side of the target so that the wind blows the boat away from the net during pursing. As the boat approaches the target, the cod-end is filled with water so that it sinks rapidly. The boat is reversed toward the target as the net is deployed over the bow; care is taken to ensure that the cod-end is clear of the purse ropes and does not interfere with them. The net is fully deployed and extends as a straight line from the bow as the boat is reversed past the target (taking care not to disturb it), before motoring forward and arching around the target to retrieve the sea anchor. While restacking the net, leave the sea anchor out to slow the boat's drift.

Investigators who plan to use nets in excess of 15 m long (e.g. Hunter & Mitchell 1967; Murphy & Clutter 1972) should consider techniques used by commercial seiners (e.g. Machii & Nose 1990; Kailola et al. 1993).

Advantages
- ✔ Ability to sample a large volume of water in small area.
- ✔ Can be used to target objects (e.g. drift algae: Kingsford 1995) as well as specific water masses.
- ✔ High quality of catch because of little damage from towing the net, so specimens are ideal for live experiments.
- ✔ In normal conditions the net can be retrieved, the catch taken and net cleaned and restacked for deployment in 10 minutes.

Disadvantages
- ✗ Calculating the volume of water sampled is difficult (though this can be kept constant by deploying the net in a similar arch at each set).
- ✗ The time required to clean the mesh of the net in a

small boat can be considerable where relatively fine drifting material (e.g. seagrass) is collected, but this is true of most nets.

✗ Difficult to use a purse seine in winds of over 20 knots.

✗ Can only sample surface waters.

Figure 9.11 Design of a plankton-mesh purse seine net. See also Figure 9.10 for method of deployment and arrangement of the rigging (Kingsford 1995).

9.2.9 Demersal plankton traps

There are strong links between assemblages in the water column and benthic assemblages. It is well known that organisms usually found just above, or near to, the substratum move into the water column at night (e.g. Alldredge & King 1977). Moreover, plankton traps are useful to record the hatching of larvae from demersal eggs (e.g. Kingsford 1985). A variety of types of equipment have been used to sample demersal zooplankton, but they generally involve an inverted net or perspex cone that is left to rest on the substratum (Youngbluth

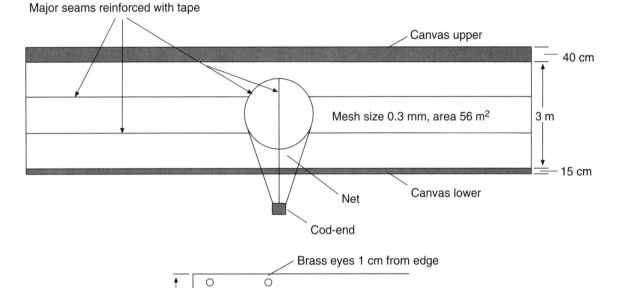

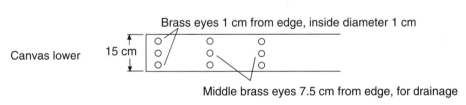

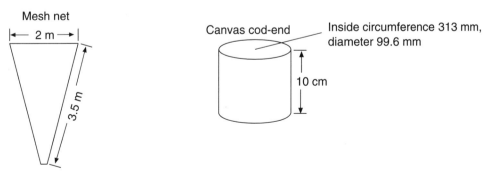

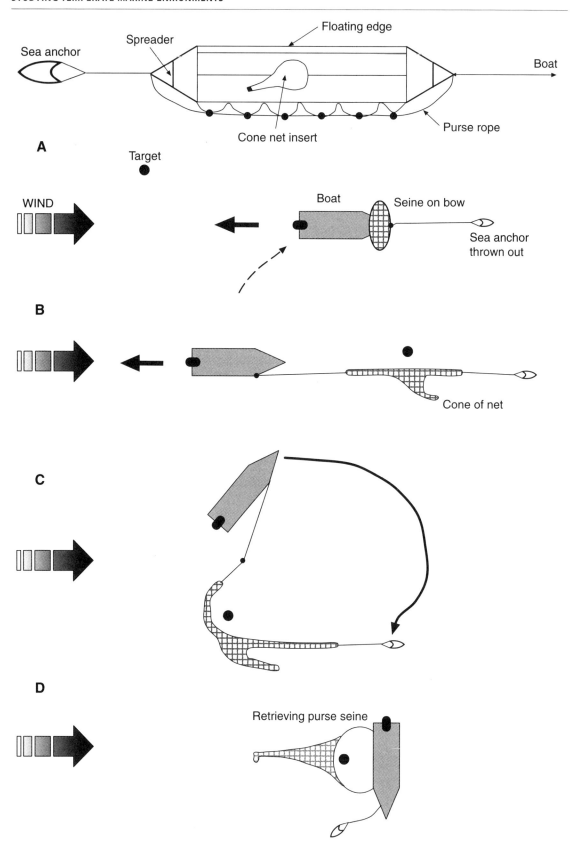

1982; Fig. 9.5). Organisms rise up and are caught in a bottle or chamber. There are various degrees of sophistication, from single traps to highly automated devices with multiple samplers (e.g. Jacoby & Greenwood 1988). In some cases, 're-entry' traps may be used to collect settling larvae or demersal zooplankters returning to the substratum from the water column (e.g. Cahoon & Tronzo 1988; Jacoby & Greenwood 1988).

Advantages

✔ Sample organisms that migrate into the water column and may be a significant contribution to local zooplankton in some environments, particularly at nights.

✔ Useful for studying the timing of migrations and for determining the taxa and or sexes of plankters that migrate.

✔ Useful for studying organisms that may be important food for consumers on reefs (e.g. fishes: Gladfelter 1979; MacDiarmid (1981); and sessile organisms such as corals.

Disadvantages

✗ Can be energy-intensive.

✗ Difficult to sample many locations at one time.

✗ Vulnerable to storms unless particularly sheltered waters are chosen.

✗ May be damaged by boat traffic in shallow water.

9.2.10 Light traps

Light traps have received special attention in recent years for the collection of the late larval and pelagic juvenile phases of reef and pelagic fishes as well as invertebrates. They have been used in scientific research for many years (e.g. Sale et al. 1976), as light attraction has been used in commercial fisheries for fish and squid (Kailola et al. 1993). Doherty (1987) described a sophisticated form of light trap with multiple chambers that has been used to capture pre-settlement reef fishes (e.g. Milicich 1994) and invertebrates (Moltschaniwskyj & Doherty 1995). The on/off sequence of lights is controlled by a timer and powered by substantial banks of batteries. (A good relationship with a battery company is a must for this type of work.) The lights in the top chamber attract fish, and once they go off, lights in deep chambers attract fishes to a collection chamber. The most recent version of the Doherty light trap has been used by many investigators (e.g. Doherty et al. 1996; see Fig. 9.13). These traps are expensive to

OPPOSITE: **Figure 9.12** The rigging of a plankton mesh purse seine net and the method for deploying it. Dark arrows = direction the boat is travelling; broken line = approach of boat to current position. Wind = speeds of less than 20 knots.

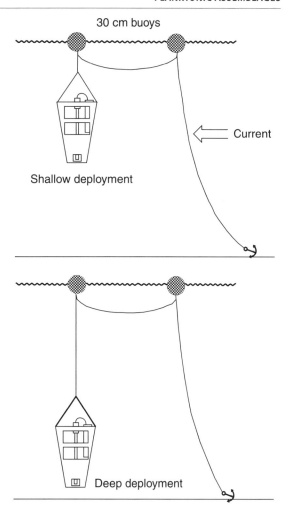

Figure 9.13 The deployment of light traps in shallow and deep water. In general, railway wheels are used to tether the heavy traps and rigging.
PETER DOHERTY

make (over A$1000 in 1996), and simpler designs may be adequate for your ecological problem. For example, Sponaugle and Cowen (1996a, b) used a light trap that was essentially a small suspended plankton net with a fluorescent light inside. The entry points of the trap were the necks of plastic soft-drink bottles sewn into the net and orientated inwards. A design of this type would cost less than $200. Light traps are generally tethered while fishing, but in some cases free-drifting traps have been used (e.g. Thorrold 1993; Doherty et al. 1996). Light traps have successfully collected late pre-settlement phases of many fishes. They appear to be particularly good at capturing clupeids, pomacentrids, serranids, lethrinids and scombrids, and poor at catching labrids, scarids and apogonids. It should be noted, however, that light traps have not been used in many parts of the world, particularly in temperate regions, and the full array of taxa and life history stages that can be collected using light traps is yet to be realised. For a comparison of light traps with other types of gear, see Choat et al. (1993).

Advantages

✔ Can fish for long periods of time.

✔ Can be placed in arrays (drifting or tethered) where all trap at the same time.

✔ High quality of catch, which can be used for experiments.

✔ Particularly good at collecting the late larval and pelagic juvenile phases of fishes (e.g. Milicich & Doherty 1994) and invertebrates (Moltschaniwsky & Doherty 1995).

✔ It is often useful to combine net tows and light traps (Brogan 1994; Doherty et al. 1996).

Disadvantages

✘ Difficult to calculate volume sampled.

✘ It has not been determined how the catchability of target organisms changes with current speed for tethered traps.

✘ Measurements of predation by fishes and invertebrates have not been estimated, which may be particularly important where traps are left for eight hours or more before the catch is processed.

✘ Catchability may change according to levels of ambient light, such as phase of the moon.

✘ Only certain taxa may enter the trap.

✘ Algal growth can rapidly diminish the power of lights at locations where fouling is a problem, though this can be alleviated by cleaning the traps.

✘ Cannot be used during the day.

9.2.11 Visual observations and collections by divers and from submarines

Visual observations have been used effectively to sample gelatinous zooplankters, macrocrustaceans and larval fishes. Hamner (1975) and others drew attention to the fact that efforts to sample the pelagic environment using nets had missed or under-represented important components of planktonic assemblages, especially gelatinous zooplankters. Studies of the taxonomy (e.g. Larson 1976), abundance and behaviour (Hamner & Hauri 1981) of these organisms have been done using visual methods and collections in jars and nets while on scuba or from submarines. These data have provided investigators with a broader perspective of the role of these important organisms in the marine environment (e.g. Zeldis 1985). Similarly, investigating macrocrustaceans in tropical and temperate waters has provided information on abundance, swarm

behaviour and trophic interrelationships of these highly mobile animals (Hamner & Carlton 1979; McFarland & Kotchian 1982; Kingsford & Choat 1989; Ritz 1994).

Data on larval fish have also been obtained through visual observation, rather than sampling by nets. Data on near-shore distribution patterns have been obtained by divers counting fish larvae in transects and using diver-controlled nets for samples and comparative estimates of abundance (Marliave 1986; Kingsford & Choat 1989). Transects may also be swum using a flowmeter (a low-speed rotor may be needed) to estimate distance while pushing a quadrat to count organisms that pass through it (most suitable for gelatinous zooplankton).

Data on behaviour and even manipulative experiments have been obtained for larvae through observations by divers (e.g. Breitberg 1991, in prep: experiments on goby larvae). Recently, Leis et al. (1996) have been following pre-settlement reef fish that have been collected from light traps (see above) and released. They have recorded the behaviour of these animals over considerable distances. One diver maintains visual contact with the fish, while the other records data on swimming depth, responses to the presence of reefs, predators, etc.

In some cases, behavioural observations of fish larvae (Kingsford & Tricklebank 1991) and estimates of abundance of plankton can be made from the surface. For example, Kingsford (1993) counted large scyphozoan jellyfish (*Catostylus mosaicus*) from the surface using transects. The transects measured 3 x 10 m, and medusae were counted to a depth of 1 m (= 30 m³). Six replicate transects were used, but the number of replicates should be determined in a pilot study (see Chapter 2). The length of transect was determined by dragging a 10 m rope behind the boat, and when the drogue passed the end of the rope, counting ceased. Transect width was measured with a light 3 m pole, held by the persons counting on the bow of the boat. For longer transects, it is possible to determine transect length by using a hand-held GPS (i.e. travelling at known speed for x min), although some correction would be needed in areas of strong currents. A particular advantage of this method is there is no need to release or pick up drogues and rope.

Advantages

✔ Collections (e.g. in jars) of delicate organisms.

✔ Can study organisms in places where it is difficult to use other types of equipment (e.g. very shallow water; topographically complex areas).

✔ Collect information on behaviour that could not be collected in any other way.

✔ Obtain estimates of abundance of organisms (e.g. medusae) that are difficult to obtain with nets.

✔ *In situ* behavioural observations of planktonic organisms.

Disadvantages

✗ For some visual methods it is difficult to sample large volumes of water.

✗ Potentially inaccurate estimates of density (though some other types of gear do not or cannot sample the organisms at all).

✗ Diver safety in rough seas and in the presence of uncomfortably large sharks.

✗ Estimates of abundance may only be done when there is good visibility.

✗ Visual counts from a boat are generally limited to calm conditions.

✗ With the exception of large, conspicuous organisms (e.g. some medusae), the accuracy of identifications may be questioned without voucher specimens.

9.2.12 Optical plankton counters

Technology for optical sampling of plankton continues to develop and is likely to become an important tool for plankton research in the future. Electronic optical plankton counters (OPCs) have been developed that count and size zooplankton, fish eggs and other marine organisms with diameters between 250 mm and 2 cm (Herman 1988, 1992). The device uses a beam of light, and occlusions of the light by plankton are measured. These instruments have been developed both for the laboratory and use at sea, to depths of up to 1000 m. The underwater versions can be attached to a range of plankton nets, and flying systems such as the Batfish (Fig. 9.6) and Nu-Shuttle (with slit sizes of 10 x 2 or 22 x 2cm). Because they can be operated continuously at speeds in excess of 8 knots, and cycled through a range of depths, large regions of the ocean can be surveyed both horizontally and vertically over relatively short periods of time. Studies of the spatial and temporal patterns of zooplankton biomass in Lake Erie (Stockwell & Sprules 1995) using an OPC, for example, provided information for developing models of predator-prey interactions and energy flow in a pelagic community. The high towing speed of an OPC also enables the abundance of large zooplankters capable of avoiding traditional nets, such as euphausiids, to be measured. These instruments therefore reduce the tedious and time-consuming process of net sampling and counting, although tows are required for reference and calibration.

Advantages

✔ Continuous data over small and large spatial scales.

✔ Rapid sampling of multiple depths.

✔ Can be deployed at high speeds, thus enabling near-synoptic high-resolution abundance data of zooplankton.

✔ Can obtain concurrent data on temperature, salinity, chlorophyll (using fluorescence) and depth of sampling.

✔ Some OPCs can be deployed from small boats.

✔ Potentially sample highly mobile plankton.

Disadvantages

✗ Expensive.

✗ Complicated equipment with great potential for technical problems.

✗ Difficult to sail over foul ground (like most gear).

✗ Most zooplankters can not be identified with any degree of confidence.

✗ Inadequate sampling of some large zooplankton (e.g. medusae).

9.2.13 Acoustics

Acoustic instrumentation can and often has been used to provide information on the size, density, spatial and depth distribution of zooplankton and micronekton. To date, acoustic techniques have been primarily applied to large crustaceans such as euphausiids (e.g. Everson & Bone 1986), but have the potential to resolve species down to 1 mm (i.e. copepods). There has been considerable development of the techniques over the past decade, and it is now recognised that split-beam and multi-frequency systems are required to provide size class and abundance/biomass information (Holliday et al. 1989; Green & Weibe 1990). These systems are currently being developed for moored arrays and use on towed profiling vehicles. Similarly, Acoustic Doppler Current Profilers (ADCPs) used for profiling currents, either mounted to a vessel or moored, can provide information on the distribution of biological material (especially zooplankton and micronekton). While it is difficult to distinguish between the effects of size and abundance from ADCP backscatter signals, these devices do provide synoptic information on the spatial distribution of biomass at the same scales as current measurements, and other physical parameters (Roe et al. 1996).

Advantages

✔ Provide size, abundance and biomass data for zooplankton and micronekton at high resolution through the water column, and enable synoptic surveys to be made.

✔ High resolution of plankton with gas-filled bladders (i.e. swim bladders of fish larvae).

Disadvantages

✗ Acoustic techniques are complex and require expensive instrumentation, traditionally used from large vessels, although moored and towed systems are now being developed.

✗ Only provide size/biomass data, and presently cannot distinguish taxonomic composition.

✗ Backscatter models are very sensitive to their input parameters, which can cause large variations in acoustic biomass estimates.

✗ Acoustic estimates often disagree with those obtained by net sampling or other techniques.

✗ Signal strength may vary with taxa: organisms with gas-filled bladders have the strongest signal.

9.2.14 Videography

Video studies of the distribution and behaviour of planktonic organisms continues to develop. Videography enables continuous distributional information on planktonic taxa to be collected in the ocean at temporal and spatial scales not possible with point-sampling methods such as nets, pumps and bottles. The video plankton recorder developed by Davis et al. (1992), for example, can provide data on plankton distribution and physical variables at scales of millimetres to hundreds of kilometres. Because this instrument samples non-invasively, it is also able to provide distributional data for fragile organisms such as gelatinous zooplankton at spatial scales unobtainable with conventional sampling equipment.

Unlike acoustic technology and optical plankton samplers, which also provide high-resolution distributional data for plankton, videography is able to provide the actual taxonomic composition of the plankton communities studied. High-resolution video systems have been deployed in a number of configurations, including ROVs (Paffenhofer et al. 1991), towed arrays (Benfield et al. 1996), fixed platforms, and a moored vertically profiling system is currently under development (Davis et al. 1996).

While much of the video footage collected to date has been examined field by field using tape editors, advances in image-analysis processing and increases in computer power will enable this aspect of the analysis to be automated in the near future. A semi-automated video analysis has already been developed that enables rapid visualisation of video data while at sea (Davis et al. 1996). Continued development of video instrumentation is likely to make this technology an essential tool in the field study of plankton dynamics in the near future.

Advantages

✔ Enables simultaneous sampling of a wide size range of planktonic organisms (from tintinnids to large

juvenile fish), and towed systems can provide high-resolution data on spatial distribution by species.

✔ A major strength of videography is that it provides the ability to observe and identify the behaviour and physiological reactions of organisms *in situ*.

✔ Able to obtain data on delicate gelatinous animals and structures (e.g. marine snow) generally destroyed by traditional sampling tools such as nets (Tiselius 1998).

Disadvantages

✗ Difficult to define the volume of water being sampled, and the relatively low tow speed of the current towed systems limits synoptic coverage to small areas.

✗ Storage, processing and telemetry of the large quantities of data is a problem with current technology.

✗ Post-processing of video footage is yet to be automated, and is currently very time-consuming.

✗ Moored systems are yet to be developed, and current systems are restricted to use from large vessels.

✗ Costs may be high.

9.2.15 Methods used to study different categories of organisms and structures

Microbes

The recognition of the role of micro-organisms in marine plankton has revolutionised our concepts of the flow of energy and nutrients through the pelagic food web. Prior to the 1970s it was thought that the majority of primary production in the oceans was carried out by organisms larger than 60 μm, such as diatoms, which were grazed upon by zooplankton, which in turn were eaten by larger organisms. Micro-organisms were rarely noted because of the sampling (e.g. using a plankton net) and preservation methods used (e.g. fixatives such as alcohol and formalin, which destroy many of the smaller fragile cells: Porter et al. 1985, Graham 1991).

The introduction of epifluorescence microscopy to count micro-organisms led to far higher estimates of their abundance than the previous, selective methods such as dilution cultures (e.g. Caron et al. 1989). Samples are fixed and filtered by gravity or minimal pressure onto a black membrane filter with a suitable pore size (0.2 μm if bacteria are to be retained). The fixative agents most used are formalin and glutaraldehyde (Sherr et al. 1993). Cells are stained with a fluorochrome such as 4, 6-diamidino-2-phenylindole (DAPI), proflavine or primulin (Porter & Feig 1980; Kuuppo-Leinikki & Kuosa 1989; Martinussen & Thingstad 1991). Organisms with photosynthetic pigments are naturally autofluorescent under these conditions and do not need to be stained. Cells are counted with an epifluorescence microscope using UV excitation. Estimates of abundance may be affected by operator subjectivity and are time-consuming. These problems may be overcome in the future by the use of computerised image analysis and flow cytometry (Bjørnsen 1986; Burkill 1987; Sieracki & Webb 1991). Further details of methodology used in the quantification of marine microorganisms can be found in Kemp et al. (1993).

In marine environments the number of bacteria ranges from 10^5 to 10^8 per ml and the number of heterotrophic nanoflagellates (measuring 2–20 μm) from 10^2 to 10^4 per ml, with abundances depending on the trophic status of the water and the depth sampled (Sanders et al. 1992). Numbers of cyanobacteria and pico-planktonic (< 2 μm) photosynthetic eukaryotes vary from 10^5 to 10^7 per ml, with their relative contribution to photosynthesis, and to some extent their abundance, decreasing as organisms from higher trophic levels become more abundant. Nanoplanktonic algae range from barely detectable abundances up to a similar level to bacteria in extreme bloom conditions in nutrient rich waters.

Phytoplankton

Phytoplankton are the predominant primary producers in the sea, and therefore the essential food for consumers such as zooplankton and fish, and a range of filter feeding benthic organisms, especially in coastal regions. Quantitative measurement of the distribution, standing stock or production rates of phytoplankton are often integral to understanding the processes associated with studies of higher trophic levels, within both the plankton and the benthos.

Samples of phytoplankton for determining abundance and biomass are generally collected by water bottles or pumps, although nets can be used for sampling large-celled phytoplankton (Table 9.6). Once the samples have been collected, there are a number of methods for quantifying phytoplankton. Standard measurements of particulate carbon, nitrogen, phosphorus, chlorophyll and various other parameters can be used (see Parsons et al. 1993 for methods), although these do not provide measures of species abundance. Automated particle counters can also be used to measure particle size and abundance (Sournia 1978). Many counters fail to differentiate phytoplankton from other planktonic components such as microzooplankton, inorganic material and detritus, although some instruments can now distinguish phytoplankton from other material by incorporating cytofluorescent technology. The pigment composition of samples, determined using high pressure liquid chromatography (HPLC), is frequently used to differentiate major phytoplankton taxa (i.e. diatoms, dinoflagellates, ciliates, etc.) within samples. New satellites (e.g. SeaWIFS, Adeos) will similarly soon be able to differentiate major taxa over large areas of sea surface using sensors that record different spectral wavelengths. Microscopic counts of preserved samples (usually fixed in Lugol's solution, but see Sournia 1978) are still the primary means of obtaining phytoplankton abundance data at the species level. Although the process is labour-intensive, efficiency gains have been achieved through the development of image-analysis systems. These data can also be used to estimate biomass or energy associated with the phytoplankton community, and is especially important in studies that focus on trophic interactions within the pelagic ecosystem.

The concentration of chlorophyll within water samples is generally used to quantify phytoplankton biomass. Fluorometric techniques are most commonly used to measure the concentrations of chlorophyll (see Parsons et al. 1993). Technological advances in the design of fluorometers over recent years has now enabled *in situ* measurements of fluorescence. Fluorometers are frequently deployed on mooring arrays and

Table 9.6 Examples of methods used to study phytoplankton based on a specific problem.
D = description or mensurative experiment.

Problem	Equipment	Volume sampled or pump rate	Sources
Distribution of different species of phytoplankton with depth (D)	Water bottles	5 litres	Chang 1983
Broad-scale measures (tens of kms) of primary production (D)	Pump and epifluorescence	125 l/min	Zeldis et al. 1995
Relationships between biomass and production of phytoplankton in areas of upwelling (D)	Bucket pump	2 litres continuous	Bradford et al. 1986

other oceanographic equipment such as CTDs and undulating vehicles, and this has enabled fluorescence to be measured over a large range of temporal and spatial scales. While fluorescence units are often used as a proxy measure for chlorophyll, laboratory measures of chlorophyll from discrete water samples are still required to relate *in situ* fluorescence to chlorophyll concentrations. The relationship between fluorescence and chlorophyll concentration is complex and depends on a range of factors (e.g. the taxonomic composition of phytoplankton, past light history, physiological state, life-history stage, etc.). It is essential that this be taken into consideration when designing studies that use this technology.

Primary productivity has traditionally been measured with the ^{14}C method (Steemann-Nielson 1952). With advances in continuous flow mass spectrometry, stable-isotope techniques are becoming increasingly popular and viable alternatives to the ^{14}C method. Slawyk (1979) was the first to use the ^{15}N method (Dugdale & Goering 1967) in conjunction with ^{13}C experiments to estimate new and total production. The use of ^{13}C simultaneously with ^{15}C not only enables a second estimate of primary productivity to be made, but also enables C:N uptake rates to be measured directly on the same water sample (Slawyk et al. 1977).

The measurement of primary production is often complex, and there is a broad literature on the merits of the different measurement techniques (e.g. Slawyk et al. 1984; Collos & Slawyk 1985; Bury et al. 1995; Mousseau et al. 1995). A thorough evaluation of the literature and consultation with appropriate experts is advised before embarking on the measurement of primary production, to ensure that appropriate techniques are used and the necessary supplementary measurements (e.g. nutrients and light) are made.

Mesoplankton include a broad range of larval forms (e.g. echinoderms, bivalves) and holoplankters (e.g. copepods; Fig. 9.14, p. 256). Many of these organisms are important consumers of phytoplankton and protists, and provide an important source of food for higher trophic groups. There has been extensive sampling of these organisms using plankton nets over a range of mesh sizes (see Table 9.7) and using almost all methods of towing (9.2.4). Mesoplankton are generally abundant and accurate estimates of abundance can generally be obtained from samples of less than 10 m³.

Fish larvae and highly mobile plankton and micronekton
Fish larvae and other highly mobile plankters (e.g.

Table 9.7 Examples of methods used to study mesoplankton based on a specific problem.
D = descriptive or mensurative experiment, E = manipulative experiment.
ND = no data provided; NA = not applicable.

Problem	Equipment method	Volume sampled	Sources
Distribution of different species of copepods with depth (D)	Net depth stratified	2–100 m³	Frost (1988)
Broadscale spatial patterns of abundance (tens–hundreds km, D)	WP2 net	12–50 m³	Bradford (1985)
Grid sampling for zooplankton related to oceanography (D)	High-speed sampler	≈ 40 m³	Murdoch (1989)
Measures of copepod biomass in and out of an area of upwelling (D)	Net pump	7–42 m³ 1–4 m³	Bradford-Grieve et al. (1993)
Survey for fish eggs to estimate biomass of fishes	Oblique tows	ND	Zeldis (1993)
The timing of diel migrations by demersal zooplankters (D)	Automated demersal plankton trap	NA	Jacoby & Greenwood (1988)
The impact of fishes on concentrations of zooplankton (D)	Net vertical tow	5–15 m³	Kingsford & MacDiarmid (1988)
Interactions between mesoplankton and fish larvae (E)	Mesocosm	NA	Cowan & Houde (1990)

chaetognaths, euphausids and mysids) pose special sampling difficulties in that many of these organisms are sufficiently mobile to be considered nekton. Many of these organisms are of intense interest to scientists and managers alike, being the young of species that are harvested commercially, those that provide a crucial source of food for higher trophic groups (e.g. Lancraft et al. 1989; Quetin & Ross 1992) or predators. For examples, see Table 9.8.

Large gelatinous zooplankters and structures
Gelatinous zooplankters (e.g. ctenophores, salps, hydromedusae, scyphomedusae) and some structures (e.g. marine snow: Alldredge & Silver 1988) pose

special sampling problems. Many of these organisms are so delicate that they disintegrate on impact with a plankton net or in preservative at a later time. Quantitative measurements of densities of gelatinous organisms should consider how robust the target organisms are (see Table 9.9).

Effective methods for sampling these organisms, and delicate structures, include collecting in jars while diving, and *in situ* photography. Quantitative measures of abundance may be obtained by divers while swimming a quadrat and using a low-speed flowmeter to estimate distance travelled; video may be used in a similar way. The behaviour of gelatinous zooplankters and interactions with predators and prey may also be observed.

Table 9.8 Examples of methods used to study highly mobile zooplankton based on a specific problem. D = descriptive or mensurative experiment, E = manipulative experiment, ND = no data provided; NA = not applicable.

Problem	Equipment method	Volume sampled	Sources
The distribution patterns of mysids in near-reef habitats (D)	Visual diver collections	ND	Ohtsuka et al. 1995
Broad-scale distribution patterns of lobster phyllosoma	Fine-mesh mid-water trawl	> 1000 m³	Booth (1994)
Spatial patterns of distribution of a conspicuous zooplankter, *Munida gregaria* (D)	Aerial survey	NA	Jillett & Zeldis (1985)
Behaviour of mysids in aggregations (D and E)	Video	NA	Ritz (1994)
How drifting algae influences the distribution of small fish and invertebrates (D and E)	Purse seine	16–44 m³	Kingsford & Choat (1985) Kingsford (1995)
Relationships between numbers of potential settlers and recruitment fish counts of reef fish (D)	Light traps	NA	Meekan et al. (1993)

Table 9.9 Examples of methods to study gelatinous and delicate zooplankton based on a specific problem. D = descriptive or mensurative experiment, E = manipulative experiment.

Problem	Equipment method	Volume sampled	Sources
Collection of hydromedusae for taxonomy (D)	≈ 50 mm mesh net, oblique tow	0.5–5 m³	Claudia Mills pers. comm.
Concentrations of *Aurelia* in windrows (D)	Purse seine 0.28 mm mesh	44 m³	Kingsford et al. (1991a)
Concentrations of medusae in aggregations (D)	Visual belt transect	30 m³	Kingsford (1993)
Collection of intact marine snow (D)	Syringe barrels	6 ml	Alldredge & McGillvary (1991)

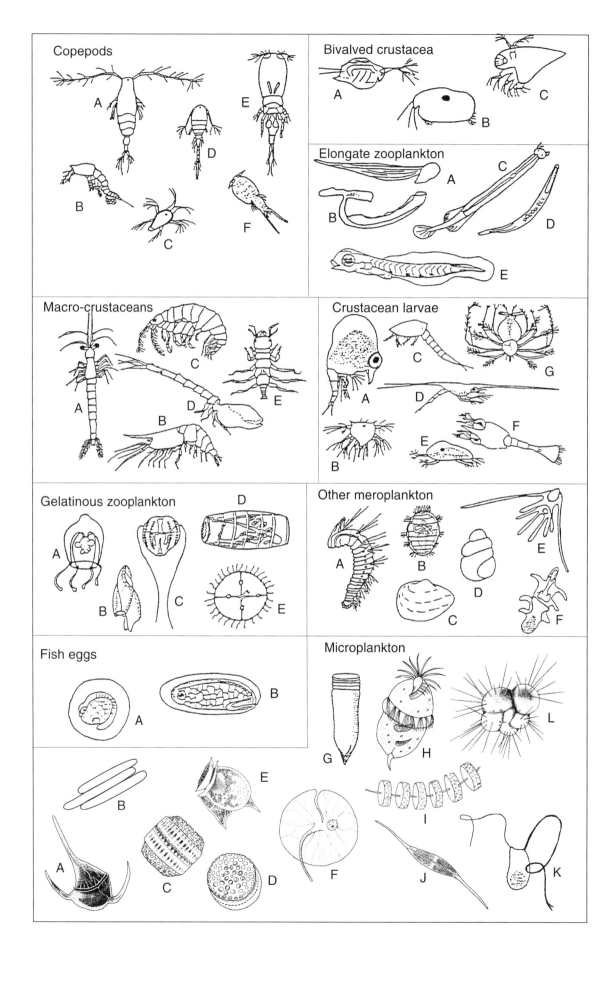

9.3 Treatment of samples

9.3.1 Recording data

Special consideration should be given to the organisation of data sheets for the sampling of plankton in the field and in the laboratory. In addition to standard details such as location, site, date, time of day, state of the tide and the time of high and low tides (see Chapter 2), other details may include GPS waypoint numbers, latitude and longitude, flowmeter readings, depth of tow, duration of tow, wind direction and strength, current direction and strength, and tidal range. In the laboratory, these details may be added to the laboratory data sheet, but in addition to counts of total samples or subsamples, additional columns may be required for number per cubic metre. In general, however, volumetric or area conversions will be easier to do once the data have been entered onto a spreadsheet or other database (see Table 3.2).

Figure 9.14 Types of planktonic organisms that have been divided into groups based on visual appearance, taxa and functional groups. Plankton of these types are found in waters throughout the world.

Copepods: A *Acartia* sp.; B *Euterpina* sp.; C copepod nauplius; D *Oithona* sp.; E *Corycaeus* sp.; F parasitic copepod.
Bivalved crustacea: A *Penilia* sp.; B ostracod; C *Evadne* sp.
Elongate zooplankton: appendicularians, A *Oikopleura* sp.; B *Fritillaria* sp.; C chaetognath, *Sagitta* sp.; D nematode; E fish larva.
Macro-crustaceans: A mysid; B decapod shrimp; C gammaridian amphipod; D cumacean; E isopod, *Paragnathia* sp.
Crustacean larvae: A brachyuran crab zoea; B barnacle nauplius; C euphausid calytopis stage; D anomuran zoea; *Petrolisthes* sp.; E barnacle cyrid, F pagrid crab larva; G lobster phyllosoma larvae.
Gelatinous zooplankton: A anthomedusa; B siphonophore; C ctenophore, *Pleurobranchia* sp.; D doliolid; E leptomedusa, *Obelia* sp.
Other meroplankton: A polychaete larva; B holothurian larva; C bivalve veliger; D gastropod larva; E ophiuroid larva; F echinoid larva (pluteus).
Fish eggs: A pilchard egg, *Sardinops neopilchardus*; B anchovy egg, *Engraulis* sp.
Microplankton: A dinoflagellate, *Ceratium* sp.; B pennate diatom, *Bacillaria* sp.; C and D diatom *Thalassiosira* sp; E dinoflagellate, *Dinophysis* sp.; F flagellate *Noctiluca* sp.; G tintinnid *Helicostomella* sp.; H oligotrichs, *Strombilidium* sp.; I diatom, *Schroderella* sp.; J diatom, *Phaedactylum* sp.; K pennate diatom *Phaeocystis* sp.; L foraminiferan, *Globergerina*.

IMAGES ADAPTED FROM DAKIN & COLEFAX 1940, NEWELL & NEWELL 1977 AND HALLEGRAEFF 1988

9.3.2 Subsampling

Subsampling is commonly required for samples of planktonic organisms before they can be counted under a microscope. Moreover, it may be necessary to have total counts of some organisms and subsamples of others, and depending upon their abundance within the samples. This is common where estimates of abundance of relatively rare organisms (e.g. fish larvae) are required as well as counts of more abundant smaller plankters. Pilot studies or good background information (see Chapter 2) are necessary to make decisions on the number of replicate subsamples that are required for an acceptable level of sampling precision, because subsampling error can be considerable (van Guelpen et al. 1982).

Folsam splitters are useful when samples are particularly large, and they have a high degree of precision when compared with other subsampling methods (van Guelpen et al. 1982). This device (Fig. 9.15, p. 258) splits the sample into two or more equal parts (relatively even splitting may take more than one attempt). Thus samples can be split into halves, quarters, eighths, sixteenths and so on. When small organisms are to be subsampled, a modified pipette is useful. The end can be broken off a 5–10 ml calibrated pipette to provide a wide mouth. A suction device (calibrated ones are useful) can then be used to subsample aliquots of known volume. For example, it may be convenient to add water until your sample equals 100 ml. Then mix, and extract subsamples of 1–5 ml, depending on concentration of plankton. Estimates of total numbers of particular plankters in the sample can then be obtained by multiplication. At least two replicate subsamples should always be taken.

Very small planktonic organisms such as protists may be subsampled as drops and/or replicate fields of view under a compound microscope.

9.3.3 Photographing samples of plankton

Silhouette photography is another technique that can be used for the qualitative and quantitative examination of zooplankton samples (Ortner et al. 1979; Lough & Potter 1983; Murdoch et al. 1990). A photographic record, either on film or photographic paper, of all or part of a plankton sample can be obtained relatively easily using flash techniques. For example, Murdoch et al. (1990) tipped a 50 ml subsample of plankton onto a 10.15 x 12.7 cm piece of Ilford Ilfospeed Grade 5 photographic paper held in a 3 cm deep tray of slightly larger dimensions for a snug fit. The flash was adapted to produce a point source of light (through a 2 mm hole). Organisms within the sample could then be identified

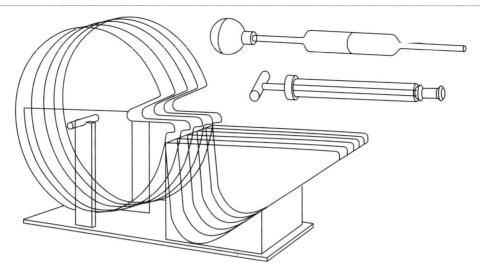

Figure 9.15 Design of a splitter and pipettes used to subsample plankton. The splitter is particularly useful for large samples and plankton that is too large to suck up with a pipette. The pipettes are usually most suitable for relatively small subsamples and are ideal for plankton in mesoplankton or below size ranges (Fig. 9.1). Replicate subsamples should always be taken.

and counted through microscopic examination of the image. This technique has a number of advantages: it provides a readily accessible long-term record of the sample, the photographic record is much easier to handle than the original sample and is amenable to subsampling with image analysis techniques; it is inexpensive, quick, and easy to perform. It is a particularly useful technique to use on unstable platforms such as research vessels, where microscope work is often difficult and samples may need to be examined in near-real time. One disadvantage of this technique is that identification beyond genera for zooplankton is sometimes difficult, particularly for copepods.

9.4 Categories of organisms and data

The taxonomic precision and presentation of data depends on the question that is being addressed.

Concentration: Data are often presented as number of organisms per unit volume. The units chosen will depend on the organisms that are sampled. For example, concentrations of meroplankton are often presented as number per m^3, whilst for fish larvae much greater volumes of water will be used (e.g. number per 500–1000 m^3). In general, it is probably most appropriate to use the average volume sampled for the presentation of data, to minimise potential distortion of data through multi-

plication or division. Rounding of the average to a multiple of 10 will make it easier for readers to compare concentrations in different studies. Catches are converted to per unit volume, based on a knowledge of the volume of water filtered (e.g. use of a flowmeter) or observed (e.g. transect methods).

Number per unit area: In studies where the objective is to determine where zooplankters of a particular species are most abundant, the depth of the water column needs to be considered. For example, concentrations of species x may be 10 per m^3 at locations with water depths of 10 and 100 m, respectively. If the water column is considered a cube that is 1 m^2 at the surface and stretches to the bottom, then 100 of species x would be found in 10 m of water and 1000 in 100 m. The expression of data in this way is common for studies of ichthyoplankton where samples are taken in nets with oblique tows; for example, number below 10 m^2 (Kendall & Picquelle 1989).

Wet weight/dry weight : Wet weight (in grams) or the volumetric displacement of plankton is often used as a crude measure of sample biomass. This measure is very crude and is of little value if the composition of the plankton varies greatly among times and locations. For example, samples of zooplankton with and without gelatinous organisms may vary considerably in wet weight, but not in total carbon. Some of these problems can be alleviated by measuring the weights of size fractions of plankton (e.g. Suthers & Frank 1990).

Dry weight/organic content : Measures of organic content are relevant where problems relate to the inputs and outputs of production from marine environments. There are acceptable relationships between the dry weight of plankton samples and the organic content of samples (grams of C). Samples are sieved and then rinsed in water to remove any interstitial salt. According to Steedman (1976), samples are dried in an oven at 70°C for 24 hours and then weighed. Pilot studies on the number of hours taken to remove most of the water from the sample should be done (i.e. times greater than 24 hours). It is generally assumed that the carbon content of samples accounts for about 42% of dry weight. Beers (1966) argued that this figure is the average carbon content from a sample of copepods, which are often the most abundant type of zooplankton in samples. This figure should be treated with caution, however, because preservation in formalin can alter dry weights and carbon content of zooplankton (Fudge 1968; Williams & Robins 1982).

Taxonomy

Planktonic organisms may be classified by taxon or size. For many biological questions, operational or functional taxonomic units are adequate (e.g, grazers; see Chapter 3). For example, in a study on the impact of fishes on zooplankton Kingsford and MacDiarmid (1988) did analyses on total zooplankton caught in a 0.28 mm mesh net and categorised zooplankton as copepods (of all orders), appendicularians (i.e. *Oikopleura* and *Fritillaria* spp.), gelatinous zooplankters (e.g. hydromedusae and salps), and macrocrustaceans (e.g. euphausids). For some questions it may be appropriate to identify species to the level of genus, species or even sex (e.g. Bradford-Grieve 1994), but this is difficult to do for many zooplankters and often requires considerable taxonomic expertise and time.

It may also be appropriate to sample and analyse plankton according to size. The size of planktonic organisms is often used as a proxy for taxonomic group or trophic status, and studies of the whole planktonic community frequently use estimates of the proportions of biomass within different planktonic size ranges to identify food web structures. Similarly, trophic models of plankton communities frequently incorporate size classes to represent trophic levels (Platt & Denman 1975; Morales et al. 1990, 1991; Kumar et al. 1991). The relationship between major taxonomic groupings and size groups is shown in Figure 9.1. While the trophic status of each size fraction overlap, various analytical techniques can be used to identify the proportion of each size-class attributable to different modes of feeding (i.e. heterotrophs versus autotrophs, carnivores versus herbivores).

Microscopy

The size of the target organisms and level of taxonomy that is appropriate for the study will determine what microscope is the most appropriate. Good-quality microscopes are essential for most studies. High-power epifluorescence microscopy is required for microbes and counts have to be made at 400–1000 x. Micro-zooplankton may have to be counted under a compound microscope at 100–200 x. Mesoplankton and fish larvae can generally be identified and enumerated at magnifications of 6–50 x. For some organisms you may require both binocular and compound microscopes. For example, although copepods can be counted easily under a binocular microscope, thoracic appendages may have to be observed under a compound microscope to key animals to the level of species or sex (e.g. Dakin & Colifax 1940; Bradford 1972). It is often useful to have top and bottom lighting in order to alternate use depending on the organisms you are observing. Fibre-optic cold-light sources for top lights are recommended, since this minimises the chances of convection currents in the observation dish.

9.5 Nutrients and physical measurements

9.5.1 Nutrients (for nitrates, nitrites, phosphorus, silica)

Measurement of mineral nutrients in the sea is often an essential component of plankton studies, especially if they focus on phytoplankton standing stock or primary production measurements. The concentration of macro-nutrients such as nitrate, phosphorus or silica often limits phytoplankton growth in the surface layers and coastal regions of the ocean, and is generally correlated to the community structure of the plankton.

Samples of sea water are collected from target depths using water bottles, and most oceanographic studies use a CTD-rosette system. Subsamples for nitrate/nitrite, reactive phosphorus and silica or ammonia are collected filtered and stored in acid-cleaned 125–250 ml polyethylene bottles rinsed with de-ionised distilled water prior to use. The bottles are rinsed three times with the sample water before final filling and capping, and care must be taken to avoid contamination. If the nutrient measurements cannot be made immediately, the bottles should be stored frozen in a clean environment to avoid contamination.

A number of studies (e.g. Ryle et al. 1981) have established that it is possible to store samples frozen and obtain high-precision estimates of nutrients. Analysis should take place within 2–4 weeks after collection. Measurements of the nutrients are generally made with an autoanalyser, and the chemical techniques are described in detail by Parsons et al.

(1993) and Grasshoff et al. (1983). The use of an auto-analyser can be expensive.

9.5.2 CTDs and data loggers

Internally recording stand-alone instruments, or data loggers, are now able to measure a wide range of parameters (e.g. light, fluorescence, temperature, conductivity and nitrate) and can be deployed in the sea for periods in excess of one year. These instruments enable variables to be measured over a range of temporal scales, from tidal and daily cycles to seasonal cycles. Because many of them are relatively cheap (under $500), arrays of them can be deployed at different depths and locations (e.g. HOBO probes: Doherty et al. 1996). Marine fouling can be a problem for the probes of some loggers if they are to be left in the water for a long period of time: they may have to be removed for cleaning on a regular basis (e.g. for meters that measure light levels and primary production).

Conductivity, temperature and depth devices (CTDs) enable more comprehensive descriptions of the physical structure of the water column, and simultaneously collect data on temperature and salinity with depth. These data can be used to obtain vertical profiles or contour plots of water-column parameters (e.g. Kingsford & Suthers 1996) on a variety of spatial scales. CTDs can also be towed slowly in surface waters to study horizontal variation in oceanography over small spatial scales (tens of metres), but care must be taken to avoid turbulence around the sensors (e.g. Kingsford & Gray 1996). Many CTDs are now small, so they can be lowered by hand from a small boat or deployed with optical plankton counters (see 9.2.12). Some caution is required to lower CTDs slowly enough to accommodate the response time of the sensors (e.g. the conductivity response takes longer than temperature). CTDs can also be fitted with a range of additional sensors, such as fluorometers, oxygen probes, nephelometers, PAR sensors, etc. Regular recalibration of the sensors is essential.

Sigma-t is used as a measure of water density. Density is determined by a complicated relationship between temperature, salinity and pressure (the latter is negligible over waters of continental shelves, deeper than 100 m). Plots of density, horizontal or vertical, are used to interpret thermohaline flow patterns. Because small changes in density can cause large changes in water circulation, water density is measured to five decimal places (e.g. 1.02677 g/cm^3). Sigma-t is an abbreviation of this, where 1 is deducted from this figure and the residue multiplied by 1000 = 26.77 (Duxbury & Duxbury 1991). In practice the function is difficult to calculate, although algorithms are available in public-domain software. A Pascal function for the calculation of sigma-t is given on p. 326 of Baker and Wolff (1987).

Temperature and salinity data is often best plotted as vertical profiles against depth. Inclusion of Sigma-t in these plots will also aid interpretation of the water-column structure. Plots of temperature versus salinity for each profile together on one plot is also a useful means of identifying different water masses, since each water mass generally has a distinctive temperature/salinity relationship. If temperature/salinity depth profiles have been collected at stations along transect lines or grids, it is possible to contour isotherms/isohalines on depth versus distance plots, or different depth strata over the grid area, respectively. Such plots enable the physical structure of the study region to be described, and are important for identifying physical features such as eddies and fronts.

9.5.3 Current measurements

Measurements of current velocities and a synoptic view of current patterns are crucial to many biological studies (Mann & Lazier 1991). Measurements may be as rudimentary as calculating current speed from drogues or current meters deployed at given locations, to complex numerical models that predict water movements over a range of temporal and spatial scales (e.g. Murdoch et al. 1990). The method chosen to measure currents again depends on the biological question. Methods can be classified loosely as tracking of water masses with drogues (Lagrangian methods) and measuring current velocity at a fixed point (Eularian methods). Lagrangian methods have been used for a wide range of applications that range from studying small-scale current patterns around reefs (e.g. Kingsford & MacDiarmid 1988; Shapiro et al. 1988), the movement of water along frontal boundaries (Kingsford & Suthers 1996), tracking cohorts of fish larvae for hours and days (Fortier & Leggett 1985) and tracking water masses over thousands of kilometres to understand the movement of planktonic and nektonic organisms. In the vicinity of reefs, the position of drogues can be marked on a map. Hand-held GPS will assist greatly with this method, especially when drogues move further from land. Waypoints can be used to plot the position of drogues and or distance and bearings from known waypoints that can be found on a chart. Drogues left to drift for hundreds to thousands of kilometres may be tracked by satellite. The use of Acoustic Doppler Current Profilers (ADCP) has now become common, and these can be attached to an oceanographic vessel or small boat to provide real-time depth profiles of current velocities while the vessel is steaming.

Eularian methods range from the use of sophisticated

current meters or ADGPs that can be left at a location for long periods of time (e.g. S4 current meters: Wolanski 1994) to diver-operated measurements. Current meters of the S4 type have many megabytes of memory and record current strength and direction as well as water depth (which will vary with state of the tide and swell for these tethered devices) and temperature. There is a wide range of mechanical flowmeters that can be used to measure the amount of water that passes a point, and when this is combined with measurements of time, current speeds can be calculated. Crude estimates of current speed are useful before you buy these devices because you often have to make decisions on the type of rotor that is used. For example, the General Oceanics 2030 flowmeter has a standard rotor (25–270 cm/sec) and a low-speed rotor (1–100 cm/sec). Flowmeters of this type are commonly used to measure the amount of water filtered by a plankton net. In some situations current speeds may be difficult to measure with a flowmeter (e.g. current speeds too low), in which case a diver may make replicate measurements of the passage of particles past a quadrat (e.g. Kingsford & MacDiarmid 1988). Quadrats with dimensions of 0.5 m or less reduce parallax problems.

In some cases a relative measure of the amount of water passing a point is required. Clod cards are often appropriate for problems of this type, although erosion of the clod in areas of high sediment load may confound interpretation of relative water movement. Clod cards are made from dental cement, which can usually be obtained from medical suppliers (e.g. EBOS in Australasia). The cement is mixed with water then placed in a small mould with a rounded surface (e.g. plastic egg containers). The dried moulds are glued to plates with an all-purpose silicon glue and labelled. Moulds are dried to extract moisture (80° C for eight hours), weighed and deployed in the field. Once retrieved, they are washed in fresh water, dried and reweighed. The weight loss of the clod can be used as a relative measure of current speed. It is advisable to use replicate clod cards at each site, because some clods may get bubbles in them (Doty 1971).

On small spatial scales, current patterns are influenced by phase of the tide, wave height, mainstream currents, freshwater inflows, wind and topographic forcing (Wolanski 1994). On large scales, contributing factors include meteorological forcing by oceanwide phenomena such as the Southern Oscillation (which influences sea levels, wind and wave height), the rotation of the earth (Coriolis forces), deep ocean currents (e.g. deep western boundary currents, polar currents), planetary waves, coastal trapped waves, internal tides, bottom friction (particularly in shallow areas) and topography.

9.5.4 Remote sensing

Remote sensing can provide a great deal of information on marine coastal environments, and information can be collected by vehicles ranging from low-altitude aircraft to spacecraft. Information is generally obtained through a process of data capture, in the form of photography or digital data and data analysis (Young 1993). The resolution of images varies greatly, depending on the altitude from which data are collected and the sensitivity and spatial resolution of the sensors (centimetres to kilometres per spatial pixel). A review of the characteristics of remote sensing systems in space is given in Young (1993). The relevance of this type of information to changes in planktonic assemblages may include weather analysis; tracking cyclones; measuring sea-surface temperatures so that climatic influences such as El Niño-Southern Oscillation (ENSO: see Allan et al. 1996) and the East Australian Current can be monitored as well as smaller-scale phenomena such as upwelling and fronts (see Plates 55, 57); sea-surface colour to measure marine biomass; sediment content (e.g. from riverine plumes: Grimes & Kingsford 1996), sea surface chlorophyll (distribution of phytoplankton, Plate 59); eddy genesis; sea-surface elevation; wind velocities and wave climate; the breakup of subpolar ice and tracks of individual icebergs; and the spread of ash clouds from volcanoes that may influence the input of compounds to the pelagic environment and ultimately affect the nature of marine sediments.

9.6 Experimental methods

It could be perceived that experiments on planktonic organisms can only be done in the laboratory. Although laboratory experiments are useful, and often the only option, field studies should be considered seriously because organisms are normally subjected to natural conditions. As with all experiments, controls are critical for the interpretation of manipulations (see section 2.6). Moreover, treatments should be considered that measure potential experimental artifacts (e.g. increases in mortality, changes in feeding behaviour, differences in the growth of organisms).

9.6.1 Laboratory experimentation

There are still a large number of measurements for planktonic organisms that can only be made under laboratory conditions. This is particularly true for behavioural (Sulkin 1990), reproductive, physiological and metabolic measurements, such as respiration rates, ingestion and egestion rates of zooplankton, egg production and larval development times, and excretion rates. *In situ* studies of the behaviour and physiology of marine pelagic organisms are difficult to undertake in the ocean. Behavioural observations of zooplankton have been made through use of video systems, divers, and submersibles, but these have generally focused on large species visible to the naked eye. Physiological/metabolic measurements and behavioural observations of pelagic organisms are essentially confined by current technology to laboratory studies of live specimens. Samples of live plankton are usually collected in water bottles or plankton nets (collecting buckets with small mesh windows only at the top will often reduce damage to specimens). Damage to fragile species, especially phytoplankton and gelatinous zooplankton, is still a problem with traditional collection techniques, however, and needs to be considered when collecting live organisms for further experimentation. Transfer of the live plankton to other containers should be done immediately, and the densities of organisms kept relatively low. Live specimens should be handled with care to minimise damage.

Once collected and maintained in culture, organisms can then be studied to establish feeding rates and habits, physiological tolerance to physical and chemical conditions, assimilation and growth efficiencies, respiration and excretion, and development and productivity (e.g. Kleppel 1992, 1993; Gifford 1993; Pavlova 1993; Bailey et al. 1994; Hwang et al. 1994; Klein-Bretler et al. 1995). Methods for collecting live zooplankton, rearing and culture techniques, and experimental methods for undertaking physiological/metabolic measurements for zooplankton are also described in Omori and Ikeda (1984).

9.6.2 Mesocosms

The manipulation of plankters of a variety of sizes is often difficult in the field, but mesocosms (mesh or part-mesh enclosures) can provide a method of tracking physical and biological conditions that plankton experience while maintaining the behavioural characteristics of the enclosed organisms. Organisms may be enclosed with other organisms to study interactions between them, or be enclosed in a mesh that retains the target organisms and allows other organisms (e.g. food) into the enclosure. The term 'mesocosm' can refer to environmental chambers that are used on land (e.g. Keller et al. 1990) as well as free-floating or tethered structures in the ocean. de Lafontaine and Leggett (1987) evaluated enclosures for field studies of larval fishes. The internal volume of the net was 3.2 m^3 (cf. Keller et al. 1990: 13 m^3) and consisted of 4.1 m cylinder sections (diameter = 1 m, 25 micron Dacron) each

separated by a metal ring to maintain shape and facilitate removal of the enclosure from the water. A wooden frame at the top of the device allowed attachment to a block and tackle so that the mesocosm could be hauled on board. A fifth section was a cone of 53 micron mesh and a cod-end. It took about an hour and a half to complete the hauling operation of each enclosure.

The same design of mesocosm was used by Cowan and Houde (1990), who attached multiple enclosures to rafts and retrieved enclosures using a winch on a portable A-frame. Where zooplankters were sampled, a submersible pump was used so that plankters could be sampled at discrete depths. de Lafontaine and Leggett concluded that 'low cost, ease of handling, environmental reproduceability and quality of replication provided by enclosures made them particularly appropriate for experimental studies on the interactions between larval fishes, their predators and prey'. Larvae and plankton were found to migrate in mesocosms as they do in the field and there was a near-instant response to changes in temperature, salinity and dissolved oxygen. Densities of phytoplankton were less variable inside the net and the size fraction of phytoplankton that entered enclosures was dependent on mesh size, as would be expected. Mesh enclosures provide a more 'natural' environment than plastic bags. For example, particles had a lower rate of settlement than had previously been found in plastic bags. Since feeding success of larvae may vary with the magnitude of disturbance, differences of this type are important.

Mesocosm technology may be suitable for studies on the influence of pollutants on zooplankters (e.g. Fairchild & Little 1993) and conditions experienced by larvae in different watermasses.

9.6.3 Tagging experiments

It may be important to tag the early life-history forms of fishes and invertebrates to validate ageing techniques or to track individual larvae or batches of larvae. Methods of tagging very small animals include the use of visible and invisible tags.

Visible tags include vital dyes such as Bismark Brown. Although these have been used to tag a variety of larvae (Levin 1990), care must be taken to check the persistence of tags in a pilot study before larvae are released. A disadvantage of visible tags is that they may alter the vulnerability of larvae to predation by making them more conspicuous. Moreover, many of these dyes are only suitable for short-term studies.

Invisible markers are attractive in that they should not alter the vulnerability of larvae to predation, but the detection of tags is more time-consuming because the specimens may have to be observed under a microscope with fluorescence, or analysed chemically. Fluorescent tagging is possible for the eggs and larvae of molluscs and fishes by batch marking in antibiotics such as calcin (e.g. Rowley 1993) and tetracycline (Geffen 1992); 0.5 g per litre is a typical dose. These tags leave a permanent mark. Jones and Milicich (pers. comm.), for example, batch tagged the eggs of reef fish (*Pomacentrus amboinensis*) and demonstrated that some of them settled back to their natal reef after 25 or more days in the plankton as larvae. Other chemical tags include the use of strontium, which can be detected chemically with technics such as Inductively Coupled Plasma Mass Spectrometry (ICP-MS: Pollard et al. in press).

Invisible microwire tags are used to identify individuals (see Chapter 5 and section 6.2.4k), but their use would generally be limited to adults and robust juveniles. It is unlikely that tags of this type could be used with delicate larvae. The advantage of these tags is that they can be detected easily with a special detector, without destroying the specimen.

9.6.4 Colonisation experiments

Some planktonic and nektonic organisms are attracted to structures in the pelagic environment. Copepods may be attracted to fragments of marine snow to feed on protists, phytoplankton and debris stuck to them (Alldredge & Silver 1988). Fish and invertebrate larvae are often found in association with objects such as drift algae and flotsam (Mansueti 1963; Hunter & Mitchell 1967; Kingsford 1993), as are larger fishes (Rountree 1989, 1990). The use of fish-attraction devices (FADs) is the focus of pelagic fisheries in many parts of the world (Boy & Smith 1984). Colonisation experiments, therefore, are relevant to studies on pelagic ecology, dispersal and applied aspects such as the use of natural or artificial structures for nursery areas and the redistribution of planktonic or nektonic organisms for collection (i.e. experiments, measurements of relative abundance) and fisheries.

Artificial structures of a variety of shapes (e.g. Mitchell & Hunter 1970; Rountree 1990) and natural structures (e.g. drift algae: Kingsford & Choat 1985; Kingsford 1992) have been used in experiments on planktonic and nektonic organisms. In some cases structures have been tethered (e.g. Edgar 1987), while others have been free-drifting (e.g. Kingsford 1992). Where structures are tethered in strong currents, some organisms may find it difficult to stay in association with objects, or not be attracted. Jars (in the case of microbes), dip nets, mesh bags or purse seine nets may be used to collect organisms from experimental objects.

When conducting colonisation experiments, you should consider using open-water controls and trialing

objects of different sizes, as this may or may not have a strong influence, depending on the range of sizes used and the behaviour of fishes (Safran & Omori 1990; Kingsford 1992). If an array of drifting objects are released, it is useful to deploy long-line flags that have a drogue and perhaps a radar reflector attached. This provides tangible reference points when retrieving and tracking the structures. Untethered experiments are difficult to follow at night, unless equipped with radio transmitters, satellite detection systems (e.g. ARGOS) and/or lights (but it is likely that lights will influence catches: see 9.2.10). The option of living on board the structure makes tracking unnecessary (e.g. Gooding & Magnuson 1967), but this is very time-intensive for multiple replicates.

9.7 Examples of sampling designs

9.7.1 Are fish larvae influenced by the proximity of reefs?

Problem: Determine whether the distribution patterns of pre-settlement reef fish are influenced by the proximity of reefs, regardless of the distance they are located across a continental shelf (Kingsford & Choat 1989)

Situation: The study was done off the northeastern coast of New Zealand. Islands are found at different distances across a continental shelf of about 70 km. It was possible that sampling down a channel without regard for the presence of islands would miss certain taxa. Preliminary studies had indicated that acceptable precision on estimates of abundance were obtained with three replicate tows.

Approach: Sampling was done at three distances from shore (inner, mid and outer; Fig. 9.16, p. 264). Within each distance, sampling was done at random locations, and at each location bays and points were sampled close to shore and at a site 3 km from shore. At each of these geographic entities three replicate oblique tows were taken with a square net of 1 m^2 mouth area, 1 mm mesh and an open-area ratio of 10:1. Oblique tows were chosen because a wide variety of taxa were being sampled. Two cruises were done that took about four days to compete. It was possible that some patterns of abundance could be confounded by time, and therefore a separate design was completed where a single location (i.e. with geographic entities of a bay, point and offshore) were sampled twice per day for three days).

A number of patterns were identified. Some groups such as tripterygiids stayed close to reefs regardless of

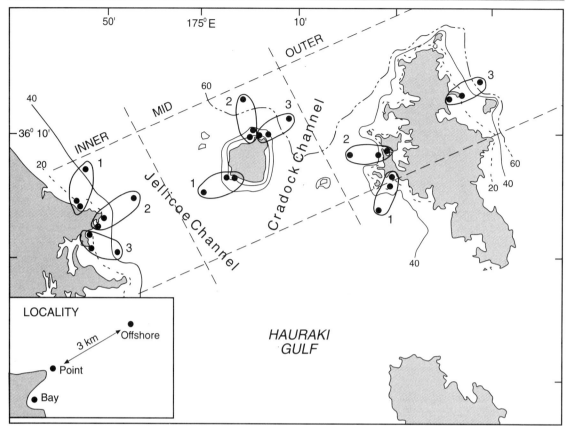

Figure 9.16 Sampling design used by Kingsford and Choat (1989) to sample ichthyoplankton near to and away from land at different distances across a continental shelf.

distance across the continental shelf, while others (e.g. snapper, *Pagrus auratus*) showed no affinity for a particular geographic entity. These conclusions corroborated with the three-day study. In addition to net tows, diver observations were made in shallow water at different states of the tide. These demonstrated that substantial numbers of tripterygiid and gobieosocid larvae were found within 10 m of the shore.

High concentrations of larvae have been found close to shore in other studies (e.g. Marliave 1986), and at times have been related to demersal modes of spawning (e.g. Leis 1991; Suthers & Frank 1991). Some caution should be taken when comparing these studies as the definition of near-shore varies greatly among authors (metres in some, kilometres in others). See Gray (1996a, b) for other studies on short-term variation in abundance.

9.7.2 The influence of physical oceanography on vertical distribution patterns of fish larvae

Problem: Do thermoclines explain the vertical distributions of larval fishes in the dynamic coastal waters of south-eastern Australia (Gray 1996c)?

Situation: Information from other parts of the world suggested that the position of the thermocline would have a great influence on the distribution patterns of fish larvae (e.g. Heath 1992). Thermoclines, therefore, could be a useful predictor of patterns of abundance of larvae in New South Wales.

Approach: The approach used was to sample the water column at multiple depths (surface, 5, 15, 30, 45, 55, and 65 m; n = 3 replicate tows) at two locations where total depth was approximately 70 m. The position and intensity of thermoclines was measured with a CTD. Samples were taken six times. On four occasions the water column was stratified by thermoclines that ranged greatly in depth and depth range. These results were compared with two times when no thermoclines were found. Great differences in the abundance of fish larvae were found with depth, in both the presence and absence of thermoclines. Some species were consistently associated with shallow, mid and deep strata regardless of oceanography. The presence of a thermocline was not considered a good predictor of the vertical distribution of larval fish. In other parts of the world, good relationships between the abundance of larvae and plankton have been found with the position of thermoclines (e.g. Frost, 1988). It was argued by Gray (1996c) that the thermocline varies greatly close to the coast of Sydney over short temporal and spatial scales as a result of highly dynamic oceanography (Middleton 1996), and this could explain the differences in patterns compared to other studies. Moreover, biotic interactions (e.g. distribution of food) and larval behaviour may have an additive effect with oceanography to depth-related patterns, or even be totally responsible.

9.7.3 Experimental investigation of the fishes associated with drifting objects

Problem: Determine the rate at which larval and pelagic juvenile fish are attracted to artificial drifting objects in temperate waters of Australia (Druce & Kingsford 1995).

Situation: Information on the colonisation of drifting objects from New Zealand (e.g. Kingsford & Choat 1985; Kingsford 1992) suggested that fishes would be quickly attracted (minutes and hours) to artificial structures.

Approach: Mopheads were used as structures. These were easy to replicate and have been used successfully in other parts of the world to attract crab megalopa (Shanks 1983). Sampling designs were as follows:

1. Open-water control, drift time = 1 minute, 1 hour, 3 hours, 5 hours; n = 3 mops.
2. Large and small mops of different colours (black and white). This was an asymmetrical design (see Chapter 3) where the orthogonal part of the design (colour versus size) was compared with open-water controls.
3. Open-water controls were compared with experimental mops and natural drift algae (*Phyllospora comosa*) that had been cut from reefs and left to drift.

Primarily pelagic juveniles were attracted to FADs of all sizes, colours and sizes. Densities around FADs were always greater than in open water controls. Some fishes were attracted to objects in drift times as short as one hour. Fishes attracted to FADs included syphaenids, mullids, blenniids, sparids and pomacentrids. No size differences in catches were recorded for FADs of different sizes, although the range of FADs used was relatively small; other studies have found major differences

in abundance of fishes and invertebrates around natural FADs of different sizes (Kingsford & Choat 1985; Safran & Omori 1990; Kingsford 1992).

The study demonstrated that manipulative experiments could be used in the pelagic environment and that measures of relative abundance could be obtained with FADs, and therefore augment information from other techniques such as ichthyoplankton nets and light traps. Finally, all of the fishes collected were in excellent condition and could have been used for descriptions of development and for experiments.

9.7.4 Scyphomedusae: pelagic-benthic links

Purpose: Determine the factors influencing patterns of abundance of the scyphozoan *Phyllorhiza punctata* (Cnidaria: Rhizostomaeae) in two estuaries in Western Australia (Rippingale & Kelly 1995).

Situation: Two estuaries superficially appeared to have great differences in numbers of jellyfish. It was well known that the life history of medusae included benthic (polypoid) phases and pelagic (medusae) phases. This has been found for all of the major orders of the class Scyphomedusae, including semestostomes (van der Veer & Oorthuysen 1985; Grondahl 1988a, b) and rhizostomes (e.g. Calder 1982). Benthic and pelagic phases were studied, thus all stages of the life history of *P. punctata* were considered.

Approach: Observations of medusae were made from bridges, wharves and jetties in one estuary. More quantitative information could have been collected with visual transects (see 9.2.11) or trawl nets. No information was collected from the other estuary, even though the absence of animals from estuary 2 was only based on anecdotal information. Pieces of rock and other hard substrata were collected from estuary 1 and placed in aquaria so they could be inspected for scyphistomae. This part of the project, however, was unsuccessful in that no scyphistomae were found on natural substrata. Sexually mature medusae were collected during the summer while adults were abundant in the estuary. They were shaken in a bucket with estuarine water to potentially release planulae (ovoid shape and 300–500 mm) held in the tentacles. Planulae were successfully obtained using this method, enabling the growth and mortality of polypoid phases to be measured in waters of different salinities (Fig. 9.17, p. 266). Asexual reproduction was also found in scyphistomae, where ciliated buds became free from the parents and behaved in ways that were indistinguishable from planulae (see generalised life history). Medusae were rare in low-salinity waters found during the winter rains. Scyphistomae

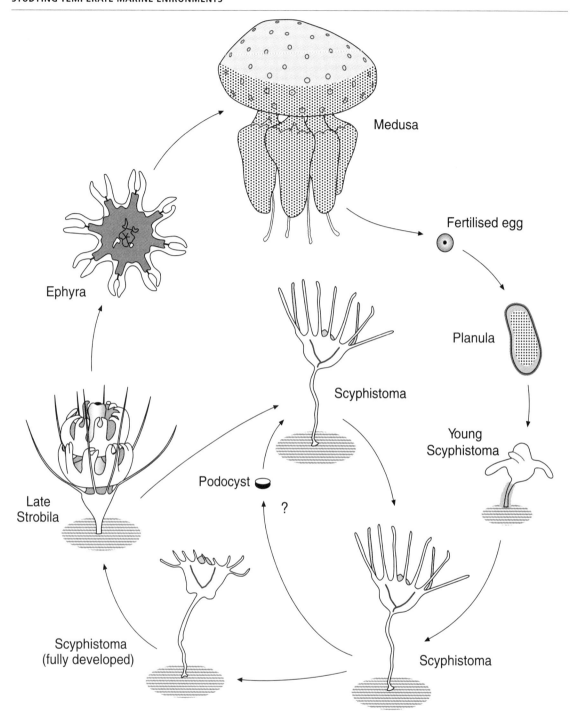

Figure 9.17 Life history of Phyllorhiza punctata, a rhizostome jellyfish, showing pelagic and benthic phases. All of the polypoid phases are attached to the substratum.
? = It is not known if this species has a podocyst stage as for other rhizostomes.

MODIFIED FROM RIPPENGALE & KELLY 1995 WITH PERMISSION

grew and fed successfully at salinities above 15‰ and temperatures above 20° C. It was argued that scyphistomae could survive in deep water of higher salinity in only one estuary. The other estuary was shallow and subject to regular flushings of low-salinity waters.

A knowledge of life history of scyphomedusae is also relevant to understanding the nature of introductions. It was noted by Rippengal and Kelly that early research into the distribution of medusae suggested it was highly likely that polypoid and perhaps ephyrae were transported by ships (see introductions in Chapter 1).

9.7.5 Salp grazing and effects on phytoplankton abundance

Purpose: Use field data and modelling to examine dynamic interactions of salps and phytoplankton biomass (Zeldis et al. 1995).

Situation: Thaliacean blooms are common in coastal waters of the world, and large numbers of these animals had previously been found in the Hauraki Gulf, New Zealand, where the study was done (Jillett 1971).

Approach: A multi-disciplinary approach was used in the field and collected data was augmented with numerical modelling. Sampling was done at stations organised as a grid (regular sampling) that encompassed most of the Hauraki Gulf, New Zealand (about 50 x 50 km). Sampling using multiple methods was done in three austral summers from 1985 to 1987. At each grid station conductivity-temperature-depth (CTD) data were collected at 1m intervals and water samples for chlorophyll *a*. Nutrients and phytoplankton were taken at a number of discrete depths (e.g. 1, 10, 15, 20, 25, 35, 45 and 55 m) using vertical profiling from a CTD hose assembly (i.e. water was pumped to the surface through a hose such as a 50 mm i.d. hose, which could supply water at 125 l/min, with a known delay and sampled at the outlet). Mesoplankton was collected with single depth-integrated/oblique tows (towed at 2–2.8 knots) that sampled to within a metre of the bottom (cylinder/cone net, mouth 0.8 m, 365 mm, 1:12 ratio). Large quantities of salps often had to be removed using a coarse sieve (3 mm). Small organisms were washed through and the volume of salps measured. The vertical distribution of salps was studied on occasions with an opening/closing net. Phytoplankton species composition was studied using whole-water samples stored in Lugol's iodine, and subsamples were examined at 200 x using an inverted microscope (picoplankton < 2 mm were beyond the resolution of the microscope and were not included in the analysis). Zeldis et al. (1995) noted that plankton in this size range typically contributed about

20% of the total chlorophyll *a* in the water column. Chlorophyll *a* was measured using *in vitro* and *in vivo* methods, and results were calibrated against known concentrations of standards. Some problems in measurement of chlorophyll *a* were encountered because of degradation of samples. The filtrates of chlorophyll *a* samples were used for analyses of nutrients. Nitrogen for all zooplankters was estimated from wet weight by using published wet weight-carbon relationships for each group. A nutrient-phytoplankton-zooplankton model was used to study the potential effects of salp introductions and subsequent bloom formation in the Hauraki Gulf. Specifically the model contrasted a salp-dominated 'community' that has high sedimentation of organic wastes with a salp-free microzooplankton-dominated community within which wastes were recycled *in situ*.

In conclusion, rates of sedimentation of waste material from the grazing of salps were greater than the mixing of re-mineralised nutrient back up through the pycnocline. This can cause a deep chlorophyll *a* maximum to persist for some time after grazing pressure dissipates. Different regimes of grazing, light and nutrients correlated with major changes in the composition of phytoplankton. The authors concluded that in coastal waters salps appeared to deepen phytoplankton distributions and reduce biomass, rather than to completely remove phytoplankton biomass from the euphotic zone as for slope and oceanic waters. Interestingly, the authors found that the survival of first feeding snapper (*Pagrus auratus*) was lowest in the presence of large numbers of salps. Although interactions between plankters and larvae have been argued to influence survivorship in other systems (Frank & Leggett 1985; Cushing 1990), hypotheses for other species and latitudes have often focused on the role of physical oceanography (Sinclair 1988). For other works on salps, see Heron (1972a, b), Heron and Benham (1984), Morris et al. 1988); and for other important roles that gelatinous zooplankters have in pelagic environments see section 9.7.

Grid, or partial grid, sampling designs are commonly used in studies on the distribution patterns of fish larvae (e.g. Crossland 1981, 1982a; Kendall & Picquelle 1989) largely because the relationships with them and variation in the abundance of other organisms (e.g. abundance of salps), geography (e.g. shelf, slope and oceanic stations) and oceanography (e.g. different currents: Robertson 1980) are not always apparent to allow for stratified sampling. Consideration should be given to analyses because there is not replication at each station. In some studies it may be appropriate to take replicate tows at a subset of stations to make stronger cases for the generality of some patterns.

SELECTED READING

Particular reference is given to temperate waters of Australia and New Zealand. The references listed are examples of research that will lead the reader to a broader literature, also see the text of the chapter. * indicates a multi-authored book or collection of proceedings; references to volumes of a journal refer to a set of relevant papers.

Benthic-pelagic coupling: Alldredge & Silver 1988; Parsons & Lalli 1988; Graf 1992.

Biological oceanography: Hamner & Hauri 1977; Alldredge & Hamner 1980; Tranter et al. 1980; Foster & Battaerd 1985; Bowman et al. 1986; Griffiths & Wadley 1986; Kingsford & Choat 1986; Wolanski & Hamner 1988; Black et al. 1990; Murdoch et al. 1990; Mann & Lazier 1991; Black 1993; Bradford-Grieve et al. 1993; Moore & Murdoch 1993; Kingsford & Suthers 1994, 1996; Bjorkstedt & Roughgarden 1997; Griffin et al. 1997.

Condition of plankton: Fraser 1989; Ferron & Leggett 1994; Lemmens 1994; Westerman & Holt 1994; Rissik & Suthers 1996.

Demersal zooplankton: Clutter 1969; Hamner 1981; Alldredge & King 1977, 1985; McWilliam et al.1981; Youngbluth 1982; Jacoby & Greenwood 1988; Ritz 1994.

Ichthyoplankton ecology: Kingsford 1988 (review); Bailey & Houde 1989; Leis 1991; Heath 1992; Kingsford 1993; Leggett & Deblois 1994.

Ichthyoplankton – reef fish: Sale 1991*; *J. Mar. Freshw.* 1996 vol. 47.

Larvae-teleplanic: Scheltema 1986, 1988.

Larval transport: Shanks 1983, 1995; Phillips 1981; Kingsford 1990; Caputi et al. 1996.

Marine protected areas: Carr & Reed 1993; see also Chapter 1.

Marine snow: Prezelin & Alldredge 1983; Alldredge & Silver 1988; Shanks & Edmondson 1990; Coombs et al. 1994.

Methodology: UNESCO 1968; Omori & Ikeda 1984.

Modelling (biological oceanography): Alexander & Roughgarden 1996.

Physical oceanography
General: Pond & Pickard 1978; Gade et al. 1983; Duxbury & Duxbury 1991; Baines 1995; Baines & Murray 1995; Summerhayes & Thorpe 1996; see also Biological oceanography.

New Zealand: Brodie 1960; Heath 1985; Harris 1990; Sharples 1997.
Australia: Mulhearn 1987; Schahinger 1987; Tate et al. 1989; Black et al. 1990; Griffin & Middleton 1991; Gibbs et al. 1991; Wolanski 1994; Middleton et al. 1996.

Plankton ecology (see also Biological oceanography)
General: Alldredge 1972, Silver et al. 1978; Goldman et al. 1979; Raymont 1983a, b; Purcell 1985; Warwick & Joint 1987; Kingsford 1993; Kiorbe 1993; Arai 1997.
New Zealand: Brodie 1960; Jillett 1971; Maddock & Taylor 1984; Murdoch 1989; Murdoch & Quigley 1994.
Australia: Heron 1972a, b; Fancett 1986, 1988; Jenkins 1988; O'Brien 1988; Clayton & King 1990; Ritz 1994.

Plankton migrations: Forward 1988; Haney 1988; Heath 1992.

Plankton and planktivores: Bray 1981; Kingsford & MacDiarmid 1988.

Pollution: Black et al. 1990; Heath 1992; Kjorsvik et al. 1990; Breitberg et al. 1994; Gray 1996a; Kingsford & Gray 1996; Kingsford et al. 1996.

Preservation: see Chapter 10.

Production and enrichment: Tranter et al. 1980; Platt et al. 1989; Behrenfield et al. 1996; Coale et al. 1996.

Protist ecology: Capriulo 1990; Margulis et al. 1990; Sherr & Sherr 1991.

Red tides and toxic blooms: Devassy 1989; Mackenzie 1991; Villarino et al. 1995.

Reproduction: Jacoby & Youngbluth 1983; see also Plankton ecology.

Sediments and turbidity: Griffiths & Glasby 1985; Grimes & Kingsford 1996.

Settlement and recruitment: Sale 1991*; Watanabe et al. 1996*.

Taxonomy: see Chapter 10.

Trophic interactions and cascades: Mazumder et al. 1990; Roman et al. 1990.

Tropical plankton: Sorokin 1993.

Water quality: Corbett et al. 1993.

TEN

IDENTIFICATION, TREATMENT AND RECORDING OF SPECIMENS

M. J. Kingsford and C. N. Battershill

10.1 Introduction

A wide diversity of species are often encountered in studies of reefs, soft substrata and planktonic assemblages. The accurate description of taxonomy is vital for characterising habitats and examining interactions and associations of target species. It is important that collection and preservation of samples and specimens for taxonomy be treated rigorously for the following reasons: (a) to gain consistent measures of species 'diversity' (number of species) and 'evenness' (relative abundance of species maintained within the assemblage over time), which will permit valid comparison with other locations; (b) uncertain taxonomy can only be checked by examination of properly preserved specimens in consultation with experts.

In some studies, new species may be discovered or range and distribution records of known species established or extended. In these cases, specimens should be sent to a museum so identification can be checked and the specimens catalogued and deposited within the museum's collection (e.g. Schiel et al. 1986; Kingsford et al. 1989). These specimens may become the 'type' for the species – the original reference material for the new species description, against which comparisons with other specimens can be made.

This chapter provides a brief guide to the correct handling and preservation of marine plant and animal specimens, and lists experts in the taxonomy of plants and animals most commonly encountered by marine researchers. These experts are internationally recognised, but specialise in southern hemisphere temperate marine biota. Major source works are also included.

As for all components of a study, the level of taxonomy required and the appropriate method of preservation will vary according to your biological question. Appropriate levels of taxonomy may be at the level of species, genus or logical groups of species that are often operational or functional taxonomic units (e.g. grazing gastropods; see section 4.5).

Examples of studies that required expertise on taxonomy and preservation are given at the end of this chapter.

10.2 Experts on taxonomy

Before any collection is made that may require the services of an expert taxonomist, it is advisable to contact a knowledgeable person to check on any particular requirements for your target organisms and to check whether they can help or collaborate. You may, for example, need information on taxonomy for field identifications; preservatives (especially where there may be special considerations for electron microscopy and other specialist tasks); and advice on how, when and where to collect organisms. If you do not know whom to consult, seek advice from local institutions. If you have access to the World Wide Web, you will be able to browse for the latest information on institutions and resident taxonomists. Examples of appropriate institutions follow for Australasia.

New Zealand
Ministry of Fisheries, the National Institute of Water and Atmospheric Research Ltd (NIWA), museums (especially the Museum of New Zealand in Wellington), universities, marine laboratories in the North Island (University of Auckland's Leigh Marine Laboratory, Victoria University of Wellington's Island Bay Marine Laboratory) and South Island (University of Canterbury's Kaikoura Marine Laboratory, University of Otago's Portobello Marine Laboratory), Department of Conservation regional offices and field centres,

regional councils, catchment authorities, port companies, marine sciences centres and marine environmental consultancies. The annual *New Zealand Marine Sciences Review* (available in most libraries) documents current activities and backgrounds most institutions and people involved in marine science in New Zealand.

Australia

Museums (e.g. Australian Museum in Sydney), Fisheries Research Centres (in all states), Conservation and Land Management Centres (CALM), Environmental Protection Authorities (EPA), marine laboratories, Institutes of Marine Science (VIMS, Victoria, and AIMS, Townsville and Karatha), CSIRO (Hobart), water boards, field offices of national parks and universities (in all states). Annual reports of the Australian Marine Sciences Association (AMSA), Australian Coral Reef Society (ACRS) and Australian Society for Fish Biology (ASFB) are useful for the names of members.

The list below provides examples of persons with current working experience in the taxonomy of species in Australian and New Zealand temperate waters. A comprehensive list of experts involved in the taxonomy of invertebrates in temperate waters is found in Cook (in press).

Algae

C. H. Hay, Cawthron Institute, 98 Halifax St (East), Private Bag 2, Nelson, New Zealand

A. Miller, Royal Botanical Gardens, Mrs MacQuaries Road, Sydney, NSW 2000, Australia

W. A. Nelson, Museum of New Zealand, PO Box 467, Wellington, New Zealand

M. J. Parsons, Lincoln University, Private Bag, Lincoln, Christchurch, New Zealand

B. Womersley, Department of Botany, University of Adelaide, SA 5000, Australia

Porifera

B. Alvarez de Glasby, NIWA, PO Box 14901, Kilbirnie, Wellington New Zealand

C. N. Battershill, NIWA, PO Box 14901, Kilbirnie, Wellington, New Zealand

P. R. Bergquist, School of Biological Sciences, University of Auckland, PO Box 92019, Auckland, New Zealand

J. Fromont, Museum of Western Australia, Perth, Western Australia 6010, Australia

M. Kelly-Borges, Unitech, PO Box 92025, Auckland, New Zealand

J. N. A. Hooper, Queensland Museum, PO Box 300, South Brisbane, Queensland 4101, Australia

Bryozoa

D. P. Gordon, NIWA, PO Box 14901, Kilbirnie, Wellington, New Zealand

Hydroids and Cnidaria (including scleractinian corals)

P. Alderslade, Darwin Museum, Darwin, Northern Territory 5794, Australia

D. Dunn (Anemones), Department of Invertebrate Zoology, California Academy of Sciences, Golden Gate Park, San Francisco, CA 94118, USA

K. R. Grange, NIWA, PO Box 893, Nelson, New Zealand

K. Miller, NIWA, PO Box 14901, Kilbirnie, Wellington, New Zealand

P. Schuchert, Zoological Institute, University of Basel, Rheinsprung 9, CH-4051, Basel, Switzerland

C. Wallace (Corals), Museum of Tropical Queensland, 84 Flinders St, Townsville, Queensland 4810, Australia

C. Veron, (Corals), AIMS, PMB No. 3, Townsville, Queensland 4810, Australia

G. Williams (Octocorals), Californian Academy of Sciences, Golden Gate Park, San Francisco, CA 94118, USA

Polychaetes

P. Hutchins, Australian Museum, PO Box A285, Sydney, NSW 2000, Australia

C. J. Glasby, NIWA, POBox 14901, Kilbirnie, Wellington, New Zealand

H. Paxton, School of Biological Sciences, MacQuarie University, NSW 2113, Australia

G. B. Read, NIWA, PO Box 14901, Kilbirnie, Wellington, New Zealand

G. Rouse, School of Biological Sciences, University of Sydney, NSW 2006, Australia

R. Wilson, Museum of Victoria, Abbotsford Campus, Melbourne, Victoria 3000, Australia

Turbellaria

L. Newman, Queensland Museum, PO Box 300, South Brisbane, Queensland 4101, Australia

Echinoderms

M. Byrne, Anderson Building, University of Sydney, NSW 2006, Australia

H. E. S. Clark, NIWA, PO Box 14901, Kilbirnie, Wellington, New Zealand

D. G. McKnight, NIWA, PO Box 14901, Kilbirnie, Wellington, New Zealand

Molluscs

R. G. Creese, University of Auckland, Leigh Marine Laboratory, PO Box 349, Warkworth, New Zealand

K. R. Grange, NIWA, PO Box 893, Nelson, New Zealand

J. Healey, Department of Zoology, University of Queensland, St Lucia, Brisbane, Queensland 4067, Australia

B. A. Marshall, Museum of New Zealand, PO Box 467, Wellington, New Zealand

W. Ponder, Australian Museum, 6–8 College St, Sydney, NSW 2000, Australia

R. C. Willan, Northern Territorty Museum, PO Box 4646, Darwin, Northern Territory 5794, Australia

Zooplankton

J. M. Bradford-Grieve, NIWA, PO Box 14901, Kilbirnie, Wellington, New Zealand.

J. Greenwood, Zoology Department, University of Queensland, St Lucia, Brisbane, Queensland 4067, Australia

Crustacea

Small epiphytal crustacea (e.g. amphipods)

G. R. F. Hicks, Museum of New Zealand, PO Box 467, Wellington, New Zealand

J. Lowry, Australian Museum, 6–8 College St, Sydney, NSW 2000, Australia

Large (e.g. lobsters)

J. D. Booth, NIWA, PO Box 14901, Kilbirnie, Wellington, New Zealand

A. B. MacDiarmid, NIWA, PO Box 14901, Kilbirnie, Wellington, New Zealand

Ascidians

C. N. Battershill, NIWA, PO Box 14901, Kilbirnie, Wellington, New Zealand

P. Kott (= P. Mather), Queensland Museum, PO Box 300, South Brisbane, Queensland 4101, Australia

L. J. Stocker, ISTP Murdoch University, Murdoch, Perth, Western Australia 6150, Australia

Fishes

Adult

M. P. Francis, Fisheries Research Institute, PO Box 297, Wellington, New Zealand

M. Gomon, Department of Ichthyology, Museum of Victoria, Melbourne, Victoria 3000, Australia

P. R. Last, CSIRO, Hobart, Tasmania 7001, Australia

C. Paulin, Museum of New Zealand, PO Box 467, Wellington, New Zealand

M. McGrouther, Australian Museum, PO Box A285, Sydney, NSW 2000, Australia

C. Roberts, Museum of New Zealand, PO Box 467, Wellington, New Zealand

B. Russell, Division of Natural Sciences, Northern Territory Museum, PO 4646, Darwin, Northern Territory 5794, Australia

Larvae

J. Leis, Australian Museum, PO Box A285, Sydney, NSW 2000, Australia

A. Miskiewicz, Sydney Water Board, Sydney, NSW 2000, Australia

P. Niera, Victoria Institute of Marine Science, Melbourne, Victoria 3225, Australia.

D. Robertson, NIWA, PO Box 14901, Kilbirnie, Wellington, New Zealand.

Birds

R. Harper, Zoology Department, Victoria University of Wellington, New Zealand.

R. Kingsford, New South Wales, National Parks and Wildlife Service, PO Box 1967, Hurstville, NSW 2220, Australia.

Marine mammals

A. N. Baker, Department of Conservation, PO Box 10420, Wellington, New Zealand

M. Bryden, Veterinary Anatomy, University of Sydney, NSW 2006, Australia

M. Cawthorn, 53 Motuhara St, Plimmerton, New Zealand

S. Dawson, School of Marine Sciences, University of Otago, PO Box 56, Dunedin, New Zealand

N. Gains, Department of Conservation, PO Box 10420, Wellington, New Zealand.

E. Slooten, School of Marine Sciences, University of Otago, PO Box 56, Dunedin, New Zealand

10.3 Collection, preservation and recording of specimens

10.3.1 Introduction

Organisms are collected and preserved for many different reasons, including laboratory collection, voucher specimens for museums, the examination of otoliths for ageing, histology of gonads, study of gut contents or resident microbes, and electron microscopy to study the impact of pollutants on the gonads of an animal. Appropriate methods of preservation vary with the type of organism and the reason for which it is collected. Consult an appropriate expert (e.g. a histologist) for advice: it may save you a lot of time, money and waste of organisms.

Because preservatives are designed to preserve tissues, safety of the person using them is an important issue. In particular, formalin and electron-microscope fixatives are suspected carcinogens and may be harmful to reproductive systems. Always use a fume hood or work in a well-ventilated area (e.g. in the field). Consult your workplace and safety adviser for advice on the use of preservatives and safe disposal of chemical waste. If you are using a microscope, it is possible to avoid bathing in the smell of formalin by rinsing in water as a temporary measure for viewing and using gentle suction-type ventilation that does not disturb the sample and minimises the chances of exposure. Always wear disposable gloves when handling preservatives. Alcohol is flammable, so care in its storage and transfer should always be exercised. It is also crucial to consider how preservatives are to be stored and transported safely to the study site and in the field.

A key text is Lincoln and Sheals (1979), which provides detailed instructions on methods of collection of almost all marine invertebrate groups and their preservation. Detailed accounts on methods of collection and preservation of different algal groups may be found in Womersley (1984) and Adams (1994).

The summary of methods below provides a quick reference for the field researcher.

10.3.2 Collection procedures

For efficiency in the field, it is important that the researcher be equipped with sufficient relabelled plastic bags and jars of various sizes to be able to collect specimens into separate containers. For underwater work, pre-labelling is vital, as it saves time and ensures cross-referencing with any photographs (*in situ* photography of specimens is always advised) and pertinent notes on location, depth, habitat type, etc. Samples should be processed immediately after collection, in a boat, laboratory or camp, by transferring them into new, labelled plastic bags that will not leak, or, better, into jars. An inventory should be compiled and cross-referenced with field notes and photographic frame numbers. All specimens should then be preserved as described below and accompanied by a preservative-proof label (pre-test your labels to make sure they will not fade or fall apart). The label should show date collected, method of collection, collector, location (with latitude and longitude if possible, or GPS waypoint, but certainly the site details should be included in field notes), depth, identification (as far as possible: even 'sponge' will do), colour, habitat. Once back at your home laboratory, a register of the collection should be completed. If the sample is to be sent to a museum or similar institution, details will be recorded on the insti-

tute's register and the sample given a unique collection number.

If samples are not collected while diving (e.g. plankton samples, cores, dredging, nets), initial preservation may be in formalin, alcohol, Lugol's iodine, glutaldehyde or by freezing, depending on the organisms and the purpose for which they are being collected.

10.3.3 Photography

It is often useful for specimens to be accompanied by *in situ* photographs, and obviously this is easiest for large organisms. Small organisms (e.g. plankton) may have to be photographed under a microscope (Chapter 9). Most organisms rapidly lose their colour on death, and particularly after preservation. Even the most vibrant-coloured cnidarians, sponges, ascidians and fishes turn brown or grey after prolonged preservation. Underwater cameras can be used to photograph organisms of a wide range of sizes (Chapter 2). Snap-on close-up lenses usually offer greatest flexibility, in that small organisms can be photographed and the close-up lens removed for larger organisms and habitats. Multiple cameras, and helpers to carry them, can provide even greater flexibility.

A bracketed sequence of two or three exposures should be taken. Photographic frames should be cross-referenced with notes and collection numbers. If a variety of colour morphs exist for any one species, these should be included if possible. It is important to photograph sedentary organisms in their extended modes, which is often difficult because many animals will retract at the slightest movement. Photography can be the only objective, but it is also an important tool for identification supported by vouchers. Good quality *in situ* photographs can be taken with modern cameras incorporating automatic light metering without much impact on diving time.

In general, films with a rating of 60–100 ASA should be used, but higher film speeds (e.g. 200 ASA) may be more appropriate for photographs of habitats in poorly lit environments. It is also worth noting that cameras with iris-diaphragm-type shutters (e.g. rangefinder cameras and twin-reflex cameras), instead of blind shutters (SLRs), have flash synchronisation at all speeds (not just 1/60 or 1/125 of a second). With a standard flash unit, such cameras offer scope for very high speeds, e.g. 1/000 of a second (M. Bradstock pers. comm.).

In addition to *in situ* photography or if underwater cameras are unavailable, 'deck' photos (photographs of the specimen taken on land) should always be taken. These are valuable in their own right because they show what the 'landed' organism looks like and provide a backup to underwater photos. The specimen should be photographed against a black background and a colour-coded reference panel with centimetre gradations and label space included. (We suggest a hexidecimal colour coding, which can be extracted from most computer draw programs.) If there is much of this work to be done, it is best to make up a framing device to support the camera and flash unit to get consistently oriented photographs. This whole assembly can be mounted on a black painted box that can be covered with an easily cleaned glass top on which to lay specimens. Photographs of pelagic invertebrates should be taken in an aquarium if at all possible, and lighting arranged to ensure the specimen is highlighted. Large fish and invertebrates or algae may need to be laid out on a clean flat surface and photographed from a stepladder or similar. The equipment available for digital photography is growing rapidly, and the quality of some images is comparable to that of conventional film. The great benefit of digitised imagery is that it can be stored and manipulated on a computer for archives, analyses (e.g. number and area of organisms) and presentations. The long-term storage of photographs as digitised images should also be considered. Conventional 35 mm slides can be scanned and stored on a CD or computer disk.

10.3.4 General preservation procedures

There are four general steps in the preparation of specimens:

1. Narcotising: some animals need relaxing/narcotisation before preservation (see 10.3.5).
2. Killing: in many cases this may be during steps 1 and 3.
3. Fixing, where preservatives are introduced to an animal's tissues to stabilise the protein constituents and prevent degradation;
4. Preservation: for long-term storage, after fixation animals may be transferred to another solution. For example, fixing in formalin (for a short period) and transferring to 70% ethanol in tap water works well for many specimens.

Alcohol (isopropyl or ethanol) and formalin are the two main preservatives used, with alcohol being favoured for long-term preservation for most organisms. In the field, and in the absence of these fixatives, other forms of alcohol may be used as a last resort (methylated spirits diluted to about 70%, and even whisky or other spirits). These are only very short-term alternatives, and samples should be refreshed with laboratory-grade alcohol (through increasing concentrations from 30 to 70%) as soon as possible on return to the laboratory or civilisation.

Formalin solution will usually be neutral if the

Dilution of formalin

Concentrated formalin is a saturated solution of formaldehyde gas (i.e. 40% formaldehyde). Formalin is diluted in fresh water (e.g. 10% formalin is a 1:10 mixture of concentrated formalin and water). You may buffer formalin with borax, but be careful, as excessive amounts of borax can clear animals and hinder identification.

If you are trying to avoid decalcification, it is best to use ethanol (final solution including animal or plant matter greater than 75%). A solution of formaldehyde and sea water is called 10% formol sea water, but the distinction between formalin and formol is rare in the literature.

---

mixture is made with seawater, but may need to be brought to pH 7 if made up with artificial sea water.

For many organisms, initial preservation in a cheaper form of alcohol will usually suffice (e.g. isopropyl alcohol). For long-term preservation, it is best to use good-quality laboratory-grade ethanol. Dilutions of alcohol are made with sea or fresh water. For a final solution of 70% alcohol, consider the size of the specimen and container, the preservative may be too dilute if a large specimen is squashed into a relatively small container.

It is important to note that careful consideration to the type and concentration of alcohol is required if you want to prevent decalcification. This, for example, is important if you want to age fish using otoliths. Increments used to age fish (annual or daily) may be obscured if preservation is poor.

For histological specimens, Bouin's fixative (or fluid) is recommended for many organisms.

Some organisms need to be relaxed before they are fixed, as they tend to contract beyond recognition if placed immediately into preservative. This is particularly the case with ascidians, worms and cnidarians. It is necessary to relax these animals in either sea water onto which menthol crystals have been sprinkled, or by using magnesium chloride (7.5% $MgCl_2$). See Lincoln and Sheals (1979) for additional methods for very delicate species requiring special care.

For long-term preservation, glass jars with airtight lids (not metal) should be used as plastic containers

---

Bouin's fixative

750 ml picric acid (saturated solution); 200–250 ml concentrated formalin in sea water; 50 ml acetic acid (Clark 1973). Material is usually fixed in 12 hours, and most specimens should be removed to fresh alcohol after this time.

Note that this fixative is extremely toxic and must not be allowed to dry out: picric acid is explosive when dry. Even storing picric acid solution in a glass-stoppered bottle is inadvisable.

273

perish. Store containers inside a larger container to minimise evaporation and leakage.

Subsamples may need to be taken for electron microscopy. If this is the case, small (1 mm cubes) pieces of the tissue of interest should be placed in the fixative recommended by the electron-microscope facility to be used. Experts listed above may also be consulted for information on the best electron-microscope fixative to be used for any particular plant or animal group or species or tissue type of interest. In most cases, the fixative is 2.5% gluteraldehyde with a 0.2M sodium cacodylate buffer. The strength of glutaraldehyde and makeup of buffer is critical to the success of fixation for electron microscopy, and is very species- and tissue-specific. Usual fixation process involves fixing in gluteraldehyde for three to four hours, followed by three washes of buffer for 10 minutes each. Store in buffer in cool conditions until staining, etc., can be carried out in an appropriate laboratory.

10.3.5 Collection and preservation of plant and animal groups

Algae
Algae should be collected with holdfasts, and as intact as possible. This may be impractical when dealing with large bull kelps and the like, in which case representative samples of the different regions of the plant should be collected. Often a positive identification is only possible if the specimen is reproductive, particularly with red algae.

Short-term storage: Leave in sea water and store in a cool dark place. Most algae will travel well if kept damp and dark.

Preservation: Use 3–5% formalin in sea water. Specimens may be dried and pressed between sheets of absorbent paper for herbarium collections (Adams 1994).

Protists
These will include flagellated forms and those with hard structures, the Foraminifera. Flagellates and pelagic forms can be collected in water samples and fine-mesh plankton nets, and forams may be collected by sieving sediments (see 9.2.15).

Preservation: Initial fixation in 50% alcohol, then refreshed with 70–90% alcohol. Alternatively, 3–5% formalin, buffered to pH 7, may be used. Lugol's iodine can be used as a quick-killing/fixing agent and provide a high degree of flagella retention in protists (see panel p. 275). Steedman (1976, p. 262) suggested that 0.4 ml of Lugol's per 100 ml of sample be placed in the

bottom of a jar before adding the sample. The sample is then mixed and formaldehyde added for a final concentration of 2%. Buffers may be added to the stock solution. Glutaraldehyde 2.5% is thought to be best for fixing flagella and cell contents.

Porifera
Sponges are grouped into three main classes characterised by their skeletal makeup: the Hexactinellida (glass sponges, with a siliceous skeleton), Demospongiae (siliceous such and spongy sponges) and the Calcarea (calcareous skeletons). They have a range of textures, morphologies and degrees of fragility. All should be treated with care to maintain the structure, as many are brittle. Avoid low-pH solutions, which may dissolve skeletons. Protect the collector from injury from spicules (which can readily penetrate skin and even gloves) or toxic chemicals. Where possible, the entire specimen should be collected, including the substratum, as there are often diagnostic spicules attached to the substrate. With large specimens, a representative portion or slice incorporating all sponge layers may be adequate. It is vital that specimens be collected into separate containers, as they may shed spicules, thereby confusing later identification if many species are collected together. It is important to record the colour of sponges on collection (preferably by photography; see 10.3.3) as well as their colour in spirit, as many change colour even on exposure to air (these changes can be diagnostic).

Immediate storage: Place the wet specimen in 70% alcohol. Sponges usually have enough moisture within them that dilution to 30–50% is readily achieved on immersion.

Preservation: After about two weeks and no longer than a month, the samples should be refreshed with 100% ethanol. It may be necessary to repeat this procedure after a year.

Cnidaria
Classes: Hydrozoa, Scyphozoa, Cubozoa, Anthozoa
Once again these should be collected with the substrate wherever possible, to avoid damage to the colony. They are all delicate animals and should be treated carefully. Medusae may be collected with plankton nets or in containers that reduce the chance of damage (e.g. use plastic bags while diving). Avoid having stiff labelling materials in the same container as delicate planktonic specimens. Steedman's solution works well with small forms.

Immediate preparation: Colonies and other fragile forms should be relaxed first in sea water, with menthol crystals

---

Steedman's solution
Stock solution for one litre: propylene phenoxetol, 50 ml; propylene glycol, 450 ml; formalin, 500 ml; can be used as a quick-killing agent. For fixing and preservation, add stock solution to a measured amount of sea water to make a final solution of 10% (Steedman 1976).

---

sprinkled on the water surface after the animals have extended themselves. The time for relaxation may be as much as six hours.

Preservation: Animals may be fixed in 10% formalin and kept long term in 5% formalin or 70% alcohol. Anthozoa with calcarous skeletal components should be preserved in alcohol, as formalin dissolves calcium. If only the skeleton is required for the true corals, these may be prepared by soaking in 50% bleach solution, rinsed and dried.

Ctenophora
These pelagic animals are extremely delicate, and much care in handling is needed. They should be collected together with a quantity of water in a large plastic bag to buffer the animal against damage during transit. It is useful to photograph the animals in the field because many of them disintegrate at the slightest touch.

Immediate preparation: Animals should be relaxed first.

Preservation: This is often species- or genus-specific: see Lincoln and Sheals (1979). Fixation (in chromic acid solution for about 15 minutes) is usually followed by alcohol washes through an increasing concentration gradient, 30–70%. Storage is generally in 70% alcohol. Steedman's solution also works well.

Bryozoa
Whole colonies should be collected wherever possible. This may be difficult with thinly encrusting species. If possible, these should be collected on their substrate. As with sponges, colour and morphology should be recorded on collection (preferably with photography), as should notes on associations. Most bryozoans are best kept dry for identification; only uncalcified or lightly calcified species need to be preserved in alcohol.

Immediate preparation: Samples should be relaxed with menthol (they may need to be left overnight until polyps no longer retract) or using methods described in Lincoln and Sheals (1979).

Preservation: Species with calcareous walls need to be washed in fresh water and preserved in 70% alcohol.

Other species can be preserved in 5% formalin solution or Steedman's solution. It is also possible to dry specimens after washing.

Brachiopoda
Preservation: Specimens should anaesthetised with a small amount of alcohol, to ensure the valves part, then fixed and preserved in 70% alcohol.

Sipuncula, Echiura, Annelida
Initial preparation: These worms should all be relaxed first. For example, the relaxation of echiurans takes a few hours in $MgCL_2$.

Preservation: When they no longer respond to stimuli, they should be fixed in 5–10% formalin for a few days. Then they should be preserved in 70% alcohol, though Steedman's solution is also good for some groups (e.g. Echiura: A. Low pers. comm.). Refer to Lincoln and Sheals (1979) for methods specific to each group.

Polychaetes are generally preserved in 10% formalin for 1–24 hours, depending on size (G. Rouse pers. comm.). Never use alcohol on small polychaetes, as it makes them too brittle (P. O'Donnell pers. comm.).

Crustacea
Initial preparation: It is necessary to kill or heavily relax specimens first: this is more humane and reduces the chances of animals shedding limbs. Large specimens can be killed by freezing and then thawed in preservative. It is vital that the colour of specimens be recorded before preservation, as this will be lost. Deck photos of large crustacea and photomicrographs of small animals are often useful in addition to *in situ* photography.

Preservation: Specimens tend to shrink if put directly into alcohol. It is best to dip the specimen in alcohol to make the exoskeleton porous, then place in 5% formalin. Although small crustaceans such as amphipods can be stored in 5% formalin, large crustaceans should generally be transferred into 70% alcohol after 24 hours. For large specimens, it may be necessary to inject formalin into internal tissues and large muscles after relaxing and killing.

Mollusca
Initial preparation: It is always necessary to relax molluscs (e.g. in cold water and menthol for 24 hours to speed the process up). Beware that decomposition can be quick, so avoid prolonged periods in aqueous solutions.

Preservation: Preserve in 5% formalin: see Lincoln and

Sheals (1979). Store in 70% alcohol to avoid damage to shells.

Echinodermata

Initial preparation: Holothurians and soft-bodied or fragile species should be relaxed first.

Preservation: Most species with hard bodies can be preserved directly. All specimens should be preserved in 10% formalin buffered to pH7. Seventy per cent alcohol can also be used. Specimens with hard bodies can also be dried.

Ascidians

Ascidians occur as solitary, social (connected solitary) and colonial forms of a wide variety of morphologies. They should be collected with their substrate wherever possible, as damage can occur to the animal if prised off surfaces. As with sponges and bryozoans, ascidians should be photographed and colour-recorded while fresh.

Initial preparation: Ascidians should be relaxed (e.g. with menthol) before preservation, otherwise zooids will contract beyond recognition.

Preservation: 70% alcohol is best, or the animals can be preserved in 5% buffered formalin. Some colonial ascidians have spicules of calcium carbonate that will dissolve in low pH solutions, thus buffering is crucial.

Other group of invertebrates

Flatworms, Nemertinea, Aschelminthes, Acanthocephala, Priapulida: see Lincoln and Sheals (1979).

Fish

Once again, good-quality *in situ* and deck photos are extremely valuable if your objective is to identify a species and collect voucher specimens. If you intend to count fin-rays, it is advisable to do this before preservation, because the spines are difficult or impossible to move after preservation; though X-rays could be used.

Short-term storage: Freeze, except when collecting for histological purposes, because tissues will be disrupted.

Long-term storage: Preserve in 10% formalin. Inject the visceral cavity of large fish and soak fish in a bath of preservative. For very long-term storage, fish are usually transferred to 70% ethanol. If you require skeletal components for ageing or other purposes, preserve with alcohol, by freezing, or remove skeletal components of interest before long-term preservation. Preservation in Boiun's fluid is generally used for the histology of gonads.

Ichthyoplankton and other zooplankton

Preserve in formalin (2–5%) in sea water. In the case of ichthyoplankton, transfer to 2% formalin and fresh water (see Tucker and Chester 1984). If the otoliths of fish larvae are to be examined, preserve the sample in 95% ethanol to avoid decalcification. Final concentration of alcohol should be greater than 75%. It is important, therefore, to avoid overpacking specimens into a jar, as this dimishes the the the concentration of alcohol.

For long-term preservation of zooplankton, concentrate into a small amount of measured sea water and add a stock solution of Steedman's solution to make a final concentration of 10%.

10.4 Examples of studies

10.4.1 Bioprospecting for biologically active compounds in marine invertebrates and algae

Purpose: Collect invertebrates (especially sessile invertebrates) and algae so they can be screened for biological activity. The aim is to provide a detailed description of every species collected, together with notes on location, taxonomy and habitat. This enables species that may be of interest to be collected and builds up biogeographic and ecological databases. Preparation of vouchers for taxonomy has to be rigorous, and subsamples may also need to be deep-frozen for chemistry (Pomponi 1988). In many instances, where the chemical interest may be associated with symbionts, electronmicroscope fixation will also need to be carried out (see below). It is also now possible to cryopreserve cells for subsequent work (Pomponi & Willoughby 1994).

Approach: Museum vouchers need to be taken of every specimen and cross-referenced to frozen material. Detailed records of sample references, locations and notes on the environment must be accurate otherwise any 'hits' may not be relocated. As many of the species targeted in marine bioprospecting can be rare (e.g nudibranchs, some sponges and hydroids), great care must be taken to cause minimal damage to the population, especially as there is usually no information about the density or distribution of most marine encrusting organisms. Assume that the population is not large and, wherever possible, take only a portion (up to a third) of colonial organisms. Colonial or modular organisms will usually heal at the site of damage and can often re-grow removed tissue. For rare invertebrates, nudibranchs and the like, the population of any species should not be sampled until it has been established that there are enough individuals in any particular location to support some sampling. In most cases, only 10–50 g of tissue is needed for chemical analyses; frequently less for fleshy species. There is no point in sampling for bioprospecting purposes a species that will be removed completely from any one location in the process, as there would be no further material to sustain re-collecting for later bioactive compound development work, not to mention the obvious serious problem of potentially bringing a local population to extinction or ecological extinction. There have already been a number of instances where marine species have been completely removed from large areas, or where very exciting pharmacological leads have been lost because the species could either not be relocated or the only known population of it had been removed.

Desirable: In situ and deck photos should be taken at the site of collection wherever possible. Full notes should be taken of the morphology, colour, texture, etc., of each specimen, together with ecological notes.

Collection: Ideally, an underwater camera should be taken on all collection dives of this nature. Thus, each species can be photographed *in situ* (with a camera fitted with a close-up lense, ruler and sample number plate). Samples will be labelled, cross-referenced underwater to the photo number, and then placed separately (by species or even individual) into pre-labelled ziplock bags. Identification (as far as possible) should also be included even at this early stage, as it acts as a check should samples be mixed for some reason (e.g. broken bags if a diver gets caught in a surge over rocks). On deck, labels (check they are resistant to preservatives) need to be made for every subsampling container, e.g. for the museum vouchers, frozen samples, reproductive samples and maybe electron-microscope samples. It is often useful to weigh wet samples prior to freezing at this point. Record-keeping and cross-referencing is vital at this point. Deck photos should also be taken as a backup, including a size scale, reference number and colour scale). Further identification is possible at this point, and scientists would have on hand a variety of taxonomic texts appropriate to the region to aid description. Quick identifications can be made of sponges using spicule mounts made after dissolving tissue in bleach, or from quick dissections of ascidians, etc. A field compound and dissecting microscope is invaluable for this process. Samples will then be treated as described above for each phyla.

Preservation and identification: Final identification will frequently involve an expert for the phylum, and this stage is paramount to all subsequent research associated with the search and development of new drugs from marine resources, as well as any research into chemical ecology and biogeography. The quality and degree of the final taxonomic assignment depends on how well the material has been fixed. The methods described above should be adopted, and for those organisms likely to require special treatment, a specialist should be consulted or even taken along. For algae, it is vital that good-quality herbarium mounts be prepared, where algae are pressed after a wash in fixative, between herbarium sheets and compressed. For sponges, each species needs to be kept separate from the time it is removed from the substrate, otherwise they shed spicules and mucous, which may confuse taxonomy and chemistry. Ascidians and cnidarians need to be well relaxed. Care should be taken with labels that might interfere with soft animals or leach chemicals into frozen samples.

10.4.2 Symbionts, chemistry, reproductive patterns and cell culture of sponges

Purpose: In ecology and in follow-up research associated with the search for and development of new chemicals from marine sources, examination of the histology and physiology of target species is necessary, particularly when reproductive phenomena may be of interest, or in the likely instance that symbionts or parasites may have an influence on the observed chemistry or ecology. In a recent case involving three species of sponge from New Zealand, the reproductive state and presence of micro-organisms needed to be examined to explain variability in the chemistry of samples. Here electron microscopy is all-important and material needs to be preserved for the variety of procedures (see above). Standard alcohol-preserved samples also need to be taken, and these are likely to include some specific staining procedure for identification of reproductive parts (e.g. Bouin's fixation followed by Malory-Heidenhain or Haemotoxylin/Eosin procedures). There is also interest in culturing either the organism or its microflora for further examination. This is possible with cryopreservation techniques, which will allow for a continuous supply of cells for work back in the home institution, eliminating the need for many costly field visits to retrieve fresh samples.

Collection: Samples of each of the three species of sponge are collected *in situ* into pre-labeled bags. Sampling is stratified by position within each sponge (inner, outer, pinacoderm, matrix); between individuals, site, depth, location, season and year.

Preservation and identification: Each sample is fixed in Bouin's fluid for light microscopy, and subsamples are fixed in gluteraldehyde for electron microscopy.

Subsamples are also cryopreserved for culture studies (Pomponi *et al.* 1997, 1998) and subsamples frozen for chemistry. It is usually possible to be confident that the same species is sampled in different areas (once the investigator has focused on a few species), but if there is any doubt, samples preserved in alcohol are useful to confirm identification. Reproductive state is determined by light microscopy after staining, and the presence of symbionts is determined with scanning and transmission electron microscopy. This is further elucidated with cell-separation techniques, whereby cells are separated in Ficoll or Percoll density gradients (synthetic isotonic sugar solutions) and identified. Each cell type is also subjected to chemical evaluation (Battershill and McLean 1994; Garson et al., 1992, 1993). Should further analysis be required, cell culture experiments are made, where separated cell lines are cultured in isolation and their chemistry tracked (Pomponi and Willoughby 1994; Battershill and McLean 1994).

10.4.3 Collection of fishes from a remote offshore island

Purpose: Describe patterns of abundance of reef fishes at Raoul Island, in the Kermadec Islands (Schiel et al. 1986). The composition of the fauna was poorly known and it was necessary to collect voucher specimens for museums and, in some cases, identify new species (e.g. *Parma kermadecensis:* Allen 1987).

Collection: Fish were speared with a hand spear. Fish were also counted using belt transects, and underwater photographs were taken of the fishes and the habitats in which they were found.

Preservation and identification: Speared fish were injected with 10% formalin and placed in a plastic bag with formalin so they were well bathed in it; a water- and formalin-proof tag was added for identification. The bag was tied and placed in a drum of formalin (with an O-ring top for a good seal). The additional formalin meant that if the bags leaked, the preservation of fish would not be compromised. In cases where samples were sent to expert ichthyologists, fish that had been soaked in formalin for a month or more were wrapped in heavy tissues soaked in formalin, then wrapped in three plastic bags to avoid leakage, and mailed in a strong box, thus minimising the chances of damage in the mail (check the legality of sending specimens this way). In the field, deck photographs were taken of some specimens, and underwater photographs were also taken. A board was used to pin fish in the erect position for deck photographs; a drop of formalin at the base of fins may help keep them upright.

10.4.4 Collection of fishes for gonad samples

Purpose: To determine if *Achoerodus viridis* changes sex, like many other labrid fishes (Gillanders 1995a).

Collection: Fish of a broad range of size range were collected at two sites (about 5 km apart) with a speargun. The fish were speared behind the head (to avoid damage to otoliths for ageing) and above the visceral cavity. The barbs of the spear were filed to minimise damage.

Preservation: The visceral cavity of fish was injected with 10% formalin within two hours of spearing, to minimise deterioration. The needle of a syringe was moved in and out of the visceral cavity to maximise the spread of formalin. Gonads were carefully removed in the laboratory and placed in Bouin's for histology. Gonads were transferred to ethanol (greater than 75%) until they were prepared for histology, which was done according to the Mallory-Heidenhain method (Humason 1989). A pilot study was done to determine where transverse sections of the gonad should be taken, and how many sections needed to be examined under a compound microscope (100–400 x) for accurate estimates of the representation of different stages of gametogenesis. For other examples, see Jones 1980 (*Notolabrus celidotus*), Cowen 1986 (*Semicossyphus pulcher*) and Nakamura et al. 1989 (*Thalassoma duperry*).

10.4.5 Analysis of the gut contents of a planktivore

Purpose: Study the impact of planktivorous fish on zooplankton (Kingsford & MacDiarmid 1988). In addition to gut sampling, fish were counted in belt transects, feeding rates were measured and information was collected on currents and densities of plankton.

Collection: Fish were speared with a handspear, high on the body to avoid damage to the guts.

Preservation and identification: The visceral cavities of the fish were injected with 10% formalin within two hours of spearing, to minimise deterioration. The needle of a syringe was moved in and out of the visceral cavity to maximise the spread of formalin. The planktivorous fish that were sampled (e.g. *Chromis dispilus*) had a discrete stomach, and this was carefully removed. The stomach was slit open and plankton was scraped out and put into a small vial that contained 5% formalin. We were careful to disperse the plankton to minimise the chances of clumping. For most samples of gut contents, two subsamples were taken to estimate abundance of each type of planter in the stomach. Although some groups could be identified to genus or species using Dakin and Colefax (1940), most analyses were done at the level of operational taxonomic units (e.g. gelatinous zooplankton, copepods, etc: see Fig. 9.12). Fish were easily identified in the field, and the taxonomy was according to Ayling and Cox (1982).

10.4.6 Patterns of abundance of ichthyoplankton

Purpose: Describe the patterns of abundance of different species of ichthyoplankton at different distances across a continental shelf. Sampling was done near to and away from land, at a range of distances across the shelf (Kingsford & Choat 1989).

Collection: Ichthyoplankton was collected in a 1 mm mesh net with a filtration efficiency of 1:10 (area of mouth:open area of mesh), towed at 2–3 knots. Oblique tows were done (3–20 minutes in duration). Samples were concentrated in the cod end of the net.

Preservation and identification: Samples were decanted from the cod end into a 1-litre jar with sufficient 10% formalin so that when the jar was filled the final concentration was 2–5%. In the laboratory, samples were poured through a 1 mm mesh sieve (a wide plastic pipe with mesh glued on the end) so that samples could be stored in small containers (100–250 ml) in a mixture of fresh water and formalin (final concentration 2–5%). Larvae were identified to family and in many cases to species. Identification was done according to Crossland (1981, 1982a), some unpublished keys and a collection of larvae of different species and sizes.

If you are involved in the identification of ichthyoplankton or other planktonic groups, a collection is crucial. You should also consider sending samples to a museum so that important species, or whole samples, can be checked by others if necessary. Note that it would not have been possible to look at the otoliths of fish larvae, because the fish were preserved in formalin: the preservative would have had to have been alcohol or selected specimens could have been frozen (Thorrold 1988; Thorrold & Milicich 1990; Nishimura 1993).

10.4.7 Impact of oil dispersant on an intertidal gastropod

Purpose: To identify the effects both lethal and sublethal of the use of new oil dispersants on intertidal gastropods.

Collection: Specimens of the gastropod *Nerita melanotragus* were collected by hand from a rocky shore and transferred into cages, which were then sprayed with various concentrations of oils and dispersants (Battershill & Bergquist 1982, 1984). At the beginning and end of the experiment (which lasted a month), samples of animals were collected for reproductive analysis and tissue analysis. All animals in cages were also collected and fixed accordingly for reproductive and other tissue analysis, after being weighed.

Preservation and identification: The shell of each animal was cracked and tissue placed in Bouin's fluid, with a subset of animals being frozen for analyses of hydrocarbons. Tissue analysis included examination of reproductive state and assessment of tissue damage. Gonad tissue was excised from both fresh and preserved specimens with a scalpel, and gill tissue were separately examined under a light microscope.

SELECTED READING

The references listed are examples of research that will lead the reader to a broader literature, also see the text of the chapter. * indicates a multi-authored book or collection of proceedings; references to volumes of a journal refer to a set of relevant papers. If you do not find a satisfactory references from the sources provided, contact a local institution, or known expert in the field (some are listed in this chapter).

General taxonomy
Australia: Shepherd & Thomas 1982, 1989; Bennett 1987; Anderson & Kautsky 1996; Edgar 1997, Anderson 1998; Note the series *Fauna of Australia*, CSIRO Publishing, Collingwood.
England and Europe: Kermack et al. 1979; Hayward and Ryland 1995.
New Zealand: Morton & Miller 1973; Doak 1979; Powell 1979; Jones 1993; Stocker 1985; Cook in press.
North America: Meinkoth 1994.

Spain: Carlos & Calvo 1995.
South Africa: Branch et al. 1994.
USA (West): Abbott & Hollenberg 1976; Morris et al. 1980.

Taxonomy by taxa
(Check some of the above general references first)

Algae: New Zealand: Bonin 1981; Fuhrer et al. 1981; Nelson & Adams 1984; Womersley 1984, 1987; Adams 1994, 1997.
Coastal plants: New Zealand: Salmon 1991; Crowe 1995. Australia: Carolin & Clarke 1991.
Sessile and mobile invertebrates: New Zealand: Powell, 1979; Walsby & Morton, 1982; Pritchard et al. 1984; Cook in press.
Australia: Shepherd & Thomas 1982, 1989; Edgar 1997; Shepherd & Davies 1997.
Porifera: Pritchard et al. 1984; see also Sessile and mobile invertebrates.

Polychaetes: Australia: Day & Hutchings 1979.
Molluscs: New Zealand: Powell 1979; Walsby & Morton 1982.
Australia: Wilson 1993a, b.
Crustacea: Jones & Morgan 1994.
Amphipoda: Barnard & Karaman 1991a, b.
Crabs: McLay 1988.
Lobsters: FAO 1991.
Brachiopods: Brunton & Curry 1979.
Ascidians: Kott 1985, 1990a, b, 1992a, b.

Plankton

General: Dakin & Colefax 1940; Yamaji 1959; Newell & Newell 1977; Smith 1977; Todd & Laverick 1991; Cook in press.
Protists: Lee et al. 1985; Tomas 1993.
Phytoplankton: Wood 1954, 1961, 1963; Crosby & Wood 1958, 1959; Wood et al. 1959; Lanigan 1972; Sournia 1978.
Cnidaria: Russell 1970; Larson 1976.
Chaetognaths: Lutschinger 1993.
Copepoda: Dakin & Colefax 1940, Rose 1970; Bradford 1972; Bradford & Jillett 1980; Bradford et al. 1983; Bradford-Grieve 1994; Dussart & Defaye 1995.
Decapod larvae: Wear & Fielder 1985; McLay 1988.
Cumacea: Jones 1963.
Bivalve larvae: Booth 1979.
Fish eggs and larvae: Robertson 1975; Russell 1976; Crossland 1981, 1982a; Moser et al. 1984; Fahay 1983; Leis & Rennis 1983 Leis & Trnski 1989; Materese et al. 1989; Okiyama 1988; Moser 1996; Neira et al. 1997.

Fishes

Australia: Russell 1988; Hutchins & Swainston 1986; Paxton et al. 1989; Randall et al 1990; Kuiter 1993; Gomon et al. 1994; Last & Stevens 1994.

Japan: Masuda et al. 1984.
New Zealand: Thompson 1981; Ayling & Cox 1982; Paul 1986; Paulin et al. 1989; McDowall 1990; Doak 1991; Paulin & Roberts 1992; Francis 1996; Cox & Francis 1997.
North Atlantic and Mediterrean: Whitehead et al. 1989; Lythgoe & Lythgoe 1991, Carlos & Calvo 1995.
South Africa: Smith & Heemstra 1986.
South America: Fowler 1941.
USA – West Coast: Eschmeyer et al. 1983; Love 1991.
USA – East Coast: Robins et al. 1986

Sea birds

Australia and New Zealand: Harper & Kinsky 1978; Chambers 1989; Dorfman & Smith 1995; Pizzy & Knight 1997.
World: Harrison 1983.

Whales and dolphins

Australia and New Zealand: Gaskin 1978; Dawson 1985; Baker 1983; Dawson & Slooten 1997; Edgar 1997.
North America: Leatherwood et al. 1988.
World: Evans 1987; Jefferson 1993.

Seals: King 1983.

Narcotisation: Lincoln & Sheals 1979; Smaldon & Lee 1979.

Preservation: Steedman 1976; Lincoln and Sheals 1979.

Staining: Burdi & Flecker 1968; Clarke 1973; Bruhbaker & Angus 1974; Humanson 1979; Ivanchenko & Ivanchenko 1973; Dinderkus 1981; Moser et al. 1984; Hyatt 1993.

X-radiography: Miller & Tucker 1979.

Accuracy – closeness to, or departure from, the true mean of a population.

Analysis of covariance (ANCOVA) – a statistical method used to analyse groups of samples that are classified according to one or more criteria (usually called factors) and individual measurements that have been made over a continuous scale (e.g. age versus length). The simplest form of analysis is a comparison of two samples. Tests for slopes and adjusted means are given in this analysis.

Analysis of variance (ANOVA) – a statistical method used to analyse groups of replicate samples that are classified according to one or more criteria (usually called factors, e.g. location, time). The simplest form of analysis is a one-way ANOVA (i.e. one factor). More complicated analyses may have two or more factors and related interactions between the factors (orthogonal designs). The factors are classed as fixed or random factors (see 3.5.3), if both fixed and random factors are in a model, it is called a 'mixed model'. Nested ANOVAs are so-called 'hierarchical designs' that do not have interactions between factors (i.e. lowest levels are nested in the factors above). Some designs are 'partially hierarchical', which means they have orthogonal and nested components in the design. Asymmetrical ANOVA refers to designs that have an orthogonal part to the design and asymmetrical components; see 'Beyond BACI'. (For examples of designs that use ANOVA, see Chapter 3.)

ANOVA – Analysis of Variance (q.v.)

ANCOVA – Analysis of Covariance (q.v.)

Assemblage – multiple species of plants and animals living in the same place and time (cf. 'Community'); see also discussion in Fauth et al. 1996.

Asymmetrical design – a design that is not orthogonal (q.v.). For example, with a single impact site and multiple control sites, the levels of certain factors are unbalanced.

Autocorrelation, spatial – the degree of similarity of samples at set distances apart and in a particular direction.

Autocorrelation, temporal – the degree of similarity of samples at set times apart.

BACI – Before After Control Impact, a term commonly used in impact studies to mean sampling at one time before and after at one impact (e.g. sewage outfall) and one control site (cf. 'BACIP', 'Beyond BACI' and 'MBACI')

BACIP Before After Control Impact Paired series, a term commonly used in impact studies to mean sampling at paired times before and after at one impact (e.g. sewage outfall) and one control site (cf. 'BACI', 'Beyond BACI' and 'MBACI'), analyses usually with t-tests.

Baseline study – data collected to define the present state of an assemblage.

Beyond BACI – a sampling design that refers to comparisons of an impact site with random multiple control locations, multiple times before and after the impact (cf. 'BACI', 'BACIP' and 'MBACI'); usually analysed with ANOVA.

Bias – direction and consistency of inaccuracy (see 'Accuracy').

Binary data – data consisting of zeros and ones. Commonly used for presence/absence data (0 = absent, 1 = present).

Brown body – a pigmented patch in the gonads of mature female fishes that is indicative of the resorption of unused ova during the spawning season.

Cartesian co-ordinates – the x, y positions of individual points on a grid.

Cascades – see 'Indirect effects'.

Census – total count of organisms. In marine systems this is difficult to do unless a very small area is searched. When total counts are presented (generally based on stratified sampling), they usually have a measure of error (e.g. 95% confidence limits). A census is often equated with measures of stock size.

Community – multiple species of plants and animals living in the same place and time (cf. 'Assemblage'). Because 'community' is sometimes used to imply that a group of organisms have 'emergent properties' that are greater than the sum of the interactions between species, we have chosen not to use this term; see also discussion in Fauth et al. 1996.

Condition – the health of organisms as measured by approaches such as body dimensions (morphometrics), length/weight relationships and quantity of biochemical energy stores.

Continuous sampling – sampling done in a swath (e.g. video, swath mapping, oceanographic undulators), rather than discrete units such as quadrats (cf. 'Discrete sampling').

Control – areas or sampling units outside of real or predicted impact, or unaffected by experimental manipulation. Controls are essential to compare changes in impact areas or experimental treatments. These are often called 'reference areas' in environmental studies.

Correlation – the degree of association between two variables (e.g. measure of topographic complexity versus number of fishes). Measures of association are often expressed as a correlation coefficient (r; may vary between 1 and – 1); cf. 'Regression'.

Crypt – a subsection of a male gonad that is full of a particular stage of gametogenesis.

Cryopreservation – preservation by supercooling. This generally involves a fast freeze to $-70°C$ and slow thaw. For example, this technique is used to transport and store sponge cells.

CSIRO – Commonwealth Scientific and Industrial Research Organisation, Australia; branch of the Australian Federal Government.

Creel survey – survey of recreational fishermen (usually of their catches).

Data matrix – a two-dimensional table of data described in terms of columns and rows.

Demersal – living on or close to the seafloor.

Demersal eggs – eggs that have been spawned onto the substratum, where they adhere for a period of hours/ days. When the eggs hatch (usually at night), larvae swim away to live in the plankton.

Dependent variable – one that varies according to independent variables and is not used as a predictor variable. For example, the size of a fish is dependent on age, so age is the independent variable (x-axis) and size the dependent variable (y-axis). In an experimental context, independent variables are the factors controlled by the investigator and the dependent variables the response of target organisms of environmental factors (cf. 'Independent variables').

Depth profile – an imaginary, or transect, line running perpendicular to the shore and down a depth gradient of a hard- or soft-bottomed environment.

df – degrees of freedom in a statistical test.

DGPS – Differential Global Positioning System, see 'GPS'.

Discrete sampling – discrete sample units such as quadrats, cores and water samples; cf . 'Continuous sampling', where sampling is done in a swath (e.g. video).

Diversity – this is usually expressed as species richness (i.e. number of species) and evenness (i.e. proportional representation of each species).

DoC – Department of Conservation, a branch of the New Zealand Government Services.

DPV – diver-propulsion vehicle.

Ecosystem – made up of interacting biotic (assemblage of micro-organisms, plants and animals) and abiotic components (e.g. includes the energy into and out of an environment).

EIA – Environmental Impact Assessment. The specific requirements of EIAs vary greatly with state and country. An EIA generally involves an overview of the potentially impacted area with relevant information on the ecology of organisms and other biology as well as information on physical attributes of the environment such as geography, geology and oceanography. Speculation is then made on the potential impacts of any development.

Elutriation – repeated rinsing of sediment with water so that sediment sinks and the animals remain in suspension, allowing the supernatant and animals to be decanted.

ENSO – El Niño-Southern Oscillation

Environment – defined physical characteristics (e.g. rocky reef, soft sediments) within which can be found a number of habitats (e.g. kelp forest, urchin-grazed barrens).

Epifauna – fauna living on a surface.

Epiphytic (or epiphytal) – organisms living on the surface of algae.

Experiment – our focus is on field experiments. There are two forms recognised in this book: mensurative and manipulative experiments.

Factor – refers to the levels of sampling. For example, times, locations, sites.

Functional group – the pooling of species for ecological questions. For example, multiple species of grazers are pooled to the following groups: limpets, urchins, turbinids. The assumption is generally that members of a functional group have similar ecological functions (see also 4.5.1).

Gametogenesis – the production of gametes (i.e. ova and sperm). This process is cyclic in most marine organisms (cf. 'Oogenesis' and 'Spermatogenesis').

Genet – an identical genetic entity that has arisen from one individual (e.g. an ascidian tadpole). This term is often used for modular organisms such as sponges and ascidians where multiple individuals (see 'Ramet') of the same genotype can result from one genet.

Gonochorism – separate sexes (= dioecious); cf. 'Hermaphroditism'.

GPS – Global Position System. Can give accurate and precise estimates of an observer's position on the planet. Differential GPS can give positions with an accuracy of ± 1 m. Without a differential capability, precision is $\pm 15–20$ m.

Halocline – oceanographic term for a rapid change in water salinity; cf. 'Thermocline' and 'Pyncnocline'.

Habitat – a characteristic biological assemblage and/or physical structure (e.g. kelp forest, shallow broken rock); cf. 'Strata'.

Hermaphroditism – individual organisms with both male and female gametes (= monecious). Some organisms are simultaneous hermaphroditites (i.e. have ripe spawn eggs and sperm in their gonads); others are sequential hermaphroditites (i.e. function as one sex before reproducing as the other sex). See also 'Protogyny' and 'Protandry'.

Hierarchical design – see 'Nested sampling design'.

Holoplankton – planktonic organisms that spend their whole life as part of the plankton.

Impact – a perturbation causing an alteration in a measured variable of a population (e.g. abundance) or assemblage (e.g. diversity). Examples of impacts include the potential positive effect of a marine protected area, and the negative effect of a source of pollution. See also 'Pulse' and 'Press'.

Importance – an indication of the result of an interaction between organisms. Interactions can be important if one organism causes a change in the distribution of some demographic characteristic (e.g. growth, reproductive output). This should not be confused with intensity (q.v.). Organisms can have an intense interaction (e.g. chases, predation), but the outcome of those interactions may not be important at the level of populations. (Adapted from Weldon & Slauson 1986.)

Inertia – in the context of impact studies, is the size of a disturbance that does not cause a response in a population (cf. 'Dependent variable').

Independent variables – variables that predict dependent variables. For example, age is independent of the size of a fish and is sometimes a good predictor of size. In an experimental context, independent variables are the factors controlled by the investigator, and the dependent variables are the response of target organisms of environmental factors (cf. 'Dependent variable').

Indirect effects – in many biological assessments it is important to think beyond direct effects. Changes in the abundance of an organism (e.g. direct effect of harvesting) may cause changes to other organisms in the same assemblage. For example, the removal of urchins may cause growth of macroalgae, which in turn influences the recruitment and survivorship of the organisms. Effects of this type are often called cascades.

Infauna – fauna that dwells in sediment or similar soft substrata.

Initial-phase male (IP male) – see 'Terminal-phase male'.

in situ – in the natural environment where subject organisms are found.

Intensity – the intensity of an interaction between organisms; see also 'Importance'.

Introductions – in this book we specifically refer to the introduction of exotic marine organisms, usually by ballast water, hitchhiking, as introduced bait or through intentional introductions (e.g. the northern Pacific sea star to Tasmania).

in vitro – biological processes occurring in isolation from the living organism.

in vivo – biological processes occurring within the living organism.

Lecithotrophic – refers to larvae that do not feed (cf. 'Planktitrophic'); nutrition is from maternal sources only.

Life history – the changes that take place during the life of an organism. For example, in fish it may include their early life history as eggs and larvae, gaining maturity, changing sex from female to male and living to a maximum age of 20 years.

Macrofauna – organisms associated with sediment and retained in a geological sieve of 0.5 mm (cf. 'Microfauna' and 'Meiofauna'; see also Chapter 8).

MBACI – a sampling design (Multiple Before After Control Impact) that refers to comparisons of multiple impact sites with random multiple control locations, at multiple times before and after the impacts (cf . 'BACI', 'BACIP' and 'MBACI'); analyses are usually done with ANOVA or repeated-measures ANOVA.

Meiofauna – very small organisms associated with sediment; pass through a geological sieve of 0.5 mm and are retained in a sieve of 0.063 mm (cf. 'Microfauna' and 'Macrofauna'; see also Chapter 8).

Meroplankton – organisms that only spend part of their lives in the plankton, usually as larvae (e.g. barnacles and fishes).

Metapopulation – a group of populations that may correspond to the geographic range of a species (cf. 'Population'); see Figure 1.3.

Microfauna – very small organisms associated with sediment and are retained in a geological sieve of 0.063 mm (cf. 'Macrofauna' and 'Meiofauna'; see also Chapter 8).

Mixed-model designs – see 'ANOVA'.

MLWM – mean low-water mark.

Modular organisms – colonial organisms that have an ability to reproduce asexually (e.g. by budding). Generally made up of individual components that can exist as groups (colonies) or as individuals (e.g. colonial ascidians) (see 'Genet' and 'Ramet').

Monitoring – sampling in time with adequate replication to detect variation over a temporal range from short to long time periods; done at more than one location.

Morphometrics – measurements of the shape of an organism. For example, many measurements are made on individual fishes for taxonomic descriptions, including total length, standard length, body depth, depth at the anus, eye diameter and head length.

Multivariate statistics – the analysis of multiple variables (e.g. species, environmental variables), which may have been measured at multiple levels (i.e. factors; e.g. times, locations, sites, habitats). You can, therefore, simultaneously analyse multiple dependent and multiple independent variables.

Natal reef – the reef from which individuals were born as eggs, larvae or juveniles.

Nested (= hierarchical) sampling design – dealing with multiple levels of variation (e.g. days within months, months within seasons, seasons within years). See Chapter 3 for nested designs using ANOVA

Neuston – planktonic organisms that live just below the surface meniscus.

NIWA – National Institute of Water and Atmospheric Research Ltd, consultancy and a branch of the New Zealand Government Services.

Non-parametric statistics – so-called distribution-free methods in which data are analysed according to rank. Often used where data violate the assumptions of parametric statistics (q.v.) or some attribute of the study organism can only be ranked (e.g. behaviour: degree of aggression).

Ontogeny –development changes in an individual during its life. Ontogenetic change is often used to mean changes that relate to age and size (e.g. diet).

Oogenesis – the production of ova (cf. 'Gametogenesis' and 'Spermatogenesis'). Cells increase in size from oogonia to hydrated vitellogenic oocytes, and 'polar bodies' of very low volume are lost at each division.

Orthogonal design – a symmetrical design that enables the combined effects of factors to be evaluated from interaction effects (cf. 'Nested sampling design' and 'Asymmetrical design').

Oviparous – broadcast spawning of eggs and sperm. Fertilisation is external and this mode of reproduction is common in the sea (e.g. many fishes, echinoderms); cf. 'Oviviparous' and 'Viviparous'.

Oviviparous – live-bearing. Young are released from the female as juveniles. While inside the female they receive little or no additional nutrition other than that provided by the yolk (e.g. some sharks); cf. 'Oviparous' and 'Viviparous'.

Partially hierarchical sampling design – see 'ANOVA'.

Parameter – mathematical or numerical values to describe some characteristic of a population (e.g. population standard deviation).

Parametric statistics –statistics of data values collected (e.g. number of fish per 100 m²); can generally be described according to parameters such as sample mean and standard deviation (SD), which are estimates of the true mean and SD of the population you have sampled (cf. 'Non-parametric statistics').

Pelagic – living in the water column; usually referring to open water.

Phenology – the study of timing, including descriptions of patterns of reproduction and the factors influencing timing (e.g. day length, water temperature, abundance of food).

Planktivorous – fishes that eat plankton.

Plankton – small (often microscopic) plants and animals that live in the sea (some species are only found in fresh water). The term normally implies that they are free-drifting and have limited powers of locomotion (e.g. copepods, larvacea, ctenophores, medusae). Most marine animals spend time as part of the plankton as larvae (= meroplankton; cf. 'Holoplankton').

Planktotrophic – a term that refers to larvae that feed on plankton as they develop (cf. 'Lecithotrophic'); nutrition is initially from the maternal yolk, then, near or after yolk absorption, the plankton (exogenous source); e.g. fish larvae.

Pleuston – organisms that rest on the surface of the water (e.g. Portugese man-of-war).

Polygyny – spawning or mating with more than one partner, not necessarily at the same time.

Population – a group of animals or plants of a single species. Usually referred to as a local population within a study area. A reproductively isolated population may be termed a mesopopulation within the entire geographic range of a species metapopulation (see Fig. 1.3).

Population dynamics – variation in numbers of organisms within a population. Important factors include recruitment, growth to maturity, mortality, reproductive output and extraction through harvesting (see Fig. 1.3).

Power of a test –the probability that it will correctly yield a statistically significant result (i.e. correct rejection of the null hypothesis); = 1 – β.

Postsettlement – after settlement, i.c. juvcnilcs and adults on reefs (cf. 'Presettlement').

Presettlement – before settlement. The planktonic stages of organisms (and in some cases pelagic juveniles) before settling on a substratum.

Press – in this book we are generally referring to a type of impact that is a sustained alteration of species densities that, in some cases, can result in the complete elimination of a particular species.

Protandric hermaphroditism – change of sex from male to female (e.g. Pisces, *Amphiprion*).

Protogynous hermaphroditism – change of sex from female to male (e.g. Pisces, Labridae).

Pseudofactorialism – two or more measurements made on each experimental unit and analysed in a single analysis, resulting in an erroneous increase in degrees of freedom (*sensu* Hurlbert & White 1993).

Pseudoreplication – the confounding of experimentally controlled effects with uncontrolled effects (*sensu* Hurlbert 1984). Pseudoreplication can be avoided at the design phase. This mistake is common in sampling designs that segregate the replicate samples of treatments from those of controls, while having only one group of each. Unless there are multiple groups of treatments and controls, an experimental effect is confounded with a spatial effect.

Pulse – in this book we are generally referring to a type of impact that is a relatively instantaneous alteration of species numbers, after which the system relaxes back to its previous state.

Pycnocline – an oceanographic term for a rapid change in water density (usually measured as Sigma-t); cf. 'Halocline' and 'Thermocline'.

Quadrat – a square sampling unit. Permanent quadrats are sampled at multiple times but remain in the same position.

Qualitative – a subjective assessment that does not necessarily involve any relevant measurements (cf. 'Quantitative').

Quantitative – actual measurements will or have been made (e.g. mean number of organisms per sampling unit); cf. 'Qualitative'.

Ramet – an individual organism in a colony. For species that reproduce asexually, it is possible to have multiple ramets of identical genotype (genet).

Random numbers – numbers that have a random distribution (within a range decided by the investigator); this can be checked with a frequency plot. Tables of random numbers are useful for random sampling procedures. Random numbers can also be generated from computer software, and other innovative methods can also be used if your resources are limited; see 3.4.2.

Random sampling – sampling units positioned in space using a randomisation procedure (e.g. at random intersections on a tape; the positions are determined *a priori* from random numbers).

Rare – in general use this term means very low in abundance. Relative rarity and potential for local or total extinction may vary according to the biogeographic range of species, habitat specificity, number of local populations and reproductive links between populations.

Reference – control (q.v.).

Regression – a functional relationship between two variables (e.g. age versus weight of an animal). This is generally expressed as r^2 (the correlation of determination, which is the explained sums of squares as a proportion of total sums of squares, proportion between 0 and 1); cf. 'Correlation'.

Replicate – the lowest level of sampling in a design; referred to in this book as sampling units. For example, 1 m^2 quadrats used to sample a kelp forest or a 13 cm diameter corer used to sample infauna. Sometimes we refer to replication at higher levels of sampling design (e.g. 'replicate locations').

Resilience – in the context of impacts, is the ability of a population to recover from a disturbance (cf. 'Inertia', 'Stability').

ROV – remotely operated vehicle.

Sampling design – detail of the allocation of sampling units in time and space.

Sampling grid – a sampling design where regularly spaced samples are taken on a grid.

Sampling unit – generally the lowest level of replication (e.g. 1 m^2 quadrat, 50 m transect, one plankton sample).

Sessile organisms – benthic organisms that are attached to the substratum (e.g. barnacles).

Sedentary organisms – benthic organisms with restricted movements (e.g. abalone).

Sigma-t – measure of the density of water.

Spatial scale – defined units of space. In general this should be defined in units of measurement, because terms such as coarse scale, mesoscale, etc. are used in different ways by investigators. If these terms are used, they should be defined (e.g. coarse scale 1 – 100 km).

Spatial variation – a general term used to describe variation of a variable (e.g. abundance of an organism) in space. This may be combined variation in space from multiple spatial scales (millimetres to thousands of kilometres; see 'Spatial scale') or one or more spatial scales that are defined in a sampling design (e.g. variation in abundance of reef fish at spatial scales of tens of metres to tens of kilometres).

Spermatogensis – the production of sperm (cf. 'Gametogenesis' and 'Oogensis'). Gametes decrease in size from large cells (spermatogonia) to sperm.

Stability – in the context of impacts, is the rate at which a population or assemblage recovers after a disturbance (cf. 'Inertia', 'Resilience').

Standing crop – biomass of material, generally measured per unit weight (e.g. kg of algae/unit area).

Statistical inference – where a null hypothesis is rejected at a predetermined level of probability (e.g. $P < 0.05$, i.e. there is less than a 5% chance that the null hypothesis is correct).

Strata – areas with environmental characteristics that are similar. For example, depth strata, shallow, mid and deep; habitat strata fringe, kelp forest, urchin-grazed barrens and sponge garden.

Stratified sampling – sampling designs where units are allocated according to predetermined strata such as habitat types. Random stratified sampling involves the placement of equal numbers of sample units in each stratum; the allocation of units for proportional stratified sampling is by area; and for optimal stratified sampling by the size of variances from and initial random stratified sampling design.

Stock – a management unit in fisheries. The determination of stock boundaries is a major area of endeavour for fisheries managers. Ideally a stock should be reproductively isolated, and this corresponds with the definition for mesopopulations (see also 'Population'); see Figure 1.3.

Subtidal – waters below the low-tide mark. Note that in some countries the term 'littoral' may extend up to 100 m from shore.

Succession – a sequence of ecological events that relate to the colonisation and persistence of species in an assemblage through time.

Survivorship – as opposed to survival, this term refers in a comparative way to the numbers surviving over a period of time.

Temporal scale – defined in units of time. In general this should be defined in units of measurement, because terms such as short term, long term, etc. are used in different ways by investigators. If these terms are used, they should be defined (e.g. longterm = years to tens of years).

Temporal variation – a general term used to describe variation of a variable (e.g. abundance of an organism) in time. This may be combined variation in time from multiple temporal scales (hours to tens of years; see also 'Temporal scale') or one or more temporal scales that are defined in a sampling design (e.g. variation in abundance on reef fish at temporal scales of days to tens of years).

Terminal-phase male (TP male) – generally refers to a male of species that are protogynous hermaphroditites. TP males are usually the largest and oldest fish in a population, and have generally spawned as females before they spawn as males. A small percentage of fish change sex from female to male early in life and never spawn as females. These and initial-phase males (q.v.) may 'steal' fertilisations before growing into TP males (if they live that long).

Thermocline – an oceanographic term that refers to a rapid change in water temperature; cf. 'Halocline' and 'Pyncnocline'.

Univariate statistics –methods that can be used to analyse one dependent variable (number per unit area of an organism). Independent measures of that variable may have been done at multiple temporal and spatial scales (cf. 'Multivariate statistics').

Variables – the measurements taken at each level of sampling; e.g. species, morphometrics of species and measures of the environment. For example, you may have estimates of density of species in your study area as well as environmental measurements (e.g. temperature, tubidity, etc.). See also 'Dependent variables' and 'Independent variables'.

Variance – the sums of squared deviations about a mean value. Although deviations are in negative and positive values, the squared values are all positive. The square root of the variance = standard deviation of a population or sample (see also Table 3.1).

Viviparous – live–bearing animals. Young are released from the female as juveniles and, in rare cases, as adults (e.g. embiotocid fishes). While young are brooded inside the female they gain nutrition from her; cf. 'Oviparous' and 'Oviviparous'.

Waypoint – a position fix to give latitude and longitude (see also 'GPS').

REFERENCES

Abbott, I. A., Hollenberg, G. J. (1976). *Marine algae of California*. Stanford University Press. 827 pp.

Abel, D. J., Williams, W. T., Williams, D. McB. (1985). A fast algorithm for large problems under the Bray-Curtis Measure. *J. Exp. Mar. Biol. Ecol.* 89: 237–245.

Acuna, J. L., Anadon, R. (1992). Appendicularian assemblages in a shelf area and their relationship with temperature. *J. Plankton Res.* 14: 1233–1250.

Adams, C., Stevely, J. M., Sweat, D. (1995). Economic feasibility of small-scale sponge farming in Pohnpei, Federated States of Micronesia. *J. World Aquacult. Soc.* 26: 132–142.

Adams, D. C., Gurevitch, J., Rosenberg, M. S. (1997). Resampling tests for meta-analysis of ecological data. *Ecology* 78: 1277–1283.

Adams, N. M. (1994). *Seaweeds of New Zealand*. Canterbury University Press, Christchurch. 360 pp.

Adams, N. M. (1997). *Common seaweeds of New Zealand*. Canterbury University Press, Christchurch. 96 pp.

Addessi, L. (1994). Human disturbance and long-term changes on a rocky intertidal community. *Ecol. Appl.* 4: 786–797.

Agardy, T. S. (1994). Advances in marine conservation: the role of marine protected areas. *TREE* 9(7): 267–270.

Agardy, T. S. (1997). *Marine protected area and ocean conservation*. Academic Press, San Diego. 244 pp.

Alexander, S. E., Roughgarden, J. (1996). Larval transport and population dynamics of intertidal barnacles: a coupled benthic/oceanic model. *Ecol. Monogr.* 66: 259–277.

Allan, R., Lindesay, J., Parker, D. (1996). *El Niño Southern Oscillation and climatic variability*. CSIRO, Collingwood. 405 pp.

Alldredge, A. L. (1972). Discarded appendicularian houses as sources of food, surface habitats, and particulate organic matter in planktonic environments. *Limnol. Oceanogr.* 21: 14–23.

Alldredge, A. L., Hamner, W. M. (1980). Recurring aggregation of zooplankton by a tidal current. *Estuar. Cstl Mar. Sci.* 10: 31–37.

Alldredge, A. L., King, J. M. (1977). Distribution, abundance, and substrate preferences of demersal reef zooplankton at Lizard Island Lagoon, Great Barrier Reef. *Mar. Biol.* 41: 317–333.

Alldredge, A. L., King, J. M. (1985). The distance demersal zooplankton migrate above the benthos: implications for predation. *Mar. Biol.* 84: 252–260.

Alldredge, A.L., McGillivary, P. (1991). The attachment probabilities of marine snow and their implications for particle coagulation in the ocean. *Deep Sea Res.* 38: 431–443.

Alldredge, A. L., Silver, M. W. (1988). Characteristics, dynamics and significance of marine snow. *Prog. Oceanogr.* 20: 41–82.

Allen, G. R. (1987). A new species of pomacentrid fish with notes on other damselfishes of the Kermadec Islands. *Rec. West. Aust. Mus.* 13: 263–273.

Ambrose, R. F., Raimondi, P. T., Engle, J. M. (1992). Final study plan for inventory of intertidal resources in Santa Barbara County. Unpublished report to the Minerals Management Service, Pacific OCS Region. 47 pp.

Anderson, D. T. (1998). *Invertebrate Zoology*. Oxford University Press, Melbourne. 467 pp.

Anderson, D. T. (1994). *Barnacles: structure, function, development and evolution*. Chapman & Hall, London. 357 pp.

Anderson, D. T., White, B. M., Egan, E. A. (1976). The larval development and metamorphosis of the ascidians *Pyura praeputialis* (Heller) and *Pyura pachydermatina* (Herdman) (Pleurogona, Family Pyuridae). *Proc. Linn. Soc. NSW.* 100: 205–217.

Anderson, M. J., Underwood, A. J. (1994). Effects of substratum on the recruitment and development of an intertidal estuarine fouling community. *J. Exp. Mar. Biol. Ecol* 184: 217–237.

Anderson, S., Kautsky, L. (1996). Copper effects on reproductive stages of Baltic Sea *Fucus vesiculosus*. *Mar. Biol.* 125: 171–176.

Anderson, T. A. (1991). Mechanisms of digestion in the marine herbivore, the luderick, *Girella tricuspidata* (Quoy and Gaimard). *J. Fish. Biol.* 39: 535–547.

Andrew, N. L. (1988). Ecological aspects of the common sea urchin. *N.Z. J. Mar. Freshwat. Res.* 22: 415–426.

Andrew, N. L. (1989). Contrasting ecological implications of food limitation in sea urchins and herbivorous gastropods. *Mar. Ecol. Prog. Ser.* 51: 189–193.

Andrew, N. L. (1991). Changes in subtidal habitat following mass mortality of sea urchins in Botany Bay, New South Wales. *Aust. J. Ecol.* 16: 353–362.

Andrew, N. L. (1993). Spatial heterogeneity, sea urchin grazing, and habitat structure on reefs in temperate Australia. *Ecology* 74: 292–302.

Andrew, N. L. (1994). Survival of kelp adjacent to areas grazed by sea urchins in New South Wales, Australia. *Aust. J. Ecol.* 16: 353–362.

Andrew, N. L., Choat, J. H. (1982). The influence of predation and conspecific adults on the abundance of juvenile *Evechinus chloroticus* (Echinoidea: Echinometridae). *Oecologia* 54: 80–87.

Andrew, N. L., Choat, J. H. (1985). Habitat related differences in the survivorship and growth of juvenile sea urchins. *Mar. Ecol. Prog. Ser.* 27: 155–161.

Andrew, N. L., Jones, G. P. (1990). Patch formation by herbivorous fish in a temperate Australian kelp forest. *Oecologia* 55: 57–68.

Andrew, N. L., MacDiarmid, A. B. (1991). Interrelationships between sea urchins and spiny lobsters in northeastern New

Zealand. *Mar. Ecol. Prog. Ser.* 70: 211–222.

Andrew, N. L., Mapstone, B. D. (1987). Sampling and the description of spatial pattern in marine ecology. *Oceanogr. Mar. Biol. Ann. Rev.* 25: 39–90.

Andrew, N. L., Pepperell, J. G. (1992). The by-catch of shrimp trawl fisheries. *Oceanogr. Mar. Biol. Ann. Rev.* 30: 527–565.

Andrew, N. L., Stocker, L. J. (1986). Dispersion and phagokinesis in the echinoid *Evechinus chloroticus*. *J. Exp. Mar. Biol. Ecol.* 100: 11–23.

Andrew, N. L., Underwood, A. J. (1989). Patterns of abundance of the sea urchin *Centrostephanus rodgersii* (Agassiz) on the central coast of New South Wales, Australia. *J. Exp. Mar. Biol. Ecol.* 131: 61–80.

Andrew, N. L., Underwood, A. J. (1993a). Density-dependent foraging in the sea urchin *Centrostephanus rodgersii* on shallow subtidal reefs in New South Wales, Australia. *Mar. Ecol. Prog. Ser.* 99: 89–98.

Andrew, N. L., Underwood, A. J. (1993b). Associations and abundance of sea urchins and abalone on shallow subtidal reefs in southern New South Wales. *Aust. J. Mar. Freshwat. Res.* 43: 1547–1559.

Andrew, T. G., Buxton, C. D., Hecht, T. (1996). Aspects of the reproductive biology of the concha wrasse, *Nelabrichthys ornatus*, at Tristan da Cunha. *Environ. Biol. Fish.* 46: 139–149.

Anon. (1992). US Joint Ocean Flux Study Program. *Oceanus* 35: 57–59.

Arai, M. N. (1997). *A functional biology of Scyphozoa*. Chapman & Hall, London. 316 pp.

Arena, G., Barea, L., Defeo, O. (1994). Theoretical evaluation of trap capture for stock assessments. *Fish. Res.* 19: 349–362.

Aronson, R. B., Heck, J. K. L. (1995). Tethering experiments and hypothesis testing in ecology. *Mar. Ecol. Prog. Ser.* 121: 307–309.

Arron, E. S., Lewis, K. B. (1993). Wellington south coast substrates 1:15,000. *N.Z. Oceanogr. Inst. Chart, Misc. Ser.* 69.

Arya, J. C., Lardner, R. W. (1979). *Mathematics for the biological sciences*. Prentice-Hall, Englewood Cliffs. 712 pp.

Asper, V. L. (1987). A review of sediment trap technique. *J. Mar. Technol. Soc.* 21: 18–25.

Ates, R. M. L. (1988). Medusivorous fishes, a review. *Zool. Meded.* 62: 29–42.

Attrill, M. J. (1997). *Rehabilitated estuarine ecosystems: the Thames Estuary, its environment and ecology*. Chapman & Hall, London. 220 pp.

Ayling, A. L. (1980). Patterns of sexuality, asexual reproduction and recruitment in some subtidal marine Demospongiae. *Biol. Bull.* 158: 271–282.

Ayling, A. L. (1983). Factors affecting the spatial distribution of thinly encrusting sponges from temperate waters. *Oecologia* 60: 412-418.

Ayling, A. M., (1978). Okakari Point to Cape Rodney Marine Reserve: a biological survey. [University of Auckland] *Leigh Lab. Bull.* 1. 98 pp.

Ayling, A. M. (1981). The role of biological disturbance in temperate subtidal encrusting communities. *Ecology* 62: 830–847.

Ayling, A. M., Cox, G. J. (1982). *Collins guide to the sea fishes of New Zealand*. Collins, Auckland. 343 pp.

Ayling, A. M., Paxton, J. R. (1983). *Odax cyanoallix*, a new species of odacid fish from northern New Zealand. *Copeia*: 95–101.

Bagenal, T. (1978). *Methods of assessment of fish production in fresh waters*. Blackwell, Oxford. 365 pp.

Bailey, K. M., Houde, E. D. (1989). Predation on eggs and larvae of marine fishes and the recruitment problem. *Adv. Mar. Biol.* 25: 1–83.

Bailey, T. G., Torres, J. J., Youngbluth, M. J., Owen, G. P. (1994). Effect of decompression on mesopelagic gelatinous zooplankton: A comparison of in situ and shipboard measurements of metabolism. *Mar. Ecol. Progr. Ser.* 113: 13–27.

Baines, P. G. (1995). *Topographic effects in stratified flows*. Cambridge University Press, Cambridge. 482 pp.

Baines, P. G., Murray, D. L. (1995). Topographic influence on the pattern of flow through Bass Strait. *Mar. Freshwat. Res.* 46: 763–767.

Baker, A. N. (1983). *Whales and dolphins of New Zealand and Australia*. Victoria University Press, Wellington. 133 pp.

Baker, J. M., Crothers, J. H. (1987). Intertidal rock. Pp. 157–197 *in* Baker, J. M., Wolff, W. J. (eds) *Biological surveys of estuaries and coasts*. Cambridge University Press, Cambridge. 449 pp.

Baker, J. M., Wolff, W. J. (1987). *Biological surveys of estuaries and coasts*. Cambridge University Press, Cambridge. 449 pp.

Ballantine, W. J. (1991). Marine Reserves for New Zealand. [University of Auckland] *Leigh Lab. Bull.* 25: 196 pp.

Barkai, A., McQuaid, C. D. (1988). Predator-prey role reversal in a marine benthic ecosystem. *Science* 242: 62–64.

Barnard, J. L., Karaman, G. S. (1991a). The families and genera of marine Gammaridean Amphipoda (except marine gammaroids) Part 1. *Rec. Aust. Mus.* 13: 1–417.

Barnard, J. L., Karaman, G. S. (1991b). The families and genera of marine Gammaridean Amphipoda (except marine gammaroids) Part 2. *Rec. Aust. Mus.* 13: 419–866.

Barrett, N. S. (1995). Aspects of the biology and ecology of six species of temperate reef fishes (Families: Labridae and Monacanthidae). PhD thesis, University of Tasmania, Hobart. 192 pp.

Barthel, D., Theede, H. (1986). A new method for the culture of marine sponges and its applicaion for experimental studies. *Ophelia* 25: 75–82.

Batschelet, E. B. (1981). *Circular statistics in biology*. Academic Press, London. 371 pp.

Battershill, C. N. (1986). The marine benthos of caves, archways, and vertical reef walls of the Poor Knights Islands. Unpublished report. Poor Knights Marine Reserve Committee. 179 pp.

Battershill, C. N., Bergquist, P. R. (1982). Responses of an intertidal gastropod to field exposure of an oil and dispersant. *Mar. Pollut. Bull.* 13: 159–162.

Battershill, C. N., Bergquist, P. R. (1984). The influence of biorhythms on sensitivity of *Nerita* to pollutants at sublethal levels. *Oil Petrochem. Pollut.* 2(1): 1–8.

Battershill, C. N., Bergquist, P. R. (1990). The influence of storms on asexual reproduction, recruitment and survivorship of sponges. Pp. 397–403 *in* Rutzler, K. (ed.), *New perspectives in sponge biology*. Smithsonian Institution Press, Washington DC.

Battershill, C. N., Kingsford, M. J.; MacDiarmid A. B. (1985). *Te Haupa Island marine habitats*. Auckland City Council, Auckland. 60 pp.

Battershill, C. N., McLean, M. G. (1994). Cell separation

experiments on *Lissodendoryx* n.sp. *National Cancer Institute Contract Report* 1994 12.

Battershill, C. N., Schiel, D. R., Jones, G. P., Creese, R. G., MacDiarmid, A. B. (Eds) (1993). Proceedings of the Second International Temperate Reef Symposium, Auckland, January 1992. NIWA Marine, Wellington. 251 pp.

Bauer, R. T., Martin, J. W. (1991). *Crustacean sexual biology.* Columbia Press, New York. 355 pp.

Bayle-Sempere, J. T., Ramos-Espla, A. A. (1993). Some population parameters as bioindicators to assess the 'reserve effect' on the fish assemblage. Pp. 189–214 *in* Boudouresque, C. F., Avon, M., Pergent-Martini, C. (eds), *Qualité du milieu marin-Indicators biologiques et physico-chimiques.* GIS posidone publ., France.

Beamish, R. J., McFarlane, G. A. (1983). The forgotten requirement of age determination in fisheries biology. *Trans. Am. Fish. Soc.* 112: 735–743.

Beck, M. W. (1995). Size-specific shelter limitation in stone crabs: a test of the demographic bottleneck hypothesis. *Ecology* 76: 968–980.

Beck, M. W. (in press). Comparison of the measurement and effects of habitat structure on gastropods in rocky intertidal and mangrove habitats. *Mar. Ecol. Prog. Ser.*

Beckley, L. E. (1985) Tidepool fishes: recolonization after experimental elimination. *J. Exp. Mar. Biol. Ecol.* 85: 287–295.

Beckley, L. E., Branch, G. M. (1992). A quantitative scuba-diving survey of the sublittoral macrobenthos at subantarctic Mariona Island. *Polar Biol.* 11: 553–563.

Beder, S. (1989). *Toxic fish and sewer surfing: How deceit and collusion are destroying our great beaches.* Allen & Unwin, Sydney. 176 pp.

Beder, S. (1990). Sun, surf, and sewage. *New Scient.* 14 July: 24–29.

Beers, J. R. (1966). Studies on the chemical composition of the major zooplankton groups in the Sargasso Sea off Bermuda. *Limnol. Oceanogr.* 11: 520–528.

Behrenfield, M. J., Bale, A. J., Kolber, Z. S., Aiken, J., Falkowski, P. G. (1996). Confirmation of iron limitation of phytoplankton photosynthesis in the equatorial Pacific Ocean. *Nature* 383: 508–511.

Bell, C. (1984). Cape Rodney to Okakari Point Marine Reserve: a visitor survey. Report prepared by the Department of Lands and Survey for the Cape Rodney to Okakari Point Marine Reserve Management Committee. 85 pp.

Bell, J. D. (1979). Observations of the diet of the red morwong, *Cheilodactylus fuscus* Castelnau (Pisces: Cheilo-dactylidae). *Aust. J. Mar. Freshwat. Res.* 30: 129–133.

Bell, J. D. (1983). Effects of depth and marine reserve fishing restrictions on the structure of a rocky reef fish assemblage in the north-western Mediterranean sea. *J. Appl. Ecol.* 20: 357–369.

Bell, J. D., Burchmore, J. J., Pollard, D. A. (1980). The food and feeding habits of the rock blackfish, *Girella elevata* Macleay (Pisces: Girellidae), from the Sydney region, New South Wales. *Aust. J. Zool.* 20: 391–405.

Bell, J. D., Craik, G. J. S., Pollard, D. A., Russell, B. C. (1985a). Estimating length frequency distributions of large reef fish underwater. *Coral Reefs* 4: 41–44.

Bell, J. D., Pollard, D. A. (1989). Ecology of fish assemblages and fisheries associated with seagrasses. Pp. 565–609 *in* Larkum, A. W. D., McComb, A. J., Shepard, S. A. (eds), *Biology of seagrasses: a treatise on the biology of seagrasses with special reference to the Australian region.* Elsevier, Amsterdam. 841 pp.

Bell, J. D., Steffe, A. S., Westoby, M. (1985b). Artificial sea-grass: How useful is it for field experiments on fish and macroinvertebrates? *J. Exp. Mar. Biol. Ecol.* 90: 171–177.

Bell, J. D, Steffe, A. S., Westoby, M. (1988). Location of sea-grass beds in estuaries: Effects of fish and decapods. *J. Exp. Mar. Biol. Ecol.* 122: 127–146.

Bell, J. D., Westoby, M. (1986a). Importance of local changes in leaf height and density to fish and decapods associated with seagrasses. *J. Exp. Mar. Biol. Ecol.* 104: 249–274.

Bell, J. D., Westoby, M. (1986b). Variation in seagrass height and density over a wide spatial scale: effects on fish and decapods. *J. Exp. Mar. Biol. Ecol.* 104: 275–295.

Bell, J. D., Westoby, M. (1986c). Abundance of macro-fauna in dense seagrass is due to habitat preference, not predation. *Oecologia (Berlin)* 68: 205–209.

Bell, J. D., Worthington, D. (1993). Links between estuaries and coastal rocky reefs in the life of fishes from South-eastern Australia. *Proc. 2nd Int. Temperate Reef Symp.* 85–92.

Bell, S. S., McCoy, E. D., Mushinsky, H. R. (1991). *Habitat structure: the physical arrangement of objects in space.* Chapman & Hall, London. 438 pp.

Bellwood, D. R., Alcala, A. C. (1988). The effect of a minimum length specification on visual estimates of density and biomass of coral reef fishes. *Coral Reefs* 7: 23–27.

Bender, E. A., Case, T. J., Gilpin, M. E. (1984). Perturbation experiments in community ecology: Theory and practise. *Ecology* 65: 1–13.

Benfield, M. C., Davis, C. S., Wiebe, P. H., Gallager, S. M., Lough, R.G.. Copley, N.J. (1996). Video plankton recorder estimates of copepod, pteropod and larvacean distributions from a stratified region of Georges Bank with comparative measurements from a MOCNESS sampler. *Deep Sea Res.* 43: 1925–1945.

Bennett, I. (1987). *W. J. Dakin's classic study: Australian Seashores.* Angus & Robertson, London. 411 pp.

Bennett, J. T., Boehlert, G. W., Turekian, K. K. (1982). Confirmation of longevity in *Sebastes diploproa* (Pisces: Scorpaenidae) from $210Pb/226Ra$ Measurements in otoliths. *Mar. Biol.* 7: 209–215.

Bergquist, P. R. (1978). *Sponges.* Hutchinson University Library, London. 268 pp.

Berkelmans, R. (1992). Video photography: a quantitative sampling method. *Reef Res.* 2: 10-11.

Beukers, J. S., Jones, G. P., Buckey, R. M. (1995). Use of implant micro-tags for studies on populations of small reef fish. *Mar. Ecol. Prog. Ser.* 125: 61–66.

Bjorkstedt, E., Roughgarden, J. (1997). Larval transport and coastal upwelling: an application of HF radar in ecological research. *Oceanography* 10: 64–67.

Bjornsen, P. K (1986). Automatic determination of bacterio-plankton biomass by image analysis. *Appl. Environ. Microbiol.* 51: 119–1204.

Blaber, S. J. M., Blaber, T. G. (1980). Factors affecting the distribution of juvenile estuarine and inshore fish. *J. Fish. Biology.* 17: 143–162.

Black, K. P. (1993). The relative importance of local retention and inter-reef dispersal of neutrally buoyant material on coral reefs. *Coral Reefs* 12: 43–53.

Black, K. P., Gay, S. L., Andrews, J. C. (1990). Residence times of neutrally-buoyant matter such as larvae, sewage or nutrients on coral reefs. *Coral Reefs* 9: 105–114.

Black, R., Fisher, K., Hill, A., McShane, P. (1979). Physical and biological conditions on a steep intertidal gradient at Rottnest Island, Western Australia. *Aust. J. Ecol.* 4: 67–74.

Blaxter, J. H. S. (1977). The effect of copper on the eggs and larvae of plaice and herring. *J. Mar. Biol. Ass. U.K.* 57: 849–858.

Bloesch, J., Burns, N. M. (1980). A critical review of sedimentation trap technique. *Schweiz. Z. Hydrol.* 42: 15–55.

Blunt, J. W., Munro, M. H. G., Battershill, C. N., Copp, B. R., McCombs, J. D., Perry, N. B., Prinsep, M., Thompson, A.M. (1990). From the Antarctic to the Antipodes; 45° of marine chemistry. *New J. Chem.* 14: 761–775.

Bodkin, J. L. (1988). Effects of kelp forest removal on associated fish assemblages in central California. *J. Exp. Mar. Biol. Ecol.* 117: 227–238.

Bohnsack, J. A. (1989). Are high densities of fishes at artificial reefs the result of habitat limitation or behavioural preference? *Bull. Mar. Sci.* 44: 631–645.

Bohnsack, J. A., Bannerot, S. P. (1986). A stationary visual census technque for quantitatively assessing community structure of coral reef fishes. *NOAA Tech. Rep.* NMFS 41: 1–15.

Bohnsack, J. A., Sutherland, D. L. (1985). Artificial reef research: a review with recommendations for future priorities. *Bull. Mar. Sci.* 37: 11–39.

Bonin, D. R. (1981). Systematics and life histories of New Zealand Bonnemaisoniaceae (Rhodophyta). Unpublished MSc thesis, University of Auckland, New Zealand. 126 pp.

Booth, D. J. (1992). Larval settlement and preferences by domino *Dasyllus albisella* Gill. *J. Exp. Mar. Biol. Ecol.* 155: 85–104.

Booth, J. D. (1979). Common bivalve larvae in New Zealand : Pteriacea, Anomaiacea, Ostreacea. *N.Z. J. Mar. Freshwat. Res.* 13: 131–139.

Booth, J. D. (1984). Movements of packhorse rocklobsters *Jasus verreauxi* tagged along the eastern coast of the North Island of New Zealand. *N.Z. J. Mar. Freshwat. Res.* 18: 275–281.

Booth, J. D. (1994). *Jasus edwardsii* larval recruitment off the east coast of New Zealand. *Crustaceana* 66: 295–317.

Booth, J. D., Phillips, B. F. (1994). Early life history of spiny lobster. *Crustaceana* 66: 271–294.

Bortone, S. A., Hastings, R. W., Oglesby, J. L. (1986). Quantification of reef fish assemblages: a comparison of several *in situ* methods. *Northwest Gulf Sci.* 8: 1–22.

Botsford, L. W., Castilla, J. C., Peterson, C. H. (1997). The management of fisheries and marine ecosystems. *Science* 277: 509–515.

Bouchon, C. 1981. Comparison of two quantitative sampling methods used in coral reef studies: the line transect and the quadrat methods. (Abstr.) *Proc. 4th Int. Coral Reef Symp.*, Manila, 2: Abstracts.

Bouwman, L. A. (1987). Meiofauna. Pp. 140–156 *in* Baker, J. M., Wolff, W. J. (eds) *Biological surveys of estuaries and coasts*. Cambridge University Press, Cambridge. 449 pp.

Bowman, M. J., Yentsch, C. M, Peterson, W. J. (1986). *Tidal mixing and plankton dynamics*. Springer-Verlag, Berlin.

Boy, R. L., Smith, B. R. (1984). Design improvements to fish attraction device (FAD) mooring systems in general use in Pacific countries. South Pacific Commission, Noumea, New Caledonia. *SPC Handbook* 24: 77 pp.

Bradford, J. M. (1972). Systematics and ecology of New Zealand central east coast plankton sampled at Kaikoura. *Memoir. N.Z. Oceanogr. Inst.* 54: 89 pp.

Bradford, J. M. (1985). Distribution of zooplankton off Westland, New Zealand, June 1979 and February 1982. *N.Z. J. Mar. Freshwat. Res.* 19: 311–326.

Bradford, J. M., Haakonssen, L., Jillett, J. B. (1983). The marine fauna of New Zealand : Pelagic calanoid copepods: Families Euchaetidae, Phaennidae, Scolecithricidae, Diaixidae, and Tharybidae. *Mem. N.Z. Oceanogr. Inst.* 90: 150 pp.

Bradford, J. M., Jillett, J. B. (1980). The marine fauna of New Zealand: Pelagic calanoid copepods: Family Aetideidae. *Mem. N.Z. Oceanogr. Inst.* 86 : 102 pp.

Bradford, J. M., Lapennas, P. P., Murtagh, R. A., Chang, F. H., Wilkinson,V. (1986). Factors controlling summer phytoplankton production in greater Cook Strait, New Zealand. *N.Z. J. Mar. Freshwat. Res.* 20: 253–279.

Bradford-Grieve, J. M. (1994). Pelagic Calanoid Copepoda: Megacalanidae, Calanidae, Paracalanidae, Mecynoceridae, Eucalanidae, Spinocalanidae, Clausiocalanidae. *Mem. N.Z. Oceanogr. Inst.* 102: 160 pp.

Bradford-Grieve, J. M., Murdoch, R. C., Chapman, B. E. (1993). Composition of macrozooplankton assemblages associated with the formation and decay of pulses within an upwelling plume in greater Cook Strait, New Zealand. *N.Z. J. Mar. Freshwat. Res.* 27: 1–22.

Bradstock, M. C. (1989). *Between the tides: New Zealand shore and estuary life*. Rev. edn. David Bateman, Auckland. 158 pp.

Branch, G. M., Barkai, A., Hockey, P. A. R., Hutchings, L. (1987). Biological interactions: Causes or effects of variability in the Benguela ecosystem. *S. Afr. J. Mar. Sci.* 5: 425–445.

Branch, G., Branch, M. (1981). *The living shores of Southern Africa*. C. Struik, Cape Town. 272 pp.

Branch, G. M., Griffiths, C. L., Branch, M. L., Beckley, L. E. (1994). *Two oceans: a guide to the marine life of Southern Africa*. David Phillip, Cape Town. 360 pp.

Brandon, S. J., Edgar, S. J., Shepherd, S. A. (1986). Reef fish populations of the Investigator group, South Australia, a comparison of two census methods. *Trans. R. Soc. S. Aust.* 101: 69–76.

Bray, R. N. (1981). Influence of water currents and zooplankton densities on daily foraging movements of blacksmith, *Chromis punctipinnis*, a plantivorous reef fish. *Fish. Bull. U.S.* 78: 829–841.

Bray, R. N., Miller, A. C., Geesey, G. G. (1981). The fish connection: a trophic link between planktonic and rocky reef communities? *Science* 214: 204–205.

Breen, P. A., McKoy, J. L. (1988). An annotated bibliography of the red rock lobster *Jasus edwardsii* in New Zealand. *N.Z. Fish. Occ. Publ.* 3: 43 pp.

Breitberg, D. L. (1984). Residual effects of grazing: inhibition of competitor recruitment by encrusting coralline algae. *Ecology* 65: 1136–1143.

Breitberg, D. L. (1991). Settlement patterns and presettlement behavor of the naked goby, *Gobiosoma bosci*, a temperate oyster reef fish. *Mar. Biol.* 109: 213–221.

Breitberg, D. L., Steinberg, N., DuBeau, S., Cooksey, C., Houde, E. D. (1994). Effects of low dissolved oxygen on predation on estuarine fish larvae. *Mar. Ecol. Prog. Ser.* 104: 235–246.

Brodie, J. W. (1960). Coastal surface currents around New Zealand. *N.Z. Jl Geol. Geophys.* 3: 235–252.

Brogan, M. W. (1994). Two methods of sampling fish larvae over reefs a comparison from the Gulf of California. *Mar. Biol.* 118: 33–45.

Brosnan, D. M., Crumrine, L. L. (1994). Effects of human trampling on marine rocky shore communities. *J. Exp. Mar. Biol. Ecol.* 177: 79–97.

Brown, A. C. (1983). The ecophysiology of sandy beach animals – a partial review. Pp. 575–605 *in* McLachlan, A., Erasmus, T. (eds) *Sandy beaches as ecosystems*. Junk, The Hague.

Brown, A. C., McLachlan, A. (1990). *Ecology of sandy shores*. Elsevier, Amsterdam. 328 pp.

Brown, D. M., Cheng, L. (1981). New net for sampling the ocean surface. *Mar. Ecol. Prog. Ser.* 5: 225–227.

Brown, P. J. 1996. Effects of visitor trampling on intertidal algae on rocky shores: Implications for marine reserve management. Unpublished MSc thesis, University of Auckland. 230 pp.

Bruhbaker, J. M., Angus, R. A. (1974). A procedure for staining fish with alizarin without causing exfoliation of scales. *Copeia* : 989–990.

Bruhn, J., Gerard, V. A. (1996). Photoinhibition and recovery of the kelp *Laminaria saccharina* at optimal and superoptimal temperatures. *Mar. Biol.* 125: 639–648.

Brunton, C. H., Curry, G. B. (1979). British brachiopods. P. 64 *in* Kermack, D. M., Barnes, R. S. K. (eds), Synopses of the British Fauna. *Synopses of the British Fauna* 17. Academic Press, London.

Buchanan, J. B. (1984). Sediment analysis. Pp. 41–65 *in* Holme, N. A. McIntyre, A. D. (eds), *Methods for the study of marine benthos*, 2nd edition. Blackwell, Oxford. 387 pp.

Buchanan, J. B., Longbottom, M. R., (1970). The determination of organic matter in marine muds: the effect of the presence of coal and the routine determination of protein. *J. Exp. Mar. Biol. Ecol.* 5: 158–169.

Buckley, R. M., Blankership, H. L. (1990). Internal intrinsic identification systems: overview of implanted wire tags, otolith marks and parasites. *Am. Fish. Soc. Symp.* 7: 173–182.

Buckley, R. M., West, J. E., Doty, D. C. (1994). Internal micro-tag systems for marking juvenile reef fishes. *Bull. Mar. Sci.* 55: 848–857.

Budd, J. T. C. (1991). Remote sensing techniques for monitoring land-cover. Pp. 33–59 *in* Goldsmith, F. B. (ed.), *Monitoring for conservation and ecology*. Chapman & Hall, London.

Buehr, M., Picton B. (1985). Basic underwater photography for divers. Pp. 6–16 *in* George, J. D., Lythgoe, G. I., Lythgoe, J.N. (eds) *Underwater photography and television for scientists*. Clarendon Press, Oxford.

Buller, A. T., McManus, J., (1979). Sediment sampling and analysis. Pp. 87–130 *in* Dyer, K. R. (ed.) *Estuarine hydrography and sedimentation*. Cambridge University Press.

Burchett, M. S. (1993). The life cycle of *Notothenia rossii* from South Georgia. *Br. Antarctic. Surv. Bull.* 61: 71–73.

Burchmore, J. J., Pollard, D. A., Bell, J. D., Middleton, M. J., Pease, B. C., Matthews, J. (1985). A ecological comparison of artificial and natural rocky reef communities in Botany Bay, New South Wales, Australia. *Bull. Mar. Sci.* 37: 70–85.

Burdi, A. R., Flecker, F. (1968). Differential staining of cartilage and bone in the intact embryonic skeleton in vitro. *Stain Tech.* 43: 47–48.

Burdick, R. K., Graybill, F. A. (1992). *Confidence intervals on variance components*. Marcel Dekker, New York. 211 pp.

Burkill, P. H. (1987). Analytical flow cytometry and its application to marine microbial ecology. Pp. 139–166 *in* Sleigh, M. (ed.) *Microbes in the sea*. Ellis Horwood, Chichester.

Burnham, K. P., Anderson, D. R., Laake, J. L. (1987). *Design and analysis methods for fish survival experiments based on release-recapture*. American Fisheries Society Monograph 5.

Bury, S. J., Owens, N. J. P. and Preston, T. (1995). ^{13}C and ^{15}N uptake by phytoplankton in the marginal ice zone of the Bellingshausen Sea. *Deep Sea Res.* 42: 1225–1252.

Bustamante, R. H., Branch, G. M., Eekhart, S. (1995). Maintenance of an exceptional grazer biomass in South Africa: a subsidy by subtidal kelps. *Ecology* 76: 2314–2329.

Butler, A. J. (1986). Recruitment of sessile invertebrates at five sites in Gulf St Vincent, South Australia. *J. Exp. Mar. Biol. Ecol.* 97: 13–36.

Butler, A. J. (1995). Subtidal rocky reefs. Pp. 83–105 *in* Underwood, A. J., Chapman, M. G. (eds) *Coastal marine ecology of temperate Australia*. UNSW Press, Sydney.

Butman, C. A. (1987). Larval settlement of soft-sediment invertebrates: the spatial scale of pattern explained by active habitat selection and the emerging role of hydrodynamical processes. *Oceanogr. Mar. Biol. Ann. Rev.* 25: 113–165.

Buxton, C. D. (1993). Life-history changes in exploited reef fishes on the east coast of South Africa. *Environ. Biol. Fish.* 36: 47–63.

Buxton, C. D., Clarke, J. R. (1992). The biology of the bronze bream, *Pachymetopon grande* (Teleostei: Sparidae) from the south-east Cape Coast, South Africa. *S. Afr. J. Zool.* 27: 21–32.

Buxton, C. D., Garratt, P. A. (1990). Alternative reproductive styles in seabraema (Pisces: Sparidae). *Environ. Biol. Fish.* 28: 113–124.

Buxton, C. D., Smale, M. J. (1989). Abundance and distribution patterns of three temperate marine reef fish (Teleostei: Sparidae) in exploited and unexploited areas off the southern Cape Coast. *J. Appl. Ecol.* 26: 441–451.

Byrne, M. (1990). Annual reproductive cycles of the commercial sea urchin *Paracentrotus lividus* from an exposed intertidal and a sheltered subtidal habitat on the west coast of Ireland. *Mar. Biol.* 104: 275–289.

Byrne, M. (1994). Class Ophiuroidea. Pp. 247–343 *in* Harrison, F., Chia, F. S. (eds), *Microanatomy of invertebrates, Vol. 14. Echinodermata*. Alan R. Liss, New York.

Byrne, M. (1995). Changes in larval morphology in the evolution of benthic development by *Patiriella exigua* (Asteroidea), a comparison with the larvae of *Patiriella* species with planktonic development. *Biol. Bull.* 188: 293–305.

Byrne, M. (1996). Starfish wanted, dead, or alive. *New Scientist*, 19 October: 53.

Caceres, C. W., Fuentes, L. S., Ojeda, F. P. (1994). Optimal feeding strategy of the temperate reef fish *Aplodactylus punctatus*: the effects of food availability on digestive and reproductive patterns. *Oecologia* 99: 118–123.

Caffey, H. M. (1985). Spatial and temporal variation in settlement and recruitment of intertidal barnacles. *Ecol. Monogr.* 55: 313–332.

Cahoon, L. B., Tronzo, C. R. (1988). A comparison of demersal zooplankton collected at Alligator Reef, Florida, using emergence and reentry traps. *Fish. Bull. U.S.* 86: 838–845.

Caillet, G. M., Radtke, R. L., Welden, B. A. (1986). Elasmo-branch age determination and verification: a review. Pp. 345–360 in Uyeno, T., Arai, T., Taniuchi, T., Matsuura, K. (eds), *Indo-Pacific Biology: Proc. 2nd Int. Conf. Indo-Pacific Fishes*. Ichthyological Society of Japan, Tokyo.

Calder, D. R. (1982). Life history of the cannonball jellyfish, *Stomolophus·meleagris* L. Agassiz, 1860 (Scyphozoa, Rhizostomida). *Biol. Bull.* 162: 149–162.

Cappo, M. (1995). The population biology of the temperate reef fish *Cheilodactylus nigripes* in an artificial reef environment. *Trans. R. Soc. S. Aust.* 119: 113–122.

Cappo, M., Brown, I. W. (1994). Evaluation of sampling methods for reef fish populations of commercial and recreational interest. Report to the Commonwealth Research Centre, Townsville, on Task 2.4.1: 45 pp.

Capriulo, G. M. (1990). *Ecology of marine protista*. Oxford University Press, Oxford.

Caputi, N., Fletcher, W. J., Pearce, A., Chubb, C. F. (1996). Effect of the Leeuwin Current on the recruitment of fish and invertebrates along the Western Australian coast. *Mar. Freshwat. Res.* 47: 147–156.

Carlos, J., Calvo, C. (1995). *El ecosistemo Marino Mediterraneo*. Murcia, Madrid. 797 pp.

Carls, M. G., Rice, S. (1989). Abnormal development and growth reductions of pollock *Theragra chalcogramma* embryos exposed to water-soluble fractions of oil. *Fish. Bull.* 88: 29-37.

Carlton, J. T. (1985). Transoceanic and interoceanic dispersal of coastal marine organisms: the biology of ballast water. *Oceanogr. Mar. Biol. Ann. Rev.* 23: 313–371.

Carlton, J. T. (1989). Man's role in changing the face of the ocean: Biological invasions and implications for conservation of near-shore environments. *Conserv. Biol.* 3: 265–273.

Carlton, J. T., Geller, J. B. (1993). Ecological roulette: the global transport of nonindigenous marine organisms. *Science* 261: 78–82.

Carlton, J. T., Hodder, J. (1995). Biogeography and dispersal of coastal organisms: Experimental studies on a replica of a 16th-century sailing vessel. *Mar. Biol.* 121: 721–730.

Carlton, J. T., Thompson, J. T., Schemel, L. E., Nichols, F. H. (1990). Remarkable invasion of San Francisco Bay (California, USA) by the Asian clam *Potmocorbula amurensis*. I. Introduction and dispersal. *Mar. Ecol. Prog. Ser.* 66: 81–94.

Carolin, R., Clarke, P. (1991). *Beach plants of southeastern Australia*. Sainty & Associates, Sydney. 119 pp.

Caron, D. A., Davis, P. G., Sieburth, J. McN. (1989). Factors responsible for the differences in cultural estimates and direct microscopical counts of populations of bacterivorous nanoflagellates. *Micrbiol. Ecol.* 18: 89–104.

Carr, M. H. (1989). Effects of macroalgal assemblages on the recruitment of temperate zone reef fishes. *J. Exp. Mar. Biol. Ecol.* 126: 59–76.

Carr, M. H. (1991). Habitat selection and recruitment of an assemblage of temperate zone reef fishes. *J. Exp. Mar. Biol. Ecol.* 146: 113–137.

Carr, M. H. (1994). Effects of macroalgal dynamics on recruit-ment of a temperate reef fish. *Ecology* 75: 1320–1333.

Carr, M. H., Reed, D. C. (1993). Conceptual issues to marine harvest refuges: Examples from temperate reef fishes. *Can. J. Fish. Aquat. Sci.* 50: 2019–2028.

Carter, L., Foster, G. A., Battershill, C. N. (1995). Cyclones,

crustal collisions and Crustacea. *Water & Atmosphere* 3(2): 12–14.

Cassie, R. M. (1954). Some uses of probability paper in the analysis of size frequency distributions. *Aust. J. Mar. Freshwat. Res.* 5: 513–522.

Cassie, R. M. (1956). Early development of the snapper, *Chrysophrys auratus* Forster. *Trans. R. Soc. N.Z.* 83: 705–713.

Castilla, J. C., Bustamante, R. H. (1989). Human exclusion from the rocky intertidal zone of Las Cruces, central Chile: the effects on *Durvillaea antarctica* (Phaeophyta, Durvilleales). *Mar. Ecol. Prog. Ser.* 50: 203–214.

Castilla, J. C., Duran, L. R. (1985). Human exclusion from the rocky intertidal zone of central Chile: the effects on *Concholepas concholepas* (Gastropoda) *Oikos* 45: 391–399.

Castilla, J. C., Fernandez, M. (1998). Small-scale benthic fisheries in Chile: on co-management and sustainable use of benthic invertebrates. *Ecol. Appl.* 8: S124–S132.

Caughley, G. (1977). *Analysis of vertebrate populations*. Wiley Interscience, London. 233 pp.

Caughley, G., Gunn, A. (1996). *Conservation biology in theory and practise*. Blackwell, Oxford. 459 pp.

CCME 1993a. Guidance manual on sampling, analysis, and data management for contaminated sites. Volume I: Main report. Canadian Council of Ministers of the Environment. Report No. CCME EPC-NCS62E. Winnipeg. 79 pp.

CCME 1993b. Guidance manual on sampling, analysis, and data management for contaminated sites. Volume II: Analytical method summaries. Canadian Council of Ministers of the Environment. Report No. CCME EPC-NCS66E. Winnipeg. 171 pp.

Chambers, S. (1989). *Birds of New Zealand: locality guide*. Arun Books, Hamilton. 511 pp.

Chang, F. H. (1983). Winter phytoplankton and microzoo-plankton populations off the coast of Westland, New Zealand, 1979. *N.Z. J. Mar. Freshwat. Res.* 17: 279–304.

Chapman, M. G. (1986). Assessment of some controls in experimental transplants of intertidal gastropods. *J. Exp. Mar. Biol. Ecol.* 103: 181–201.

Chapman, M. G., Underwood, A. J. (1994). Dispersal of the intertidal snail, *Nodilittorina pyramidalis*, in response to the topographic complexity of the substratum. *J. Exp. Mar. Biol. Ecol.* 179: 145–169.

Chapman, M. G., Underwood, A. J., Skilletter, G. A. (1995). Variability at different spatial scales between a subtidal assemblage exposed to discharge of sewage and two control sites. *J. Exp. Mar. Biol. Ecol.* 189: 103–122.

Chapman, V. J. (1979). *Mangrove vegetation*. J. Cramer, Leutershausen. 447 pp.

Chatfield, C. (1980). *The analysis of time series: an introduction*. Chapman & Hall, London. 266 pp.

Checkley Jr, D. M., Ortner, B., Settle, L. R., Cummings, S. R. (1997). A continous, underway fish egg sampler. *Fish. Oceanogr.* 6: 58–73 .

Chen, Y., Jackson, D. A., Harvey, H. H. (1992). A comparison of von Bertalanffy and polynomial functions in modelling fish growth data. *Can. J. Fish. Aquat. Sci.* 49: 1297–1304.

Cheshire, A. C., Hallam, N. D. (1989). Methods for assessing the age composition of native stands of subtidal macro-algae: a case study on *Durvillaea potatorum*. *Bot. Mar.* 32: 199–204.

Chick, R. (1993). The influence of oceanography on the vertical distribution of plankton. Unpublished Hons thesis, University of Sydney. 90 pp.

Choat, J. H. (1982). Fish feeding and the structure of benthic communities in temperate waters. *Ann. Rev. Ecol. Syst.* 13: 423–449.

Choat, J. H. (1991). The biology of herbivorous fishes on coral reefs. Pp. 120–155 *in* Sale, P. F. (ed.) *The ecology of fishes on coral reefs*. Academic Press, San Diego.

Choat, J. H., Axe, L. M. (1996). Growth and longevity in acanthurid fishes: An analysis of otolith increments. *Mar. Ecol. Prog. Ser.* 134: 15–26.

Choat, J. H., Ayling, A. M. (1987). The relationship between habitat structure and fish faunas on New Zealand reefs. *J. Exp. Mar. Biol. Ecol.* 110: 257–284.

Choat, J. H., Ayling, A. M., Schiel, D. R. (1988). Temporal and spatial variation in an island fish fauna. *J. Exp. Mar. Biol. Ecol.* 121: 91–111.

Choat, J. H., Bellwood, D. R. (1985). Interactions amongst herbivorous fishes on a coral reef: the influence of spatial variation. *Mar. Biol.* 89: 221–234.

Choat, J. H., Clements, K. D. (1992). Diet of odacid and aplodactylid fishes from Australia and New Zealand. *Aust. J. Mar. Freshwat. Res.* 43: 1451–1459.

Choat, J. H., Clements, K. D. (1993). Daily feeding rates in herbivorous labroid fishes. *Mar. Biol.* 117: 205–213.

Choat, J. H., Doherty, P. J., Kerrigan, B. A., Leis, J. M. (1993). A comparison of towed nets, purse seine, and light-aggregation devices for sampling larvae and pelagic juveniles of coral reef fishes. *Fish. Bull. U.S.* 91: 195–209.

Choat, J. H., Kingett, P. D. (1982). The influence of fish predation on the abundance cycles of an algal turf invertebrate fauna. *Oecologia* 54: 88–95.

Choat, J. H., Schiel, D. R. (1982). Patterns of distribution and abundance of large brown algae and invertebrate hervibores in subtidal regions of northern New Zealand. *J. Exp. Mar. Biol. Ecol.* 60: 129–162.

Christensen, N. L., Bartuska, A. M., Brown, J. H., Carpenter, S., D'Antonio, C., Francis, R., Franklin, J. F., MacMahon, J. A., Noss, R. F., Parsons, D. J., Peterson, D. J., Turner, M. G., Woodmasee, R. G. (1996). The report of the ecological society of America committee on the scientific basis for ecosystem management. *Ecol. Appl.* 6: 665–691.

Christie, H., Evans, R. A., Sandnes, O. K. (1985). Methods for *in situ* subtidal hard bottom studies. Pp. 37–47 *in* George, J. D., Lythgoe, G. I., Lythgoe, J. N. (eds), *Underwater photography and television for scientists*. Clarendon Press, Oxford.

Cintron, G., Novelli, Y. S. (1984). Methods for studying mangrove structure. Pp. 91–113 *in* Snedaker, S. C., Snedaker, J.G. (eds), *The mangrove ecosystem: Research methods*. UNESCO, Paris.

Clapin, G., Evans, D. R. (1995). Status of the introduced marine fanworm *Sabella spallanzani* in Western Australia. July 1995. Centre for Research on Introduced Marine Pests, CSIRO Australia, Division of Fisheries, *Tech. Rep.* No. 1. 34 pp.

Clark, C. W. (1996). Marine reserves and the precautionary management of fisheries. *Ecol. Appl.* 6: 369–370.

Clark, R. B. (1989). *Marine pollution*. Clarendon Press, Oxford. 220 pp.

Clarke, G. (1973). *Staining procedures used by the biological stain commission*. Williams & Wilkins, Baltimore. 418 pp.

Clarke, J. R., Buxton, C. D. (1989). A survey of the recreational rock-angling fishery at Port Elizabeth, on the south-east coast of South Africa. *S. Afr. J. Mar. Sci.* 8: 183–194.

Clarke, K. R. (1993). Non-parametric multivariate analyses of changes in community structure. *Aust. J. Ecol.* 18: 117–143.

Clarke, P. J., Allaway, W. G. (1993). The regeneration niche of the grey mangrove (*Avicennia marina*): effects of salinity, light and sediment factors on establishment, growth and survival in the field. *Oecologia* 93: 548–557.

Clarke, P. J., Myerscough, P. J. (1993). The intertidal limits of *Avicennia marina* in southeastern Australia: the effects of physical conditions, interspecific competition and predation on establishment and survival. *Aust. J. Ecol.* 18: 307–315.

Clayton, M. N., King, R. J. (1990). *Biology of marine plants*. Longman Cheshire, Melbourne. 501 pp.

Clements, K. D. (1996). Fermentation and gastrointestinal microorganisms in fishes. Pp. 156–198 *in* Mackie, R. I., White, B. A. (eds), *Gastrointestinal microbiology: Gastro-intestinal ecosystems and fermentation*. Chapman & Hall, New York.

Clements, K. D., Gleeson, V. P., Slaytor, M. (1994). Short-chain fatty acid metabolism in temperate marine herbivorous fish. *J. Comp. Physiol. B.* 164: 372–377.

Clutter, R. I. (1969). The microdistribution and social behaviour of some pelagic mysid shrimps. *J. Exp. Mar. Biol. Ecol.* 3: 125–155.

Coale, K. H., Johnson, K. S., and 17 others. (1996). A massive phytoplankton bloom induced by an ecosystem-scale iron fertilisation experiment in the equatorial Pacific Ocean. *Nature* 383: 495–501.

Coates, D. (1980). Prey-size intake in Humbug damselfish *Dascyllus aruanus* (Pisces, Pomacentridae) living within social groups. *J. Anim. Ecol.* 49: 335–340.

Coates, M., McKillup, S. C. (1995). Role of recruitment and growth in determining the upper limit of distribution of the intertidal barnacle *Hexaminius popeiana*. *Mar. Freshwat. Res.* 46: 1065–1070.

Cochran, W. G. (1977). *Sampling techniques*. John Wiley & Sons, New York. 428 pp.

Cohen, A. N., Carlton, J. T., Fountain, M. C. (1995). Introduction, dispersal and potential impacts of the green crab *Carcinus maenas* in San Francisco Bay, California. *Mar. Biol.* 122: 225–238.

Cohen, J. (1988). *Statistical power analysis for the behavioural sciences*. 2nd edn, Lawrence Erlbaum Hillsdale, New Jersey. 567 pp.

Cole, R. G. (1994). Abundance, size structure, and diver-orientated behaviour of three large benthic carnivorous fishes in a marine reserve in northeastern New Zealand. *Biol. Conserv.* 70: 93–99.

Cole, R. G., Ayling, T. M., Creese, R. G. (1990). Effects of marine reserve protection at Goat Island, northern New Zealand. *N.Z. J. Mar. Freshwat. Res.* 24: 197–210.

Cole, R. G., Creese, R. G., Grace, R. V., Irving, P., Jackson, B. R. (1992). Abundance patterns of subtidal benthic invertebrates and fishes at the Kermadec Islands. *Mar. Ecol. Prog. Ser.* 82: 207–218.

Cole, R. G., Immenga, D., Foster, D. M. (1995). Population biology of the venerid bivalve, *Tawera spissa*, off Tauranga, New Zealand. (Abstr.) *N.Z. Mar. Sci. Soc. Ann. Review* 38: 8.

Colebrook, J. M. (1982). Continuous plankton record: seasonal variations in the distribution and abundance of plankton in the North Atlantic and North Sea. *J. Plankton Res.* 4: 435–462.

Colebrook, J. M., Robinson, G. A., Hunt, H. G., John, A. W. G., Bottrell, H. H., Lingley, J. A., Collins, N. R., Halliday, N. C.

(1984). Continuous plankton record: A possible reversal in the downward trend in the abundance of plankton in the North Sea and the North east Atlantic. *J. Cons. perm. int. Explor. Mer.* 41: 301–306.

Collos, Y., Slawyk, G. (1985). On the compatability of carbon uptake rates calculated from stable and radioactive isotope data: implications for the design of experimental protocols in aquatic primary productivity. *J. Plankton Res.* 7: 595–603.

Connell, J. H. (1974). Ecology: Field experiments in marine ecology. Pp. 21–54 *in* Mariscal, R. (ed.), *Experimental marine biology*. Academic Press, New York.

Connell, J. H. (1983). On the prevalence and relative importance of interspecific competition: Evidence from field experiments. *Am. Nat.* 122: 661–696.

Connell, J. H., Keough, M. J. (1983). Disturbance and patch dynamics of subtidal marine animals on hard substrata. Pp. 125–152 *in* Pickett, S. T. A., White, P. S. (eds), *The ecology of natural disturbances and patch dynamics: an evolutionary perspective*. Academic Press, London.

Connell, J. H., Keough, M. J. (1985). Disturbances and patch dynamics of subtidal marine animals on hard substrata. Pp. 1225–152 *in* Pickett, S. T. A., White, P. S. (eds), *The ecology of natural disturbances and patch dynamics*. Academic Press, London.

Connell, S. D. (1994). The contribution of large predatory fish to the mortality and abundance of juvenile coral reef fishes. Unpublished PhD thesis, University of Sydney. 111 pp.

Connell, S. D. (1996). Variations in mortality of a coral-reef fish: Links with predator abundance. *Mar. Biol.* 126: 347–352.

Connell, S. D. (1997). Exclusion of predatory fish on a coral reef: the anticipation, pre-emption and evaluation of some caging artifacts. *J. Exp. Mar. Biol. Ecol.* 213: 181–198.

Connell, S. D., Jones, G. P. (1991). The influence of habitat complexity on postrecruitment processes in a temperate reef fish population. *J. Exp. Mar. Biol. Ecol.* 151: 271–294.

Connolly, R. M. (1994a). Removal of seagrass canopy: effects on small fish and their prey. *J. Exp. Mar. Biol. Ecol.* 184: 99–110.

Connolly, R. M. (1994b). A comparison of fish assemblages from seagrass and unvegetated areas of a southern Australian estuary. *Aust. J. Mar,. Freshwat. Res.* 45: 1033–1044.

Conover, W. J. (1980). *Practical nonparametric statistics*. Wiley, New York. 493 pp.

Constable, A. J. (1993). The role of sutures in shrinking of the test in *Heliocidaris erythrogramma* (Echinoidea: Echinometridae). *Mar. Biol.* 117: 423–430.

Cook, S. de C. (ed.) (in press). *New Zealand coastal marine invertebrates*. 2 vols. Canterbury University Press, Christchurch.

Coombs, S. H., Robins, D. B., Conway, D. V. P., Halliday, N. C., Pomroy, A. J. (1994). Suspended particles in the Irish Sea and feeding conditions for fish larvae. *Mar. Biol.* 118: 7–16.

Corbett, S. J., Rubin, G. L., Curry, G. K., Kleinbaum, D. G. (1993). The health effects of swimming at Sydney beaches. *Am. J. Public Health.* 83: 1701–1706.

Cormack, R. M. (1994). Unification of mark-recapture analyses by loglinear modelling. Pp. 191–132 *in* Fletcher, D. J., Manly, F. J. (eds) *Statistics in ecology and environmental monitoring*. University of Otago Press, Dunedin.

Cortis, J. N., Risk, M. J. (1985). A reef under siltation stress: Cahuita, Costa Rica. *Bull. Mar. Sci.* 36: 339–356.

Cortez, T., Castro, B. G., Guerra, A. (1995a). Feeding dynamics of *Octopus mimus* (Mollusca: Cephalopoda) in northern Chile waters. *Mar. Biol.* 123: 497–504.

Cortez, T., Castro, B. G., Guerra, A. (1995b). Reproduction and condition of female *Octopus mimus* (Mollusca: Cephalopoda). *Mar. Biol.* 123: 505–510.

Coull, B. C. (1985). Long term variability of estuarine meiobenthos: An 11-year study. *Mar. Ecol. Prog. Ser.* 24: 205–218.

Coull, B. C., Wells, J. B. J. (1983). Refuges from fish predation: experiments with phytal meiofauna from the New Zealand rocky intertidal. *Ecology.* 64: 1599–1609.

Cowan, J. H., Houde, E. D. (1990). Growth and survival of bay anchovy *Anchoa mitchilli* larvae in mesocosm enclosures. *Mar. Ecol. Prog. Ser.* 68: 47–57.

Cowen, R. K. (1983). The effect of sheephead (*Semicossyphus pulcher*) predation on red sea urchin (*Strongylocentrotus franciscanus*) populations: An experimental analysis. *Oecologia* 58: 249–255.

Cowen, R. K. (1985). Large scale patterns of recruitment by the labrid, *Semicossyphus pulcher*: Causes and implications. *J. Mar. Res.* 43: 719–742.

Cowen, R. K. (1986). Site-specific differences in the feeding ecology of California sheephead, *Semicossyphus pulcher* (Labridae). *Environ. Biol. Fish.* 16: 193–203.

Cowen, R. K. (1990). Sex change and life history patterns of the labrid, *Semicossyphus pulcher*, across an environmental gradient. *Copeia*: 787–795.

Cox, G., Francis, M. (1997). *Sharks and rays of New Zealand*. Canterbury University Press, Christchurch. 68 pp.

Coyer, J., Witman, J. (1990). *The underwater catalog: a guide to methods in underwater research*. Shoals Marine Laboratory, Ithaca. 72 pp.

Creese, R. G. (1981). Patterns of growth, longevity and recruitment of intertidal limpets in N.S.W. *J. Exp. Mar. Biol. Ecol.* 51: 145–171.

Creese, R. G. (1988). Ecology of molluscan grazers and their interactions with marine algae in north-eastern New Zealand: a review. *N.Z. J. Mar. Freshwat. Res.* 22: 427–444.

Creese, R. G., Ballantine, W. J. (1986). Rocky shores of exposed islands in north-eastern New Zealand. Pp. 85–102 *in* Wright, A. E., Beever, R. E. (eds), *The offshore islands of northern New Zealand. Dept of Lands & Survey, Inform. Ser.* 16.

Creese, R. G., O'Neill, M. H. B. (1987). *Chiton aorangi* n. sp., a brooding chiton (Mollusca: Polyplacophora) from northern New Zealand. *N.Z. J. Zool.* 14: 89–93.

Creese, R. G., Schiel, D. R., Kingsford, M. J. (1990). Sex change in a giant endemic limpet, *Patella kermadecensis*, from the Kermadec Islands. *Mar. Biol.* 104: 419–426.

Creese, R. G., Underwood, A. J. (1982). Analysis of inter- and intra-specific competition amongst intertidal limpets with different methods of feeding. *Oecologia* 53: 337–347.

Crisp, P., Daniel, L., Tortell, P. (1990). *Mangroves in New Zealand: trees in the tide*. Nature Conservation Council, Wellington. 69 pp.

Cronan, G., Hay, M. E. (1996). Induction of seaweed chemical defenses by amphipod grazing. *Ecology* 77: 2287–2301.

Crosby, L. A., Wood, E. J. F. (1958). Studies on Australian and New Zealand diatoms. I. Planktonic and allied species. *Trans. R. Soc. N.Z.* 88: 483–530.

Crosby, L. A., Wood, E. J. F. (1959). Studies on Australian and New Zealand diatoms. II. Normally epontic and benthic genera. *Trans. R. Soc. N.Z.* 89: 1–58.

Crossland, J. (1976). Fish trapping experiments in northern New Zealand waters. *N.Z. J. Mar.Freshwat. Res.* 10: 511–516.

Crossland, J. (1980). The number of *Chrysophrys auratus* (Forster), in the Hauraki Gulf, New Zealand, based on egg surveys in 1974–75 and 1975–76. *Fish. Res. Bull. N.Z.* 22: 1–38.

Crossland, J. (1981). Fish eggs and larvae of the Hauraki Gulf, New Zealand. *Fish. Res. Bull. N.Z.* 23: 1–61.

Crossland, J. (1982a). Distribution and abundance of fish eggs and larvae from the spring-summer plankton of northeast New Zealand, 1976–78. *Fish. Res. Bull. N.Z.* 24: 1–59.

Crossland, J. (1982b). Movements of tagged snapper in the Hauraki Gulf. *Fish. Res. Occ. Publ. (N.Z.)* 35: 1–15.

Crowe, A. (1995). *Which coastal plant? A simple guide to the identification of New Zealand's common coastal plants.* Viking, Auckland. 64 pp.

Cullinan, V. I., Thomas, J. M. (1992). A comparison of quantititative methods for examining landscape patterns and scale. *Land. Ecol.* 7: 211–227.

Culotta, E. (1994). Is marine biodiversity at risk? *Science* 26: 918–920.

Cushing, D. (1975). *Marine ecology and fisheries.* Cambridge University Press, Oxford. 278 pp.

Cushing, D. H. (1990). Plankton production and year-class strength in fish populations: an update of the match/ mismatch hypothesis. *Adv. Mar. Biol.* 26: 250–293.

Dahl, A. L. (1978). *Coral reef monitoring handbook.* South Pacific Commission Publications Bureau, Sydney. 21 pp.

Dakin, W. J., Colefax, A. N. (1940). *The plankton of the Australian coastal waters off New South Wales. Part I.* Australian Publishing Co., Sydney. 215 pp.

Dalby, D. H., Wolff, W. J. (1987). Remote Sensing. Pp. 27–37 in Baker, J. M., Wolff, W. J. (eds) *Biological surveys of estuaries and coasts.* Cambridge University Press, Cambridge. 449 pp.

Dalzell, P., Adams, T. J. H., Polunin, N. V. C. (1996). Coastal fisheries in the Pacific Islands. *Oceanogr. Mar. Biol. Ann. Rev.* 34: 395–531.

Darwall, W. R. T., Costello, M. J., Donnelly, R., Lysaght, S. (1992). Implications of life-history strategies for a new wrasse fishery. *J. Fish. Biol.* 41(B): 111–123.

Davies, C. R. (1989). The effectiveness of non-destructive sampling of coral reef fish populations with fish traps. Unpublished BSc (Hons) thesis, James Cook University, Townsville.

Davis, A. R. (1987). Variation in recruitment of the subtidal colonial ascidian *Podoclavella cylindrica* (Quoy & Gaimard): the role of substratum choice and early survival. *J. Exp. Mar. Biol. Ecol.* 106: 57–71.

Davis, A. R. (1989). Effects of variation in initial settlement of the subtidal colonial ascidian *Podoclavella cylindrica* (Quoy & Gaimard): the role of substratum choice and early survival. *J. Exp. Mar. Biol. Ecol.* 117: 157–167.

Davis, A. R. (1996). Association among ascidians: facilitation of recruitment in *Pyura spinifera. Mar. Biol.* 126: 35–41.

Davis, C. S., Gallager, S. M., Berman, M. S., Haury, L. R., Stickler, J. R. (1992). The video plankton recorder (VPR): design and initial results. *Arch. Hydrobiol. Limnol.* 36: 67–81.

Davis, C. S., Gallager, S. M., Marra, M., Stewart, W. K. (1996). Rapid visualisation of plankton abundance and taxanomic composition using the video plankton recorder. *Deep Sea Res.* 43: 1947–1970.

Davis, G. E., Anderson, T. W. (1989). Population estimates of four kelp forest fishes and an evaluation of three *in situ* assessment techniques. *Bull. Mar. Sci.* 44: 1138–1151.

Davis, R. A. (1991). *Oceanography: An introduction to the marine environment.* WCB Publishers, Dubuque. 434 pp.

Davis, T. L. O. (1977). Age determination and growth of the freshwater catfish, *Tandana tandanus* Mitchell, in the Gwydir River, Australia. *Aust. J. Mar. Freshwat. Res.* 28: 119–137.

Dawson, S. (1985). *The New Zealand whale and dolphin digest.* Brick Row Publishing, Auckland. 130 pp.

Dawson, S., Slooten, E. (1997). *Down-under dolphins.* Canterbury University Press, Christchurch. 64 pp.

Day, J. H., Hutchings, P. A. (1979). An annotated checklist of Australian and New Zealand Polychaeta: Archiannelida and Myzostomida. *Rec. Aust. Mus.* 32: 80–161.

Day, J. W., Hall, C. A. S., Kemp, W. M., Yanez-Arancibia, A. (1989). *Estuarine Ecology.* Wiley, New York. 558p.

Day, R. W., Quinn, G. P. (1989). Comparisons of treatments following an analysis of variance. *Ecol. Monogr.* 59: 433–463.

Day, R. W., Williams, M. C, Hawkes, G. P. (1995). A comparison of fluorochromes for marking abalone shells. *Mar. Freshwat. Res.* 46: 599–605.

Dayton, P. K. (1985). Ecology of kelp communities. *Ann. Rev. Ecol. Syst.* 16: 215–245.

Dayton, P. K., Currie, V., Gerrodette, T., Keller, B. D., Rosenthal, R., Tresca, D.V. (1984). Patch dynamics and stability of some California kelp communities. *Ecol. Monogr.* 54: 253–289.

Dayton, P. K., Oliver, J. S. (1980). An evaluation of experimental analysis of population and community patterns in benthic marine environments. Pp. 93–120 *in* Tenore, K. R., Coull, B. C. (eds), *Marine benthos dynamics. No. 11.* Belle W. Baruch Institute for Marine Biology and Coastal Research by the University of South Carolina, Georgetown.

Dayton, P. K., Tegner, M. J. (1984). The importance of scale in community ecology: a kelp forest example with terrestrial analogues. Pp. 457–481 in Price, P. W., Slobodchikoff, C. N., Gaud W. S. (eds). *A new ecology: novel approaches to interactive systems. Wiley, New York.*

Dayton, P. K., Tegner, M. J., Parnell, P. E., Edwards, P. B. (1992). Temporal and spatial patterns of disturbance and recovery in a kelp forest community. *Ecol. Monogr.* 62: 421–445.

de Boer, B.A. (1978). Factors influencing the distribution of the damselfish *Chromis cyanea* (Poey), Pomacentridae, on a reef at Curacao, Netherlands Antilles. *Bull. Mar. Sci.* 28: 550–565.

de Lafontaine, Y., Leggett, W. C. (1987). Evaluation of *in situ* enclosures for larval fish studies. *Can. J. Fish. Aquat. Sci.* 44: 54–65.

De Martini, E. E., Roberts, D. A. (1982). An empirical test of biases in the rapid visual techniques for species-time censuses of reef fish assemblages. *Mar. Biol.* 70: 129–134.

De Martini, E. E., Roberts, D. A. (1990). Effects of giant kelp (*Macrocystis*) on the density and abundance of fishes in a cobble-bottom kelp forest. *Bull. Mar. Sci.* 46: 289–300.

de Ruyter van Steveninck, E. D., Admiraal, W., Breebaart, L., Tubbing, G. M. J., van Zanten, B. (1992). Plankton in the river Rhine: Structural and functional changes observed during downstream transport. *J. Plankton Res.* 14: 1351–1368.

De Vries, D. A., Grimes, C. B., Lang, K. L., White, D. B. (1990). Age and growth of king and spanish mackerel larvae and juveniles from the Gulf of Mexico and U.S. South Atlantic Bight. *Environ. Biol. Fish.* 29: 135–143.

De Vries, M. C., Tankersley, R. A., Forward Jr, R. B., Kirby-Smith, W. W., Luettich, R. A. (1994). Abundance of estuarine crab larvae is associated with tidal hydrographic variables. *Mar. Biol.* 118: 403–414.

De Witt, T. H., Morrisey, D. J., Roper, D. S., Nipper, M. G. (1996). Fact or artifact: the need for appropriate controls in ecotoxicological field experiments. *SETAC News* (Newsletter of the Society for Environmental Toxicology and Chemistry*), September 1996*: 22–23.

Denman, K. L., Powell, T. M. (1984). Effects of physical processes on planktonic ecosystems in the coastal ocean. *Oceanogr. Mar. Biol. Ann. Rev.* 26: 487–551.

Dennis, B. (1996). Discussion: Should ecologists become Bayesians? *Ecol. Appl.* 6: 1095–1103.

Dethier, M. N. (1994). The ecology of intertidal algal crusts: variations within a functional group. *J. Exp. Mar. Biol. Ecol.* 177: 37–71.

Dethier, M. N., Graham, E. S., Cohen, S., Tear, L. M. (1993). Visual versus random-point percent cover estimations: 'objective' is not always better. *Mar. Ecol. Prog. Ser.* 96: 93–100.

Devassy, V. P. (1989). Red tide discolouration and its impact on fisheries. *In* Okaichi, T., Anderson, D. M., Nemoto, T. (eds), *Red tides: Biology, Environmental Science, and Toxicology.* Elsevier, New York, pp. 57–60.

Dewey, M. R., Zigler, S. L. (1996). An evaluation of fluorescent elastomer for marking bluegills in experimental studies. *Prog. Fish. Cult.* 58: 219–220.

Dexter, D. M. (1984). Temporal and spatial variability in the community structure of the fauna of four sandy beaches in south-eastern New South Wales. *Aust. J. Mar. Freshwat. Res.* 35: 663–672.

Digby, P. G. N., Kempton, R. A. (1987). *Multivariate analysis of ecological communities.* Chapman & Hall, London. 206 pp.

Dillon, W. R., Goldstein, M. (1984). *Multivariate analysis methods and application.* Harper & Row Publishers, New York. 887 pp.

Dinderkus, G. (1981). The use of various alcohols for the alcian blue in toto staining of cartilage. *Stain Tech.* 56: 128–129.

Dix, T. G. (1970). Biology of *Evechinus chloroticus* (Echinometridae) from different localities 3. Reproduction. *N.Z. J. Mar. Freshwat. Res.* 4: 385–405.

Dix, T. G. (1977). Reproduction in Tasmanian populations of *Heliocidaris erythrogramma* (Echinodermata: Echinometridae). *Aust. J. Mar. Freshwat. Res.* 28: 509–520.

Doak, W. (1972). *Fishes of the New Zealand region.* Hodder & Stoughton, Auckland. 132 pp.

Doak, W. (1979). *The cliff dwellers. An undersea community.* Hodder & Stoughton, Auckland. 80 pp.

Doak, W. (1991). *Wade Doak's world of New Zealand fishes.* Hodder & Stoughton, Auckland. 222 pp.

Dobson, J. E., Bright, E. A., Ferguson, R. L., Field, D. W., Wood, L. L., Haddad, K. D., Iredale, H., Jensen, J. R., Klemas, V. V., Orth, R. J., Thomas, J. P., (1995). NOAA coastal change analysis program (C-CAP): Guidance for regional implementation. *NOAA Tech. Rep. NMFS* 123. 92 pp.

DOC (1994). A draft status list for New Zealand's marine flora and fauna. In Waghorn, E. J. (comp.), McCrone, A. (ed.). Department of Conservation, Wellington. 59 pp.

DOC (1995). *Marine reserves.* Information Paper. Department of Conservation, Wellington. 21 pp.

Dodge, R. E., Logan, A., Antonius, A. (1982). Quantitative reef assessment studies in Bermuda: a comparison of methods and preliminary results. *Bull. Mar. Sci.* 32: 745–760.

Doherty, P. J. (1987). Light-traps: Selective but useful devices for quantifying the distributions and abundance of larval fishes. *Bull. Mar. Sci.* 41: 423–431.

Doherty, P. J. (1991). Spatial and temporal patterns in recruitment. Pp. 261–293 *in* Sale, P. F. (ed.), *The ecology of fishes on coral reefs.* Academic Press, San Diego.

Doherty, P. J., Kingsford, M. J., Booth, D., Carlton, J. (1996). Habitat selection before settlement by *Pomacentrus coelestis. Mar. Freshwat. Res.* 47: 391–400.

Doherty, P. J., McIlwain, J. (1996). Monitoring larval fluxes through the surf zones of Australian coral reefs. *Mar. Freshwat. Res.* 47: 383–390.

Doherty, P. J., Sale, P. F. (1985). Predation on juvenile coral reef fishes: an exclusion experiment. *Coral Reefs* 4: 225–234.

Doherty, P. J., Williams, D. McB. (1988). The replenishment of coral reef fish populations. *Oceanogr. Mar. Biol. Ann. Rev.* 26: 487–551.

D'Olier, B. (1979). Side-scan sonar and reflection seismic profiling. Pp. 57–86 *in* Dyer, K.R. (ed.) *Estuarine hydrography and sedimentation.* Cambridge University Press, Cambridge.

Done, T. J. (1981). Photogrammetry in coral ecology: a technique for the study of change in coral communities. *Proc. 4th Int. Coral Reef Symp., Manila* 2: 315–320.

Dorfman, E. J., Smith, G. (1995). Marine birds and mammals. Pp. 254–262 *in* Underwood, A. J., Chapman, M. (eds), *Coastal ecology in temperate Australia.* Chapman & Hall, Sydney.

Doty, M. S. (1971). Measurements of water movement in reference to benthic algal growth. *Botanica Mar.* 14: 32–35.

Druce, B. E., Kingsford, M. J. (1995). An experimental investigation on the fishes associated with drifting objects in coastal waters of temperate Australia. *Bull. Mar. Sci.* 57: 378–392.

Drummond, A. E. (1995). Reproduction of the sea urchins *Echinometra mathaei* and *Diadema savignyi* on the South African eastern coast. *Mar. Freshwat. Res.* 46: 751–758.

Duckworth, A. R., Battershill, C. N., Bergquist, P. R. (1997). Influence of explant procedures and environmental factors on culture success of three sponges. *Aquaculture* 156: 251–267.

Dudgeon, S. R., Kubler, J. E., Vadas, R. L., Davison, I.R. (1995). Physiological reponses to environmental variation in intertidal red algae: does thallus morphology matter? *Mar. Ecol. Prog. Ser.* 117: 193–206.

Duffy, J. E., Hay, M. E. (1990). Seaweed adaptations to herbivory. *Bioscience* 40: 368–375.

Dufour, V., Riclet, E., Loyat, A. (1996). Colonisation of reef fishes at Moorea Island, French Polynesia. *Marine and Freshwater Research.* 47: 413–422.

Dugan, J. E., Davis, G. E. (1993). Applications of marine refugia to coastal fisheries management. *Can. J. Fish. Aquat. Sci.* 50: 2029–2042.

Dugan, J. E., Hubbard, D. M. (1997). Local variations in populations of the sand crab *Emerita analoga,* on the California coast. *Revta Chil. Hist. Nat.* 69: 577–588.

Dugdale, R. C., Goering, J. J. (1967). Uptake of new and regenerated forms of nitrogen in primary productivity. *Limnol. Oceanogr.* 12: 196–206.

Dussart, B. H., Defayne, D. (1995). *Copepoda: introduction to the Copepoda. Vol. 7.* SPB Academic Publishing, Amsterdam. 277 pp.

Duxbury, A. J., Duxbury, A. R. (1991). *An introduction to the world's oceans.* Wm. C. Brown, Dubuque. 446 pp.

Dye, A. H. (1983). A method for the quantitative estimation of bacteria from mangrove sediments. *Estuar. Cstl Shelf Sci.* 17: 207–215.

Dye, A. H. (1992). Experimental studies of succession and stability in rocky intertidal communities subject to artisanal shellfish gathering. *Netherlands J. Sea Res.* 30: 209–217.

Ebeling, A. W., Hixon, M. A. (1991). Tropical and temperate reef fishes: Comparison of community structures. Pp. 509–563 *in* Sale, P. F. (ed.), *The ecology of fishes on coral reefs.* Academic Press, San Diego.

Ebeling, A. W., Larson, R. J., Alevizon, W. S., Bray, R. N. (1980). Annual variability of reef-fish assemblages in kelp forests off Santa Barbara, California. *Fish. Bull. U.S.* 78: 361–377.

Ebeling, A. W., Laur, D. R. (1988). Fish populations in kelp forests without sea otters: Effects of severe storm damage and destructive sea urchin grazing. Pp. 169–191 *in* van Blaricom, G., Estes, J. (eds), *Community ecology of sea otters.* Springer-Verlag, Berlin.

Ebeling, A. W., Laur, D. R., Rowley, R. J. (1985). Severe storm disturbances and reversal of community structure in a southern California kelp forest. *Mar. Biol.* 84: 287–294.

Ebert, T. A. (1982). Longevity, life history and relative body wall size in sea urchins. *Ecol. Monogr.* 52: 353–394.

Ebert, T. A., Russell, M. P. (1992). Growth and mortality estimates for red sea urchin *Strongylocentrotus franciscanus* from San Nicolas Island, California. *Mar. Ecol. Prog. Ser.* 81: 31–41.

Edgar, G. J. (1983a). The ecology of the south-east Tasmanian phytal algal communities. I. Spatial organisation on a local scale. *J. Exp. Mar. Biol. Ecol.* 70: 129–157.

Edgar, G. J. (1983b). The ecology of south-east Tasmanian phytal animal communities. II. Seasonal change in plant animal populations. *J. Exp. Mar. Biol. Ecol.* 70: 159–179.

Edgar, G. J. (1987). Dispersal of faunal and floral propagules associated with drifting *Macrocystis pyrifera* plants. *Mar. Biol.* 95: 599–610.

Edgar, G. J. (1991). Artificial algae as habitats for mobile epifauna: factors affecting colonisation in a Japanese *Sargassum* bed. *Hydrobiologia* 226: 111–118.

Edgar, G. J. (1997). *Australian marine life: the plants and animals of temperate waters.* Reed Kew, Victoria. 544 pp.

Edgar, G. J., Moore, P. G. (1986). Macro-algae as habitats for motile macrofauna. *Monogr. Biol.* 4: 255–277.

Edgar, G. J., Shaw, C. (1995). The production and trophic ecology of shallow-water fish assemblages in southern Australia. I. Species richness, size-structure and production of fishes and epibenthic invertebrates in Western Port, Victoria. *J. Exp. Mar. Biol.Ecol.* 194: 53–81.

Edgar, G. J., Watson, G., Hammond, L. S., Shaw, C. (1994). Comparisons of species richness, size-structure and production of benthos in vegetated and unvegetated habitats in Western Port, Victoria. *J. Exp. Mar. Biol. Ecol.* 176: 201–226.

Edgar, G. J., Barrett, N. (1997). Short term monitoring of biotic change in Tasmanian marine reserves. *J. Exp. Mar. Biol. Ecol.* 213: 261–279.

Edgar, G. J., Moverley, N. S., Barrett, N. S., Peters, D., Reed, C. (1997). The conservation-related benefits of a systematic marine biological sampling programme: the Tasmanian reef

bioregionalisation as a case study. *Biological Conservation.* 79: 227–240.

Efron, B., Tibshirani, R. J. (1993). *An introduction to the bootstrap.* Chapman & Hall, New York. 436 pp.

Eggleston, D. R., Lipcius, R., Miller, D., Coba-Centina, L. (1990). Shelter scaling regulates survival of juvenile spiny lobster, *Panulirus argus. Mar. Ecol. Prog. Ser.* 62: 79–88.

Eleftheriou, A., Holme, N. A. (1984). Macrofauna techniques. Pp. 140–216 *in* Holme, N. A., McIntyre, A. D. (eds), *Methods for the study of marine benthos*, 2nd edn, Blackwell, Oxford. 387 pp.

Elner, R. W., Vadas, R. L. (1990). Inference in ecology: the sea urchin phenomenon in the northwestern Atlantic. *Am. Nat.* 136: 108–125.

El-Sabh, M., Aung, T. H., Murty, T. S. (1997). Physical processes in inverse estuarine systems. *Oceanogr. Mar. Biol. Ann. Rev.* 35: 1–69.

Emlet, R. B., Hoegh-Guldberg, O. (1997). Effects of egg size on postlarval performance: Experimental evidence from a sea urchin. *Evolution.* 51: 141–152.

Emmett, B., Jamieson, G. S. (1989). An experimental transplant of pre-recruit abalone in Barkley Sound, British Columbia. *Fish. Bull.* 87: 95–104.

English, S. A., Wilkinson, C., Baker, V. J. (1994). *Survey manual for tropical marine resources.* Australian Institute of Marine Science (AIDAB), Townsville. 368 pp.

Epifanio, C. E. (1995). Transport of the blue crab (*Callinestes sapidus*) larvae in the waters off mid-Atlantic states. *Bull. Mar. Sci.* 57: 713–725.

Erasmus, J., Cook, P. A., Sweijd, N. (1994). Internal shell structure and growth lines in the shell of the abalone *Haliotis midae. J. Shellfish Res.* 13: 493–501.

Eschymeyer, W. N., Herald, E., Hammann, H. (1983). *A field guide to Pacific coast fishes North America.* Houghton Mifflin, Boston. 336 pp.

Estes, J. A., Duggins, D. O. (1995). Sea otters and kelp forests in Alaska: generality and variation in a community paridigm. *Ecol. Monogr.* 65: 75–100.

Ettinger-Epstein, P. (1996). Interrelationships between habitat type and abundance of *Turbo torquatus* (Gastropoda) Hons thesis, University of Sydney. 76 pp.

Evans, P. G. H. (1987). *The natural history of whales and dolphins.* Facts on File Publications, New York. 343 pp.

Everson, W. E., Bone, D. G. (1986). Effectiveness of the RMT8 system for sampling krill (*Euphausia superba*) swarms. *Polar Biology* 6: 83–90.

Fabens, G. (1965). Properties of the von Bertalanffy growth curve. *Growth* 29: 265–289.

Fagan, P., Miskiewicz, A. G., Tate, P. M. (1992). An approach to monitoring sewage outfalls. A case study on the Sydney Deepwater Sewage Outfalls. *Mar. Pollut. Bull.* 25: 5–8.

Fahay, M. P. (1983). Guide to the early life stages of marine fishes occurring in the western Atlantic Ocean, Cape Hatteras to the Southern Scotian Shelf. *J. Nwest Atlant. Fish. Sci.* 4: 3–423

Fairchild, J. F., Little, E. E. (1993). Use of mesocosm studies to examine direct and indirect impacts of water quality on early life history stages of fishes. *Am. Fish. Soc. Symp.* 14: 95–105.

Fairweather, P. G. (1988). Predation creates halos of bare space among among prey on rocky seashores in New South Wales. *Aust. J. Ecol.* 13: 401–409.

Fairweather, P. G. (1990a). Is predation capable of interacting with other community processes on rocky reefs? *Aust. J. Ecol.* 15: 453–464.

Fairweather, P. G. (1990b). Sewage and the biota on seashores: assessment of impact in relation to natural variability. *Environmtl Monit. Assessmt* 14: 197–210.

Fairweather, P. G. (1991a). A conceptual framework for ecological studies of coastal resources: an example of a tunicate collected for bait on Australian seashores. *Ocean Shoreline Mgmt* 15: 125–142.

Fairweather, P. G. (1991b). Statistical power and design requirements for environmental monitoring. *Aust. J. Mar. Freshwat. Res.* 42: 555–567.

Fairweather, P. G. (1993). Links between ecology and ecophilosophy, ethics and the requirements of environmental management. *Aust. J. Ecol.* 18: 3–20.

Faith, D. P., Minchin, P. R., Belbin, L. (1987). Compositional dissimilarity as a robust measure of ecological distance. *Vegetatio*. 69: 57–68.

Falcon, J. M., Bortone, S. A., Brito, A., Bundrick, C. M. (1996). Structure of an relationships within and between the littoral, rock-substrate fish communities off four islands in the Canarian Achipelago. *Mar. Biol.* 125: 215–231.

Fancett, M. S. (1986). Species composition and abundance of scyphomedusae in Port Phillip Bay, Victoria. *Aust. J. Mar. Freshwat. Res.* 37: 379–384.

Fancett, M. S. (1988). Diet and prey selectivity of scyphomedusae from Port Phillip Bay, Australia. *Mar. Biol.* 98: 503–509.

Fanelli, G., Piraino, S., Belmonte, G., Geraci, S., Boero, F. (1994). Human predation along Apulian rocky coasts (SE Italy): Desertification caused by *Lithophaga lithophaga* (Mollusca) fisheries. *Mar. Ecol. Prog. Ser.* 110: 1–8.

FAO (1991). FAO Species Calalogue: Vol. 13. *Marine lobsters of the world. FAO Fisheries Synopsis No.* 125, FAO, Rome. 292 pp.

Farnsworth, E. J., Ellison, A. M. (1996). Scale-dependent spatial and temporal variability in biogeography of mangrove root epibiont communities. *Ecol. Monogr.* 66: 45–67.

Fauth, J. E., Bernado, J., Camara, M., Resetaritis, J. W. J., Buskirk, J. V., McCollum, S. A. (1996). Simplifying the jargon of community ecology: a conceptual approach. *Am. Nat.* 147: 282–286.

Fenchel, T. (1988). Marine plankton and food chains. *Ann. Rev. Ecol. Syst.* 19: 19–38.

Fernandes, L. (1990). Effect of the distribution and density of benthic target organisms on manta tow estimates of their abundance. *Coral Reefs* 9: 161–165.

Fernandes, L., Marsh, H., Moran, P. J., Sinclair, D. (1990). Bias in manta tow surveys of *Acanthaster planci. Coral Reefs* 9: 155–160.

Ferrell, D. J., Bell, J. D. (1991). Differences among assemblages of fish associated with *Zostera capricorni* and bare sand over a large spatial scale. *Mar. Ecol. Prog. Ser.* 72: 15–24.

Ferrell, D. J., Henry, G. W., Bell, J. D., Quartararo, N. (1992). Validation of annual marks in the otoliths of young snapper, *Pagrus auratus* (Sparidae). *Aust. J. Mar. Freshwat. Res.* 43: 179–184.

Ferron, A., Leggett, W. C. (1994). An appraisal of condition measures for fish larvae. *Adv. Mar. Biol.* 30: 217–303.

Fielding, P. J. (1995). A preliminary investigation of abalone *Haliotis midae* resource along the Transkei coast, South Africa. *S. Afr. J. Mar. Sci.* 15: 253–262.

Findlay, R. H., Kirk, R. M. (1988). Post-1947 changes in the Avon-Heathcote Estuary, Christchurch: a study of the effect of urban development around a tidal estuary. *N.Z. J. Mar. Freshwat. Res.* 22: 101–127.

Finn, M. D., Kingsford, M. J. (1996). Two-phase recruitment of a coral reef fish (Pisces: Apogonidae) on the Great Barrier Reef. *Mar. Freshwat. Res.* 47: 423–432.

Fisher, R. A., Yates, F. (1963). *Statistical tables for biological, agricultural and medical research.* Harlow: Longman, London. 146 pp.

Fletcher, D. J. (1994). A mark-recapture model in which sightings probablity depends on the number of sightings on the previous occasion. Pp. 105–110 *in* Fletcher, D. J., Manly, F. J. (eds), *Statistics in ecology and environmental monitoring.* University of Otago Press, Dunedin.

Fletcher, D. J., Manly, B. J. (1994). *Statistics in ecology and environmental monitoring.* University of Otago Press, Dunedin. 269 pp.

Fletcher, W. J. (1987). Interactions among subtidal Australian sea urchins, gastropods, and algae: effects of experimental removals. *Ecol. Monogr.* 57: 89–109.

Fletcher, W. J., Creese, R. G. (1985). Competitive interactions between co-occurring species of herbivorous gastropods. *Mar. Biol.* 86: 183–191.

Floderus, S., Pihl, L. (1990). Resuspension in the Kattegat: impact of variation in wind and climate and fishery. *Estuar. Cstl Shelf Sci.* 31: 487–498.

Floodgate, G. D., Gareth Jones, E. B, (1987). Bacteria and fungi. Pp. 238–279 *in* Baker, J. M., Wolff, W. J. (eds), *Biological surveys of estuaries and coasts.* Cambridge University Press, Cambridge. 449 pp.

Folk, R. L. (1974). *Petrology of sedimentary rocks.* Hemphill Publishing Co., Austin, Texas. 182 pp.

Forrester, G. E. (1990). Factors influencing the juvenile demography of a coral reef fish. *Ecology* 71: 1666–1681.

Forrester, G. E. (1991). Social rank, individual size and group composition as determinants of food consumption by humbug damselfish, *Dascyllus aruanus. Anim. Behav.* 42: 701–711.

Förstner, U. (1979). Metal pollution assessment from sediment analysis. Pp. 111–196 *in* Förstner, U., Wittman, G. T. W. (eds), *Metal pollution in the aquatic environment.* Springer-Verlag, Berlin.

Fortier, L., Leggett, W. C. (1985). A drift study of larval fish survival. *Mar. Ecol. Prog. Ser.* 25: 245–257.

Forward, R. B. (1988). Diel vertical migration: zooplankton photobiology and behaviour. *Oceanogr. Mar. Biol. Ann. Rev.* 26: 361–393.

Foster, B. A., Battaerd, W. R. (1985). Distribution of zoo-plankton in a coastal upwelling in New Zealand. *N.Z. J. Mar. Freshwat. Res.* 19: 213–226.

Foster, M. S. (1972). The algal turf community in the nest of the ocean gold fish *Hysypops rubicunda. Proc. Int. Seaweed Symp.* 7: 55–60.

Foster, M. S. (1990). Organisation of macroalgal assemblages in the Northwest Pacific: the assumption of homogeneity and the illusion of generality. *Hydrobiologia.* 192: 21–34.

Foster, M. S., Harrold, C., Hardin, D. D. (1991). Point vs. photo quadrat estimates of the cover of sessile marine organisms. *J. Exp. Mar. Biol. Ecol.* 146: 193–203.

Foster, M. S., Schiel, D. R. (1985). The ecology of giant kelp forests in California: a community perspective. *U.S. Fish and Wildlife Service Biological Report* 85 (7.2): 152 pp.

Foster, M. S., Schiel, D. R. (1993). Zonation, El Niño disturbance, and the dynamics of subtidal vegetation along a 30 m depth gradient in two giant kelp forests. *Proc. 2nd Int. Temperate Reef Symp.* : 151–162.

Fowler, A. J. (1990). Validation of annual growth increments in the otoliths of a small, tropical coral reef fish. *Mar. Ecol. Prog. Ser.* 64: 25–38.

Fowler, H. N. (1941). Fishes of Chile. Systematic catalog. *Revta chil. Hist. nat.* 45: 22–57.

Francis, M. P. (1981). Age and growth of moki, *Latridopsis ciliaris* (Teleostei: Latridae). *N.Z. J. Mar. Freshwat. Res.* 15: 47–49.

Francis, M. P. (1988). *Coastal fishes of New Zealand*. Heinemann Reed, Auckland. 63 pp.

Francis, M. P. (1993). Does water temperature determine year class strength in New Zealand snapper (*Pagrus auratus*)? *Fish. Oceanogr.* 2: 65–72.

Francis, M. P. (1995). Spatial and seasonal variation in the abun-dance of juvenile snapper (*Pagrus auratus*) in the north-western Hauraki Gulf. *N.Z. J. Mar. Freshwat. Res.* 29: 565–579.

Francis, M. P. (1996). Geographic distribution of marine fishes in the New Zealand Region. *N.Z. J. Mar. Freshwat. Res.* 30: 35–55.

Francis, M. P., Randall, J. E. (1993). Further additions to the fish faunas of Lord Howe and Norfolk Islands, Southwest Pacific Ocean. *Pacific Science* 47: 118–135.

Francis, M. P., Grace, R. V., Paulin, C. D. (1987). Coastal fishes of the Kermadec Islands. *N.Z. J. Mar. Freshwat. Res.* 21: 1–13.

Francis, M. P., Williams, M. W., Pryce, A. C., Pollard, S., Scott, S. G. (1992). Daily increments in otoliths of juvenile snapper, *Pagrus auratus* (Sparidae). *Aust. J. Mar. Freshwat. Res.* 43: 143–160.

Francis, R. I. C. C., Mattlin, R. H. (1986). A possible pitfall in the morphometric application of disciminant analysis: measurement bias. *Mar. Biol.* 93: 311–313.

Frank, K. T., Leggett, W. C. (1982). Coastal water mass replacement: Its effect on zooplankton dynamics and the predator-prey complex associated with larval capelin (*Mallotus villosus*). *Can. J. Fish. Aquat. Sci.* 39: 991–1003.

Frank, K. T., Leggett, W. C. (1985). Reciprocal oscillations in densities of larval fish and predators. *Can. J. Fish. Aquat. Sci.* 42: 1841–1849.

Fraser, A. (1989). Triacylglycerol content as a condition index for fish, bivalve and crustacean larvae. *Can. J. Fish. Aquat. Sci.* 46: 1868–1873.

Fricke, H., Fricke, S. (1977). Monogamy and sex change by aggressive dominance in coral reef fish. *Nature* 266: 830–832.

Frost, B. W. (1988). Variability and possible adaptive signif-icance of diel vertical migration in *Calanus pacificus*, a planktonic marine copepod. *Bull. Mar. Sci.* 43: 675–694.

Fudge, H. (1968). Biochemical analysis of preserved zoo-plankton. *Nature* 219: 380–381.

Fuhrer, B., Christianson, I. G., Allender, B. M. (1981). *Seaweeds of Australia*. Reed, Sydney. 112 pp.

Fulton, M. R. (1996). The digital stopwatch as a source of random numbers. *Bull. Ecol. Soc. Am.* 77: 217–218.

Gade, H. G., Edwards, A., Svendsen, H. (1983). *Coastal Oceanography*. Plenum Publishing Corp. 582 pp.

Gaines, S. D., Bertness, M. (1993). Dispersal of juveniles and variable recruitment in sessile marine species. *Nature* 360: 579–580.

Gaines, S. D., Lubchenco, J. (1982). Unified approach to marine plant-herbivore interactions. II. Biogeography. *Ann. Rev. Ecol. Syst.* 13: 111–138.

Gaines, S. D., Roughgarden, J. (1985). Larval settlement rate: a leading determinant of structure in an ecological community of the marine intertidal zone. *Proc. Natn Acad. Sci. USA* 83: 4707–4711.

Gaines, S. D., Roughgarden, J. (1987). Fish on offshore kelp forests affect recruitment to intertidal barnacle populations. *Science* 235: 479–481.

Gallahar, N. K., Kingsford, M. J. (1992). Patterns of increment width and Sr/Ca ratios in otoliths of juvenile rock blackfish, *Girella elevata*. *J. Fish. Biol.* 41: 749–763.

Gallahar, N. K., Kingsford, M. J. (1993). Depth distribution patterns of mobile epiphytic invertebrates on *Ecklonia radiata*. *Proc. 2nd Int. Temperate Reef Symp.*: 163–168.

Gallahar, N. K., Kingsford, M. J. (1996). Factors influencing Sr/Ca ratios in *Girella elevata*: an experimental approach. *J. Fish. Biol.* 48: 174–186.

Gamble, J. C. (1984). Diving. Pp. 99–139 *in* Holme, N. A., Mcintyre, A. D. (eds), *Methods for the study of marine benthos*, 2nd edn, Blackwell, Oxford. 387 pp.

Gardner, S. N., Mangel, M. (1996). Mark-resight population estimation with imperfect observations. *Ecology* 77: 880–884.

Garilov, L. A. (1994). Diatom kills by flagellates. *Nature* 367: 520.

Garson, M. J., Thompson, J. E., Larsen, R. M., Battershill, C. N., Murphy, P. T., Bergquist, P. R. (1992). Terpenes in sponge cell membranes: Cell separation and membrane fractionation studies with the tropical marine sponge *Amphimedon* sp. *Lipids* 27: 378–388.

Garson, M. J., Zimmermann, M. P., Hoberg, M., Larsen, R. M., Battershill, C. N., Burphy, P. T. (1993). Isolation of brominated long chain fatty acids from the tropical marine sponge *Amphimedon* sp. *Lipids* 28: 1011–1014.

Gaskin, D. E. (1968). The New Zealand Cetacea. *Fish. Res. Bull. N.Z.* 1: 1–92.

GBRMPA (1985a). Workshop on reef fish assessment and monitoring. Great Barrier Reef Marine Park Authority, Townsville. *GBRMPA Wkshop Ser.* 2. 64 pp.

GBRMPA (1985b). Workshop on coral trout assessment techniques (1979). Great Barrier Reef Marine Park Authority, Townsville. *GBRMPA Wkshop Ser.* 3. 85 pp.

Geffen, A. J. (1992). Validation of otolith increment deposition rate. Pp. 101–113 *in* Stevenson, D. K., Campana, S. E. (eds), *Otolith microstructure examination and analysis. Can. Spec. Publ. Fish. Aquat. Sci.* 117.

George, J. D., Lythgoe, G. I., Lythgoe, J. N. (eds) (1985). *Underwater photography and television for scientists*. Clarendon Press, Oxford, 184 pp.

Gibbs, C. F., Arnott, G. H., Longmore, A. R., Marchant, J. W. (1991). Nutrient and plankton distribution near a shelf break front in the region of the Bass Strait Cascade. *Aust. J. Mar. Freshwat. Res.* 42: 201–217.

Giese, A. C., Pearse, J. S. (1974). *Reproduction of marine invertebrates: Acoelomate and pseudocoelomate metazoans*. Academic Press, New York. Vol. 1. 540 pp.

Gifford, D. J. (1993). Protozoa in the diets of *Neocalanus spp.* in the oceanic Subarctic Pacific Ocean. *Prog. Oceanogr. 32*: 1–4, pp. 223–237.

Gill, J. L., Hafs, H. D. (1971). Analysis of repeated measurements of animals. *J. Anim. Sci.* 33: 331–336.

Gillanders, B. M. (1995a). The reproductive biology of the protogynous hermaphrodite, *Achoerodus viridis* (Labridae) from South-eastern Australia. *Mar. Freshwat. Res.* 46: 999–1008.

Gillanders, B. M. (1995b). Feeding ecology of the temperate marine fish *Achoerodus viridis* (Labridae): size, seasonal and site specific differences. *Mar. Freshwat. Res.* 46: 1009–1020.

Gillanders, B. M. (1997a). Patterns of abundance and size structure in the blue groper, *Achoerodus viridis* (Pisces, Labridae): Evidence of links between estuaries and coastal reefs. *Environ. Biol. Fish.* 49: 153–173.

Gillanders, B. M. (1997b). Comparison of growth rates between estuarine and coastal reef populations of *Achoerodus viridis* (Pisces: Labridae). *Mar. Ecol. Prog. Ser.* 146: 283–287.

Gillanders, B. M., Brown, M. T. (1994). The chemical composition of *Xiphophora gladiata* (Phaeophyceae: Fucales): Seasonal and within plant variation. *Botanica. Mar.* 37: 483–490.

Gillanders, B. M., Kingsford, M. J. (1993). Abundance patterns of *Achoerodus viridis* (Labridae) on estuarine and exposed rocky reefs: possible linkages. *Proc. 2nd Int. Temperate Reef Symp.*: 93–98.

Gillanders, B. M., Kingsford, M. J. (1996). Elements in otoliths may elucidate the contribution of estuarine recruitment to sustaining populations of reef fish. *Mar. Ecol. Prog. Ser.* 141: 13–20.

Gillanders, B. M., Kingsford, M. J. (1998). The influence of habitat on the abundance and size structure of a large labrid. *Mar. Biol.* (in press).

Gittins, R. (1985). *Canonical analysis: a review with application in ecology.* Springer-Verlag, Berlin. 351 pp.

Gladfelter, W. B. (1979). Twilight migrations and foraging activities of the copper sweeper *Pempheris schombirgki* (Teleostei: Pempheridae). *Mar. Biol.* 50: 109–119.

Gladstone, W., Westoby, M. (1988). Growth and reproduction in *Canthigaster valentini* (Pisces, Tetradontidae): a comparison of a toxic reef fish with other reef fishes. *Environ. Biol. Fish.* 21: 207–221.

Glasby, G. P. (1991). A review of the concept of sustainable management as applied to New Zealand. *Jl R. Soc. N.Z.* 21: 61–81.

Glasby, T. M. (1997). Analysing data from post-impact studies using asymmetrical analyses of variance. *Aust. J. Ecol.* 22: 448–459.

Glasby, T. M., Kingsford, M. J. (1994). *Atypichthys strigatus* (Pisces: Scorpididae): An opportunistic planktivore that responds to benthic disturbances and cleans other fishes. *Aust. J. Ecol.* 19: 385–394.

Glasby, T. M., Underwood, A. J. (1996). Sampling to different-iate between pulse and press perturbations. *Environ. Monit. Assmt* 42: 241–252.

Glynn, W. (1988). El Niño-Southern Oscillation 1982–1983: Nearshore population, and ecosystem reponses. *Ann. Rev. Ecol. Syst.* 19: 309–346.

Godfriaux, B. L., (1973). Sediment variability in offshore sampling areas. *N.Z. J. Mar. Freshwat. Res.* 7: 323–329.

Goldman, J. C., McCarthy, J. J., Peavey, D. G. (1979). Growth rate influence on the chemical composition of phytoplankton in oceanic waters. *Nature* 279: 210–215.

Gomon, M. F., Glover, J. C. M., Kuiter, R. H. (1994). *The fishes of Australia's south coast.* State Print, Adelaide. 992 pp.

Gooding, R., Magnuson, J. J. (1967). Ecological significance of a drifting object to pelagic fishes. *Pacif. Sci.* 21: 486–497.

Gordon, D. P., Ballantine, W. J. (1976). Cape Rodney to Okakari Point Marine Reserve: a review of knowledge and bibliography to December 1976. *Tane* 22 (suppl.): 146 pp.

Gordon, I. (1993). Precopulatory behaviour of captive sand-tiger, *Carcharinus tauruas. Environ. Biol. Fish.* 38: 159–164.

Goshima, S., Fujiwara, H. (1994). Distribution and abundance of cultured scallop *Patinopecten yessoensis* in extensive sea beds as assessed by underwater camera. *Mar. Ecol. Prog. Ser.* 110: 151–158.

Gosliner, T. M. (1995). Introduction and spread of *Philine auriformis* (Gastropoda: Opisthobranchia) from New Zealand to San Francisco Bay and Bodega Bay. *Mar. Biol.* 122: 249–256.

Gowing, L. (1994). Environmental assessment of landfill leachate on estuarine benthos. Unpublished MSc thesis, University of Auckland. 103 pp.

Graf, G. (1992). Benthic-pelagic coupling: a benthic view. *Oceanogr. Mar. Biol. Ann. Rev.* 30: 149–190.

Graham, J. M. (1991). Symposium introductory remarks: a brief history of aquatic microbial ecology. *J. Protozool.* 38: 66–69.

Graham, K. R., Sebens, K. P. (1996). The distribution of marine invertebrate larvae near vertical surfaces in the rocky subtidal zone. *Ecology* 77: 933–949.

Grange, K. R. (1979). Soft-bottom communities of Manukau Harbour, New Zealand. *N.Z. J. Mar. Freshwat. Res.* 13: 315–329.

Grange, K. R. (1989). Monitoring of littoral macrobenthos and surface sediments in Manukau Harbour, New Zealand. *Misc. Publ. N.Z. Oceanogr. Inst.* 102 : 49 pp.

Grange, K. R. (1990): Unique marine habitats in the New Zealand fiords: a case for preservation. Report prepared for Department of Conservation, Wellington. *NZOI Contract Report* 90/7: 70 pp.

Grange, K. R., Anderson, P. W. (1976). A soft-sediment sampler for the collection of biological specimens. *NZOI Rec.* 3: 9–16.

Grange, K. R., Singleton, R. J. (1988). Population structure of black coral, *Antipathes aperta*, in the southern fiords of New Zealand. *NZ J. Zool.* 15: 481–487.

Grant, A., Morgan, P. J., Olive, P. J. W. (1987). Use made in marine ecology of methods for estimating demographic parameters from size/frequency data. *Mar. Biol.* 95: 201–208.

Grassoff, K., Ehrhardt, M., Kremling, K. 1983. *Methods of sea water analysis* (2nd edn). Basel, Verlag Chemie. 419 pp.

Gray, C. A. (1996a). Intrusions of surface sewage plumes into continental shelf waters: interactions with larval and pre-settlement juvenile fishes. *Mar. Ecol. Prog. Ser.* 139: 31–45.

Gray, C. A. (1996b). Small-scale temporal variability in assemblages of larval fishes: implications for sampling. *J. Plankton Res.* 18: 1643–1657.

Gray, C. A. (1996c). Do thermoclines explain the vertical distributions of larval fishes in the dynamic coastal waters of south-eastern Australia? *Mar. Freshwat. Res.* 47: 183–190.

Gray, C. A., Bell, J. D. (1986). Consequences of two common techniques for sampling vagile macrofauna associated with the seagrass *Zostera capricorni. Mar. Ecol. Prog. Ser.* 28: 43–48.

Gray, C. A., Otway, N. M., Laurenson, F. A., Miskiewicz, A. G. (1992). Distribution and abundance of ichthyoplankton in relation to effluent plumes from sewage outfalls and depth of water. *Mar. Biol.* 113: 549–559.

Green, C. H., Weibe, P. H. (1990). Bioacoustical oceanography: New tools for zooplankton and micronekton research in the 1990s. *Oceanography* 3: 12–17.

Green, R. H. (1979). *Sampling design and statistical methods for environmental biologists*. Wiley-Interscience, New York. 257 pp.

Green, R. H. (1989). Power analysis and practical strategies for environmental monitoring. *Environ. Res.* 50: 195–205.

Green, R. H. (1993a). Application of repeated measures designs in environmental impact and monitoring studies. *Aust. J. Earth Sci.* 18: 81–98.

Green, R. H. (1993b). Relating two set of variables in environmental studies. Pp. 151–165 *in* Rao, C. R. (ed.), *Multivariate analysis: future directions*. Elsevier, Amsterdam.

Green, R. H. (1994). Aspects of power analysis in environmental monitoring. Pp. 173–182 *in* Fletcher, D. J., Manly, F. J. (eds), *Statistics in ecology and environmental monitoring*. University of Otago Press, Dunedin.

Green, R. H., Boyd, J. M., MacDonald, J. S. (1993). Relating sets of variables in environmental studies: the sediment quality triad as a paridigm. *Environmetrics* 4: 439–457.

Greenacre, M. J. (1984). *Theory and application of correspondence analysis*. Academic Press, New York.

Griffin, D. A., Middleton, J. H. (1991). Local and remote wind forcing of New South Wales inner shelf currents and sea level. *J. Phys. Oceanogr.* 21: 304–322.

Griffin, D. A., Thompson, P. A., Bax, N. J., Bradford, R. W., Hallegraeff, G. M. (1997). The 1995 mass mortality of pilchard: no role found for physical or biological oceanographic factors in Australia. *Marine and Freshwater Research.* 48: 27–42.

Griffiths, C. A., Glasby, G. P. (1985). Input of river-derived sediment to the New Zealand continental shelf: I. Mass. *Estuar. Cstl Shelf Sci.* 21: 773–787.

Griffiths, C. L., Hockey, P. A. R., Van Erkom Schurink, C., Roux, P. J. (1992). Marine aliens on South African shores, implications for community structure and trophic functioning. *S. Afr. J. Mar. Sci.* 12: 713–772.

Griffiths, F. B., Wadley, V. A. (1986). A synoptic comparison of fishes and crustaceans from a warm-core eddy, the East Australian Current, the Coral Sea and the Tasman Sea. *Deep Sea Res.* 33: 1907–1922.

Grimes, C. B., Kingsford, M. J. (1996). How do estuarine and riverine plumes of different sizes influence fish larvae: do they enhance recruitment? *Mar. Freshwat. Res.* 47: 191–208.

Groenveld, J. C., Rossouw, G. J. (1995). Breeding period and size in the South Coast lobster *Palinurus gilchristi* (Decapoda: Palinuridae). *S. Afr. J. Mar. Sci.* 15: 17–24.

Grondahl, F. (1988a). A comparative ecological study on the scyphozoans *Aurelia aurita, Cyanea capillata* and *C. lamarckii* in the Gullmar Fjord, western Sweden, 1982–1986. *Mar. Biol.* 97: 541–550.

Grondahl, F. (1988b). Interactions between polyps of *Aurelia aurita* and planktonic larvae of scyphozoans: an experimental study. *Mar. Ecol. Prog. Ser.* 45: 87–93.

Grondahl, F., Hernroth, L. (1987). Release and growth of *Cyanea capillata* (L.) ephyrae in the Gullmar Fjord, Western Sweden. *J. Exp. Mar. Biol. Ecol.* 106: 91–101.

Grossman, G. D., Nickerson, D. M., Freeman, M. C. (1991). Principal component analysis of assemblage structure data: utility of tests based on eigenvalues. *Ecology* 72: 341–347.

Grutter, A. S. (1997). Spatiotemporal variation and selectivity in the diet of the cleaner fish *Labroides dimidiatus. Copeia*: 346–355.

Gubbay, S. (1995). *Marine protected areas: Principles and techniques for management*. Chapman & Hall, London. 232 pp.

Gulland, J. A. (1988). *Fish population dynamics*. Wiley, New York. 422 pp.

Haddon, M., Willis, T. J. (1995). Morphometric and meristic comparison of orange roughy (*Hoplostethus atlanticus*: Trachichthyidae) from the Puysegur Bank and Lord Howe Rise, New Zealand, and its implication for stock structure. *Mar. Biol.* 123: 19–27.

Haddon, M., Willis, T. J., Wear, R. G., Anderlini, V. C. (1996). Biomass and distribution of surf clams off an exposed west coast North Island beach, New Zealand. *J. Shellfish Res.* 15: 331–339.

Hagen, N. T. (1996). Tagging sea urchins: a new technique for individual identification. *Aquaculture* 139: 271–284.

Hair, C., Bell, J., Kingsford, M. J. (1994). Effects of position in the water column, vertical movement and shade on settlement to artificial habitats. *Bull. Mar. Sci.* 55: 434–444.

Hairston Sr, N. G. (1989). *Ecological experiments*. Cambridge University Press, Cambridge. 370 pp.

Hall, S. J., Raffaelli, D., Turrell, W. R. (1990). Predation caging experiments in marine systems: a re-evaluation of their value. *Am. Nat.* 136: 657–672.

Hallegraeff, G. M. (1988). *Plankton : a microscopic world*. CSIRO in association with E.J. Brill/Robert Brown & Associates, Bathurst. 111 pp.

Hammond, L. S., Synnot, R. N. (1994). *Marine biology*. Longman Cheshire, Melbourne. 518 pp.

Hamner, R. M. (1981). Day-night differences in the emergence of demersal zooplankton from a sand substrate in a kelp forest. *Mar. Biol.* 62: 275–280.

Hamner, W. M. (1975). Underwater observations of blue-water plankton: Logistics, techniques, and safety procedures for divers at sea. *Limnol. Oceanogr.* 20: 1045–1051.

Hamner, W. M., Carlton, J. H. (1979). Copepod swarms: Attributes and role in coral reef ecosystems. *Limnol. Oceanogr.* 24: 1–14.

Hamner, W. M., Hamner, P. P., Strand, S. W. (1994). Sun-compass migration by *Aurelia aurita* (Scyphozoa): Population retention and reproduction in Saanich Inlet, British Columbia. *Mar. Biol.* 119: 347–356.

Hamner, W. M., Hauri, I. R. (1977). Fine-scale currents in the Whitsunday Islands, Queensland, Australia: Effect of tide and topography. *Aust. J. Mar. Freshwat. Res.* 28: 333–359.

Hamner, W. M., Hauri, I. R. (1981). Long-distance horizontal migrations of zooplankton (Scyphomedusae: *Mastigias*). *Limnol. Oceanogr.* 26: 414–423.

Hamner, W. M., Jones, M. S., Carlton, J. H., Hauri, I. R., Williams, D. McB. (1988). Zooplankters, planktivorous fish, and water currents on a windward reef face: Great Barrier Reef, Australia. *Bull. Mar. Sci.* 42: 459–479.

Hamner, W. M., Madin, L. P., Alldredge, A. L., Gilmer, R. W., Hamner, P. P. (1975). Underwater observations of gelatinous zooplankton: Sampling problems, feeding biology, and behaviour. *Limnol. Oceanogr.* 20: 907–917.

Hancock, D. A. (1994). Recreational fishing: What's the catch? Proceedings of the Australian Society for Fish Biology Workshop. Australian Government Publishing Service, Canberra. 274 pp.

Haney, J. F. (1988). Diel patterns of zooplankton behavior. *Bull. Mar. Sci.* 43: 583–603.

Hansen, J. A., Skilleter, G. A. (1994). Effects of the gastropod *Rhinoclavis aspera* (Linnaeus, 1758) on microbial biomass and productivity in coral-reef-flat sediments. *Aust. J. Mar. Freshwat. Res.* 45: 569–584.

Haq, S. M. (1972). Breeding of *Euterpina acutifons*, a harpacticoid copepod, with special reference to dimorphic males. *Mar. Biol.* 15: 221–235.

Harmelian-Vivien, M. L., Francour, P. (1992). Trawling or visual censuses? Methodological bias in the assessment of fish populations in seagrass beds. *Mar. Ecol.* 13: 41–51.

Harper, P. C., Kinsky, F. C. (1978). *Southern albatrosses and petrels*. Price Milburn, Wellington. 116 pp.

Harris, T. F. W. (1990) *Greater Cook Strait – form and flow*. DSIR Marine and Freshwater, Wellington. 212 pp.

Harrison, P. (1983). *Seabirds: An identification guide*. Houghton Mifflin, Boston. 448 pp.

Hartley, J. P., Dicks, B. (1987). Macrofauna of subtidal sediments using remote sampling. Pp. 106–130 *in* Baker, J. M., Wolff, W. J. (eds), *Biological surveys of estuaries and coasts*. Cambridge University Press, Cambridge.

Hartley, J. P., Dicks, B., Wolff, W. J. (1987). Processing sediment macrofauna samples. Pp. 131–139 *in* Baker, J. M., Wolff, W. J. (eds), *Biological surveys of estuaries and coasts*. Cambridge University Press, Cambridge.

Harvey, E., Shortis, M. (1996). A system for stereo-video measurement of subtidal organisms. *J. Mar. Technol. Soc.* 29: 10–22.

Harvey, N. (1998). *Environmental impact assessment: procedures, practise, and prospects in Australia*. Oxford University Press, Oxford. 233 pp.

Hastings, A., Harrison, S. (1994). Metapopulation dynamics and genetics. *Ann. Rev. Ecol. Syst.* 25: 167–188.

Hatcher, B. G., Kirkman, H., Wood, W. F. (1987). Growth of the kelp *Ecklonia radiata* near the northern limit of its range in Western Australia. *Mar. Biol.* 95: 63–73.

Hauri, I. R., McGowan, J. A., Wiebe, P. H. (1978). Patterns and processes in the time-space scales of plankton distributions. Pp. 277–337 *in* Steele, J. (ed.*) Spatial pattern plankton communities*. Plenum Press, New York.

Hawkins, S. J., Hartnoll, R. G. (1983). Grazing of intertidal algae by marine invertebrates. *Oceanogr. Mar. Biol. Ann. Rev.* 21: 195–282.

Hay, C. H. (1990). The distribution of *Macrocystis* (Phaeophyta, Laminariales) as a biological indicator of cool sea surface temperature, with special reference to New Zealand waters. *Jl R. Soc. N.Z.* 20: 313–336.

Hay, M. E. (1991). Fish-seaweed interactions on coral reef: Effects of herbivorous fishes and adaptations of their prey. Pp. 96–119 *in* Sale, P. F. (ed.) *The ecology of fishes on coral reefs*. Acadmic Press, San Diego.

Hay, M. E. (1996). Marine chemical ecology: What's known and what's next? *J. Exp. Mar. Biol. Ecol.* 200: 103–134.

Hay, M. E. (1997). Synchronous spawning: When timing is everything. *Science*. 275: 1080–1081.

Hay, M. E., Fenical, W. (1988). Marine plant-herbivore interactions: the ecology of chemical defense. *Ann. Rev. Ecol. Syst.* 19: 111–145.

Hay, C. H., South, G. R. (1979). Experimental ecology with particular reference to proposed commercial harvesting of *Durvillaea* (Phaeophyta, Durvilleales) in New Zealand. *Bot. Mar.* XXII: 431–436.

Haynes, D., Quinn, G. P. (1995). Temporal and spatial variability in community structure of a sandy intertidal beach, Cape Paterson, Victoria, Australia. *Mar. Freshwat. Res.* 46: 931–942.

Hayward, B. W., Hollis, C. J., Grenfell, H. (1994b). Foraminiferal associations in Port Pegasus, Stewart Island, New Zealand. *N.Z. J. Mar. Freshwat. Res.* 28: 69–95.

Hayward, P. J., Ryland, J. S. (1995). *Handbook of the marine fauna of North-West Europe*. Oxford University Press, Oxford. 800 pp.

Hayward, B. W., Stephenson, A. B., Riley, J., Motu, G., Grenfell, H., Morley, M., Blom, W. (1994a). Changes in benthic communities of the Waitemata Harbour, 1930s to 1990s. (Abstr.) *N.Z. Mar. Sci. Soc. Ann. Review* 37: 22.

Hayward, B. W., Stephenson, A. B., Riley, J. L., Blom, W. M., Grenfell, H. R., Morley, M. (1995). Soft-bottom macro-benthos of the Waitemata Harbour. (Abstr.) *N.Z. Mar. Sci. Soc. Ann. Review* 38: 41.

Heath, M. R. (1992). Field investigation of the early life history stages of marine fish. *Adv. Mar. Biol.* 28: 1–174.

Heath, M. R., Rankine, P. (1988). Growth and advection of larval herring (*Clupea harengus* L.) in the vicinity of the Orkney Isles. *Estuar. Cst Shelf Sci.* 27: 547–565.

Heath, R. A. (1985). A review of the physical oceanography of the sea around New Zealand – 1982. *N.Z. J. Mar.Freshwat. Res.* 19: 79–124.

Heck, K. L., Wilson, K. A. (1990). Epifauna and infauna: Biomass and abundance. Pp. 125–128 *in* Phillips, R. C, McRoy, C. P. (eds), *Seagrass research method*. UNESCO, Paris.

Hegge, B., Eliot, I., Hsu, J. (1996). Sheltered sandy beaches of southwestern Australia. *J. Cstl Res.* 12: 748–760.

Heip, C. H. R., Goosen, N. K., Herman, P. M. J., Soetaert, K. (1995). Production and consumption of biological particles in temperate tidal estuaries. *Oceanogr. Mar. Biol. Ann. Rev.* 33: 1–149.

Herendeen, R. A. (1998). *Ecological numeracy: quantitative analysis of environmental issues*. Wiley, New York. 331pp.

Herman, A. W. (1988). Simultaneous measurement of zoo-plankton and light attenuance with a new optical plankton counter. *Continental Shelf Sci* 8: 205–221.

Herman, A.W., Mitchell, M.R., Young, S.W. (1984). A continous pump sampler for profiling copepods and chlorophyll in the upper oceanic layers. *Deep Sea Res.* 31: 439–450.

Herman, A. W. (1992). Design and calibration of a new optical plankton counter capable of sizing small zooplankton. *Deep Sea Res.* 39: 395–415.

Heron, A. C. (1972a). Population ecology of a colonying species: the pelagic tunicate *Thalia democratica*. II. Population growth rate. *Oecologia* 10: 294–312.

Heron, A. C. (1972b). Population ecology of a colonying species: the pelagic tunicate *Thalia democratica*. I. Individual growth rate and generation time. *Oecologia* 10: 269–293.

Heron, A. C., Benham, E. E. (1984). Individual growth rates of salps in three populations. *J. Plankton Res.* 6: 811–828.

Herrnkind, W. F., Butlet, M. J., Hunt, J. H. (1997). Can

artificial habitats that mimic natural structures enhance recruitment of Caribbean spiny lobster. *Fisheries* 22: 24–27.

Hickford, M., Schiel, D. R. (1996). Catch versus count: effects of gill netting on reef fish populations in southern New Zealand. *J. Exp. Mar. Biol. Ecol.* 188: 215–232.

Hilborn, R., Medley, P. (1989). Tuna purse-seine fishing with fish-aggregating devices (FADS): models of tuna FAD interactions. *Can. J. Fish. Aquat. Sci.* 46: 28–32.

Hilborn, R., Walters, C. J. (1992). *Quantitative Fisheries Stock Assessment*. Chapman & Hall, London. 570 pp.

Hilborn, R., Walters, C. J., Ludwig, D. (1995). Sustainable exploitation of renewable resources. *Ann. Rev. Ecol. Syst.* 26: 45–67.

Hilderman, W. H., Johnston, I. S., Jokiel, P. L. (1979). Immunicompetence in the lowest metazoan phylu: transplantation immunity in sponges. *Science* 204: 420–422.

Hiscock, K. (1987). Subtidal rock and shallow sediments using diving. Pp. 198–237 in Baker, J. M., Wolff, W. J. (eds), *Biological surveys of estuaries and coasts*. Cambridge University Press, Cambridge. 449 pp.

Hixon, M. A. (1980). Competitive interactions between California reef fishes of the genus *Embiotoca*. *Ecology* 61: 918–931.

Hixon, M. A. (1981). An experimental analysis of territoriality in the California reef fish *Embiotoca jacksoni* (Embiotocidae). *Copeia* 1981: 653–665.

Hixon, M. A. (1991). Predation as a process structuring coral-reef fish communities. Pp. 475–508 *In* Sale, P. F. (ed.) *The ecology of fishes on coral reefs*. Academic Press, London.

Hixon, M. A., Beets, J. P. (1989). Shelter characteristics and Caribbean fish assemblages: experiments with artificial reefs. *Bull. Mar. Sci.* 44: 666–680.

Hixon, M. A., Beets, J. P. (1993). Predation, prey refuges, and the structure of coral-reef fish assemblages. *Ecol. Monogr.* 63: 77–101.

Hixon, M. A., Brostoff, W. N. (1996). Succession and herbivory: effects of differential fish grazing on Hawaiian coral-reef algae. *Ecol. Monogr.* 66: 67–90.

Hixon, M. A., Carr, M. H. (1997). Synergistic predation, density dependence, and population regulation in marine fish. *Science* 277: 946–949.

Hobson, E. S. (1974). Feeding relationships of teleostean fishes on coral reefs in Kona, Hawaii. *Fish. Bull. U.S.* 72: 915–1031.

Hobson, E. S. (1991). Trophic relationships of fishes specialized to feed on zooplankters above coral reefs. Pp. 69–95 *in* Sale, P. F. (ed.), *The ecology of fishes on coral reefs*. Academic Press, San Diego.

Hobson, E. S., Chess, J. R. (1976). Trophic interactions among fishes and zooplankters near shore at Santa Catalina Island, California. *Fish. Bull. U.S.* 74: 567–598.

Hobson, E. S., Chess, J. R. (1978). Trophic relationships among fishes and plankton in the lagoon at Enewetak Atoll, Marshall Islands. *Fish. Bull. U.S.* 76: 133–153.

Hockey, P. A. R., Bosman, A. L. (1987). Man as an intertidal predator in Transkei: Disturbance, community convergence and the management of a natural food resource. *Oikos* 46: 3–14.

Hoffman, K. S., Grau, E. G. (1989). Daytime changes in oocyte development with relation to the tide for the Hawaiian saddleback wrasse, *Thalassoma duperry*. *J. Fish. Biology*. 34: 529–546.

Holbrook, S. J., Kingsford, M. J., Schmitt, R. J., Stephens, J. J. S. (1994). Spatial patterns of marine reef fish assemblages. *Am. Zool.* 34: 463–475.

Holbrook, S. J., Schmitt, R. J. (1984). Experimental analyses of patch selection by foraging black surfperch (*Embiotoca jacksoni* Agazzi). *J. Exp. Mar. Biol. Ecol.* 79: 39–64.

Holbrook, S. J., Schmitt, R. J. (1989). Resource overlap, prey dynamics and the strength of competition. *Ecology* 70: 1943–1953.

Holbrook, S. J., Schmitt, R. J. (1992). Dietary specialization in surfperches. *Ecology* 73: 391–401.

Holbrook, S. J., Schmitt, R. J. (1996). On the dynamics and structure of reef fish communities: are resources tracked? Pp. 9–48 *in* Cody, M. L., Smallwood, J. A. (eds), *Long-term studies of vertebrate communities*. Academis Press, London.

Holbrook, S. J., Schmitt, R. J. (1998). Have field experiments aided in the understanding of abundance and dynamics of reef fishes? Pp. 152–169 *in* Resetarits, W. J., Bernado, J. (eds), *Issues and perspectives in experimental ecology*. Oxford University Press, Oxford.

Holbrook, S. J., Schmitt, R. J., Ambrose, R. F. (1990). Biogenic habitat structure and characteristics of temperate reef fish assemblages. *Aust. J. Ecol.* 15: 489–503.

Holdway, P., Maddock, L. (1983). Neustonic distributions. *Mar. Biol.* 77: 207–214.

Holliday, D. V., Pieper, R. E., Kleppel, G. S. (1989). Determination of zooplankton size and distribution with multi-frequency acoustic techology. *J. Cons. Int. Explor. Mer.* 46: 226–238.

Holme, N. A. (1985). Use of photographic and television cameras on the continental shelf. Pp. 88–99 *in* George, J. D., Lythgoe, G. I., Lythgoe, J. N. (eds), *Underwater photography and television for scientists*. Clarendon Press, Oxford. 184 pp.

Holme, N. A., McIntyre, A. D. (1984). *Methods for the study of marine benthos*, 2nd edn. Blackwell, Oxford, 387 pp.

Holmes, N. Saenger, P. (1995). Australia. Ch. 22, Pp. 291–305 *in* Hotta, K., Dutton, I. (eds), *Coastal management in the Asia-Pacific region: Issues and approaches*. JIMSTEF, Tokyo. 421 pp.

Hooker, S. H. (1995). Life history and demography of the pipi *Paphies australis* (Bivalvia: Mesodesmatidae) in northeastern New Zealand. Unpublished PhD thesis, University of Auckland.

Hooker, S. H., Creese, R. G. (1995). Reproduction of paua, *Haliotis iris* Gmelin 1791 (Mollusca: Gastropoda), in northeastern New Zealand. *Mar. Freshwat. Res.* 46: 617–622.

Hooper, D. J. (1979). Hydrographic surveying. Pp. 41–56 *in* Dyer, K. R. (ed.), *Estuarine hydrography and sedimentation*. Cambridge University Press, Cambridge.

Hooper, J. N. A. (1994). Coral reef sponges of the Sahul Shelf – a case for habitat preservation. *Mem. Qd Mus.* 36: 93–106.

Horn, M. H. (1989). Biology of marine herbivorous fishes. *Oceanogr. Mar. Biol. Ann. Rev.* 27: 167–272.

Horn, M. H. (1992). Herbivorous fishes: Feeding and digestive mechanisms. Pp. 339–362 *in* John, D. M., Hawkins, S. J., Price, J. H. (eds), *Plant-animal interactions in the marine benthos*. Vol. 46. Clarendon Press, Oxford.

Hostetter, E. B., Munroe, T. A. (1993). Age, growth, and reproduction of tautog *Tautoga onitis* (Labridae: Perciformes) from coastal waters of Virginia. *Fish. Bull. U.S.* 91: 45–64.

Houde, E. D., Lovdal, J. A. (1984). Seasonality of occurrence, foods and food preferences of ichthyoplankton in Biscayne Bay, Florida. *Estuar. Cstl Shelf Sci.* 18: 403–419.

Howard, R. K. (1989). The structure of nearshore fish community of western Australia: diel patterns and the habitat role of limestone reef. *Environ. Biol. Fish.* 24: 93–104.

Howell, B. R., Mokeness, E., Svasand, T. (in press). *Stock enhancement and sea ranching.* Blackwell, Oxford.

Hubbard, D. M. (1996). Tidal cycle distortion in Carpinteria Salt Marsh, California. *Bull. Sthn Calif. Acad. Sci.* 95: 88–98.

Hughes, T. P. (1994). Catastrophes, phase shifts, and large-scale degradation of a Caribbean reef. *Science* 265: 1547–1551.

Huitema, B. E. (1980). *The analysis of covariance and alternatives.* Wiley, New York. 445 pp.

Humason, G. L. (1979). *Animal tissue techniques.* Freeman, London. 661 pp.

Hunter, J., Mitchell, C. T. (1967). Association of fishes with flotsam in the offshore water of central America. *Fish. Bull. U.S.* 66: 13–29.

Hurlbert, S. H. (1984). Pseudoreplication and the design of ecological field experiments. *Ecol. Monogr.* 54: 187–211.

Hurlbert, S. H., White, M. D. (1993). Experiments with freshwater invertebrate zooplanktivores: Quality of statistical analyses. *Bull. Mar. Sci.* 53: 128–153.

Hutchings, P. A. (1988). A survey of Elizabeth and Middleton Reefs for the Australian National Parks and Wildlfe Service. Australian Museum, Sydney, 229 pp.

Hutchins, B., Swainston, R. (1986). *Sea fishes of southern Australia.* Swainston, Perth. 180 pp.

Huynh, H., Feldt, L. S. (1970). Conditions under which mean square ratios in repeated measurement designs have exact F-distributions. *J. Am. Stats. Ass.* 65: 1582–1589.

Hwang, J. S., Costello, J. H., Strickler, J. R. (1994). Copepod grazing in turbulent flow: Elevated foraging behaviour and habituation of escape responses. *J. Plankton Res.* 16: 421–431.

Hyatt, M. A. (1993). *Stains and cytochemical methods.* Plenum Press, New York. 455 pp.

Hyslop, E. J. (1980). Stomach contents analysis – a review of methods and their application. *J. Fish. Biol.* 17: 411–429.

Ianora, A., Poulet, S. A., Miralto, A., Grottoli, R. (1996). The diatom *Thalassiosira rotula* affects reproductive success in the copepod *Acartia clausi. Mar. Biol.* 125: 279–286.

Ibarra-Obando, S. E., Boudouresque, C. H. (1994). An improvement on the Zieman leaf marking technique for *Zostera marina* growth and production assessment. *Aquatic Bot.* 47: 293–302.

Inglis, G. J. (1992). Population ecology and epibiosis of the Sydney cockle *Anadara trapezia.* Unpublished PhD thesis, University of Sydney. 197 pp.

Inglis, G. J. (1993). Ambiguities in the identification and selection of representative marine reserves. *Proc. 2nd Int. Temperate Reef Symp.*: 23–28.

Inglis, G. J., Lincoln-Smith, M. P. (1995). An examination of observer bias as a source of error in surveys of seagrass shoots. *Aust. J. Ecol.* 20: 273–281.

Ingolfsson, A. (1995). Floating clumps of seaweed around Iceland: Natural microcosms and a means of dispersal for fauna. *Mar. Biol.* 122: 13–22.

Ivanchenko, O. F., Ivanchenko, L. A. (1973). The use of alcian blue to stain the cartilagenous skeleton of fish larvae and fingerlings in whole mounts. *J. Icthyol.* 13: 794–796.

Ivanovici, A. M. (1984). *Inventory of declared marine and estuarine protected areas in Australian waters.* Australian National Parks and Wildlife Service, Special Publication 12. Commonwealth of Australia Press, Canberra. Vols 1 & 2. 433 pp.

Jacoby, C. A., Greenwood, J. G. (1988). Spatial, temporal, and behavioural patterns in emergence of zooplankton in the lagoon of Heron Reef, Great Barrier Reef, Australia. *Mar. Biol.* 97: 309–328.

Jacoby, C. A., Youngbluth, M. J. (1983). Mating in three species of *Pseudodiaptomus* (Copepoda: Calanoida). *Mar. Biol.* 76: 77–86.

Jahn, A. E., Lavenberg, R. J. (1986). Fine-scale distribution of nearshore, suprabenthic fish larvae. *Mar. Ecol. Prog. Ser.* 31: 223–231.

James, F. C., McCulloch, C. E. (1990). Multivariate analysis in ecology and systematics: panacea or Pandora's box? *Ann. Rev. Ecol. Syst.* 21: 129–166.

James, R. J., Fairweather, P. G. (1995). Comparison of rapid methods for sampling the pipi, *Donax deltoides* (Bivalvia: Donacidae), on ocean beaches. *Mar. Freshwat. Res.* 46: 1093–1099.

James, R. J., Lincoln-Smith, M. P., Fairweather, P. G. (1995). Sieve mesh-size and taxonomic resolution needed to describe natural spatial variation of marine macrofauna. *Mar. Ecol. Prog. Ser.* 118: 187–198.

Jaramillo, E., McLachlan, A., Dugan, J. E. (1995). Total sample area and estimates of species richness in exposed sandy beaches. *Mar. Ecol. Prog. Ser.* 119: 311–314.

Jefferson, T. A. (1993). *Marine mammals of the world.* FAO Species Identification Guide, Rome. 320 pp.

Jeffs, A. (1986). The demostraphy of *Haliotis iris.* Unpublished report. Leigh Marine Reserve Committee, Auckland. 22 pp.

Jenkins, G. P. (1987). Age and growth of co-occurring larvae of two flounder species, *Rhombolsolea tapirina* and *Ammotretis rostratus. Mar. Biol.* 95: 157–166.

Jenkins, G. P. (1988). Micro- and fine scale distribution of microplankton in the feeding environment of larval flounder. *Mar. Ecol. Prog. Ser.* 43: 233–244.

Jenkins, G. P., May, H. M. A. (1994). Variation in settlement and larval duration of King George Whiting, *Sillaginoides punctata* (Sillaginidae) in Swan Bay, Victoria, Australia. *Bull. Mar. Sci.* 54: 281–296.

Jernakoff, P. (1983). Factors affecting the recruitment of algae in a midshore region dominated by barnacles. *J. Exp. Mar. Biol. Ecol.* 67: 17–31.

Jernakoff, P. (1985). An experimental evaluation of the influence of barnacles, crevices and seasonal patterns of grazers on algal diversity and cover in an intertidal barnacle zone. *J. Exp. Mar. Biol. Ecol.* 88: 287–302.

Jernakoff, P., Brearley, A., Nielsen, J. (1996). Factors affecting grazer-epiphyte interactions in temperate seagrass meadows. *Oceanogr. Mar. Biol. Ann. Rev.* 34: 109–162.

Jernakoff, P., Fairweather, P. G. (1985). An experimental analysis of interactions among several intertidal organisms. *J. Exp. Mar. Biol. Ecol.* 94: 71–88.

Jernakoff, P., Phillips, B. F. (1988). Effect of a baited trap on the foraging movements of juvenile western rock lobsters *Panulirus cygnus* George. *Aust. J. Mar. Freshwat. Res.* 39: 185–192.

Jillett, J. B. (1971). Zooplankton and hydrology of Hauraki Gulf, New Zealand. *Mem. N.Z. Oceanogr. Inst.* 53: 103 pp.

Jillett, J. B. (1976). Zooplankton associations off Otago Peninsula, south-eastern New Zealand, related to different water masses. *N.Z. J. Mar. Freshwat. Res.* 10: 543–557.

Jillett, J. B., Zeldis, J. R. (1985). Aerial observations of surface patchiness of a planktonic crustacean. *Bull. Mar. Sci.* 37: 609–619.

Johannes, R. E. (1978). Reproductive strategies of coastal marine fishes in the tropics. *Environ. Biol. Fish.* 3: 65–84.

Johnson, C. R., Mann, K. H. (1988). Diversity patterns of adaptation, and stability of Nova Scotia kelp beds. *Ecol. Monogr.* 58: 129–154.

Johnson, D. (1994). Seastar fight gains momentum: update on the northern Pacific seastar '*Asterias amurensis*'. *Aust. Fish.* January 1994: 25–27.

Johnson, D. H. (1995). Statistical sirens: the allure of nonparametrics. *Ecology* 76: 1998–2000.

Johnson, L. E. (1992). Potential and peril of field experimentation: the use of copper to manipulate molluscan herbivores. *J. Exp. Mar. Biol. Ecol.* 160: 251–262.

Joll, L. M. (1980). Reproductive biology of two species of Turbinidae (Mollusca: Gastropoda). *Aust. J. Mar. Freshwat. Res.* 31: 319–336.

Jones, C. M., Robson, D. S. (1991). Improving precision in angler surveys: traditional access design versus bus route design. *Am. Fish. Soc. Symp.* 12: 177–188.

Jones, D. S., Morgan, G. J. (1994). *A field guide to crustaceans of Australian waters.* Reed, Sydney. 216 pp.

Jones, G. P. (1980). Growth and reproduction in the protogynous hermaphrodite *Pseudolabrus celidotus* (Pisces: Labridae) in New Zealand. *Copeia.* 660–675.

Jones, G. P. (1981). Spawning-site choice by female *Pseudolabrus celidotus* (Pisces:Labridae) and its influence on the mating system. *Behav. Ecol. Sociobiol.* 8: 129–142.

Jones, G. P. (1983). Relationship between density and behaviour in juvenile *Pseudolabrus celidotus* (Pisces: Labridae). *Anim. Behav.* 31: 729–735.

Jones, G. P. (1984a). Population ecology of the temperate reef fish *Pseudolabrus celidotus* Bloch & Schneider (Pisces: Labridae) I. Factors influencing recruitment. *J. Exp. Mar. Biol. Ecol.* 75: 257–276.

Jones, G. P. (1984b). Population ecology of a temperate reef fish *Pseudolabrus celidotus* Bloch & Schneider (Pisces: Labridae). II. Factors influencing adult density. *J. Exp. Mar. Biol. Ecol.* 75: 277–303.

Jones, G. P. (1984c). The influence of habitat and behavioural interactions on the local distribution of the wrasse, *Pseudolabrus celidotus*. *Environ. Biol. Fish.* 10: 43–58.

Jones, G. P. (1987). Competitive interactions among adults and juveniles in a coral reef fish. *Ecology* 68: 1534–1547.

Jones, G. P. (1988a). Ecology of rocky reef fish of north-eastern New Zealand. *N.Z. J. Mar. Freshwat. Res.* 22: 445–462.

Jones, G. P. (1988b). Experimental evaluation of the effects of habitat structure and competitive interactions on the juveniles of two coral reef fishes. *J. Exp. Mar. Biol. Ecol.* 123: 115–126.

Jones, G. P. (1991). Postrecruitment processes in the ecology of coral reef fish populations: a multifactorial perspective. Pp. 294–328 *in* Sale, P. F. (ed.), *The ecology of fishes on coral reefs.* Academic Press, San Diego.

Jones, G. P., Andrew, N. L. (1990). Herbivory and patch dynamics on rocky reefs in temperate Australasia: the roles of fish and sea urchins. *Aust. J. Ecol.* 15: 505–520.

Jones, G. P., Andrew, N. L. (1993). Temperate reefs and the scope of seascape ecology. *Proc. 2nd Int. Temperate Reef Symp.*: 63–76.

Jones, G. P., Cole, R. C., Battershill, C. N. (1993). Marine reserves: Do they work? *Proc. 2nd Int. Temperate Reef Symp.*: 29–45.

Jones, G. P., Ferrell, D. J., Sale, P. F. (1990). Spatial pattern in the abundance and structure of mollusc populations in the soft sediments of a coral reef lagoon. *Mar. Ecol. Prog. Ser.* 62: 109–120.

Jones G. P., Ferrell, D. J., Sale, P. F. (1992). Fish feeding and dynamics of soft-sediment mollusc populations in a coral reef lagoon. *Mar. Ecol. Prog. Ser.* 80: 175–190.

Jones, G. P., Norman, M. D. (1986). Feeding selectivity in relation to territory size in a herbivorous reef fish. *Oecologia* 68: 549–556.

Jones, G. P., Thompson, S. M. (1980). Social inhibition of maturation in females of the temperate wrasse *Pseudolabrus celidotus* and a comparison with the blennioid *Tripterygion varium*. *Mar. Biol.* 59: 247–256.

Jones, J. B. (1992). Environmental impact of trawling on the seabed: a review. *N.Z. J. Mar. Freshwat. Res.* 26: 59–67.

Jones, N. S. (1963). The marine fauna of New Zealand: crustaceans of the order Cumacea. *Mem. N.Z. Oceanogr. Inst.* 23: 80 pp.

Jones, P. J. S. (1994). A review and analysis of the objectives of marine nature reserves. *Ocean Coast. Mgmt.* 24: 149–178.

Jones, R. S. (1968). A suggested method for quantifying gut contents in herbivorous fishes. *Micronesica* 4: 369–371.

Jones, R. S., Thompson, M. J. (1978). Comparison of Florida reef fish assemblages using a rapid visual technique. *Bull. Mar. Sci.* 28: 159–172.

Jones, W. E., Bennell, S., Beveridge, S., McConnell, B., Mack-Smith, S., Mitchell, J., Fletcher, A. (1980). Methods of data collection and processing in rocky intertidal monitoring. Pp. 137–170 *in* Price, J. H., Irvine, D. E. G., Farnham, W. F. (eds), *The shore environment, Vol 1: Methods.* Academic Press, London. 321 pp.

Jumars, P. A., Thistle, D., Jones, M. L. (1977). Detecting two-dimensional spatial structure in biological data. *Oecologia* 28: 109–123.

Kaehler, S., Williams, G. A. (1996). Distribution of algae on tropical rocky shores: Spatial and temporal patterns of non-coralline encrusting algae in Hong Kong. *Mar. Biol.* 125: 177–188.

Kailola, P. J., Williams, M. J., Stewart, P. C., Reichelt, R. E., McNee, A., Grieve, C. (1993). *Australian Fisheries Resources.* Bureau of Resource Sciences and the Fisheries Research and Development Corporation, Canberra. 422 pp.

Kalish, J. M. (1990). Use of otolith microchemistry to distinguish the progeny of sympatric anadromous and non-anadromous salmonids. *Fish. Bull. U.S.* 88: 657–666.

Kaly, U. L. (1988). Distribution, abundance and size of mangrove and saltmarsh gastropods. Unpublished PhD thesis, University of Sydney.

Karas, P., Neuman, E., Sandstrom, O. (1991). Effects of a pulp mill effluent on the population dynamics of perch, *Perca fluviatis*. *Can. J. Fish. Aquat. Sci.* 48: 28–34.

Kaseloo, P. A., Weatherly, A. H., Lotimer, J., Farina, M. D. (1992). A biotelemetry system recording fish activity. *J. Fish. Biol.* 40: 165–179.

Kato, M., Sudo, H., Azeta, M., Matsumiya, Y. (1991). Field experiments of iridium marking for red sea bream in Shijiki

Bay. *Bull. Nat. Res. Far Seas Fish.* 28: 21–45.

Kaufman, K. W. (1981). Fitting and using growth curves. *Oecologia* 49: 293–299.

Kay, A. M., Keough, M. J. (1981). Occupation of patches by the epifaunal communities on pier pilings and the bivalve *Pinna bicolor* at Edithburgh, South Australia. *Oecologia* 48: 125–130.

Keats, D. W. (1991). Refugial *Laminaria* abundance and reduction in urchin grasing in communities in the northwest Atlantic. *J. Mar. Biol. Ass. U.K.* 71: 867–876.

Keestra, B. (1987). An experimental assessment of the distribution patterns of subtidal gastropods; especially *Cookia sulcata*. Unpublished MSc thesis, University of Auckland, New Zealand.

Kelleher, G., Kenchington, R. A. (1992). Guidelines for establishing marine protected areas. A marine conservation and development report. IUCN, Gland, Switzerland. 79 pp.

Keller, A. A., Doering, P. H., Kelly, S. P., Sullivan, B. K. (1990). Growth of juveile Atlantic menhaden, *Brevoortia tyrannus* (Pisces: Clupeidae) in MERL mesocosms : Effects of eutrophication. *Limnol. Oceanogr.* 35: 109–122.

Kelley, J. C. (1976). Sampling the sea. Pp. 361–387 *in The ecology of the seas.* Blackwell, Oxford.

Kelly, M. (1983). A bibliography and literature review for the Poor Knights Islands Marine Reserve. Poor Knights Islands Marine Reserve Management Committee, Auckland. 126 pp.

Kelly, M. G. (1980). Remote sensing of seagrass beds. Pp. 69–85 *in* Phillips, R. C., McRoy, C. P. (eds) *Handbook of seagrass biology: an ecosystem perspective.* Garland STMP Press, New York.

Kelly, S. (1997). Offshore movements of lobster in the Cape Rodney to Okakari Point Marine Reserve. Science for Conservation. Unpublished report, Department of Conservation, Te Papa Atawhai, Wellington. 58 pp.

Kelly-Borges, M.; Bergquist, P. R. (1988). Success in a shallow reef environment : Sponge recruitment by fragmentation with a different. *Proc. 6th Int. Coral Reef Symp.* 2: 757–762.

Kemp, P. F., Sherr, B. F., Sherr, E. B., Coale, J. J. (1993). *Handbook of methods in aquatic microbial ecology.* Lewis Publishers, Boca Raton.

Kenchington, R. A. (1978). Large area surveys of coral reefs. *In* Stoddart, D. R., Johannes, R. E. (eds.), *Coral reefs: research methods.* UNESCO, Paris. 581 pp.

Kenchington, R. A. (1990). *Managing marine environments.* Taylor & Francis, New York. 248 pp.

Kendall, A. W., Picquelle, S. J. (1989). Egg and larval distributions of the walleye pollock *Theragra chalcogramma* in Shelkof Strait, Gulf of Alaska. *Fish. Bull. U.S.* 88: 133–154.

Kennelly, S. J. (1983). An experimental approach to the study of factors affecting algal colonisation in a sublittoral kelp forest. *J. Exp. Mar. Biol. Ecol.* 68: 257–276.

Kennelly, S. J. (1987a). Inhibition of kelp recruitment by turfing algae and consequences for an Australian kelp community. *J. Exp. Mar. Biol. Ecol.* 112: 49–60.

Kennelly, S. J. (1987b). Physical disturbances in an Australian kelp community. I. Temporal effects. *Mar. Ecol. Prog. Ser.* 40: 145–153.

Kennelly, S. J. (1987c). Physical disturbances in an Australian kelp community. II. Effects on understorey species due to

differences in kelp cover. *Mar. Ecol. Prog. Ser.* 40: 155–165.

Kennelly, S. J. (1991). Caging experiments to examine the effects of fishes on understorey species in a sublittoral kelp community. *J. Exp. Mar. Biol. Ecol.* 147: 207–230.

Kennelly, S. J. (1995). Kelp beds. Pp. 106–120 *in* Underwood, A. J., Chapman, M. G. (eds), *Coastal marine ecology of temperate Australia.* UNSW Press, Sydney.

Kennelly, S. J., Craig, J. R. (1989). Effects of trap design, independence of traps and bait on sampling population of spanner crabs. *Mar. Ecol. Prog. Ser.* 51: 49–56.

Kennelly, S. J., Larkum, A. W. D. (1983). A preliminary study of temporal variation in the colonization of the subtidal algae in an *Ecklonia radiata* community. *Aquat. Bot.* 17: 275–282.

Kennelly, S. J., Underwood, A. J. (1985). Sampling of small invertebrates on hard substrata in a sublittoral kelp forest. *J. Exp. Mar. Biol. Ecol.* 89: 55–67.

Kennelly, S. J., Underwood, A. J. (1992). Fluctuations in the distributions and abundances of species in sublittoral kelp forests in New South Wales. *Aust. J. Earth Sci.* 17: 367–382.

Kent, M., Coker, P. (1992). *Vegetation description and analysis.* CRC Press, Boca Raton. 363 pp.

Keough, M. J. (1983). Patterns of recruitment of sessile invertebrates in two subtidal habitats. *J. Exp. Mar. Biol. Ecol.* 66: 213–245.

Keough, M. J. (1984a). Effects of patch size on the abundance of sessile marine invertebrates. *Ecology* 65: 423–437.

Keough, M. J. (1984b). Dynamics of the epifauna of the bivalve *Pinna bicolor*: Interactions among recruitment, predation and competition. *Ecology* 65: 677–688.

Keough, M. J., Downes, B. J. (1982). Recruitment of marine invertebrates: the role of active larval choices and early mortality. *Oecologia* 54: 348–352.

Keough, M. J., Mapstone, B. D. (1995). Protocols for designing marine ecological monitoring programs associated with BEK mills. *National Pulp Mills Research Program, Tech. Rep.* 11. CSIRO, Canberra. 185 pp.

Keough, M. J., Mapstone, B. D. (1997). Designing environmental monitoring for pulp mills in Australia. *Wat. Sci. Tech.* 35: 397–404.

Keough, M. J., Quinn, G., King, A. (1990). Ecology of temperate reefs. *Aust. J. Ecol.* 15: 361–531.

Keough, M. J., Quinn, G. P., King, A. (1993). Correlations between human collecting and intertidal mollusc populations on rocky shores. *Conserv. Biol.* 7: 378–390.

Keough, M. J., Raimondi, P. T. (1995). Responses of settling invertebrate larvae to organic films: effects of different types of films. *J. Exp. Mar. Biol. Ecol.* 185: 235–253.

Keough, M. J., Raimondi, P. T. (1996). Responses of settling invertebrate larvae to organic films: a comparison of the effects of local and 'foreign' films. *J. Exp. Mar. Biol. Ecol.* 207: 59–78.

Kermack, D. M., Barnes, R. S. K., Crothers, J. H. (1979). *Synopses of the British Fauna (New Series).* 1–49 vols. Academic Press, London.

Kerrigan, B. A. (1987). Comparisons between populations of *Evechinus chloroticus* in subtidal and intertidal habitats. Unpublished MSc thesis, University of Auckland. 80 pp.

Key, S. (1988). Distribution of living planktonic foraminifera (Protozoa) of eastern Australia. *Aust. J. Mar. Freshwat. Res.* 39: 71–85.

Kiirikki, M., Bloomster, J. (1996). Wind induced upwelling as a possible explanation for mass occurrences of epiphytic

Ectocarpus siliculosus (Phaeophyta) in the northern Baltic Proper. *Mar. Biol.* 127: 353–358.

Kimmel, J. J. (1985). A new species-time method for visual assessment of fishes and its comparison with established methods. *Environ. Biol. Fish.* 12: 23–32.

Kimmerer, W. J. (1984). A further improvement in the operation of opening/closing nets. *J. Plankton Res.* 6: 527–529.

King, C. K., Hoegh-Guldberg, O., Byrne, M. (1994). Reproductive cycle of *Centrostephanus rodgersii* (Echinodea), with recommendations for the establishment of a sea urchin fishery in New South Wales. *Mar. Biol.* 120: 95–106.

King, J. E. (1983). *Seals of the world.* Cornell University Press, New York. 240 pp.

Kingsford, M. J. (1980). Interrelationships between spawning and recruitment of *Chromis dispilus* (Pisces: Pomacentridae). Unpublished MSc thesis, University of Auckland. 79 pp.

Kingsford, M. J. (1985). The demersal eggs and planktonic larvae of *Chromis dispilus* (Teleostei: Pomacentridae) in northeastern New Zealand coastal waters. *N.Z. J. Mar. Freshwat. Res.* 19: 429–438.

Kingsford, M. J. (1986). Biological assessments of marine protected areas: a handbook of methodology for rocky reef environments. Unpublished report commissioned by the Poor Knights Management Committee of the New Zealand Ministry of Agriculture and Fisheries, Auckland, 127 pp.

Kingsford, M. J. (1988). The early life history of fish in coastal waters of northern New Zealand: a review. *N.Z. J. Mar. Freshwat. Res.* 22: 463–479.

Kingsford, M. J. (1989). Distribution patterns of planktivorous reef fish along the coast of northeastern New Zealand. *Mar. Ecol. Prog. Ser.* 54: 13–24.

Kingsford, M. J. (1990). Linear oceanographic features: a focus for research on recruitment processes. *Aust. J. Ecol.* 15: 391–401.

Kingsford, M. J. (1992). Drift algae and small fish in coastal waters of northeastern New Zealand. *Mar. Ecol. Prog. Ser.* 80: 41–55.

Kingsford, M. J. (1993). Biotic and abiotic structure in the pelagic environment: importance to small fish. *Bull. Mar. Sci.* 53: 393–415.

Kingsford, M. J. (1995). Drift algae: a contribution to nearshore habitat complexity in the pelagic environment and an attractant for fish. *Mar. Ecol. Prog. Ser.* 116: 297–301.

Kingsford, M. J., Atkinson, M. H. (1994). Increments in otoliths and scales: how they relate to the age and early development of reared and wild larval and juvenile snapper *Pagrus auratus* (Sparidae). *Aust. J. Mar. Freshwat. Res.* 45: 1007–1021.

Kingsford, M. J., Choat, J. H. (1985). The fauna associated with drift algae captured with a plankton-mesh purse-seine net. *Limnol. Oceanogr.* 30: 618–630.

Kingsford, M. J., Choat, J. H. (1986). Influence of surface slicks on the distribution and onshore movements of small fish. *Mar. Biol.* 91: 161–171.

Kingsford, M. J., Choat, J. H. (1989). Horizontal distribution patterns of presettlement reef fish: are they influenced by the proximity of reefs? *Mar. Biol.* 100: 285–297.

Kingsford, M. J., Gray, C. A. (1996). Influence of pollutants and oceanography on abundance and deformities of wild fish larvae. Pp. 233–253 *in* Schmitt, R. J., Osenberg, C. W. (eds), *Detecting ecological impacts: concepts and applications in coastal habitats.* Academic Press, Santa Barbara.

Kingsford, M. J., MacDiarmid, A. B. (1988). Interrelations between planktivorous reef fish and zooplankton in temperate waters. *Mar. Ecol. Prog. Ser.* 48: 103–117.

Kingsford, M.J., Milicich, M.J. (1987). Presettlement phase of *Parika scaber* (Pisces: Monacanthidae), a temperate reef fish. *Mar. Ecol. Prog. Ser.* 36: 65–79.

Kingsford, M. J., Schiel, D. R., Battershill, C. N. (1989). Distribution and abundance of fish in a rocky reef environment at the subantarctic Auckland Islands, New Zealand. *Polar Biol.* 9: 179–186.

Kingsford, M. J., Suthers, I. M. (1994). Dynamic estuarine plumes and fronts: Importance to small fish and plankton in coastal waters of NSW, Australia. *Continental Shelf Res.* 14: 655–672.

Kingsford, M. J., Suthers, I. M. (1996). The influence of phase of the tide on patterns of ichthyoplankton abundance in the vicinity of an estuarine front, Botany Bay, Australia. *Estuar. Cstl Shelf Sci.* 43: 33–54.

Kingsford, M. J., Suthers, I. M., Gray, C. A. (1996). Potential degradation of larval habitat and the incidence of deformities in larval fishes. *Mar. Poll. Bull.* 33: 201–212.

Kingsford, M. J., Tricklebank, K. A. (1991). Ontogeny and behavior of *Aldrichetta forsteri* (Teleostei: Mugilidae). *Copeia:* 9–16.

Kingsford, M. J., Underwood, A. J., Kennelly, S. J. (1991a). Humans as predators on rocky reefs in New South Wales, Australia. *Mar. Ecol. Prog. Ser.* 72: 1–14.

Kingsford, M. J., Wolanski, E., Choat, J. H. (1991b). Influence of tidally induced fronts and Langmuir circulations on distribution and movements of presettlement fishes around a coral reef. *Mar. Biol.* 109: 167–180.

Kiorbe, T. (1993). Turbulence, phytoplankton cell size, and the structure of pelagic food webs. *Adv. Mar. Biol.* 29: 1–72.

Kirk, R. E. (1995). *Experimental design: Procedures for the behavioural sciences.* 3rd edn. Brooks/Cole Publishing, Boston. 921 pp.

Kirkman, H. (1985). Community structure in seagrasses in southern Western Australia. *Aquat. Bot.* 21: 363–375.

Kirkman, H. (1990). Seagrass distribution and mapping. Pp. 19–25 *in* Phillips, R. C., McRoy, C. P. (eds), *Seagrass research methods* UNESCO, Paris.

Kittaka, J., MacDiarmid, A. B. (1994). Breeding. Pp. 384–401 *in* Phillips, B. F., Cobb, J. S., Kittaka, J. (eds), *Spiny lobster management.* Blackwell, Oxford. 550 pp.

Kjorsvik, E., Mangor-Jensen, A., Holmefjord, I. (1990). Egg quality in fishes. *Adv. Mar. Biol.* 26: 71–113.

Klein-Breteler, W. C. M.; Gonzalez, S. R.; Schogt, N. (1995). Development of *Psuedocalanus elongatus* (Copepoda, Calanoida) cultured at different temperature and food conditions. *Mar. Ecol. Prog. Ser.* 119: 99–110.

Kleppel, G. S. (1992). Environmental regulation of feeding and egg production by *Acartia tonsa* off southern California. *Mar. Biol.* 112: 57–65.

Kleppel, G. S. (1993). On the diets of calanoid copepods. *Mar. Ecol. Prog. Ser.* 99: 183–195.

Koslow, J. A., Brault, S., Dugas, J., Fourier, R. O., Hughes, P. (1985). Condition of larval cod (*Gadus morhua*) off southwest Nova Scotia in 1983 in relation to plankton abundance and temperature. *Mar. Biol.* 86: 113–121.

Klimly, A. P., Brown, S. T. (1983). Stereophotography for the field biologist: measurement of lengths and three-dimensional positions of free-swimming sharks. *Environ. Biol. Fish.* 12: 23–32.

Knox, G. A. (1969). *The natural history of Canterbury*. Reed, Wellington. 620pp.

Koslow, J. A., Hanley, F., Wickland, R. (1988). Effects of fishing on reef fish communities at Pedro Bank and Port Royal Cays, Jamaica. *Mar. Ecol. Prog. Ser.* 43: 201–212.

Kott, P. (1985). The Australian Ascidiacea, Part 1. Phlebobranchia and Stolidobranchia. *Mem. Qld Mus.* 23 : 1–438.

Kott, P. (1990a). The Australian Ascidiacea, Part 2. Aplousobranchia (1). *Mem. Qld Mus.* 29(1): 1–266.

Kott, P. (1990b). The Australian Ascidiacea, Part 1. Phlebobranchia and Stolidobranchia. Supplement. *Mem. Qld Mus.* 29(1):267–298.

Kott, P. (1992a). The Australian Ascidiacea, Part 3. Aplousobranchia (2). *Mem. Qld Mus.* 32(2): 375–620.

Kott, P. (1992b). The Australian Ascidiacea. Supplement 2. *Mem. Qld Mus.* 32(2): 621–655.

Krebs, C. J. (1989). *Ecological methodology*. Harper & Row, New York. 654 pp.

Krogh, M. (1994). Spatial, seasonal and biological analysis of sharks caught in the New South Wales protective beach meshing programme. *Aust. J. Mar. Freshwat. Res.* 45: 1087–1106.

Kuiter, R. H. (1993). *Coastal fishes of south-eastern Australia*. Crawford House Press, Bathhurst. 437 pp.

Kumar, S. K., Vincent, W. F., Austin, P. C., Wake, G. C. (1991). Picoplankton and marine food chain dynamics in a variable mixed-layer: a reaction-diffusion model. *Ecol. Model.* 57: 193–219.

Kuo, J., Phillips, R. C., Walker, D. I., Kirkman, H. (1996). *Seagrass biology*. Proceedings of an International Workshop, Rottnest Island, Western Australia, 25–29 January 1996. Faculty of Science, University of Western Australia, Perth. 385 pp.

Kuuppo, Leinikki, P., Kuosa, H. (1989). Preservation of picoplanktonic cyanobacteria and heterotrophic nanoflagellates for epifluorescence microscopy. *Arch. Hydrobiol.* 144: 631–636.

Laedsgaard, P., Byrne, M., Anderson, D. T. (1991). Reproduction of sympatric populations of *Heliocidaris erythrogramma* and *H. tuberculata* (Echinodermata) in New South Wales. *Mar. Biol.* 110: 359–374.

Lafferty, K. D., Kuris, M. (1996). Biological control: Recent developments. *Ecology* 77: 1989–2000.

Lagler, K. F., Bardach, J. E., Miller, R. R., Passino, D. R. (1977). *Ichthyology*. Wiley, New York. 506 pp.

Lancraft, T. M., Torres, J. J., Hopkins, T. L. (1989). Micronekton and macrozooplankton in the open waters near Antarctic Ice Edge Zones (AMERIEZ 1983 and 1986). *Polar Biol.* 9: 225–233.

Lande, R. (1996). Statistics and partitioning of species diversity, and similarity among multiple communities. *Oikos* 76: 70–82.

Lanigan, K. A. (1972). Phytoplankton from the Hauraki Gulf. *Tane* 18: 169–176.

Larkum, A. W. D., Collett, L. C., Williams, R. J. (1984). The standing stock, growth and shoot production of *Zostera capricorni* Aschers in Botany Bay, New South Wales, Australia. *Aquatic Bot.* 19: 307–327.

Larkum, A. W. D., McComb, A. J., Shepherd, S. A. (eds) (1989*). Biology of Seagrasses: a treatise on the biology of seagrasses with special reference to the Australian region.* Elsevier, Amsterdam. 265 pp.

Larkum, A. W. D., West, R. J. (1990). Long-term changes of seagrass meadows in Botany Bay, Australia. *Aquat. Bot.* 37: 55–70.

Larson, R. J. (1976). Marine flora and fauna of the northeastern United States. Cnidaria: Scyphozoa. *NOAA Tech. Rep. NMFS Circ.* 397.

Larson, R. J. (1980a). Territorial behaviour of the black and yellow rockfish and gopher rockfish (Scorpaenidae, *Sebastes*). *Mar. Biol.* 58: 111–122.

Larson, R. J. (1980b). Influence of territoriality on adult density in two rockfishes of the genus *Sebastes*. *Mar. Biol.* 58: 123–132.

Larson, R. J. (1980c). Competition, habitat selection, and the bathymetric segregation of two rockfish (*Sebastes*) species. *Ecol. Monogr.* 50: 221–239.

Larson, R. J., DeMartini, E. E. (1984). Abundance and vertical distribution of fishes in a cobble-bottom kelp forest off San Onofre, California. *Fish. Bull. U.S.* 82: 37–53.

Lasker, R. (1975). Field criteria for survival of anchovy larvae: the relation between inshore chlorophyll maximum layers and successful first feeding. *Fish. Bull. U.S.* 73: 453–463.

Lassig, B., Engelhardt, U. (1994). COTS COMMS. *Reef Research*. Newsletter of the Research and Monitoring Section of the Great Barrier Reef Marine Park Authority 4, 17–24.

Last, P. R., Stevens, J. D. (1994). *Sharks and Rays of Australia*. CSIRO, Melbourne. 513 pp.

Lazzaro, X. (1987). A review of planktivorous fishes: their evolution, feeding behaviours, selectivities, and impacts. *Hydrobiologia* 146: 97–167.

Leatherwood, S., Reeves, R. R., Perris, W. F., Evans, W. E. (1988). *Whales, dolphins, and porpoises of the eastern North Pacific and adjacent arctic waters*. Dover Publications, New York. 245 pp.

Lechowicz, M. J. (1982). The sampling characteristics of electivity indices. *Oecologia* 52: 22–30.

Lee Long, W. J., Mellors, J. E., Cole, R. G. (1993). Seagrasses between Cape York and Hervey Bay, Queensland, Australia. *Aust. J. Mar. Freshwat. Res.* 44: 19–31.

Lee, J. J., Hutner, S. H., Bovee, E. C. (1985). *An illustrated guide to the Protozoa*. Allen Press, Laurence. 629 pp.

Legendre, L., Demers, S. (1984). Towards dynamic biological oceanography. *Can. J. Fish. Aquat. Sci.* 41: 2–19.

Legendre, P., Fortin, M-J. (1989). Spatial pattern and ecological analysis. *Vegetatio*. 80: 107–138.

Legendre, L., Legendre, P. (1994). *Numerical ecology*. 2nd English edition, Elsevier, Amsterdam. 585 pp.

Legendre, P. (1993). Spatial autocorrelation: trouble of new paridigm? *Ecology* 74: 1659–1674.

Legendre, P., Fortin, M-J. (1989). Spatial pattern and ecological analysis. *Vegetatio* 80: 107–138.

Leggett, W. C., Deblois, E. (1994). Recruitment in marine fishes: is it regulated by starvation and predation in the egg and larval stages? *Neth. J. Sea Res.* 32: 119–134.

Lehodey, P., Grandperrin, R., Marchal, P. (1997). Reproductive biology and ecology of the deep-demersal fish alfonsino *Berynx splendens*, over seamounts off New Caledonia. *Mar. Biol.* 128: 17–27.

Leis, J. M. (1986). Vertical and horizontal distribution of fish larvae near coral reefs at Lizard Island, Great Barrier Reef. *Mar. Biol.* 90: 505–516.

Leis, J. M. (1991). The pelagic stage of reef fishes: the larval biology of coral reef fishes. Pp. 183–230 *in* Sale, P. F. (ed.), *The ecology of fishes on coral reefs*. Academic Press, San Diego.

Leis, J. M., Carson-Ewart, B. M. (1997). In situ swimming speeds of the late pelagic larvae of some Indo-Pacific coral-reef fishes. *Mar. Ecol. Prog. Ser.* 159: 165–174.

Leis, J. M., Rennis, D. S. (1983). *The larvae of Indo-Pacific coral reef fishes*. UNSW Press, Sydney. 269 pp.

Leis, J. M., Sweatman, P. A., Reader, S. E. (1996). What pelagic stages of coral reef fishes are doing in blue water: daytime field observation of larval behavioural capablities. *Mar. Freshwat. Res.* 47: 401–412.

Leis, J. M., Trnski, T. (1989). *The larvae of Indo-Pacific shorefishes*. UNSW Press, Sydney. 371 pp.

Lemmens, J. W. T. J. (1994). Biochemical evidence for absence of feeding in puerulus larvae of the Western rock lobster *Panuliris cygnus* (Decapoda: Palinuridae). *Mar. Biol.* 118: 383–392.

Leonard, G. H., Clark, R. P. (1993). Point quadrat versus transect estimates of the cover of benthic red algae. *Mar. Ecol. Prog. Ser.* 101: 203–208.

Lepage, R., Billard, L. (1992). *Exploring the limits of bootstrap*. Wiley, New York.

Lessios, H. A. (1988). Mass mortality of *Diadema antillarum* in the Caribbean: what have we learned? *Ann. Rev. Ecol. Syst.* 19: 371–393.

Leum, L., Choat, J. H. (1980). Density and distribution patterns of the temperate marine fish *Cheilodactylus spectabilis* (Cheilodactylidae) in a reef environment. *Mar. Biol.* 57: 327–337.

Levin, L. A. (1990). A review of methods for labelling and tracking marine invertebrate larvae. *Ophelia* 32: 115–144.

Levin, P. S. (1991). Effects of microhabitat on recruitment variation in a Gulf of Maine reef fish. *Mar. Ecol. Prog. Ser.* 75: 183–189.

Levin, P. S. (1993). Habitat structure, conspecifc presence and spatial variation in the recruitment of a temperate reef fish. *Oecologia* 94: 176–185.

Levin, P. S. (1994). Small-scale recruitment variation in temperate fish: the roles of macrophytes and food supply. *Environ. Biol. Fish.* 40: 271–281.

Levin, P. S., Hay, M. E. (1996). Responses of temperate reef fishes to alterations in algal structure and species composition. *Mar. Ecol. Prog. Ser.* 134: 37–47.

Levin, S. A. (1992). The problem of pattern and scale in ecology. *Ecology* 73: 1943–1983.

Lewis, F. G., Stoner, A. W. (1981). An examination of methods for sampling macrobenthos in seagrass meadows. *Bull. Mar. Sci.* 31: 116–129.

Lewis, K. B. (1994). New maps of the seabed. *Wat. Atmos.* 2: 8–10.

Liang, D., Uye, S. (1996). Population dynamics and production of the planktonic copepods in a eutrophic inlet of the inland sea of Japan. II. *Acartia omorii. Mar. Biol.* 125: 109–118.

Lincoln, R. J., Sheals, J. (1979). *Invertebrate animals: collection and preservation*. British Museum, London. 150 pp.

Lincoln-Smith, M. P. (1988a). Improving multispecies rocky reef fish censuses by counting different groups of species using different procedures. *Environ. Biol. Fish.* 26: 29–37.

Lincoln-Smith, M. P. (1988b). Effects of observer swimming speed on sample counts of temperate rocky reef fish assemblages. *Mar. Ecol. Prog. Ser.* 43: 223–231.

Lincoln-Smith, M. (1998). Guidelines for assessment of aquatic ecology in EIA. New South Wales, Australia. Department of Urban Affairs and Planning, 80 pp.

Lincoln-Smith, M. P., Bell, J. D., Hair, C. A. (1991). Spatial variation in abundance of recently settled rocky reef fish in southeastern Australia: Implications for detecting change. *Mar. Ecol. Prog. Ser.* 77: 95–103.

Lindquist, N., Hay, M. E. (1996). Palatability and chemical defense of marine invertebrate larvae. *Ecol. Monogr.* 66: 431–451.

Littler, M. M. (1980). Southern California rocky intertidal ecosystems: Methods, community structure and variability. Pp. 565–608 *in* Price, J. H., Irvine, D. E. G., Farnham, W. F. (eds), *The shore environment, Vol. 2: Ecosystems*. Academic Press, London.

Littler, M. M., Murray, S. N. (1975). Impact of sewage on the distribution, abundance and community structure of rocky intertidal macro-organisms. *Mar. Biol.* 30: 277–291.

Lively, C. M., Raimondi, P. T. (1987). Desiccation, predation, and mussel–barnacle interactions in the northern Gulf of California. *Oecologia* 74: 304–309.

Lively, C. M., Raimondi, P. T., Delph, L. F. (1993). Intertidal community structure: Space-time interactions in the northern Gulf of California. *Ecology* 74: 162–173.

Lloyd, M. (1967). Mean crowding. *J. Anim. Ecol.* 36: 1–30.

Lobban, C. S., Harrison, P. J. (1996). *Seaweed ecology and physiology*. Cambridge University Press, Cambridge. 384 pp.

Long, B. G., Skewes, T. D., Poiner, I. R. (1994). An efficient method for estimating seagrass biomass. *Aquat. Bot.* 47: 277–291.

Lotrich, V. A., Meredith, W. H. (1974). A technique and the effectiveness of various acrylic colors for subcutaneous marking of fish. *Trans. Am. Fish. Soc.* 103: 140–142.

Lough, R. G., Potter, D. C. (1983). Rapid shipboard identific-ation and enumeration of zooplankton samples. *J. Plankton Res.* 5: 755–782.

Love, M. R. (1991). *Probably more than you want to know about the fishes of the Pacific Coast*. Really Big Press, Santa Barbara. 215 pp.

Love, M. S., Brooks, A., Busatto, D., Stephens, J., Gregory, P. A. (1996). Aspects of the life histories of the kelp bass, *Paralabrax clathratus*, and barred sand bass, *P. nebulifer*, from the southern California Bight. *Fish. Bull.* 94: 472–481.

Love, M. S., Carr, M. H., Haldorson, L. J. (1991). The ecology of substrate-associated juveniles of the genus *Sebastes*. *Environ. Biol. Fish.* 30: 225–243.

Lowe, I. (1996). *Australia: state of the environment 1996*. CSIRO Publishing, Collingwood. 533 pp.

Lubchenco, J., Gaines, S. D. (1981). Unified approach to marine plant-herbivore interactions. I. Populations and communities. *Ann. Rev. Ecol. Syst.* 12: 405–437.

Luckens, P. A. (1976). Settlement and succession on rocky shores at Auckland, North Island, New Zealand. *NZOI Memoir.* 70: 64.

Luckens, P. A. (1991). Distribution, growth rate, and death from octopod and gastropod predation of *Tawera bollonsi* (Bivalvia: Veneridae) at the Auckland Islands. *N.Z. J. Mar. Freshwat. Res.* 25: 255–268.

Luckhurst, B. E., Luckhurst, K. (1978). Analysis of the influence of substratum variables on coral reef communities. *Mar. Biol.* 49: 317–323.

Luoma, S. N., Cain, D. J., Brown, C., Axtman, E. V. (1991). Trace metals in clams (*Macoma balthica*) and sediments at the Palo Alto mudflat in south San Francisco Bay; April 1990–April 1991. *U.S. Geol. Surv. Rep.* 91–460. 47 pp.

Lutschinger, S. (1993). The marine fauna of New Zealand: Chaetognatha (Arrow Worms). *Mem. N.Z. Oceanogr. Inst.* 101: 61 pp.

Lythgoe, J., Lythgoe, G. (1991). *Fishes of the sea: the North Atlantic and Mediterranean.* MIT Press, Cambridge, Mass. 256 pp.

McAllister, M. K., Pikitch, E. K., Punt, A. E., Hilborn, R. (1994). A baysian approach to stock assessment and harvest decisions using the sampling importance resampling algorithm. *Can. J. Fish. Aquat. Sci.* 51: 2673–2687.

McArdle, B. H., Blackwell, R. G. (1989). Measurement of density variability in the bivalve *Chione stutchburyi* using spatial autocorrelation. *Mar. Ecol. Prog. Ser.* 52: 245–252.

McArdle, B. H., Pawley, M. D. M. (1994). Cost benefit analysis in the design of biological monitoring programs: is it worth the effort? Pp. 237–253 *in* Fletcher, D. J., Manly, B. F. J. (eds), *Statistics in ecology and environmental monitoring,* University of Otago Press, Dunedin.

Machii, T., Nose, Y. (1990). Mechanical properties of a rectangular purse seine in a normal operation. *Nip. Sui. Gakk.* 56: 173–178.

McClatchie, S., Millar, R. B., Webster, F., Lester, P. J., Hurst, R., Bagley, N. (1996). Demersal fish community diversity off New Zealand is related to depth, latitude and regional surface phytoplankton. *Deep Sea Res.* 44: 647–668.

McClintock, J. B. (1987). Investigation of the relationship between invertebrate predation and biochemical composition, energy content, spicule armament and toxicity of benthic sponges at McMurdo Sound, Antarctica. *Mar. Biol.* 94: 479–487.

McCloskey, L. R., Muscatine, L., Wilkerson, F. P. (1994). Daily photosynthesis, respiration and carbon budgets in a tropical marine jellyfish (*Mastigias* sp.). *Mar. Biol.* 119: 13–22.

McCormick, M. I. (1989). Spatio-temporal patterns in the abundance and population structure of a large temperate reef fish. *Mar. Ecol. Prog. Ser.* 53: 215–225.

McCormick, M. I. (1990). Handbook for stock assessment of agar seaweed *Pterocladia lucida*; with a comparison of survey techniques. *N.Z. Fish. Tech. Rep.* 24: 36.

McCormick, M. I. (1994a). Variability in age and size at settlement of the tropical goatfish *Upeneus tragula* (Mullidae) in the northern Great Barrier Reef Lagoon. *Mar. Ecol. Prog. Ser.* 103: 1–15.

McCormick, M. I. (1994b). Comparison of field methods for measuring surface topography and their associations with a tropical reef fish assemblage. *Mar. Ecol. Prog. Ser.* 112: 87–96.

McCormick, M. I., Choat, J. H. (1987). Estimating total abundance of a large temperate-reef fish using visual strip transects. *Mar. Biol.* 96: 469–478.

McCormick, T. B., Herbinson, K., Mill, T. S., Altick, J. (1994). A review of abalone seeding, possible significance and a new seeding device. *Bull. Mar. Sci.* 55: 680–693.

MacDiarmid, A. B. (1981). Factors influencing the distribution and abundance of two temperate planktivorous reef fish, *Pempheris adspersa* and *Scorpis violaceus*. Unpublished MSc thesis, University of Auckland. 101 pp.

MacDiarmid, A. B. (1985). Sunrise release of larvae from the palinurid rock lobster *Jasus edwardsii*. *Mar. Ecol. Prog. Ser.* 21: 313–315.

MacDiarmid, A. B. (1987). The ecology of *Jasus edwardsii* (Hutton) (Crustacea: Palinuridae). Unpublished PhD thesis, University of Auckland. 169 pp.

MacDiarmid, A. B. (1988). Experimental confirmation of external fertilization in the southern temperate rock lobster *Jasus edwardsii* (Hutton) (Decapoda, Palinuridae). *J. Exp. Mar. Biol. Ecol.* 120: 277–285.

MacDiarmid, A. B. (1989a). Moulting and reproduction of the spiny lobster *Jasus edwardsii* (Decapoda: Palinuridae). *Mar. Biol.* 103: 303–310.

MacDiarmid, A. B. (1989b). Size at onset of maturity and size-dependent reproductive output of female and male spiny lobsters *Jasus edwardsii* (Hutton) (Decapoda, Palinuridae) in northern New Zealand. *J. Exp. Mar. Biol. Ecol.* 127: 229–243.

MacDiarmid, A. B. (1991). Seasonal changes in depth distribution, sex ratio and size frequency of spiny lobsters *Jasus edwardsii* (Decapoda, Palinuridae) on a coastal reef in northern New Zealand. *Mar. Ecol. Prog. Ser.* 70: 129–141.

MacDiarmid, A. B. (1994). Cohabitation in the spiny lobster *Jasus edwardsii* (Hutton, 1897). *Crustaceana* 66: 341–355.

MacDiarmid, A. B., Breen, P. A. (1993). Spiny lobster population change in a marine reserve. *Proc. 2nd Int. Temperate Reef Symp.*: 47–56.

MacDiarmid, A. B., Hickey, B., Maller, R. A. (1991). Daily movement patterns of spiny lobster *Jasus edwardsii* (Hutton) on a shallow reef in northern New Zealand. *J. Exp. Mar. Biol. Ecol.* 147: 185–205.

McDonald, L. L., Erickson, W. P. (1994). Testing for bioequivalence in field studies: has a disturbed site been adequately reclaimed? Pp. 183–197 *in* Fletcher, D. J., Manly, B. F. J. (eds), *Statistics in ecology and environmental monitoring,* University of Otago Press, Dunedin.

McDowall, R.M. (1990). *New Zealand freshwater fishes: a natural history guide.* Heinemann Reed, Auckland. 553pp.

McDowall, R. M. (1996). *Freshwater fishes of southeastern Australia.* Reed, Sydney. 247 pp.

McFarland, W. N., Kotchian, N. M. (1982). Interactions between schools of fish and mysids. *Behav. Ecol. Sociobiol.* 11: 71–76.

McGuinness, K. A. (1987). Disturbance and organisms on boulders. I. Patterns in the environment and the community. *Oecologia* 71: 409–419.

McGuinness, K. A. (1988). Explaining patterns in abundance of organisms on boulders: the failure of 'natural experiments'. *Mar. Ecol. Prog. Ser.* 48: 199–204.

McGuinness, K. A. (1989). Effects of some natural and artificial substrata on sessile marine organisms at Galeta Reef, Panama. *Mar. Ecol. Prog. Ser.* 52: 201–208.

McGuinness, K. A. (1990). Effects of oil spills on macroinvertebrates of saltmarshes and mangrove forests in Botany Bay, New South Wales. *J. Exp. Mar. Biol. Ecol.* 142: 121–135.

McGuinness, K. A., Underwood, A. J. (1986). Habitat structure and the nature of communities on intertidal boulders. *J. Exp. Mar. Biol. Ecol.* 104: 97–123.

McIntyre, A. D., Warwick, R. M. (1984). Meiofauna techniques. Pp. 217–244 *in* Holme, N. A., McIntyre, A. D. (eds), *Methods for the study of marine benthos,* 2nd edition, Blackwell, Oxford. 387 pp.

MacKenzie, L. (1991). Toxic and noxoius phytoplankton in Big Glory Bay, Stewart Island, New Zealand. *J. Appl. Phycology.* 3: 19–34.

McKillup, S. C., Butler, A. J. (1979). Cessation of hole-digging by the crab *Helograpsus haswellianus*: a resource-conserving adaptation. *Mar. Biol.* 50: 157–161.

McKoy, J. L., Esterman, D. B. (1981). Growth of rock lobsters (*Jasus edwardsii*) in the Gisborne region, New Zealand. *N.Z. J. Mar. Freshwat. Res.* 15: 121–136.

McKoy, J. L., Leachman, A. (1982). Aggregations of ovigerous female rock lobsters *Jasus edwardsii* (Decapoda: Palinurudae). *N.Z. J. Mar. Freshwat. Res.* 16: 141–146.

McLachlan, A. (1983). Sandy beach ecology – a review. Pp. 321–380 *in* McLachlan, A., Erasmus, T. (eds), *Sandy beaches as ecosystems*. Junk, The Hague.

McLachlan, A., Jamarillo, E. (1995). Zonation on sandy beaches. *Oceanogr. Mar. Biol. Ann. Rev.* 33: 305–336.

McLachlan, A., Dugan, J. E., Defeo, O., Ansell, A. D., Hubbard, D. M., Taramillo, E., Penchazadoh, P. E. (1996). Beach clam fisheries. *Oceanogr. Mar. Biol. Ann. Rev.* 34: 163–232.

McLaughlin, R. H., O'Gower, A. K. (1971). Life history and underwater studies of a heterodont shark. *Ecol. Monogr.* 41: 271–289.

McLay, C. L. (1988). Crabs of New Zealand. *Leigh Lab. Bull.* 22. 463 pp.

McLean, C., Miskiewicz, A. G., Roberts, E. A. (1991). Effect of three primary treatment sewage outfalls on metal concentrations in the fish *Cheilodactylus fuscus* collected along the coast of Sydney, Australia. *Mar. Pollut. Bull.* 22: 134–140.

McNeill, S. E. (1994). The selection and design of marine protected areas: Australia as a case study. *Biodiversity Conserv.* 3: 586–605.

McNeill, S.E., Bell, J.D. (1992). Comparison of beam trawls for sampling macrofauna of *Posidonia* seagrass. *Estuaries* 15: 360-367.

McNeill, S. E., Worthington, D. G., Ferrell, D. J., Bell, J. D. (1992). Consistently outstanding recruitment of five species of fish to a seagrass bed in Botany Bay, NSW. *Aust. J. Ecol.* 17: 359–365.

MacPherson, K., Biagi, F., Francour, P., Garcia-Rubies, A. G., Harmelin, J., Harmelin-Vivien, L., Jouvenel, J. Y., Planes, S., Viliola, L., Tunesi, L. (1997). Mortality of juvenile fishes of the genus *Diplodus* in protected and unprotected areas in the western Mediterranean Sea. *Mar. Ecol. Prog. Ser.* 160: 135–147.

McQuaid, C. D. (1996). Family Littorinidae II role in the ecology of intertidal shallow marine ecosystems. *Oceanogr. Mar. Biol. Ann. Rev.* 34: 263–302.

McShane, P. E. (1994). Estimating the abundance of stocks of abalone (*Haliotis* spp.): Examples from Victoria and New Zealand. *Fish. Res.* 19: 379–394.

McShane, P.E. (1995). Estimating the abundance of abalone: the importance of patch size. *Mar. Freshwat. Res.* 46: 657–662

McShane, P. E. (1996). Recruitment processes in abalone (*Haliotis* spp.). Pp. 315–324 *in* Watanabe, Y., Yamashita, Y., Oozeki, Y. (eds), *Survival strategies in early life stages of marine resources*. A. A. Balkema, Rotterdam.

McShane, P. E., Naylor, J. R. (1991). A survey of kina populations (*Evechinus chloroticus*) in Dusky Sound and Chalky Inlet, southwestern New Zealand. *N.Z. Fish. Assmt Res. Doc.* 91/17: 21 pp.

McShane, P. E., Naylor, J. R., Anderson, O. A., Gerring, P., Stewart, R. (1993). Pre-fishing surveys of kina (*Evechinus chloroticus*) in Dusky Sound, southwestern New Zealand. *N.Z. Fish. Assmt Res. Doc.* 93/8: 28 pp.

McShane, P. E., Smith, M. G. (1988). Measuring abundance of juvenile abalone *Haliotis rubra* Leach (Gastropoda: Haliotidae) : Comparison of a novel method with two other methods. *Aust. J. Mar. Freshwat. Res.* 39: 331–336.

McWilliam, P. S., Sale, P. F., Anderson, D. T. (1981). Seasonal changes in resident zooplankton sampled by emergence traps in One Tree Lagoon, Great Barrier Reef. *J. Exp. Mar. Biol. Ecol.* 52: 185–203.

Maddock, L., Taylor, F. J. (1984). The application of multivariate statistical methods to phytoplankton counts from the Hauraki Gulf, New Zealand. *N.Z. J. Mar. Freshwat. Res.* 18: 309–322.

Madhupratap, M., Nehring, S., Lenz, J. (1996). Resting eggs of zooplankton (Copepoda and Cladocera) from Kiel Bay and adjacent waters (southwestern Baltic). *Mar. Biol.* 125: 77–88.

Maller, R. A., de Boer, E. S. (1988). An analysis of two methods of fitting the von Bertalanffy curve to capture-recapture data. *Aust. J. Mar. Freshwat. Res.* 39: 459–466.

Manly, B. F. J. (1985a). A test of Jackson's method for separating death and emigration with mark-recapture data. *Res. Popul. Ecol.* 27: 99–109.

Manly, B. F. J. (1985b). *The statistics of natural selection on animal populations*. Chapman & Hall, London. 484 pp.

Manly, B. F. J. (1991). *Randomisation and Monte Carlo methods in biology*. Chapman & Hall, New York. 281 pp.

Manly, B. F. J. (1994a). CUSUM methods for detecting changes in monoitoring environmental variables. Pp. 235–238 *in* Fletcher, D. J., Manly, F. J. (eds), *Statistics in ecology and environmental monitoring*. University of Otago Press, Dunedin.

Manly, B. F. J. (1994b). *Multivariate statistical methods: a primer*. Chapman & Hall, London. 215 pp.

Manly, B. F. J. (1997). *Randomisation, bootstrap and Monte Carlo methods in biology*. Chapman & Hall, London. 399 pp.

Mann, K. H., Lazier, J. R. N. (1991). *Dynamics of marine ecosystems: biological–physical interactions in the ocean*. Blackwell, Oxford. 466 pp.

Mansueti, R. (1963). Symbiotic behaviour between small fishes and jellyfishes, with new data on that between the stromateid, *Peprilus alepidotus*, and the Scyphomedusa, *Chrysaora quinquecirrha*. *Copeia*: 40–80.

Mapstone, B. D. (1995). Scalable decision rules for environmental impact studies: Effect size, Type I, and Type II errors. *Ecol. Appl.* 5: 410–410.

Mapstone, B. D., Ayling, A. M. (1993). An investigation of optimum methods and unit sizes for the visual estimation of abundances of some coral reef organisms. A report to the Great Barrier Reef Marine Park Authority. 71 pp.

Marcus, N. H. (1995). Seasonal study of planktonic copepods and their benthic resting eggs in northern California coastal waters. *Mar. Biol.* 123: 459–466.

Margulis, L., Corliss, J. O., Melkonian, M., Chapman, D. J. (1990). *Handbook of Protoctista: the structure, cultivation, habitats and life histories of the eukaryotic microorganisms and their descendants exclusive of animals, plants and fungi*. Jones & Bartlett Publishers. Boston. 914 pp.

Marliave, J. B. (1986). Lack of planktonic dispersal of rocky intertidal fish larvae. *Trans. Am. Fish. Soc.* 115: 149–154.

Marshall, P. A., Keough, M. J. (1994). Assymmetry in intra-specific competion in the limpet *Cellana tramoserica* (Sowerby). *J. Exp. Mar. Biol. Ecol.* 177: 121–138.

Martinussen, I., Thingstad, T. F. (1991). A simple double staining technique for simultaneous quantification of auto- and heterotrophic nano- and picoplankton. *Mar. Microbiol. Food Webs* 5: 5–11.

Masuda, H., Amaoka, K., Araga, C., Uyeno, T., Yoshino, T. (1984). *The fishes of the Japanese Archipelago*. Tokai University Press, Tokyo. 448 pp.

Matarese, A. C., Kendall Jr., A. W., Blood, D. M., Vinter, B. M. (1989). Laboratory guide to early life history stages of Northeast Pacific fishes. *NOAA Tech. Mem.* 650 pp.

Matthews, K. R. (1990). An experimental study of the habitat preferences and movement patterns of copper, quillback and brown rockfishes (*Sebastes* spp.). *Environ. Biol. Fish.* 29: 161–178.

Mazumder, A., Taylor, W. D., McQueen, D. J., Lean, D. R. S. (1990). Effects of fish and plankton on lake temperature and mixing level. *Science* 247: 312–315.

Meadows, P. S., Meadows, A. (1991). The geotechnical and geochemical implications of bioturbation in marine sedimentary ecosystems. Pp. 157–181 *in* Meadows, P. S., Meadows, A. (eds), *The environmental impact of burrowing animals and animal burrows*. Clarendon Press, Oxford.

Meekan, M. G., Choat, J. H. (1997). Latitudinal variation in abundance of herbivorous fishes: a comparison of temperate and tropical reefs. *Mar. Biol.* 128: 373–384.

Meekan, M. G., Fortier, L. (1996). Selection for fast growth during the larval life of Atlantic cod *Gadus morhua* on the Scotian Shelf. *Mar. Ecol. Prog. Ser.* 137: 25–37.

Meekan, M. G., Milicich, M. J., Doherty, P. J. (1993). Larval production drives temporal patterns of larval supply and recruitment of a coral reef damselfish. *Mar. Ecol. Prog. Ser.* 93: 217–225.

Meinkoth, N. A. (1994). *The Audubon Society: Field guide to North American sea shore creatures*. Knopf, New York. 813 pp.

Mellors, J. E. (1991). An evaluation of a rapid visual technique for estimating seagrass biomass. *Aquat Bot.* 42: 67–73.

Menge, B. A. (1991). Generalizing from experiments: is predation strong or weak in the New England rocky intertidal? *Oecologia.* 88: 1–8.

Menge, B. A., Berlow, E. L., Blanchette, C. A. (1994). The keystone species concept: variation in interaction strength in a rocky intertidal habitat. *Ecol. Monogr.* 64: 249–286.

Merrick, J. R., Schmida, G. E. (1984). *Australian Freshwater Fishes: Biology and management*. Griffin Press, Netley. 409 pp.

Middleton, J. H., Cox, D., Tate, P. (1996). The oceanography of the Sydney region. *Mar. Poll. Bull.* 33: 124–131.

Milicich, M. J. (1994). Dynamic coupling of reef fish replenishment and oceanographic processes. *Mar. Ecol. Prog. Ser.* 110: 135–144.

Milicich, M. J., Doherty, P. J. (1994). Larval supply of coral reef fish populations: Magnitude and synchrony of replenishment to Lizard Island, Great Barrier Reef. *Mar. Ecol. Prog. Ser.* 110: 121–134.

Millar, A. J. K., Kraft, G. T. (1993). Catalogue of marine and freshwater red algae (Rhodophyta) of New South Wales, including Lord Howe Island, southwestern Pacific. *Aust. Syst. Bot.* 6: 1–90.

Millar, A. J. K., Kraft, G. T. (1994). Catalogue of marine benthic green algae (Chlorophyta) of New South Wales, including Lord Howe Island, southwestern Pacific. *Aust. Syst. Bot.* 7: 419–453..

Miller, C. D., Judkins, D. C. (1981). Design of pumping systems for sampling zooplankton, with descriptions of two high-capacity samplers for coastal studies. *Biol. Oceanogr.* 1: 29–56.

Miller, J. W., Tucker, J. W. (1979). X-radiography of larval and juvenile fishes. *Copeia*: 535–538.

Miller, J. Z. (1993). Factors determining community structure within patches of *Mesochaetopterus sagittarius* (Polychaeta: Chaetopteridae). Unpublished MSc thesis, University of Sydney.

Miller, M. W., Hay, M. E. (1996). Coral-seaweed-grazer-nutrient interactions on temperate reefs. *Ecol. Monogr.* 66: 323–344.

Miller, P. J. (1979). *Fish Phenology: anabolic adaptiveness in teleosts*. Academic Press, London. 447 pp.

Miller, R. J. (1983). Considerations for conducting field experiments with baited traps. *Fisheries* 8: 14–17.

Miller, R. J. (1990). Effectiveness of crab and lobster traps. *Can. J. Fish. Aquat. Sci.* 47: 1228–1251.

Miller, R. J., Hunte, W. (1987). Effective area fishes by Antillean fish traps. *Bull. Mar. Sci.* 40: 484–493.

Minchin, P. R. (1987). An evaluation of the relative robustness of techniques for ecological ordination. *Vegetatio.* 69: 89–107.

Miskiewicz, A. G. (1992). *Assessment of the distribution, impacts and bioaccumulation of contaminants in aquatic environments*. Proceedings of a Bioaccumulation Workshop. Water Board and Australian Marine Sciences Assn, Sydney. 334 pp.

Mitchell, C., Hunter, J. R. (1970). Fishes associated with drifting kelp, *Macrocystis pyrifera*, off the coast of southern California and northern Baja California. *Calif. Dept. Fish. Game Bull.* 56: 288–297.

Mittelbach, G. G., Turney, A. M., Hall, D. J., Rettig, J. E., Osenberg, W. (1995). Perturbation and resilience: a long-term, whole-lake study of predator extinction and reintroduction. *Ecology* 76: 2347–2360.

Moller, H. (1984). Reduction of a larval herring population by a jellyfish predator. *Science* 224: 621-622.

Moltschaniwskyj, N. A., Doherty, P. J. (1995). Cross-shelf distribution patterns of tropical cephalopods sampled with light traps. *Mar. Freshwat. Res.* 46: 707–714.

Montgomery, D. C. (1991). *Design and analysis of experiments*. Wiley, New York. 649 pp.

Mooi, R., Telford, M. (1998). *Echinoderms: San Francisco*. A. A. Balkema, Rotterdam. 923 pp.

Moore, M. I., Murdoch, R. C. (1993). Physical and biological observations of coastal squirts under non-upwelling conditions. *J. Geophys. Res.* 98: 20,043–20,061.

Morales, C. E., Bautista, B., Harris, R. P. (1990). Estimates of ingestion in copepod assemblages: Gut fluorescence in relation to body size. Pp. 565–577 *in* Barnes, M., Gibson, R. N. (eds), *Trophic relationships in the marine environment*. Aberdeen University Press.

Morales, C. E., Bedo, A., Harris, R. P., Tranter, P. R. G. (1991). Grazing of copepod assemblages in the north-east Atlantic: the importance of small size fraction. *J. Plankton Res.* 13: 455–472.

Moran, A. L. (1997). Spawning and development of the black turban snail *Tegula funebralis* (Prosobranchia: Trochidae). *Mar. Biol.* 128: 107–114.

Moran, M. J., Sale, P. F. (1977). Seasonal variation in territorial response, and other aspects of the ecology of the Australian temperate pomacentrid fish *Parma microlepis*. *Mar. Biol.* 39: 121–128.

Moran, P. J., Johnson, D. B., Miller-Smith, B. A., Mundy, C. N., Bass, D. K., Davidson, J., Miller, I. R., Thompson, A. A. (1989). *A guide to the AIMS manta tow technique*. Australian Institute of Marine Science, Townsville. 20 pp.

Moreno, C. A., Jara, H. F. (1984). Ecological studies on fish fauna associated with *Macrocytis pyrifera* belts in the Fueguian Islands, Chile. *Mar. Ecol. Prog. Ser.* 15: 99–107.

Moreno, C. A., Sutherland, J. P., Jara, H. F. (1984). Man as a predator in the intertidal zone of southern Chile. *Oikos* 42: 155–160.

Morgan, G. R. (1974). Aspects of the population dynamics of the western rock lobster, *Palinurus cygnus*, George. II. Seasonal changes in the catchability coefficient. *Aust. J. Mar. Freshwat. Res.* 25: 249–259.

Morris, R. H., Abbot, D. P., Haderlie, E. C. (1980). *Intertidal invertebrates of California*. 2nd edn. Stanford University Press. 626 pp.

Morris, R. J., Bone, Q., Head, R., Braconnot, J. C., Nival, P. (1988). Role of salps in the flux of organic matter to the bottom of the Ligurian Sea. *Mar. Biol.* 97: 237–241.

Morrisey, D. J., Howitt, L., Underwood, A. J., Stark, J. S. (1992a). Spatial variation in soft-sediment benthos. *Mar. Ecol. Prog. Ser.* 81: 197–204.

Morrisey, D. J., Stark, J. S., Howitt, L., Underwood, A. J. (1994a). Spatial variation in concentrations of heavy metals on marine sediments. *Aust. J. Mar. Freshwat. Res.* 45: 177–184.

Morrisey, D. J., Underwood, A. J. Howitt, L., Stark, J. S. (1992b). Temporal variation in soft sediment benthos. *J. Exp. Mar. Biol. Ecol.* 164: 235–245.

Morrisey, D. J., Underwood, A. J., Howitt, L. (1996). Effects of copper on the faunas of marine soft sediments: an experimental field study. *Mar. Biol.* 125: 199–213.

Morrisey, D. J., Underwood, A. J., Stark, J. S., Howitt, L. (1994b). Temporal variation in concentrations of heavy metals in marine sediments. *Estuar. Cstl Shelf Sci.* 38: 271–282.

Morse, D. E. (1990). Recent progress in larval settlement and metamophosis: closing the gaps between molecular biology and ecology. *Bull. Mar. Sci.* 46: 465–483.

Morton, J., Miller, M. (1973). *The New Zealand sea shore*. Collins, Auckland. 653 pp.

Moser, H. G. (1996). *The early stages of fishes in the California Current region*. Allen Press, Kansas. 1505 pp.

Moser, H. G., Richards, W. J., Cohen, D. M., Fahay, M., Kendall, J. A. W., Richardson, S. L. (1984). *Ontogeny and systematics of fishes*. Allen Press, Lawrence. 760 pp.

Mousseau L., Dauchez S., Legendre L., Fortier L. (1995). Photosynthetic carbon uptake by marine phytoplankton: Comparison of the stable (^{13}C) and radioactive (^{14}C) isotope methods. *J. Plankton Res.* 17:1449–1460.

Mulhearn, P. J. (1987). The Tasman front: a study using satellite infrared imagery. *J. Phys. Oceanogr.* 17: 1148–1155.

Mundy, C. N. (1990). Field and laboratory investigations of the line intercept transect technique for monitoring the effects of the crown-of-thorns starfish, *Acanthaster planci*. Australian Institute of Marine Science, Townsville. 42 pp.

Munday, P. L., Wilson, S. K. (1997). Comparative efficacy of clove oil and other chemicals in anaesthetization of *Poma-*

centrus ambionensis, a coral reef fish. *J. Fish. Biology.* 51: 931–938.

Munk, P. (1988). Catching large herring larvae: gear applicability and larval distribution. *J. Cons. int. Explor. Mer.* 45: 97–104.

Munro, M. H. G., Blunt, J. W., Blake, R. J., Litaudon, M., Battershill, C. N., Page, M. J. (1993). From seabed to sick bed: what are the prospects? Pp. 473–484 *in* Van Soost, R., Brekman, R. (eds), *Sponges in time and space*. Proceedings of the 4th International Symposium on Sponge Biology. Balkema, Amsterdam.

Murdoch, A., MacKnight, S. D. (1991). *CRC handbook of techniques for aquatic sediment sampling*. CRC Press, Boca Raton. 210 pp.

Murdoch, R.C. (1989). The effects of a headland eddy on surface macro-zooplankton assemblages north of Otago Peninsula, New Zealand. *Estuar. Cstl Shelf Sci.* 29: 361–383.

Murdoch, R. C., Chapman, B. E. (1989). Occurrence of hoki (*Macruronus novaezelandiae*) eggs and larvae in eastern Cook Strait. *N.Z. J. Mar. Freshwat. Res.* 23: 61–67.

Murdoch, R. C., Grange, K. R. (1989). Video survey of the seabed at selected sites along the proposed HVDC power cable route. Report prepared for Transpower, Wellington. *NZOI Contract Report 1989/8*: 6 pp. + video.

Murdoch, R. C., Proctor, R., Jillett, J. B., Zeldis, J. R. (1990). Evidence for an eddy over the continental shelf in the downstream lee of Otago Peninsula, New Zealand. *Estuar. Cstl Shelf Sci.* 30: 489–507.

Murdoch, R. C., Quigley, B. (1994). Patch study of mortality, growth and feeding of the larvae of the southern gadoid *Macruronus novaezelandiae*. *Mar. Biol.* 121: 23–34.

Murdoch, W. (1994). Population regulation in theory and practice. *Ecology* 75: 271–287.

Murphy, G. I., Clutter, R. I. (1972). Sampling anchovy larvae with a plankton-mesh purse seine. *Fish. Bull. U.S.* 70: 789–798.

Mushinsky, H. R., Gibson, D. J. (1991). The influence of fire on periodicity of habitat structure. Pp. 237–252 *in* Bell, S. S., McCoy, E. D., Mushinsky, H. R. (eds), *Habitat structure: the physical arrangement of objects in space*. Chapman & Hall, London. 438 pp.

Nakamura, M., Hourigan, T. F., Yamauchi, K., Nagahama, Y., Grau, E. G. (1989). Histological and ultrastructural evidence for the role of gonandal steroid hormones in sex change in the protogynous wrasse *Thalassoma duperry*. *Environ. Biol. Fish.* 24: 117–136.

Navarrete, S. A., Menge, B. A. (1996). Keystone predation and interaction strength: interactive effects of predators on their main prey. *Ecol. Monogr.* 66: 409–430.

Neira, F. J., Miskiewicz, A. G., Trnski, T. (1997). *Larvae of temperate Australian fishes: laboratory guide for larval fish identification*. UWA Press, Perth. 436 pp.

Neira, F. J., Potter, I. C. (1992a). Movement of larval fishes through the entrance channel of a seasonally open estuary in Western Australia. *Estuar. Cstl Shelf Sci.* 35: 213–224.

Neira, F. J., Potter, I. C. (1992b). The ichthyoplankton of a seasonally closed estuary in temperate Australia: does an extended period of opening influence species composition? *J. Fish. Biol.* 41: 935–953.

Nelson, V. (1996). Monitoring of reef benthos: fixed v. random transect. *Reef Res.* 6: 12–14.

Nelson, W. A., Adams, N. M. (1984). Marine algae of the

Kermadec Islands. *Natn. Mus. N.Z. Occ. Ser.* 10: 1–29.

Newell, G. E., Newell, R. C. (1977). *Marine plankton: a practical guide.* Hutchinson, Essex. 244 pp.

Nicol, S., Endo, Y. (1997). Krill fisheries of the world. FAO Fisheries Technical Paper 367, FAO, Rome. 100pp.

Nisbet, R. M., Gurney, W. S. (1982). *Modelling fluctuating populations.* Wiley, London. 379 pp.

Nishimura, A. (1993). Occurrence of a check in otoliths of reared and sea-caught larval walleye pollock *Theragra chalcogramma* (Pallas) and its relationship to events in early-life history. *J. Exp. Mar. Biol. Ecol.* 166: 175–183.

NOAA (1990). The potential of marine fishery reserves for reef fish management in the US Southern Altantic. *NOAA Technical Memorandum NMFS-SEFC-261.* Prepared by the Plan Development Team (Contact J. A. Bohnsack) April 1990. 40 pp.

NOAA (1993). Sampling and analytical methods of the National Status and Trends Program National Benthic Surveillance and Mussel Watch Projects 1984–1992. Volume I. Overview and summary of methods. *NOAA Tech. Memo. NOS ORCA* 71: 117 pp. (NOAA, Silver Spring, Maryland)

NOAA (1995). Review of the use of marine fishery reserves in the US southeastern Atlantic. *NOAA Technical Memorandum NMFS–SEFSC–376;* Proceedings of a Symposium at the American Fisheries Society 125th Annual Meeting, 28–29 August 1995, Tampa, Florida.

Nonaka, M., Fushimi, H. (1994). Restocking. Pp. 446–460 *in* Phillips, B. F., Cobb, J. S., Kittaka, J. (eds), *Spiny lobster management.* Blackwell, Oxford.

Norcross, B. L. (1991). Estuarine recruitment mechanisms of larval Atlantic croakers. *Trans. Am. Fish. Soc.* 120: 673–683.

Norcross, B. L., Shaw, R. F. (1984). Oceanic and estuarine transport of fish eggs and larvae: a review. *Trans. Am. Fish. Soc.* 113: 153–165.

Norman, M. D., Jones, G. P. (1984). Determinants of territory size in the pomacentrid reef fish, *Parma victoriae. Oecologia* 61: 60–69.

Norse, E. A. (1993). *Global marine diversity.* Island Press, Washington. 383 pp.

Nowak, B. F. (1996). Health of red morwong, *Cheilodactylus fuscus,* and rock cale *Crinodus lophodon,* from Sydney cliff-face sewage outfalls. *Mar. Poll. Bull.* 33: 7–12.

NRCC (1988). *Biologically available metals in sediments.* National Research Council of Canada, Associate Committee on Scientific Criteria for Environmental Quality, *NRCC Publ.* 27694: 296 pp.

NZFIB (1994). *The New Zealand Seafood Industry Economic Review 1993.* N.Z. Fishing Industry Board, Wellington. 66 pp.

O'Brien, D. P. (1988). Surface schooling behaviour of the coastal krill *Nyctiphanes australia* (Crustacea: Euphausiacea) off Tasmania, Australia. *Mar. Ecol. Prog. Ser.* 42: 219–233.

Ochi, H. (1989). Mating behaviour and sex change of the anemonefish, *Amphiprion clarkii,* in the temperate waters of southern Japan. *Environ. Biol. Fish.* 26: 257–275.

Odum, E. P., Kuenzler, E. J. (1955). Measurement of territory and home-range size in birds. *Auk* 72: 128–137.

Ohlhorst, S. L., Liddell, W. D., Taylor, R. J., Taylor, J. M. (1988). Evaluation of reef census techniques. *Proc. 6th Int. Coral Reef Symp.* 2: 319–324.

Ohtsuka, S., Inagaki, H., Onbe, T., Gushima, K., Yoon, Y. H. (1995). Direct observations of groups of mysids in

shallow coastal waters of western Japan and southern Korea. *Mar. Ecol. Prog. Ser.* 123: 33–44.

Okamoto, M. (1989). Ability of a small obervation ROV to observe fish fauna around artificial reefs in comparison with diving observations. *Bull. Jpn. Soc. Sci. Fish.* 55: 1539–1546.

Okiyama, M. (1988). *An atlas of the early stage fishes in Japan.* Tokai University Press, Tokyo. 1154 pp.

Olla, B. L., Samet, C., Studholme, A. L. (1981). Correlates between number of mates, shelter availability and reproductive behaviour in the Tautog *Tautoga onitis. Mar. Biol.* 62: 239–248.

Olsen, R. R. (1985). The consequences of short-distance larval dispersal in a sessile marine invertebrate. *Ecology* 66: 30–39.

Olsgard, F., Gray, J. S. (1995). A comprehensive analysis of the effects of offshore oil and gas exploration and production on the benthic communities of the Norwegian continental shelf. *Mar. Ecol. Prog. Ser.* 122, 277–306.

Omori, M., Ikeda, T. (1984). *Methods in marine zooplankton ecology.* Wiley, New York. 332 pp.

Orth, R.J. (1976). The demise and recovery of eelgrass, *Zostera marina,* in the Chesapeake Bay, Virginia. *Aquat. Bot.* 2: 141–159.

Orth, R. J., Ferguson, R. L., Haddad, K. D. (1991). Monitoring seagrass distribution and abundance patterns. Pp. 281–300 in *Coastal Wetlands Coastal Zone '91 Conference-ASCE,* California.

Ortner, P. B., Cummings, S. R., Afring, R. P., Edgerton, H. E. (1979). Silhoutte photography of oceanic zooplankton. *Nature* 277: 50–51.

Osunkoya, O. O., Creese, R. G. (1997). Population structure, spatial pattern and seedling establishment in the New Zealand mangrove, *Avicennia marina* var. *australasica. Aust. J. Bot.* 45: 707–725.

Packham, J. R., Willis, A. J. (1997). *Ecology of dunes, salt marsh and shingle.* Chapman & Hall, London. 335 pp.

Paffenhoffer, G. A., Stewart, T. B., Youngbluth, M. J., Bailey, T. G. (1991). High–resolution vertical profiles of pelagic tunicates. *J. Plankton Res.* 13: 971–981.

Page, H. M., Dugan, J. E., Hubbard, D. M. (1992). Comparative effects of infaunal bivalves on an epibenthic microalgal community. *J. Exp. Mar. Biol. Ecol.* 157: 247–262.

Page, L. M., Burr, B. M. (1991). *A field guide to freshwater fishes.* Houghton Mifflin, Boston. 432 pp.

Page, M. J., Battershill, C. N. (in press). Ecology and impact of harvesting the deep water sponge *Lissodendoryx* n.sp. 1 on the continental shelf edge adjacent to the Kaikoura peninsula, New Zealand. *Mar. Ecol. Prog. Ser.*

Paine, R. T. (1994). *Marine rocky shores and community ecology: an experimentalist's perspective.* Ecology Institute, Oldendorf/Luhe. 152 pp.

Paine, R. T., Ruesink, J. L., Sun, A., Soulanille, E. L., Wonham, M. J., Harley, C. D. G., Brumbaugh, D. R., Secord, D. (1996). Trouble on oiled waters: Lessons from the *Exxon Valdez* oil spill. *Ann. Rev. Ecol. Syst.* 27: 197–234.

Palmer, M. W. (1993). Putting things in better order: the advantages of canonical correspondence analysis. *Ecology* 74: 2215–2230.

Pankhurst, N. W. (1995). Hormones and reproductive behaviour in male damselfish. *Bull. Mar. Sci.* 57: 569–581.

Pankhurst, N. W., Conroy, A. M. (1987). Seasonal changes in reproductive condition and plasma levels of sex steroids in the blue cod, *Parapercis colias* (Bloch & Schneider)

(Mugiloididae). *Fish Physiol. Biochem.* 4: 15–26.

Paradis, A. R., Pepin, P., Brown, J. A. (1996). Vulnerability of fish eggs and larvae to predation: review of the influence of the relative size of prey and predator. *Can. J. Fish. Aquat. Sci.* 53: 1226–1235.

Parker, N. C., Giorgi, A. E., Heidinger, R. C. (1990). Fish-marking techniques. *Am. Fish. Soc. Symp.* 7. Bethesda. 879 pp.

Parker, N. R., Mladenov, P. V., Grange, K. R. (1997). Reproductive biology of the antipatharian black coral *Antipathes fiordensis* in doubtful sound, Fiordland, New Zealand. *Mar. Biol.* 130: 11–22.

Parsons, T. R., Lalli, C. M. (1988). Comparative oceanic ecology of the plankton communities of the subarctic Atlantic and Pacific Oceans. *Oceanogr. Mar. Biol. Ann. Rev.* 26: 317–359.

Parsons, T. R., Maita, Y., Lalli, C. M. (1993). *A manual of chemical and biological methods for seawater analysis.* Pergamon Press, Oxford. 172 pp.

Pasqualini, V., Pergent-Martini, C. (1996). Monitoring of *Posidonia oceanica* meadows using image processing. Pp. 351–358 in Kuo, J., Phillips, R. C., Walker, D. I., Kirkman, H. (eds), *Seagrass biology: Proceedings of an international workshop*, Faculty of Sciences, University of Western Australia.

Paterson, D. M. (1989). Short-term changes in the erodibility of intertidal cohesive sediments related to the migratory behaviour of epipelic diatoms. *Limnol. Oceanogr.* 34: 223–234.

Paul, L. J. (1976). A study on age, growth, and population structure of the snapper, *Chrysophrys auratus* (Forster), in the Hauraki Gulf, New Zealand. *Fish. Res. Bull. N.Z.* 13. 62 pp.

Paul, L. J. (1986). *New Zealand fishes: An identification guide.* Reed Methuen, Auckland. 184 pp.

Paul, L. J. (1992). Age and growth studies of New Zealand Marine Fishes, 1921–90: A review and bibliography. *Aust. J. Mar. Freshw. Res.* 43: 879–912.

Paulin, C., Roberts, C. (1992). *The rockpool fishes of New Zealand.* Museum of New Zealand Te Papa Tongarewa, Wellington. 177 pp.

Paulin, C., Stewart, A., Roberts, C., McMillan P. (1989). *New Zealand fish: a complete guide.* Museum of New Zealand Te Papa Tongarewa, Wellington. 179 pp.

Pavlova, E.V. (1993). Diel changes in copepod respiration rates. Pp. 292–293, 333–339 in Ferrari, F. D., Bradley, B. P. (eds), *Ecology and morpology of copepods.* International Conference on Copepoda, Baltimore. 6–13 June.

Pawlik, J. R. (1986). Chemical induction of larval settlement and metamorphosis in the reef–building tube worm, *Phragmatopoma california* (Savbellariidae: Polychaeta). *Mar. Biol.* 91: 59–68.

Paxton, J. R., Eschmeyer, W. N. (1994). *Encylopedia of fishes.* UNSW Press, Sydney. 240 pp.

Paxton, J. R., Hoese, D. F., Allen, G. R., Hanley, J. E. (1989). *Zoological catalogue of Australia. Volume 7. Pisces. Petromyzontidae to Carangidae.* Australian Government Publishing Service, Canberra. 665 pp.

Pearce, A. F., Phillips, B. F. (1988). ENSO events, the Leeuwin current, and larval recruitment of the western rock lobster. *J. Cons. int. Explor. Mer.* 45: 13–21.

Pearse, J. S., Cameron, A. (1991). Echinodermata: Echinoidea. Pp. 513–662 in Pearse, J. S., Pearse, V. B. (eds.), *Reproduction of marine invertebrates.* Vol. VI. *Echinoderms and Lophophorates.* Boxwood Press, Pacific Grove.

Pearse, J. S., Pearse, V. B., Davis, K. K. (1986). Photoperiodic regulation of gametogensis and growth in the sea urchin *Stronglocentrotus purpuratus. J. Exp. Zool.* 237: 107–118.

Peres, J. M., Picard, J. (1964). Noveau manuel de bionomie benthique de la mer Mediterranée. *Rec. Trav. Stn mar. Endoume Fr.* 31: 1–136.

Peterman, R. M. (1990). Statistical power analysis can improve fisheries research and management. *Can. J. Fish. Aquat. Sci.* 47: 2–634.

Peterson, C. H. (1982). The importance of predation and intra– and interspecific competition in the population biology of two infaunal suspension-feeding bivalves, *Protothaca staminea* and *Chione undatella. Ecol. Monogr.* 52: 437–475.

Peterson, C. H., Andre, S. V. (1980). An experimental analysis of interspecific competition among marine filter feeders in a soft-sediment environment. *Ecology* 61: 129–139.

Peterson, C. H., Black, R. (1993). Experimental tests of the advantages and disadvantages of high density for two coexisting cockles in a Southern Ocean lagoon. *J. Anim. Ecol.* 62: 614–633.

Peterson, C. H., Black, R. (1994). An experimentalist's challenge: When artifacts of intervention interact with treatments. *Mar. Ecol. Prog. Ser.* 111: 289–297.

Petraitis, P. S. (1990). Direct and indirect effects of predation, herbivory and surface rugosity on mussel recruitment. *Oecologia* 83: 405–413.

Petraitis, P. S. (1995). The role of growth in maintaining spatial dominance by mussels (*Mytilus edulis*). *Ecology* 76: 1337–1346.

Petraitis, P. S., Latham, R. E., Niesenbaum, R. A. (1989). The maintenance of species diversity by disturbance. *Q. Rev. Biol.* 64: 393–418.

Pfister, C. A. (1996). The role and importance of recruitment variability to a guild of tide pool fishes. *Ecology* 77: 1928–1941.

Pham, T., Greenwood, J. (1998). Push net technique: a minimal-interference system for sampling ichthyoplankton. *Proc. Roy. Soc. Qld.*, in press

Phillips, B. F. (1981). The circulation of the southeastern Indian Ocean and planktonic life of the western rock lobster. *Oceanogr. Mar. Biol. Ann. Rev.* 19: 11–39.

Phillips, B. F., Cobb, J. S, Kittaka, J. (1994). *Spiny lobster management: Perspectives.* Blackwell, Oxford. 550 pp.

Phillips, D. J. H. (1995). The chemistries and environmental fates of trace metals and organochlorines in aquatic ecosystems. *Mar. Pollut. Bull.* 31: 193–200.

Phillips, D. J. H., Rainbow, P. S. (1993). *Biomonitoring of trace aquatic contaminants.* Elsevier Applied Science, London. 371 pp.

Phillips, R. C., McRoy, C. P. (1980). *Handbook of seagrass biology: an ecosystem perspective.* Garland Publishing, New York. 353 pp.

Phillips, R. C., McRoy, C. P. (1990). *Seagrass research methods. Monographs on oceanographic methodology* 9. UNESCO. 210 pp.

Pickett, S. T. A., White, P. S. (eds) (1985). *The ecology of natural disturbances and patch dynamics.* Academic Press, London.

Pielou, E. C. (1984). *The interpretation of ecological data: a primer on classification and ordination.* Wiley, New York. 263 pp.

Pimental, R. A. (1979). *Morphometrics: The multivariate analysis of biological data*. Kendall/Hunt Publishing, Dubuque. 276 pp.

Pineda, J. (1991). Predictable upwelling and the shoreward transport of planktonic larvae by internal bores. *Science* 253: 548–551.

Pirker, J. G. (1992). Growth, shell-ring deposition, and mortality of paua (*Haliotis iris* Martyn) in the Kaikoura region. Unpublished MSc thesis, University of Canterbury, Christchurch. 166 pp.

Pirker, J. G., Schiel, D. R. (1993). Tetracycline as a fluorescent shell-marker in the abalone *Haliotis iris*. *Mar. Biol.* 116: 81–86.

Pitcher, C. R., Skewes, T. D., Dennis, D. M., Prescott, J. H. (1992). Estimation of the abundance of the tropical lobster *Panulirus ornatus* in Torres Strait, using visual transect methods. *Mar. Biol.* 113: 57–64.

Pizzy, G., Knight, F. (1997). *The field guide to birds of Australia*. HarperCollins, Sydney. 576 pp.

Platt, T., Denman, K. L. (1975). Spectral analysis in ecology. *Ann. Rev. Ecol. Syst.* 6: 189–210.

Platt, T., Harrison, W. G., Lewis, M. R., Li, W. K. W., Sathyendranath, S., Smith, R. E. (1989). Biological production of the oceans: the case for a consensus. *Mar. Ecol. Prog. Ser.* 52: 77–88.

Polderman, P. J. G. (1980). The permanent quadrat method, a means of investigating the dynamics of saltmarsh algal vegetation. Pp. 193–212 *in* Price, J. H., Irvine, D. E. G., Farnham, W. F. (eds), *The shore environment, Vol 1: Methods*. Academic Press, London. 321 pp.

Polivka, K. M., Chotkowski, M. A. (1998). Recolonization of experimentally defaunated tidepools by Northwest Pacific intertidal fishes. *Copeia*: 456–462.

Pollard, D. A. (1984). A review of ecological studies on sea-grass communities, with particular reference to recent studies in Australia. *Aquat. Bot.* 18: 3–42.

Pollard, D. A., Lincoln-Smith, M. P., Smith, A. K. (1996). The biology and conservation of the grey nurse shark (*Carcharias taurus* Rafinesque 1810) in New South Wales, Australia. *Aquat. Conserv.: Mar. Freshwat. Ecosys.* 6: 1–20.

Pollard, D. A., Mathews, J. (1985). Experience in the construction and siting of artificial reefs and fish aggregation devices in Australian waters (abstract), with notes on and a bibliography of Australian studies. *Bull. Mar. Sci.* 37: 299–304.

Pollard, M. J., Kingsford, M. J., Batteglene, S. C. (in press). Chemical marking of juvenile snapper (*Pagrus auratus*: Sparidae) by the incorporation of strontium into the dorsal spines. *Fish. Bull. U.S.*

Pomeroy, L. R., Wiegert, R. G. (1981). *The ecology of a salt marsh*. Ecological Studies 38. Springer-Verlag, Berlin. 271 pp.

Pomponi, S. A. (1988). Maximising the potential of marine organsm collections for both pharmacological and systematic studies. *Biomedical importance of marine organisms*. 13: 7–11.

Pomponi, S. A., Willoughby, R. (1994). Sponge cell culture for production of bioactive metabolites. *In* van Soest, R. W. G., van Kempen, T. H. M. G., Brackman, J. C. (eds), Sponges in time and space: proceedings of the 4th International Porifera Congress, Amsterdam, Netherlands, April 1993. A. A. Balkema, Rotterdam, pp. 395–400.

Pomponi, S. A., Willoughby, R., Kaighn, M. E., Wright, A. E. (1997). Development of technique for *in vitro* production of bioactive natural products from marine sponges. Invited chapter *in* Maramorosch, K., Mitsuhashi, J. (eds), *Invertebrate cell culture: novel directions and biotechnology applications*.

Pomponi, S. A., Willoughby, R., Kelly-Borges, M. (1998). Sponge cell culture. Pp. 423–433 *in* Cooksey, K. (ed.), *Molecular approaches to the study of the ocean*. Chapman & Hall, London.

Pond, S., Pickard, G. L. (1978). *Introductory dynamic oceanography*. Pergamon Press, Oxford. 241 pp.

Poore, G. C. B. (1973). Ecology of New Zealand abalones, *Haliotis* species (Mollusca: Gastropoda). 4. Reproduction. *N.Z. J. Mar. Freshwat. Res.* 7: 67–84.

Porter, K. G., Feig, Y. S. (1980). The use of DAPI for identifying and counting aquatic microflora. *Limnol. Oceanogr.* 25: 943–948.

Porter, K. G., Sherr, E. B., Sherr, B. F., Pace, M., Sanders, R. W. (1985). Protozoa in planktonic food webs. *J. Protozool.* 32: 409–415.

Potts, G. W., Wood, J. W., Edwards, J. M. (1987). Scuba diver-operated low-light-level video system for use in underwater research and survey. *J. Mar. Biol. Ass. U.K.* 67: 299–306.

Povey, A., Keough, M. J. (1991). Effects of trampling on plant and animal populations on rocky shores. *Oikos* 61: 355–368.

Powell, A. W. B. (1937). Animal communities of the sea-bottom in Auckland and Manukau Harbours. *Trans. R. Soc. N.Z.* 66: 354–401.

Powell, A. W. B. (1979). *New Zealand Mollusca*. Collins, Auckland. 500 pp.

Powell, T. M., Steele, J. H. (1995). *Ecological Time Series*. Chapman & Hall, London. 496 pp.

Prezelin, B. B., Alldredge, A. L. (1983). Primary production of marine snow during and after an upwelling event. *Limnol. Oceanogr.* 28: 1156–1167.

Price, J. H., Irvine, D. E. G., Farnham, W. F. (1980). *The shore environment. Vol. 1: Methods*. Academic Press, London. 321 pp.

Pridmore, R. D., Thrush, S. F., Wilcox, T. J., Smith, T. J., Hewitt, J. E., Cummings, V. J. (1991). Effect of the organo-chlorine pesticide technical chlordane on the population structure of suspension and deposit feeding bivalves. *Mar. Ecol. Prog. Ser.* 76: 261–271.

Prince, E. D., Lee, D. W., Cort, J. L., McFarlane, G. A., Wild, A. (1995). Age validation evidence for two-tag-recaptured Atlantic albacore, *Thunnus alalunga*, based on dorsal anal and pectoral finrays, vertebrae and otoliths. Pp. 375–396 *in* Secor, D. H., Dean, J. M., Campana, S. E. (eds), *Recent developments in fish otolith research*. The Belle Library in Marine Science 19. University of South Carolina Press, Columbia. 735 pp.

Prince, J. D., Sellers, T. L., Ford, W. B., Talbot, S. R. (1987). Experimental evidence for limited dispersal of haliotid larvae (genus *Haliotis*: Mollusca: Gastropoda). *J. Exp. Mar. Biol. Ecol.* 106: 243–263.

Pritchard, K., Ward, V., Battershill, C. N., Bergquist, P. R. (1984). Marine sponges : Forty-six sponges of northern New Zealand. *Leigh Lab Bull.* 14. 149 pp.

Probert, P. K., Anderson, P. W. (1986). Quantitative distribution of benthic macrofauna off New Zealand, with particular reference to the west coast of the South Island. *N.Z. J. Mar. Freshwat. Res.* 20: 281–290.

Purcell, J. E. (1985). Predation on fish eggs and larvae by pelagic cnidarians and ctenophores. *Bull. Mar. Sci.* 37: 739–755.

Quammen, M. L. (1984). Predation by shorebirds, fish and crabs on invertebrates in intertidal mudflats: an experimental test. *Ecology.* 65: 529–537.

Quetin, L. B., Ross, R. M. (1992). A long-term ecological research strategy for polar environmental research. *Mar. Pollut. Bull.* 25: 233–238.

Quinn, G. P., Ryan, N. R. (1989). Competitive interactions between two species of intertidal herbivorous gastropods from Victoria, Australia. *J. Exp. Mar. Biol. Ecol.* 125: 1–12.

Rafaelli, D., Hawkins, S. (1996). *Intertidal ecology.* Chapman & Hall, London. 368 pp.

Raimondi, P. T. (1990). Patterns, mechanisms, consequences of variability in settlement and recruitment of an intertidal barnacle. *Ecol. Monogr.* 60: 283–309.

Rainer, S. F. (1995). Potential for the introduction and translocation of exotic species by hull fouling: a preliminary assessment. Centre for Research on Introduced Marine Pests, CSIRO Australia, Division of Fisheries. *Tech. Rep.* 1. 20 pp.

Ralston, S. (1982). Influence of hook size in the Hawaiian deep-sea handline fishery. *Can. J. Fish. Aquat. Sci.* 39: 1297–1302.

Ralston, S., Gooding, R. M., Ludwig, G. M. (1986). An ecological survey and comparison of bottom fish resource assessments (submersible versus handline fishing) at Johnston Atoll. *Fish. Bull. U.S.* 84: 141–156.

Ramos, A. A. (1985). *La Reserva marina de la Isla Plana o Nueva Tabarca (Alicante).* Ayuntamiento de Alicante, Universidad de Alicante. 196 pp.

Ramos-Espla, A. A., McNeill, S. E. (1994). The status of marine conservation in Spain. *Ocean Coast. Mgmt* 24: 125–138.

Randall, J. E., Allen, G. R., Steene, R. C. (1990). *Fishes of the Great Barrier Reef and Coral Sea.* Crawford House Press, Bathhurst. 507 pp.

Ray, G. C. (1996). Coastal marine discontinuitues and synergisms: Implications for biodiversity conservation. *Biodiversity Conserv.* 5: 1095–1108.

Raymont, J. E. G. (1983a). *Plankton and productivity in the oceans: Vol. 1 Phytoplankton.* 2nd edn. Pergamon Press, Oxford. 489 pp.

Raymont, J. E. G. (1983b). *Plankton and productivity in the oceans: Vol. 2 Zooplankton.* 2nd edn. Pergamon Press, Oxford. 824 pp.

Reed, D. C., Amsler, C. D., Ebeling, A. W. (1992). Dispersal in kelps: factors affecting spore swimming and competency. *Ecology* 73: 1577–1586.

Reed, D. C., Foster, M. S. (1984). The effects of canopy shading on algal recruitment and growth in giant kelp forest. *Ecology* 65: 937–948.

Reed, D. C., Laur, D. R., Ebeling, A. W. (1988). Variation in algal dispersal and recruitment: the importance of episodic events. *Ecol. Monogr.* 58: 937–948.

Reed, D. C., Lewis, R. J. (1994). Effects of an oil and gas production effluent on the colonisation potential of giant kelp (*Macrocystis pyrifera*) zoospores. *Mar. Biol.* 119: 277–284.

Reed, D. C., Lewis, R. J., Anghera, M. (1994). Effects of an open-coast oil-production outfall on patterns of giant kelp (*Macrocystis pyrifera*) recruitment. *Mar. Biol.* 120: 25–332.

Rhoads, D. C., Young, D. K. (1970). The influence of deposit feeding organisms on sediment stability and community trophic structure. *Mar. Biol.* 11: 255–261.

Ricker, W. E. (1987). *Computation and interpretation of biological statistics of fish populations.* Department of Fisheries and Oceans Bulletin 191, Canadian Government Publishing Centre Supply and Services, Ottawa. 382 pp.

Riddle, M. J. (1988). Cyclone and bioturbation on sediments from coral reef lagoons. *Estuar. Cstl Shelf Sci.* 27: 687–695.

Rimmer, D. W., Wiebe, W. J. (1987). Fermentative microbial digestion in herbivorous fishes. *J. Fish. Biol.* 31: 229–236.

Ripley, B. D. (1981). *Spatial statistics.* Wiley, New York. 252 pp.

Rippengale, R. J., Kelly, S. J. (1995). Reproduction and survival of *Phyllorhiza punctata* (Cnidaria: Rhizotomeae) in a seasonally fluctuating salinity regime in Western Australia. *Mar. Freshwat. Res.* 46: 1145–1151.

Rissik, D., Suthers, I. M. (1996). The nutritional significance of an estuarine plume front: Multispecies larval fish feeding. *Mar. Biol.* 125: 233–240.

Ritz, D. A. (1994). Social aggregation in pelagic invertebrates. *Adv. Mar. Biol.* 30: 156–216.

Robertson, D. A. (1975). A key to the planktonic eggs of some New Zealand marine telosts. *Fish. Res. Div. [N.Z.] Occ. Publ.* 9: 1–19.

Robertson, D. A. (1980). Hydrology and the quantitative distribution of planktonic eggs of some fishes of the Otago coast, southeastern New Zealand. *Fish. Res. Bull. N.Z.* 21: 1–69.

Robertson, D. R. (1991). The role of adult biology in the timing of spawning of tropical reef fishes. Pp. 356–386 *in* Sale, P. F. (ed.), *The ecology of fishes on coral reefs.* Academic Press, San Diego.

Robertson, D. R. (1982). Fish feces as food on a Pacific coral reef. *Mar. Ecol. Prog. Ser.* 7: 253–265.

Robertson, D. R. (1996). Interspecific competition controls abundance and habitat use of territorial Caribbean damselfishes. *Ecology* 77: 885–899.

Robins, C. R., Ray, G. C., Douglass, J., Freund, R. (1986). *A field guide to Atlantic coast fishes of North America.* Houghton Mifflin, Boston. 354 pp.

Roe, H. S. J., Griffiths, G., Hartman, M., Crisp, N. (1996). Viariability in biological distributions and hydrography from concurrent Acoustic Doppler Current Profiler and SeaSoar surveys. *ICES Jl Mar. Sci.* 53: 131–138.

Rogers-Bennett, L., Bennett, W., Fastenau, H.C., Dewees, C. M. (1995). Spatial variation in red sea urchin reproduction and morphology: Implications for harvest refugia. *Ecol. Appl.* 5: 1171–1180.

Rohlf, F. J. (1974). Methods of comparing classifications. *Ann. Rev. Ecol. Syst.* 5: 101–113.

Rohlf, F. J. (1990). Morphometrics. *Ann. Rev. Ecol. Syst.* 21: 299–316.

Rohlf, F. J., Sokal, R. R. (1981). *Statistical tables.* Freeman, New York. 219 pp.

Roman, M. R., Furnas, M. J., Mullin, M. M. (1990). Zooplankton abundance and grazing at Davies Reef, Great Barrier Reef. *Mar. Biol.* 105: 73–82.

Roper, D. S. (1986). Occurrence and recruitment of fish larvae in a northern New Zealand estuary. *Estuar. Cstl Shelf Sci.* 22: 705–717.

Roper, D. S. (1990). Benthos associated with an estuarine outfall, Tauranga Harbour, New Zealand. *N.Z. J. Mar. Freshwat. Res.* 24: 487–498.

Roper, D. S., Thrush, S. F., Smith, D. G. (1988). The influence of runoff on intertidal mudflat benthic communities. *Mar. Environ. Res.* 26: 1–18.

Rose, M. (1970). Copépodes pélagique. *Faune de France* 26. 374 pp.

Rotenberry, J. T., Wiens, J. A. (1985). Statistical power analysis and community-wide patterns. *Am. Nat.* 125: 164–168.

Rothlisberg, P. C., Pearcy, W. G. (1976). An epibenthic sampler to study the ontogeny of vertical migrations of *Pandalus jordani* (Decapoda, Caridea). *Fish. Bull.* 74: 994–997.

Round, F. E., Hickman, M. (1984). Phytobenthos sampling and estimation of primary production. Meiofauna techniques. Pp. 245–283 in Holme, N. A., McIntyre, A. D. (eds), *Methods for the study of marine benthos*. 2nd edn. Blackwell, Oxford. 387 pp.

Rountree, R. A. (1989). Association of fishes with fish aggregation devices: Effects of structure size on fish abundance. *Bull. Mar. Sci.* 44: 960–972.

Rountree, R. A. (1990). Community structure of fishes attracted to shallow water fish aggregation devices off South Carolina, U.S.A. *Environ. Biol. Fish.* 29: 241–262.

Rowe, F. W. E., Gates, J. (1995). *Zoological catalogue of Australia: Echinodermata*. Vol. 33. CSIRO, Melbourne. 510 pp.

Rowley, R. J. (1992). Impacts of marine reserves on fisheries: a report and review of the literature. Department of Conservation Te Papa Atawhai, Wellington. *Sci. Res. Ser.* 51: 50 pp.

Rowley, R. J. (1993). Fluorescent marking of larvae for field study of larval transport. *Proc. 2nd Int. Temperate Reef Symp.*: 208.

Rozas, L., Minello, T. J. (1997). Estimating densities of small fishes and decapod crustaceans in shallow estuarine habitats: a review of sampling design with focus on gear selection. *Estuaries* 20: 199–213.

Rumohr, H. (1995). Monitoring the marine environment with imaging methods. *Scientia Mar.* 59: 129–138.

Russ, G. (1984a). Distribution and abundance of herbivorous grazing fishes in the central Great Barrier Reef. I. Levels of variability across the entire continental shelf. *Mar. Ecol. Prog. Ser.* 20: 23–34.

Russ, G. (1984b). Distribution and abundance of herbivorous grazing fishes in the central Great Barrier Reef. II. Patterns of zonation of mid–shelf and outershelf reefs. *Mar. Ecol. Prog. Ser.* 20: 35–44.

Russ, G. R., Alcala, A. C. (1989). Effects of intense fishing pressure on an assemblage of coral reef fish. *Mar. Ecol. Prog. Ser.* 56: 13–27.

Russell, B. C. (1977). Population and standing crop estimates for rocky reef fishes of northeastern New Zealand. *N.Z. J. Mar. Freshwat. Res.* 11: 23–36.

Russell, B. C. (1983). The food and feeding habits of rocky reef fish of northeastern New Zealand. *N.Z. J. Mar. Freshwat. Res.* 17: 121–145.

Russell, B. C. (1988). Revision of the labrid genus *Pseudo-labrus* and allied species. *Rec. Aust. Mus.* 9: 2–66.

Russell, B. C., Talbot, F. H., Anderson, G. R. V., Goldman, B. (1978). Collection and sampling of reef fishes. *In* Stoddart, D. R., Johannes, R. E. (eds), *Coral reefs: research methods*. UNESCO, Paris. 581 pp.

Russell, F. S. (1970). *The medusae of the British Isles. II Pelagic Scyphozoa*. Cambridge University Press, Cambridge. 284 pp.

Russell, F. S. (1976). *The eggs and planktonic stages of British marine fishes*. Academic Press, London. 524 pp.

Russell, G. (1991). Vertical distribution. Pp. 43–67 in Mathieson, A. C., Nienhuis, P. H. (eds), *Ecosystems of the world 24: intertidal and littoral ecosystems*. Elsevier, Amsterdam.

Rutzler, K., Ferraris, J. D., Larson, R. J. (1980). A new plankton sampler for coral reefs. *Mar. Ecol.* 1: 65–71.

Ryle, V. D.; Mueller, H. R.; Gentien, P. (1981). Automated analysis of nutrients in tropical sea waters. *AIMS Tech. Bull., Oceanogr. Ser.* 3.

Safran, P., Omori, M. (1990). Some ecological observations on fishes associated with drifting seaweed off Tohoku coast, Japan. *Mar. Biol.* 105: 395–402.

Sainsbury, J. C. (1996). *Commercial fishing methods: An introduction of vessels and gears*. Fishing News Books, Oxford. 359 pp.

Sainsbury, K. J. (1982). Population dynamics and fisheries management of the paua *Haliotis iris*. I. Population structure, growth, reproduction and mortality. *N.Z. J. Mar. Freshwat. Res.* 16: 147–161.

Sainsbury, K. J. (1988). The ecological basis of multispecies fisheries, and management of a demersal fishery in tropical Australia. Pp. 349–382 in Gulland, J. A. (ed.), *Fish population dynamics*. Wiley, New York.

St John, J., Russ, G. R., Gladsone, W. (1990). Accuracy and bias of visual estimates of numbers, size struture and biomass of a coral reef fish. *Mar. Ecol. Prog. Ser.* 64: 253–262.

Sakuri, M., Nakazono, A. (1995). Twilight migrations of the temperate Japanese surfperch *Neoditrema ransonneti* (Embiotocidae*)*. *Jap. J. Ichthyol.* 42: 261–267.

Sale, P. F. (1991). Reef fish communities: Open nonequilibrial systems. Pp. 564–600 in Sale, P. F. (ed.) *The ecology of fishes on coral reefs*. Academic Press, San Diego.

Sale, P. F., Ferrell, D. (1988). Early life survivorship of coral reef fishes. *Coral Reefs* 7: 117–124.

Sale, P. F., Sharp, B. J. (1983). Correction for bias in visual transect censuses of coral reef fishes. *Coral Reefs* 2: 37–42.

Sale, P. F., McWilliam, P. S., Anderson, D. T. (1976). Composition of near-reef zooplankton at Heron Reef, Great Barrier Reef. *Mar. Biol.* 34: 59–66.

Salm, R. V., Clark, J. R. (1989). *Marine and coastal protected areas: a guide for planners and managers*. IUCN, Gland. 302 pp.

Salmon, J. T. (1991). *Native New Zealand flowering plants*. Reed, Auckland. 254 pp.

Sameoto, D. D., Jaroszynski, L. O. (1969). Otter surface sampler: a new neuston net. *J. Fish. Res. Bd Can.* 25: 2240–2244.

Sanchez-Jerez, P. (in press). Comparing the abundance of fish populations and assemblages by visual counts: problems with sampling designs and analyses. *Il Nat. Sicil.*

Sanchez-Jerez, P., Ramos Espla, A. A. (1996). Detection of environmental impacts by bottom trawling on *Posidonia oceania* (L.) Delile meadows: Sensitivity of fish and macro-invertebrate communities. *J. Aquat. Ecos. Health* 5: 239–253.

Sanders, R. W., Caron, D. A., Berninger, U-G. (1992). Relation-ships between bacteria and heterotrophic nanoplankton in marine and fresh waters: an inter-ecosystem comparison. *Mar. Ecol. Prog. Ser.* 86: 1–14.

Sano, M., Moyer, J. T. (1985). Bathymetric distribution and feeding habits of two sympatric Cheilodactylid fishes at Miake-jima, Japan. *Jap. J. Ichthyol.* 32: 239–247.

Santelices, B. (1990). Patterns of reproduction and recruitment in seaweeds. *Oceanogr. Mar. Biol. Ann. Rev.* 28: 177–276.

Santurtun, M. (1995). Catch structure and dynamics of

Antillean Z-traps on a coral reef: interaction between trap characteristics and fish behaviour. Unpublished MSc thesis, James Cook University, Townsville. 73 pp.

Sasekumar, A. (1984). Methods for the study of mangrove fauna. Pp. 145–161 *in* Snedaker, S. C., Snedaker, J. G. (eds), *The mangrove ecosystem: research methods*. UNESCO, Paris.

Scanes, P. (1993). Trace metal uptake in cockles *Anadara trapezia* from Lake Macquarie, New South Wales. *Mar. Ecol. Prog. Ser.* 102: 135–142.

Schahinger, R. B. (1987). Structure of coastal upwelling events observed off the south coast of south Australia during February 1983–April 1984. *Aust. J. Mar. Freshwat. Res.* 38: 439–459.

Scheiner, S. M., Gurevitch, J. (1993). *Design and analysis of ecological experiments*. Chapman & Hall, London. 445 pp.

Scheltema, R. S. (1986). On dispersal and planktonic larvae of benthic invertebrates: an electic overview and summary of problems. *Bull. Mar. Sci.* 39: 290–322.

Scheltema, R. S. (1988). Initial evidence for the transport of teleplanic larvae of benthic invertebrates across the east Pacific barrier. *Biol. Bull.* 174: 145–156.

Schiel, D. R. (1984). Poor Knights Islands Marine Reserve: a biological survey of subtidal reefs. *Leigh Mar. Lab. Rep.* 93 pp.

Schiel, D. R. (1985a). Growth, survival and reproduction of two species of marine algae at different densities in natural stands. *J. Ecol.* 75: 199–217.

Schiel, D. R. (1985b). A short-term demographic study of *Cystoseira osmundacea* (Fucales: Cystoseiraceae) in central California. *J. Phycol.* 21: 99–106.

Schiel, D. R. (1988). Algal interaction on subtidal reefs in northern New Zealand: a review. *N.Z. J. Mar. Freshwat. Res.* 22: 481–489.

Schiel, D. R. (1990). Macroalgal assemblages in New Zealand: Structure, interactions and demography. *Hydrobiologia* 192: 59–76.

Schiel, D. R. (1993). Experimental evaluation of commercial-scale enhancement of abalone *Haliotis iris* populations in New Zealand. *Mar. Ecol. Prog. Ser.* 97: 167–181.

Schiel, D. R. (1997). Review of abalone culture and research in New Zealand. *Moll. Res.* 18: 289–298.

Schiel, D. R., Andrew, N. L., Foster, M. S. (1995). The structure of subtidal algal and invertebrate assemblages at the Chatham Islands, New Zealand. *Mar. Biol.* 69: 355–368.

Schiel, D. R., Breen, P. A. (1991). Population structure, ageing and fishing mortality of the New Zealand abalone *Haliotis iris*. *Fish. Bull.* 89: 681–691.

Schiel, D. R., Foster, M. S. (1986). The structure of subtidal algal stands in temperate waters. *Oceanogr. Mar. Biol. Ann. Rev.* 24: 265–307.

Schiel, D. R., Kingsford, M. J., Choat, J. H. (1986). Depth distribution and abundance of benthic organisms and fishes at the subtropical Kermadec Islands. *N.Z. J. Mar. Freshwat. Res.* 20: 521–535.

Schiel, D. R., Nelson, W. A. (1990). The harvesting of macro-algae in New Zealand. *Hydrobiologia*. 204/205: 23–33.

Schlacher, T. A., Wooldridge, T. H. (1996a). How accurately can retention of benthic macrofauna by a particular mesh size be predicted from body size of organisms? *Hydrobiologia* 323: 149–154.

Schlacher, T. A., Wooldridge, T. H. (1996b). How sieve mesh size affects sample estimates of estuarine benthic macrofauna. *J. Exp. Mar. Biol. Ecol.* 201: 159–171.

Schmidt, G. H. (1982). Random and aggretative settlement in some sessile marine invertebrates. *Mar. Ecol. Prog. Ser.* 9: 97–100.

Schmidt, G. H., Warner, G. F. (1984). Effects of caging on the development of a sessile epifaunal community. *Mar. Ecol. Prog. Ser.* 34: 305–307.

Schmidt, K. F. (1997). 'No-take' zones spark fisheries debate. *Science* 277: 489–490.

Schmitt, R. J. (1982). Consequences of dissimilar defenses against predation in a subtidal marine community. *Ecology* 63: 1588–1601.

Schmitt, R. J. (1985). Competitive interactions of two mobile prey species in a patchy environment. *Ecology* 66: 950–958.

Schmitt, R. J. (1987). Indirect interactions between prey: apparent competition, predator aggregation, and habitat segregation. *Ecology* 68: 1887–1897.

Schmitt, R. J. (1996). Exploitation competition in mobile grazers: trade-offs in use of a limited resource. *Ecology* 77: 408–425.

Schmitt, R. J., Holbrook, S. J. (1985). Patch selection by juvenile black surfperch (Embiotocidae) under variable risk: interactive influence of food quality and structural complexity. *J. Exp. Mar. Biol. Ecol.* 85: 269–285.

Schmitt, R. J., Holbrook, S. J. (1986). Seasonally fluctuating resources and temporal variability of interspecific competition. *Oecologia* 69: 1–11.

Schmitt, R. J., Holbrook, S. J. (1990). Population responses of surfperch released from competition. *Ecology* 71: 1653–1665.

Schmitt, R. J., Osenberg, C. W. (1996). *Detecting ecological impacts: concepts and application in coastal habitats* Academic Press, London. 401 pp.

Schnute, J. (1981). A verstile growth model with statistically stable parameters. *Can. J. Fish. Aquat. Sci.* 38: 1128–1140.

Schnute, J. T., Richards, L. J. (1990). A unified approach to the analysis of fish growth, maturity and survivorship data. *Can. J. Fish. Aquat. Sci.* 47: 24–40.

Schroeder, A., Lowry, M., Suthers, I. (1994). Sexual dimorphism in the red morwong, *Cheilodactylis fuscus*. *Aust. J. Mar. Freshwat. Res.* 45: 1173–1180.

Schultz, E. T., Warner, R. R. (1989). Phenotypic plasticity in life-history traits of female *Thalassoma bifasciatum* (Pisces; Labridae). 1. Manipulation of social structure in tests for adaptive shifts of life-history allocations. *Evolution* 43: 1497–1506.

Scott, F. J., Russ, G. R. (1987). Effects of grazing on species composition of the epilithic algal community on coral reef of the central Great Barrier Reef. *Mar. Ecol. Prog. Ser.* 39: 293–304.

Sebens, K. P. (1991). Habitat structure and community dynamics in marine benthic systems. Pp. 211–236 *in* Bell, S. S., McCoy, E. D., Mushinsky, H. R. (eds), *Habitat structure: the physical arrangement of objects in space*. Chapman & Hall, London.

Seki, T., Taniguchi, K. (1996). Factors critical to the survival of herbivorous animals during settlement and metamor-phosis. Pp. 341–354 *in* Watanabe, Y., Yamashita, Y., Oozeki, Y. (eds), *Survival strategies in early life stages of marine resources*. A. A. Balkema, Rotterdam.

Shanks, A. L. (1983). Surface slicks associated with tidally forced internal waves may transport pelagic larvae of benthic invertebrates and fishes shoreward. *Mar. Ecol. Prog. Ser.* 13: 311–315.

Shanks, A. L. (1995). Orientated swimming by megalopae of several eastern North Pacific crab species and its potential role in their onshore migration. *J. Exp. Mar. Biol. Ecol.* 186: 1–16.

Shanks, A. L., del Carmen, K. A. (1997). Larval polychaetes are strongly associated with marine snow. *Mar. Ecol. Prog. Ser.* 154: 211–221.

Shanks, A. L., Edmondson, E. W. (1990). The vertical flux of metazoans (holoplankton, meiofauna, and larval invertebrates) due to their association with marine snow. *Limnol. Oceanogr.* 35: 455–463.

Shanks, A. L., Graham, W. M. (1988). Chemical defense in a scyphomedusae. *Mar. Ecol. Prog. Ser.* 45: 81–86.

Shanks, A. L., Wright, W. G. (1987). Internal wave-mediated shoreward transport of cyprids, megalopae, and gammarids and correlated longshore differences in the settling rate of intertidal barnacles. *J. Exp. Mar. Biol. Ecol.* 114: 1–13.

Shapiro, D. Y., Hensley, D. A., Appledoorn, R. S. (1988). Pelagic spawning and egg transport in coral-reef fishes: a skeptical overview. *Environ. Biol. Fish.* 22: 3–14.

Sharples, J. (1997). Cross-shelf intrusion of subtropical water into the coastal zone of northeast New Zealand. *Conti. Shelf Res.* 17: 835–857.

Sheehy, M. R. J., Greenwood, J. G., Fielder, D. R. (1994). More accurate chronological age determination of crustaceans from field situations using the physiological age marker, lipofuscin. *Mar. Biol.* 121: 237–246.

Shenker, J. M., Maddox, E. D., Wishinski, E., Pearl, A., Thorrold, S., Smith, N. (1993). Onshore transport of settlement-stage Nassau grouper *Epinephelus striatus* and other fishes in Exuma Sound, Bahamas. *Mar. Ecol. Prog. Ser.* 98: 31–43.

Shepherd, S. A. (1981). Ecological strategies in a deep water red algal community. *Bot. Mar.* 24: 457–463.

Shepherd, S.A. (1985). Power and efficiency of a research diver, with a description of a rapid underwater measuring gauge: Their use in measuring recruitment and density of an abalone population. Pp. 263–272 in Mitchell, C. T. (ed.), *Diving for science*. American Academy of Underwater Science, La Jolla.

Shepherd, S. A. (1986). Studies on the southern Australian abalone (genus *Haliotis*). VII. Aggregative behaviour of *H. laevigata* in relation to spawning. *Mar. Biol.* 90: 231–236.

Shepherd, S. A. (1988). Studies on the southern Australian abalone (genus *Haliotis*). VIII. Growth of juvenile *H. laevigata*. *Aust. J. Mar. Freshwat. Res.* 39: 177–183.

Shepherd, S. A. (1990). Studies on the southern Australian abalone (genus *Haliotis*). XI. Long–term recruitment and mortality dynamics of an unfished population. *Aust. J. Mar. Freshwat. Res.* 41: 475–492.

Shepherd, S. A., Avalos-Borja, M., Quintanilla, O. M. (1995). Toward a chronology of *Haliotis fulgens*, with a review of abalone shell microstructure. *Mar. Freshwat. Res.* 46: 607–615.

Shepherd, S. A., Daume, S. (1996). Ecology and survival of juvenile abalone in crustose coralline habitat. Pp. 297–314 in Watanabe, Y., Yamashita, Y., Oozeki, Y. (eds), *Survival strategies in early life stages of marine resources*. Balkema, Rotterdam.

Shepherd, S. A., Kirkwood, G. P., Sandland, R. L. (1982). Studies on the southern Australian abalone (genus *Haliotis*). III. Mortality of *H. laevigata* and *H. ruber*. *Aust. J. Mar.*

Freshwat. Res. 33: 265–272.

Shepherd, S. A., Laws, H. M. (1974). Studies on the southern Australian abalone (genus *Haliotis*). II. Reproduction of five species. *Aust. J. Mar. Freshwat. Res.* 25: 49-62.

Shepherd, S. A., Lowe, D., Partington, D. (1992a). Studies on southern Australian abalone (genus *Haliotis*) XIII: larval dispersal and recruitment. *J. Exp. Mar. Biol. Ecol.* 164: 247–260.

Shepherd, S. A., Tegner, M. J., Guzman del Proo, S. A. (1992b). *Abalone of the world: Biology, fisheries and culture.* Fishing News Books and Blackwell, Oxford. 608 pp.

Shepherd, S. A., Thomas, I. M. (1982). *Marine invertebrates of southern Australia. Part I.* Government Printer, Adelaide. 491 pp.

Shepherd, S. A., Thomas, I. M. (1989). *Marine invertebrates of southern Australia. Part II.* South Australian Government Printing Division, Adelaide. 900 pp.

Shepherd, S. A., Davies, M. (1997). *Marine invertebrates of southern Australia: Part III.* Southern Research and Development Institute and the Flora & Fauna of South Australia Handbooks Committee, Adelaide. 352 pp.

Sherr, E. B., Caron, D. A., Sherr, B. F. (1993). Staining of heterotrophic protists for visualisation via epifluorescence microscopy. Pp. 213–228 in Kemp, P. F., Sherr, B. F., Sherr, E. B., Coale, J. J. (eds), *Handbook of methods in aquatic microbial ecology*. Lewis, Boca Raton.

Sherr, E. B., Sherr, B. F. (1991). Planktonic microbes: Tiny cells at the base of the ocean's food webs. *TREE* 6: 50–54.

Short, A. D., Wright, L. D. (1983) Physical variability of sandy beaches. Pp. 133–144 in McLachlan, A., Erasmus, T. (eds), *Sandy beaches as ecosystems*. Junk, The Hague.

Shreeves, K. (1991). Special activities: Underwater photography. Pp. 5.17–5.30 in Hornsby, A. (ed.-in-chief), *The encyclopedia of recreational diving*. International PADI, Bristol.

Shumway, S. E., Cembella, A. D. (1994). Toxic algal blooms: potential hazards to scallop culture and fisheries. *Mem. Qld Mus.* 36: 361–372.

Siegel, S., Castellan, N. J. J. (1988). *Non-parametric statistics for the behavioural sciences*. McGraw-Hill, New York. 399 pp.

Siegfried, W. R. (1994). *Rocky shores: exploitation in Chile and South Africa. Ecological Studies* 103. Springer-Verlag, Berlin.

Sieracki, M. E., Webb, K. L. (1991). Applications of image analysed fluorescence microscopy for quantifying and characterising planktonic protist communities. Pp. 77–100 in Reid, P. C, Turley, C. M., Burkill, P. H. (eds), *Protozoa and their role in marine processes*. Springer-Verlag, Berlin.

Sikkel, P. C. (1995). Effects of nest quality on male courtship and female spawning site choice in an algal-nesting damselfish. *Bull. Mar. Sci.* 57: 682–689.

Silver, M. W., Shanks, A. L., Trent, J. D. (1978). Marine snow: microplankton habitat and source of small-scale patchiness in pelagic populations. *Science* 201: 371–373.

Sinclair, M. (1988). *Marine populations: An essay on population regulation and speciation*. University of Washington Press, Seattle. 252 pp.

Sindermann, C. J. (1996). *Ocean Pollution: Effects on living resources and humans*. CRC Press, New York. 265 pp.

Singleton, R. J. (1992). Community analysis package, version 3.1. N.Z. Oceanographic Institute, NIWA, Wellington. 12 pp.

Siniff, D. B., Skoog, R. O. (1964). Aerial censusing of caribou using stratified random sampling. *J. Wildl. Mgmt* 28: 391–401.

Sissenwine, M. P. (1984). Why do fish communites vary? Pp.

50–94 *in* May, R. M. (ed.), *Exploitation of marine communites*. Springer-Verlag, Berlin.

Skerman, T. M. (1960). Ship-fouling in New Zealand waters: a survey of marine fouling organisms from vessels of the coastal and overseas trade. *N.Z. J. Sci.* 3: 620–648.

Skilleter, G. A. (1996). An experimental test of artifacts from repeated sampling in soft-sediments. *J. Exp. Mar. Biol. Ecol.* 205: 137–148.

Skilleter, G.A., Underwood, A.J. (1993). Intra- and inter-specific competition for food in infaunal coral reef gastropods. *J. Exp. Mar. Biol. Ecol.* 173: 29–55.

Slawyk, G. (1979). ^{13}C and ^{13}N uptake by phytoplankton in the Antarctic upwelling area: results from the *Antipod I* cruise in the Indian Ocean sector. *Aust. J. Mar. Freshwat. Res.* 30: 431–448.

Slawyk, G., Collos, Y., Auclair, J. C. (1977). The use of the ^{13}C and ^{13}N isotopes for the simultaneous measurement of carbon and nitrogen turnover rates in marine phytoplankton. *Limnol. Oceanogr.* 22: 925–932.

Slawyk, G., Minas, M., Collos, Y., Legendre, L., Roy, S (1984). Comparison of radioactive and stable isotope tracer techniques for measuring photosynthesis: ^{13}C and ^{14}C uptake by marine phytoplankton. *J. Plankton Res.* 6: 249–257.

Sloan, N. A. (1986). World jellyfish and tunicate fisheries, and the northeast Pacific echinoderm fishery. *In* Jamieson, G. S., Bourne, N. (eds), North Pacific Workshop on stock assessment and management of invertebrates. *Can. Spec. Publ. Fish. Aquat. Sci.* 92.

Smaldon, G., Lee, E. W. (1979). *A synopsis of methods for the narcotisation of marine invertebrates*. Royal Socttish Museum, Edinburgh. 96 pp.

Smith, A. K., Pollard, D. A. (1996). The best available information – some case studies from NSW, Australia, of conservation-related management responses which impact on recreational fishers. *Mar. Policy* 20: 261–267.

Smith, B., Wilson, J. B. (1996). A consumer's guide to evenness indices. *Oikos* 76: 70–82.

Smith, D. C. (1992). Age determination and growth in fish and other aquatic animals. *Aust. J. Mar. Freshwat. Res.* 43:

Smith, D. L. (1977). *A guide to marine coastal plankton and marine invertebrate larvae*. Kendall Hunt, Iowa. 82 pp.

Smith, L. C., Hildeman, W. H. (1986). Allograft rejection, autograft fusion and inflammatory responses to injury in *Callyspongia diffusa* (Porifera: Demospongia). *Proc. R. Soc. Lond.* B 226: 445–464.

Smith, M. M., Heemstra, P. C. (1986). *Smith's sea fishes*. Spinger-Verlag, Berlin. 1047 pp.

Snedaker, S. C., Snedaker, J. G. (1984). *The mangrove ecosystem: research methods*. UNESCO, Paris.

Snedecor, G. W., Cochran, W. G. (1980). *Statistical methods*. Iowa State University Press, Ames. 507 pp.

Snow, J. 1995. The population ecology of intertidal *Ulva* in the Bay of Plenty. Unpublished MSc thesis, University of Auckland. 123 pp.

Sokal, R. R., Rohlf, F. J. (1995). *Biometry*. Freeman, New York. 887 pp.

Sokal, R. S., Oden, N. L. (1978). Spatial autocorrelation in biology. 1. Methodology. *Biol. J. Linn. Soc.* 10: 199–228.

Somerfield, P. J., Gee, J. M., Warwick, R. M. (1994). Soft sediment meiofaunal community structure in relation to a long-term heavy metal gradient in the Fal estuary system. *Mar. Ecol. Prog. Ser.* 105: 79–88.

Sorokin, Y. I. (1993). Plankton in coral-reef waters. Pp. 71–126 *in* Sorokin, Y. I. (ed.), *Coral reef ecology*. Springer-Verlag, Berlin.

Sournia, A. (1978). *Phytoplankton manual. Monographs on oceanographic methodology 6*. UNESCO, Paris. 337 pp.

Sousa, W. P. (1985). Disturbance and patch dynamics on rocky intertidal shores. Pp. 101–124 *in* Pickett, S. T. A., White, P. S. (eds), *The ecology of natural disturbances and patch dynamics*. Academic Press, London.

Spanier, E., Cobb, J. S., Clancy, M. (1994). Impacts of remotely operated vehicles (ROVs) on the behaviour of marine animals: An example using American lobsters. *Mar. Ecol. Prog. Ser.* 104: 257–266.

Sponaugle, S., Cowen, R. K. (1996a). Larval supply and patterns of recruitment for two Caribbean reef fishes, *Stegastes partitus* and. *Acanthurus bahianus. Mar. Freshwat. Res.* 47: 433–447.

Sponaugle, S., Cowen, R. K. (1996b). Nearshore patterns of coral reef fish larval supply to Barbados, West Indies. *Mar. Ecol. Prog. Ser.* 133: 13–28.

Stace, G. (1991). The elusive toheroa. *N.Z. Geographic* 9: 18–34.

Staines, V. E. (1997). Demography and restoration of grey mangrove populations in North Auckland. Unpublished MSc thesis, University of Auckland. 127 pp.

Steedman, H. F. (1976). *Zooplankton fixation and preservation*. UNESCO, Paris. 350 pp.

Steele, J. H. (1976). Patchiness. Pp. 98–115 *in* Cushing, D. H., Walsh, J. J. (eds), *The ecology of the seas*. Blackwell, Oxford.

Steele, M. A. (1997). The relative importance of processes affecting recruitment of two temperate reef fishes. *Ecology* 78: 129–145.

Steemann-Nielsen, E. (1952). The use of radioactive carbon (^{14}C) for measuring organic production in the sea. *J. Cons. Int. Explor. Mer* 18: 117–140.

Steinberg, P. D. (1994). Lack of short-term induction of phorotannins in the Australian algae *Ecklonia radiata* and *Sargassum vestitum. Mar. Ecol. Prog. Ser.* 112: 129–133.

Steinberg, P. D. (1995). Interaction between the canopy dwelling echinoid *Holopneustes purpurescens* and its host kelp *Ecklonia radiata. Mar. Ecol. Prog. Ser.* 127: 169–181.

Steneck, R. S., Dethier, M. N. (1994). A functional group approach to the structures of algal-dominated communities. *Oikos* 69: 476–498.

Stephens, J. J. S., Jordan, G. A., Morris, P. A., Singer, M. M., McGowen, G. E. (1986). Can we relate larval fish abundance to recruitment or population stability? a preliminary analysis of recruitment to a temperate rocky reef. *Cal. COFI Rep.* 27: 65–83.

Stephenson, T. A., Stephenson, A. (1972). *Life between tidemarks on rocky shores*. Freeman, San Francisco. 425 pp.

Sterba, G. (1962). *Freshwater fishes of the world*. Vista Books, London. 878 pp.

Stevenson, D. K., Campana, S. E. (1992). Otolith Micro-structure examination and analysis. *Can. Spec. Publ. Fish. Aquat. Sci.* 117: 126 pp.

Stewart, C., de Mora, S. J., Jones M. R. L., Miller, M. C. (1992). Imposex in New Zealand dogwhelks. *Mar. Pollut. Bull.* 24: 204–209.

Stewart-Oaten, A., Murdoch, W. M., Parker, K. R. (1986). Environmental impact assessment: 'pseudoreplication' in time? *Ecology*. 67: 929–940.

Stobutzki, I. C., Bellwood, D. R. (1994). An analysis of

sustained smimming abilities of pre- and postsettlement coral reef fishes. *J. Exp. Mar. Biol. Ecol.* 175: 275–286.

Stocker, L. J. (1985). Identification guide to some common New Zealand ascidians. Unpublished report. Leigh Marine Laboratory, Auckland.

Stocker, L. J. (1986). Artifactual effects of caging on the recruitment and survivorship of a subtidal colonial invertebrate. *Mar. Ecol. Prog. Ser.* 34: 305–307.

Stocker, L. J., Battershill, C. N. (in press). Ascideacea. *In* Cook, S. de C. (ed.), *New Zealand marine coastal invertebrates*. Canterbury University Press, Christchurch.

Stocker, L. J., Bergquist, P. R. (1986). Seasonal cycles, extrinsic factors, and the variable effects of turfing algae on the abundance of a subtidal colonial invertebrate. *J. Exp. Mar. Biol. Ecol.* 102: 1–22.

Stocker, L. J., Bergquist, P. R. (1987). Importance of algal turf, grazers, and spatial variability in the recruitment of a subtidal colonial invertebrate. *Mar. Ecol. Prog. Ser.* 39: 285–291.

Stockwell, J. D., Sprules, W. G. (1995). Spatial and temporal patterns of zooplankton biomass in Lake Erie. *ICES J. Mar. Sci.* 52: 557–564.

Stoddart, D. R., Johannes, R. E. (1978). *Coral reefs: research methods. Monographs on oceanographic methodology 5.* UNESCO, Paris. 581 pp.

Storr, J. F. (1964). Ecology of the Gulf of Mexico commercial sponges and its relation to the fishery. U.S. Department of the Interior, Fish and Wildlife Service, *Spec. Scient. Report, Fish.* 466. 73 pp.

Strathmann, M. F. (1987). *Reproduction and development of marine invertebrates of the northern Pacific coast.* University of Washington Press, Seattle. 670 pp.

Strickland, J. D. H. Parsons, T. R. (1972). A practical handbook of seawater analysis. *Fish. Res. Bd Can. Bull.* 167.

Suarez de Vivero, J. L., Frieyro, M. C. (1994). Spanish marine policy: Role of marine protected areas. *Mar. Policy* 18: 345–352.

Sulkin, S. D. (1990). Larval orientation mechanisms: the power of controlled experiments. *Ophelia* 32: 49–62.

Summerhayes, C. P., Thorpe, S. A. (1996). *Oceanography: an illustrated guide.* Manson Publishing, Woods Hole. 352 pp.

Sundstroem, B., Edler, L., Graneli, E. (1990). The global distribution of harmful effects of phytoplankton. Pp. 537–541 in Graneli, E., Sundstroem, B., Elder, L., Anderson, D. M. (eds), Toxic Marine Phytoplankton. International Conference on Toxic Marine Phytoplankton, Lund, 26–30 June 1989.

Suthers, I. M., Frank, K. T. (1990). Zooplankton biomass gradient off south-western Nova Scotia, nearshore ctenophore predation or hydrographic separation. *J. Plankton Res.* 112: 831–850.

Suthers, I. M., Frank, K. T. (1991). Comparative persistence of marine fish larvae from pelagic versus demersal eggs of southwestern Nova-Scotia, Canada. *Mar. Biol.* 108: 175–184.

Svane, I., Lundalv, T. (1982). Persistence stability in ascidian populations: Long-term population dynamics and reproductive pattern of *Pyura tessellata* (Forbes). *Sarsia* 67: 249–257.

Svane, I. B., Young, C. M. (1989). The ecology and behaviour of ascidian larvae. *Oceanogr. Mar. Biol. Ann. Rev.* 27: 45–90.

Sweatman, H. P. A. (1984). A field study of the predatory behaviour and feeding rate of the piscivorous coral reef fish, the lizardfish *Synodus englemani. Copeia*: 187–193.

Sweatman, H. P. A. (1985). The influence of adults of some coral reef fishes on larval recruitment. *Ecol. Monogr.* 55: 469–485.

Swihart, R. K., Slade, N. A. (1986). The importance of statistical power when testing for independence in animal movements. *Ecology* 67: 255–258.

Syms, C. (1995). Multi-scale analysis of habitat association in a guild of blennioid fishes. *Mar. Ecol. Prog. Ser.* 125: 31–43.

Tabachnik, B. G., Fidell, L. S. (1983). *Using multivariate statistics.* Harper & Row, New York. 509 pp.

Tabachnik, B. G., Fidell, L. S. (1996). *Using multivariate statistics.* HarperCollins College Publishers, New York. 880 pp.

Tankersley, R. A., McKelvey, L. M., Forward, R. B. J. (1995). Responses of estuarine crab megalopae to pressure, salinity and light: implications for flood-tide transport. *Mar. Biol.* 122: 391–400.

Tanner, J. E., Hughes, T. P., Connell, J. H. (1994). Species, coexistence, keystone species and succession: a sensitivity analysis. *Ecology* 75: 2204–2219.

Tate, P. M., Jones, I. S. F., Hamon, B. V. (1989). Time and space scales of surface temperature in the Tasman Sea, from satellite data. *Deep Sea Res.* 36: 419–430.

Taylor, F. J., Taylor, N. J., Walsby, J. R. (1985). A bloom of the planktonic diatom, *Ceratualina pelagica*, off the coast of northeastern New Zealand in 1982–83, and its contribution to an associated mortality of fish and benthic fauna. *Int. Rev. Ges. Hydrobiol.* 70: 773–795.

Taylor, R. B., Cole, R. C. (1994). Mobile epifauna on subtidal brown seaweeds in northeastern New Zealand. *Mar. Ecol. Prog. Ser.* 115: 271–282.

Taylor, R. B., Willis, T. J. (1998). Relationships amongst length, weight and growth of north-eastern New Zealand fishes. *Marine and Freshwater Research.* 49: 255–260.

Tegner, M. J., Breen, P. A., Lennert, C. E. (1989). Population biology of red abalones, *Haliotis rufescens,* in southern California and management of the red and pink, *H. corrugata,* abalone fisheries. *Fish. Bull. U.S.* 87: 313–339.

Tegner, M. J., Dayton, P. K. (1977). Sea urchin recruitment patterns and implications of commercial fishing. *Science* 196: 324–326.

Tegner, M. J., Dayton, P. K. (1987). El Niño effects on southern California kelp forest communities. *Adv. Ecol. Res.* 47: 243–279.

ter Braak, C. J. F. (1987). The analysis of vegetation-environment relationships by canonical correspondence analysis. *Ecology.* 67: 1167–1179.

Thackway, R. (1996). Developing Australia's representative system of marine protected areas: Criteria and guideline for identification and selection. Ocean Rescue 2000 Workshop Series. Proceedings of a technical meeting held at the South Australian Aquatic Sciences Centre, West Beach, Adelaide, 22–23 April 1996. 148 pp.

Thackway, R., Cresswell, I. D. (1995). An interim marine and coastal regionalisation for Australia, Stage I – the nearshore component: a framework for establishing the national system of marine protected areas. Draft version 1.1. Australian Nature Conservation Agency, Canberra.

Thayer, G. W. (1992). *Restoring the nation's marine environment.* Maryland Sea Grant College, College Park, Maryland. 716 pp.

Thomas, L., Krebs, C. J. (1997). A review of statistical power analysis software. *Bull. Ecol. Soc. Am.* 78: 126–139.

Thompson, S. (1979). Ecological and behavioural factors

influ-encing the distribution and abundance patterns of tripterygiid fishes with particular reference to *Tripterygion varium*. Unpublished MSc thesis, University of Auckland. 80 pp.

Thompson, S. (1981). Fish of the marine reserve. *Leigh Laboratory Bulletin 3*. 364 pp.

Thompson, S. (1983). Homing in a territorial reef fish. *Copeia*: 832–834.

Thompson, S. (1986). Male spawning success and female choice in the mottled triplefin, *Fosterygion varium* (Pisces: Tripterygiidae). *Anim. Behav.* 34: 580–589.

Thompson, S. M., Jones, G. P. (1983). Interspecific territoriality and competition for food between reef fishes *Fosterygion varium* and *Pseudolabrus celidotus*. *Mar. Biol.* 76: 95–104.

Thorrold, S. R. (1988). Estimating some early life history parameters in a tropical clupeid, *Herklosichthys castelnaui*, from daily growth increments in otoliths. *Fish. Bull. U.S.* 87: 73–83.

Thorrold, S. R. (1993). Post-larval and juvenile scombrids captured in light traps: Preliminary results from the central Great Barrier Reef Lagoon. *Bull. Mar. Sci.* 52: 631–641.

Thorrold, S. R., Milicich, M. J. (1990). Comparison of larval duration and pre- and post-settlement growth in two species of damselfish, *Chromis atripectoralis* and *Pomacentris coelestis*. *Mar. Biol.* 105: 375–384.

Thresher, R. E. (1984). *Reproduction in reef fishes*. TFH, Neptune City. 399 pp.

Thresher, R. E., Gunn, J. S. (1986). Comparative analysis of visual census techniques for highly mobile, reef associated piscivores (Carangidae). *Environ. Biol. Fish.* 17: 93–116.

Thresher, R. E., Proctor, C. H., Gunn, J. S. (1994). An evaluation of electron-probe microanalysis of otliths for stock delineation and identification of nursery areas in a temperate groundfish, *Nemadactylus macropterus*. *Fish. Bull.* 92: 817–840.

Thrush, S. F., Hewitt, J. E., Cummings, V. J., Dayton, P. K. (1995). The impact of habitat disturbance by scallop dredging on marine benthic communites: What can be predicted from the results of experiments? *Mar. Ecol. Prog. Ser.* 129: 141–150.

Thrush, S. F., Hewitt, J. E., Pridmore, R. D. (1989). Patterns in the spatial arrangements of polychaetes and bivalves in intertidal sandflats. *Mar. Biol.* 102: 529–535.

Thrush, S. F., Pridmore, R. D., Hewitt, J. E. (1994). Impacts on soft sediment macrofauna: the effects of spatial variation on temporal trends. *Ecol. Appl.* 4: 31–41.

Thrush, S. F., Pridmore, R. D., Hewitt, J. E. (1996b). Impacts on soft-sediment macrofauna: the effects of spatial variation on temporal trends. Pp. 49–57 *in* Schmitt, R. J., Osenberg, C. W. (eds), *Detecting ecological impacts: concepts and applications in coastal habitats*. Academic Press, San Diego.

Thrush, S. F., Whitlatch, R. B., Pridmore, R. D., Hewitt, V. J., Cummings, V. J., Wilkinson, M. R. (1996a). Scale-dependent recolonisation: the role of sediment stability in a dynamic sandflat habitat. *Ecology* 77: 2472–2487.

Tilzey, R. D. J. (1994). *The South East Fishery*. Bureau of Resource Sciences, Australia Parkes, ACT 2600. 360 pp.

Tiselius, P. (1998). An *in situ* video camera for plankton studies: design and preliminary observations. *Mar. Ecol. Prog. Ser.* 164: 293–299

Todd, C. D., Keough, M. J. (1994). Larval settlement in hard sustratum assemblages: a manipulative field study of the effects of substratum filming and the presence of incumbents. *J. Exp. Mar. Biol. Ecol.* 181: 159–187.

Todd, C. D., Laverick, M. S. (1991). *Coastal marine zooplankton: a practical guide for students*. Cambridge University Press, Cambridge. 106 pp.

Toft, C. A., Shea, P. J. (1983). Detecting community-wide patterns: estimating power strengthens statistical inference. *Am. Nat.* 122: 618–625.

Tomas, C. R. (1993). *Marine Phytoplankton: a guide to naked flagellates and coccolithophorids*. Academic Press, New York. 263 pp.

Tranter, D. J., Parker, R.R., Cresswell, G. R. (1980). Are warm core eddies unproductive? *Nature*. 284: 540–542.

Trenery, D. R. (1986). Harvesting experiments and survey of the laminarian alga *Ecklonia radiata* with reference to possible commercial utilisation. Unpublished report commissioned by Fisheries Research Division, MAF, New Zealand. 43 pp.

Trexler, J. C., Travis, J. (1993). Nontraditional regression analysis. *Ecology* 74: 1629–1638.

Tricklebank, K. A., Jacoby, C. A., Montgomery, J. C. (1992). Composition, distribution and abundance of neustonic ichthyoplankton off northeastern New Zealand. *Estuar. Cstl Shelf Sci.* 34: 263–275.

Tsukamoto, K., Hirokawa, J., Oya, M., Sekiya, S., Fujimoto, H., Imaizum, K. (1989). Size-dependent mortality of red sea bream, *Pagrus major*, juveniles released with fluorescent otolith-tags in News Bay, Japan. *J. Fish. Biology*. 35: 59–69.

Tucker, Jnr, J. W., Chester, A. J. (1984). Effects of salinity, formalin concentration and buffer on quality of preservation of southern flounder (*Paralichthys lethostigma*) larvae. *Copeia*: 981–988.

Tully, O., O'Ceidigh, P. (1989a). The ichthyoneuston of Galway Bay (Ireland). *Mar. Biol.* 101: 27–41.

Tully, O., O'Ceidigh, P. (1989b). The ichthyoneuston of Galway Bay (west of Ireland) II. Food of post-larval and juvenile neustonic and pseudoneustonic fish. *Mar. Ecol. Prog. Ser.* 51: 301–310.

Turchin, P. (1996). Fractal analysis of animal movement: a critique. *Ecology* 77: 2086–2090.

Tutschulte, T. C., Connell, J. H. (1981). Reproductive biology of three species of abalones (*Haliotis*) in southern California. *Veliger* 23: 195–206.

Tzioumis, V., Kingsford, M. J. (1995). Periodicity of spawning of two temperate damselfishes: *Parma microlepis* and *Chromis dispilus*. *Bull. Mar. Sci.* 57: 596–609.

Underwood, A. J. (1974). The reproductive cycles and geographical distribution of some common eastern Australian prosobranches (Mollusca: Gastropoda). *Aust. J. Mar. Freshwar. Res.* 25: 63–88.

Underwood, A. J. (1977). Movements of intertidal gastropods. *J. Exp. Mar. Biol. Ecol.* 26: 191–201.

Underwood, A. J. (1979). The ecology of intertidal gastropods. *Adv. Mar. Biol.* 16: 111–210.

Underwood, A. J. (1980). The effects of grazing by gastropods and physical factors on the upper limits of distribution of intertidal macroalgae. *Oecologia* 46: 201–213.

Underwood, A. J. (1981). Techniques of analysis of variance in experimental marine biology and ecology. *Oceanogr. Mar. Biol. Ann. Rev.* 19: 513–605.

Underwood, A. J. (1985). Physical factors and biological

interactions: the necessity and nature of ecological experiments. Pp. 372–390 *in* Moore, P. G., Seed, R. (eds), *The ecology of rocky coasts.* Hodder & Stoughton, London.

Underwood, A. J. (1986). The analysis of competition by field experiments. Pp. 240–268 *in* Kikkawa, J., Anderson, D. J. (eds), *Community ecology: Pattern and process.* Blackwell, Melbourne.

Underwood, A. J. (1989). The analysis of stress in natural populations. *Biol. J. Linn. Soc.* 37: 51–78.

Underwood, A. J. (1990). Experiments in ecology and management: their logics, functions and interpretations. *Aust. J. Ecol.* 15: 365–389.

Underwood, A. J. (1991). Beyond BACI: Experimental designs for detecting human environmental impacts on temporal variations in natural populations. *Aust. J. Mar. Freshwat. Res.* 42: 569–587.

Underwood, A. J. (1992). Beyond BACI: the detection of environmental impacts on populations in the real, but variable world. *J. Exp. Mar. Biol. Ecol.* 161: 145–178.

Underwood, A. J. (1993a). Field experiments in intertidal ecology. *Proc. 2nd Int. Temperate Reef Symp.*: 7–13.

Underwood, A. J. (1993b). The mechanics of spatially replicated sampling programmes to detect environmental impacts in a variable world. *Aust. J. Ecol.* 18: 99–116.

Underwood, A. J. (1994a). On beyond BACI: Sampling designs that might reliably detect environmental disturbances. *Ecol. Appl.* 4: 3–15.

Underwood, A. J. (1994b). Rocky intertidal shores. Pp. 272–297 *in* Hammond, L. S., Synott, R. N. (eds), *Marine Biology.* Longman Cheshire, Melbourne.

Underwood, A. J. (1994c). Things environmental scientists (and statisticians) need to know to receive (and give) better statistical advice. P. 269 *in* Fletcher, D. J., Manly, B. F. J. (eds.), Statistics in ecology and environmental monitoring. University of Otago Press, Dunedin.

Underwood, A. J. (1997). *Experiments in ecology: their logical design and interpretation using analysis of variance.* Cambridge University Press, Cambridge. 504 pp.

Underwood, A. J., Atkinson, M. H. (1993). Beyond the baseline: where should we go from here? *Proc. 2nd Int. Temperate Reef Symp.*: 115–120.

Underwood, A. J., Chapman, M. G. (1985). Multifactorial analyses of directions of movement of animals. *J. Exp. Mar. Biol. Ecol.* 91: 17–43.

Underwood, A. J., Chapman, M. G. (1994). *Coastal marine ecology of temperate Australia.* UNSW Press, Sydney. 341 pp.

Underwood, A. J., Chapman, M. G. (1995). Rocky Shores. Pp. 55–82 *in* Underwood, A. J., Chapman, M. G. (eds), *Coastal marine ecology of temperate Australia.* University of New South Wales Press, Sydney. 341 pp.

Underwood, A. J., Chapman, M. G. (1996). Scales of spatial patterns of distribution of intertidal invertebrates. *Oecologia* 107: 212–224.

Underwood, A. J., Fairweather, P. W. (1989). Supply-side ecology and benthic marine assemblages. *Trends Ecol. Evol.* 4: 16–19.

Underwood, A. J., Jernakoff, P. (1981). Effects of interactions between algae and grasing gastropods on the structure of low-shore intertidal algal community. *Oecologia.* 48: 221–233.

Underwood, A. J., Kennelly, S. J. (1990a). Pilot studies for designs of surveys of human disturbance of intertidal habitats in New South Wales. *Aust. J. Mar. Freshwat. Res.* 41: 165–173.

Underwood, A. J., Kennelly, S. J. (1990b). Ecology of marine algae on rocky shores and subtidal reefs in temperate Australia. *Hydrobiologia* 192: 3–20.

Underwood A. J., Kingsford M. J., Andrew N. L. (1991). Patterns in shallow subtidal marine assemblages along the coast of New South Wales. *Aust. J. Ecol.* 6: 231–249.

Underwood, A., Peterson, C. H. (1988). Towards an ecological framework for investigating pollution. *Mar. Ecol. Prog. Ser.* 46: 227–234.

Underwood, A. J., Petraitis, P. S. (1993). Structure of intertidal assemblages in different locations: how can local processes be compared? Pp. 38–51 *in* Ricklefs, R., Schulter, D. (eds), *Species diversity in ecological communities.* University of Chicago Press.

Underwood, A. J., Skilleter, G. A. (1996). Effects of patch-size on the structure of assemblages in rock pools. *J. Exp. Mar. Biol. Ecol.* 197: 63–90.

UNESCO (1968). *Monographs on oceanographic methodology: Zooplankton sampling. Vol. 2.* UNESCO, Paris. 174 pp.

UNESCO (1978) *Coral Reefs: Research methods. Monographs on oceanographic methodology 5.* UNESCO, Paris. 581 pp.

UNESCO (1984). Comparing coral reef survey methods. *UNESCO Rep. Mar. Sci. 21*: 1–170.

US EPA (1995). QA/QC guidance for sampling and analysis of sediments, water, and tissues for dredged material evaluations. U.S. Environmental Protection Agency, Office of Water, Report No. *EPA Rep.* 823–B–95–001: 103 pp.

Vadas, R. L., Wright, W. A., Miller, S. L. (1990). Recruitment of *Ascophyllum nodosum*: wave action as a source of mortality. *Mar. Ecol. Prog. Ser.* 61: 263–272.

Vanderploeg, H. (1978). Two electivity indices with special reference to zooplankton grazing. *J. Fish. Res. Brd. Can.* 36: 362–365.

van der Veer, H. W., and Oorthysen, W. (1985). Abundance, growth and food demand of the scyphomedusa *Aurelia aurita* in the Western Wadden Sea. *Neth. J. Sea Res.* 19: 38–44.

van Guelpen, L., Markle, D. F., Duggan, D. J. (1982). An evaluation of accuracy, precision and speed of several zooplankton subsampling techniques. *J. Cons. int. Explor. Mer.* 40: 226–236.

van Rooij, J. M., Vidler, J. J. (1996). A simple method for stereo-photographic length measurement of free swimming fish: merits and constraints. *J. Exp. Mar. Biol. Ecol.* 195: 237–249.

Victor, B. C. (1982). Daily otolith increments and recruitment in two coral-reef wrasses, *Thalassoma bifasciatum* and *Halichoeres bivittatus*. *Mar. Biol.* 71: 203–208.

Victor, B. C. (1986a). Duration of the planktonic larval stage of one hundred species of Pacific and Atlantic wrasses (family Labridae). *Mar. Biol.* 90: 317–326.

Vieira, J. P. (1991). Juvenile mullets (Pisces, Mugilidae) in the estuary of Lagoa-Dos-Patos, RS, Brazil. *Copeia*: 409–418.

Villarino, M. L., Figueiras, F. G., Jones, K. J., Alvarez-Salgado, X. A., Richard, J., Edwards, A. (1995). Evidence of in situ vertical migration of a red-tide microplankton species in Ria de Vigo (NW Spain). *Mar. Biol.* 123: 607–618.

Wahle, R. A., Steneck, R. S. (1992). Habitat restrictions in early benthic life: Experiments on habitat selection and in situ predation with the American lobster. *J. Exp. Mar. Biol. Ecol.* 157: 91–114.

Wainwright, P. C. (1988). Morphology and ecology: functional basis of feeding constraints in Caribbean labrid fishes. *Ecology* 69: 635–645.

Walker, M. M. (1982). Reproductive periodicity in *Evechinus chloroticus* in the Hauraki Gulf. *N.Z. J. Mar. Freshwat. Res.* 16: 19–25.

Wallace, R. K. (1981). An assessment of diet overlap indexes. *Trans. Am. Fish. Soc.* 110: 72–76.

Wallace, R. K., Ramsey, J. S. (1983). Reliability in measuring diet overlap. *Can. J. Fish. Aquat. Sci.* 40: 347–351.

Walls, K. (1995a). The New Zealand experience in developing a marine biogeographic regionalisation. Pp. 33–48 *in* Muldoon, J. (ed.), *Towards a marine regionalisation for Australia*. Proceedings of a workshop held in Sydney, New South Wales, 4–6 March 1994. Great Barrier Reef Marine Park Authority, Australia.

Walls, K. (1995b). New Zealand. Pp. 307–317 *in* Hotta, K., Dutton, I. (eds), *Coastal management in the Asia-Pacific region: Issues and approaches*. JIMSTEF, Tokyo.

Walls, K., Dingwall, P. (1995). Marine Region 18 – Australia/ New Zealand, Part B. Pp. 154–199 *in* Kelleher et al. (eds), *A global representative system of marine protected areas, Vols I–IV*. Great Barrier Reef Marine Park Authority. World Bank and IUCN.

Walls, K., McAlpine, G. (1993). Developing a strategy for a network of marine reserves around New Zealand – a manager's perpective. *Proc. 2nd Int. Temperate Reef Symp.*: 57–62.

Walsby, J. R., Morton, J. E. (1982). Marine molluscs. Part 1. Archaeogastropods and chitons. *Leigh Lab. Bull.* 4: 89 pp.

Ward, J. (1993). Indicators of the state of the coastal environment and management practices. Centre for Resource Management and Environment, Lincoln University. *Information Pap.* 48.

Ward, J., Hedley, S. (1994). Monitoring the New Zealand Coastal Policy Statement: a preliminary assessment. Centre for Resource Management and Environment, Lincoln University. *Information Pap.* 51. 17 pp.

Ward, T. J., Jacoby, C. A. (1992). A strategy for assessment and management of marine ecosystems: baseline and monitoring studies in Jervis Bay, a temperate Australian embayment. *Mar. Pollut. Bull.* 25: 163–171.

Warner, R. R. (1988). Sex change in fishes: hypotheses, evidence, and objections. *Environ. Biol. Fish.* 22: 81–90.

Warner, R. R. (1990). Resource assessment versus tradition in mating-site determination. *Am. Nat.* 135: 205–217.

Warner, R. R., Hughes, T. P. (1988). The population dynamics of reef fishes. *Proc. 6th. Int. Coral. Reef Symp.* 1: 149–155.

Warren, J. H. (1990). The use of open burrows to estimate abundances of intertidal estuarine crabs. *Aust. J. Ecol.* 15: 277–280.

Warwick, R. M., Joint, I. R. (1987). The size distribution of organisms in the Celtic Sea: from bacteria to metazoa. *Oecologia* 73: 185–191.

Watanabe, Y., Yamashita, Y., Oozenki, Y. (1996). *Survival strategies in early life history stages of marine resources*. A. A. Balkema, Rotterdam. 367 pp.

Wear, R. G., Fielder, D. R. (1985). The marine fauna of New Zealand: larvae of Brachyura (Crustacea, Decapoda). *Mem. N.Z. Oceanogr. Inst.* 92. 89 pp.

Webb, R. O., Kingsford, M. J. (1992). Protogynous hermaphroditism in the half-banded sea perch, *Hypoplectrodes maccullochi (Serranidae). J. Fish. Biol.* 40: 951–961.

Weinberg, S. (1981). A comparison of coral reef survey methods. *Bijdr. Dierk.* 51: 199–218.

Weldon, C. W., Slauson, W. L. (1986). The intensity of competition versus its importance: an overlooked distinction and some implications. *Q. Rev. Biol.* 61: 23–44.

Werner, E. E., Gillium, J. F. (1984). The ontogenetic niches and species interactions in size-structured populations. *Ann. Rev. Ecol. Syst.* 15: 393–425.

West, G. (1990). Methods of assessing ovarian development in fishes: a review. *Aust. J. Mar. Freshwat. Res.* 41: 199–222.

West, R. J. (1990). Depth-related structural and morphological variations in an Australian *Posidonia* seagrass bed. *J. Exp. Mar. Biol. Ecol.* 36: 153–166.

West, R. J., Gordon, R. J. (1994). Commercial and recreational harvest of fish from two Australian coastal rivers. *Aust. J. Mar. Freshwat. Res.* 45: 1259–1279.

Westerman, M., Holt, G. J. (1994). RNA:DNA ratio during the critical period and early larval growth of the red drum *Sciaenops ocellatus. Mar. Biol.* 121: 1–10.

Wethey, D. S. 1984. Spatial pattern in barnacle settlement: day to day changes during the settlement season. *J. Mar. Biol. Ass. U.K.* 64: 687–697.

White, G. C., Garrott, R. A. (1990). *Analysis of wildlife radio-tracking data*. Academic Press, New York.

Whitehead, P. J. P., Bauchot, M-L., Hureau, J-C., Nielsen, J., Tortonese, E. (1989). *Fishes of the Northeastern Atlantic and the Mediterranean*. Vols 1–3. UNESCO, Paris. 1473 pp.

Wiebe, P. H., Burt, K. H., Boyd, S. H., Morton, A. W. (1976). A multiple opening/closing net and environmental sensing system for sampling zooplankton. *J. Mar. Res.* 34: 313–326.

Wiens, J. (1989). Spatial scaling in ecology. *Functional Ecol.* 3: 385–397.

Wilkins, H. K. A., Myers, A. A. (1992). Microhabitat utilisation by an assemblage of temperate Gobiidae (Pisces: Teleostei). *Mar. Ecol. Prog. Ser.* 90: 103–112.

Wilkinson, L. (1996). *SYSTAT 6.0 for Windows*. SPSS, Chicago.

Willan, R. C., Dollimore, J. M., Nicholson, J. (1979). A survey of fish populations at Karikari Peninsula, Northland, by scuba diving. *N.Z. J. Mar. Freshwat. Res.* 13: 447–458.

Williams, D. H. C., Anderson, D. T. (1975). The reproductive system, embryological development and metamorphosis of the sea urchin *Heliocidaris erythrogramma* (Val.) Echinoida, Echinodermata. *Aust. J. Zool.* 23: 371–403.

Williams, D. McB. (1982). Patterns of distribution of fish communities across the central Great Barrier Reef lagoon. *Coral Reefs 1*: 35–43.

Williams, D. McB. (1991). Patterns and processes in the distribution of coral reef fishes. Pp. 437–474 *in* Sale, P. F. (ed.), *The ecology of fishes on coral reefs*. Academic Press, San Diego.

Williams, D. McB. (1994). Annual recruitment surveys of coral reef fish are a good indicator of patterns of settlement. *Bull. Mar. Sci.* 54: 314–331.

Williams, D. McB., Dixon, P., English, S. (1988). Cross-shelf distribution of copepods and fish larvae across the central Great Barrier Reef. *Mar. Biol.* 99: 577–589.

Williams, D. McB., Hatcher, A. I. (1983). Structure of fish communities on outer slopes of inshore, midshelf and outer shelf reefs of the Great Barrier Reef. *Mar. Ecol. Prog. Ser.* 10: 239–250.

Williams, M. R. (1994). Use of principal component biplots to detect environmental impact. Pp. 263–269 *in* Fletcher, D.J., Manly, F. J. (eds) Statistics in ecology and environmental monitoring. University of Otago Press, Dunedin.

Williams, R., Robins, D. B. (1982). Effects of preservation on wet weight, dry weight, nitrogen and carbon contents of *Calanus helgolandicus* (Crustacea: Copepoda). *Mar. Biol.* 71: 271–281.

Williamson, J. E. (1992). Distribution patterns and life-history features of the intertidal encrusting alga *Pseudolitho-derma* sp. Unpublished MSc thesis, University of Auckland.

Williamson, J. E., Creese, R. G. (1996). Colonisation and persistence of patches of the crustose brown alga *Pseudolitho-derma* sp. (Ralfsiaceae). *J. Exp. Mar. Biol. Ecol* 203: 191–208.

Wilson, B. (1993a). *Australian marine shells: Prosobranch Gastropods.* Vol. 1. Odessey, Kallaroo. 408 pp.

Wilson, B. (1993b). *Australian marine shells: Prosobranch Gastropods.* Vol. 2. Odessey, Kallaroo. 370 pp.

Wilson, N. H. F., Schiel, D. R. (1995). Reproduction in two species of abalone (*Haliotis iris* and *H. australis*). *Mar. Freshwat. Res.* 46: 629–637.

Wilson, W. H. (1991). Competition and predation in marine soft-sediment communities. *Ann. Rev. Ecol. Syst.* 21: 221–241.

Winer, B. J. (1971). *Statistical principles in experimental design.* McGraw-Hill Kogakusha, Tokyo. 907 pp.

Winer, B. J., Brown, D. R., Michels, K. M. (1991). *Statistical principles in experimental design.* McGraw-Hill, New York. 1057 pp.

Wolanski, E. (1994). *Physical oceanographic processes of the Great Barrier Reef.* CRC Press, Boca Raton. 194 pp.

Wolanski, E., Hamner, W. M. (1988). Topographically controlled fronts in the ocean and their biological influence. *Science* 241: 177–181.

Wolff, W. J. (1987). Flora and macrofauna of intertidal sediments. Pp. 81–105 *in* Baker, J. M., Wolff, W. J. (eds), *Biological surveys of estuaries and coasts.* Cambridge University Press, Cambridge. 449 pp.

Womesley, H. B. S. (1984). *The marine flora of southern Australia Part I.* Government Printer, Adelaide. 329 pp.

Womesley, H.B.S. (1987). *The marine flora of southern Australia Part II.* Government Printer, Adelaide. 484 pp.

Wood, E. J. F. (1954). Dinoflagellates in the Australian region. *Aust. J. Mar. Freshwat. Res.* 2(2): 171–351.

Wood, E. J. F. (1961). Studies on Australian and New Zealand diatoms. IV. Descriptions of further sedentary species. *Trans. R. Soc. N.Z.* 88: 669–698.

Wood, E. J. F. (1963). Studies on Australian and New Zealand diatoms. VI. Tropical and subtropical species. *Trans. R. Soc. N.Z.(Bot.)* 2: 189–218.

Wood, E. J .F., Crosby, L. A., Cassie, V. (1959). Studies on Australian and New Zealand diatoms. III. Descriptions of further discoid species. *Trans. R. Soc. N.Z.* 87: 211–219.

Wootton, J. T. (1994). The nature and consequences of indirect effects in ecological communities. *Ann. Rev. Ecol. Syst.* 25: 443–466.

Worthington, D. G., Andrew, N. A. (198). Small scale variation in demography and its implications for alternative size limits in the fishery for blacklip abalone (*Haliotis rubra*) in NSW, Australia. *Can. Spec. Publ. Fish. Aquat. Sci.* 125: 341–348.

Worthington, D. G., Andrew, N. L., Hamer, G. (1995). Covariation between growth and morphology suggests

alternative size limits for the blacklip abalone, *Haliotis rubra,* in New South Wales, Australia. *Fish. Bull.* 93: 551–561.

Worthington, D. G., Fairweather, P. G. (1989). Shelter and food: interactions between *Turbo undulatum* (Archaeo-gastropoda: Turbinidae) and coralline algae on rocky seashores in New South Wales. *J. Exp. Mar. Biol. Ecol.* 129: 61–79.

Worthington, D. G., Ferrell, D. J., McNeill, S. E., Bell, J. D. (1992a). Effects of the shoot density of seagrass on fish and decapods: Are correlations evident over large spatial scales? *Mar. Biol.* 112: 139–146.

Worthington, D. G., Ferrell, D. J., McNeill, S. E., Bell, J. D. (1992b). Growth of four species of juvenile fish associated with the seagrass, *Zostera capricorni,* in Botany Bay, New South Wales. *Aust. J. Mar. Freshwat. Res.* 43: 1189–1198.

Wulff, J. L. (1986). Variation in clone structure of fragmenting coral reef sponges. *Biol. J. Linn. Soc.* 27(4): 311–330.

Yamaji, I. (1959). *The plankton of Japanese coastal waters.* Hoikusha, Osaka. 229 pp.

Young, B. M., Harvey, L. E. (1996). A spatial analysis of the relationship between mangrove (*Avicennia marina* var. *australasica*) physiognomy and sediment accretion in the Hauraki Plains, New Zealand. *Estuar. Cstl Shelf Sci.* 42: 231–243.

Young, C. M. (1990). Larval ecology of marine invertebrates: a sesquicentennial history. *Ophelia* 32: 1–48.

Young, P. C. (1993). *Concise encyclopedia of environmental systems.* Pergamon Press, Oxford. 769 pp.

Youngbluth, M. J. (1982). Sampling demersal zooplankton: a comparison of field collections using three different emergence traps. *J. Exp. Mar. Biol. Ecol.* 61: 111–164.

Zann, L. P. (1996). The state of the marine environment. Department of the Environment, Sport and Territories, Ocean Rescue 2000, Canberra. 531 pp.

Zavodnik, D. (1987). Spatial aggregations of the swarming jellyfish *Pelagia noctiluca* (Scyphozoa). *Mar. Biol.* 94: 265–269.

Zedler, J. B., Nordby, C. S. (1986). The ecology of Tijuana Estuary: an estuarine profile. *U.S. Fish and Wildlife Serv. Biol. Rep.* 85 (7.5): 104 pp.

Zeldis, J. R. (1985). Ecology of *Munida gregaria* (Decapoda, Anomura): distribution and abundance, population dynamics and fisheries. *Mar. Ecol. Prog. Ser.* 22: 77–99.

Zeldis, J. R. (1993). Applicability of egg surveys for spawning stock biomass estimation of snapper and hoki in New Zealand. *Bull. Mar. Sci.* 53: 864–890.

Zeldis, J. R., Davis, C. B., James, M. R., Ballara, S. L., Booth, W. E., Chang, F. H. (1995). Salp grazing: effects on phyto-plankton abundance, vertical distribution and taxonomic composition in a coastal habitat. *Mar. Ecol. Prog. Ser.* 126: 267–283.

Zeldis, J., Sharples, J., Uddstrom, M., Pickmere, S. (1998). Fertilising the continental shelf: biological oceanographic studies on the northeastern New Zealand continental margin. *NIWA Water & Atmosphere.* 6: 13–16.

Zieman, J. C. (1974). Methods for the study of the growth and production of turtle grass, *Thalassia testudinium* König. *Aquaculture* 4: 139–143.

INDEX